Total Construction Management

A convergence of lean management and quality management thinking has taken place in organizations across many industries, including construction. Practices in procurement, design management and construction management are all evolving constantly and understanding these changes and how to react is essential to successful management. This book provides valuable insights for owners, designers and constructors in the construction sector.

Starting by introducing the language of total quality, lean and operational excellence, this book takes the reader right up to the latest industry practice in this sector, and demonstrates the best way to manage change. Written by two of the world's leading experts, *Total Construction Management: Lean quality in construction project delivery* offers a clearly structured introduction to the most important management concepts and practices used in the global construction industry today.

This authoritative book covers issues such as procurement, BIM, all forms of waste, construction safety, and design and construction management, all explained with international case studies. It is a perfect guide for managers in all parts of the industry, and ideal for those preparing to enter the industry.

John S. Oakland is both Chairman and Head of Research and Education at Oakland Consulting LLP. He is also Emeritus Professor of Business Excellence and Quality Management at Leeds University Business School, UK. He is the author of numerous texts on Quality Management.

Marton Marosszeky retired from a full-time professorship at UNSW, Australia, in 2006 and has been working as a lean consultant since then. Between 2007 and 2012 he was the leader of the lean consulting service line within Evans and Peck. He has worked with major project teams and company executives supporting them in developing and implementing lean/quality-based management strategies in the infrastructure (road and rail), building construction, and oil and gas industries across four continents.

'This new edition from John and Marton provides an excellent "one stop shop" for anyone wanting to make a difference to quality in the construction industry. The case studies really bring the topic to life and help the reader to transition from the theory through to how total quality management can work in real life situations.'

Ian Mitchell, *Network Rail UK and Chair of the Chartered Quality Institute, UK*

'This book is a must read for those who live and work in a construction industry that is in many ways broken. Merging lean and quality in the concept "lean quality", the authors construct on that foundation a compelling diagnosis of construction industry ills and equally excellent treatments. Readers can see the desired future for the construction industry in detailed case studies of advanced practitioners, ranging from worker empowerment at J.B. Henderson Constructors to Boulder Associates' implementation of lean into their architectural practice. The authors give us hope that industry transformation is possible and is actually underway.'

Glenn Ballard, *University of California Berkeley, USA*

'Oakland and Marosszeky have successfully presented in this book the practical application of Lean quality principles within the construction industry with great examples one can relate to the day to day business environment. Engaging reading I would recommend to both quality and non quality professionals interested to enhance effectiveness in their organisations.'

Carlos Vazquez Travieso, *Head of Quality at Transport for London, UK*

'John Oakland and Marton Marosszeky have proved that significant improvements in the way we design and construct can only come through looking at what we do through the lens of safety and quality, as opposed to focusing first on cost, schedule and workflow or the use of information technologies. They do this by describing and explaining the value realized by applying Lean thinking, management and methods, BIM and project integration from the perspective of producing well designed and safely built products. *Total Construction Management: Lean quality in construction project delivery* is hard to put down because it is so well written and full of insights. It is a must read for owners, designers, construction managers and constructors wherever they work.'

Dean Reed, *DPR Construction Director for Lean Construction and co-author of* Integrating Project Delivery

'It's rare for a book to be able to combine big picture thinking about the major productivity issues facing the construction industry with in depth analysis of the various initiatives being adopted around the world. The authors' integrated view of Lean Processes, BIM, Collaborative Contracting (IPD) and Quality Management make this compelling reading for those interested in industry reform and competitive advantage. The case studies included clearly evidence the analysis and demonstrate the significant advances that clear thinking leadership can achieve.'

Richard Morwood, *Industry Director - Integrated Project Delivery, AECOM, Australia and New Zealand*

Total Construction Management

Lean quality in construction project delivery

**John S. Oakland and
Marton Marosszeky**

Routledge
Taylor & Francis Group

LONDON AND NEW YORK

First published 2017
by Routledge
2 Park Square, Milton Park, Abingdon, Oxon OX14 4RN

and by Routledge
711 Third Avenue, New York, NY 10017

Routledge is an imprint of the Taylor & Francis Group, an informa business

British Library Cataloguing-in-Publication Data
A catalogue record for this book is available from the British Library

Library of Congress Cataloging in Publication Data
Names: Oakland, John S., author. | Marosszeky, Marton, author.
Title: Total construction management : lean quality in construction project delivery / John S. Oakland and Marton Marosszeky.
Description: Abingdon, Oxon ; New York, NY : Routledge is an imprint of the Taylor & Francis Group, an Informa Business, [2017] | Includes bibliographical references and index.
Identifiers: LCCN 2016034978| ISBN 9781138908536 (hbk : alk. paper) | ISBN 9781138908543 (pbk. : alk. paper) | ISBN 9781315694351 (ebk : alk. paper)
Subjects: LCSH: Construction industry—Quality control. | Building—Cost control. | Project management—Case studies.
Classification: LCC TH438.2 .O35 2017 | DDC 624.068/4—dc23
LC record available at https://lccn.loc.gov/2016034978

ISBN: 978-1-138-90853-6 (hbk)
ISBN: 978-1-138-90854-3 (pbk)
ISBN: 978-1-315-69435-1 (ebk)

Typeset in Palatino and Univers
by Florence Production Ltd, Stoodleigh, Devon, UK

MIX
Paper from responsible sources
FSC® C013985
www.fsc.org

Printed in the United Kingdom
by Henry Ling Limited

CONTENTS

PART II: PLANNING 81

PART III: PERFORMANCE 169

FIGURES

TABLES

PREFACE

INDUSTRY CHALLENGES AND SOLUTIONS

All industries are undergoing rapid change under the pressure of technological change and changing client needs. The construction sector is no exception. The past ten years have seen accelerating globalization, a demand for larger and more complex projects, and a requirement for them to be delivered in ever shorter time frames. Meanwhile, clients of the industry are increasingly concerned that this sector is not keeping pace with the rates of improvement seen in other sectors of the economy. In addition, in this sector, the rate and cost of errors in quality and safety have been too slow to improve.

This book explores the challenges faced by the sector as well as reasons for its slow response to the pressures for change and improvement. As a part of this overview, it charts the changing roles of the professions and discusses the demands placed on design and construction managers by an ever more fragmented industry structure. Finally, it explores the potential of lean quality-based thinking for the management of organizations and projects, and for improved outcomes and value for clients.

In today's construction industry, many among clients, designers and contractors are seeing Building Information Modelling (BIM) as the silver bullet that will transform the industry. We are convinced that this position is misguided. Although BIM provides the basis for improved communication within the design team and with external stakeholders, and it provides support for solution optimization in both the design and construction stages of projects, it is no more than a very powerful enabling technology. The framework for lean quality developed in this book provides a foundation for organizational excellence across entire supply chains; it offers a powerful new perspective for policy makers, and helps to create the organizational prerequisites necessary for the effective deployment of technologies such as BIM.

In the ten years since our first book was published, the lean construction movement has matured globally, and has earned the respect of the traditional associations, which represent public and private owners, constructors, designers and construction managers. Leading owners in high technology manufacturing, petrochemical processing, healthcare and government are concerned that labour productivity in construction is not keeping pace with the rate of improvement in other sectors and are turning to lean quality frameworks to inform their thinking about the way that they engage with their supply chains and derive improved value from them.

Within the construction sector, the genesis of lean practice was at the site level, improving planning reliability and collaboration. With time, organizations are realizing that broader strategic issues need to be addressed, such as the relationship between general contractors and their supplier chains, in between projects. This book provides the basis for companies to take lean quality practices from the project level to the enterprise level. We believe that lean quality thinking provides the basis for transforming the construction sector globally and significantly improving productivity and increasing the industry's potential for value creation for its customers. The book also posits a much greater potential role for repeat industry clients to drive change in this all-important industry.

LEAN QUALITY AND THE OPPORTUNITY FOR IMPROVED OUTCOMES

When John Oakland wrote the first editions of his books on *Statistical Process Control* and *Total Quality Management* nearly 30 years ago, there were very few books on these subjects. Since their publication, interest in business performance improvement generally has exploded. There are now many texts on the subject and its various aspects, including business/operational excellence, business process management, six sigma and lean systems-based approaches.

So much has been learned during the last 30 years of implementation of these approaches that it has been necessary to rewrite these two books and revise them several times. John has done several editions since *TQM* was first published in 1989, and the content and case studies in the last edition were changed substantially to reflect the develop-ments, current understanding and experience gained of *TQM and Operational Excellence*, with new material on Lean Systems included.

John's work was relatively unknown in the construction sector until Marton Marosszeky, a specialist in construction process improvement in the areas of quality, safety and production, talked John into joining forces, and in 2006 they published *Total Quality in the Construction Supply Chain*. Pressure from clients and governments as well as commercial competitive pressures have continued to force leading organizations in the sector to differentiate themselves on the basis of customer focus, overall product and process quality, cost of products and services, and value creation for clients.

In response to these pressures, senior management in leading design and construction organizations worldwide are embracing the philosophy and principles of what we have now called *lean quality*. Often approaching the overall task from different perspectives, some have adopted frameworks of performance measurement and benchmarking, others have used the goal of continuous improvement while others have chosen to follow the values and concepts of lean construction. We see these as different perspectives through different lenses of the same broad objective – improving performance in all the activities of a business.

Traditionally in conversations about quality, the building and construction sector has had a natural orientation towards product quality. Given the complexity of its organizational relationships and traditional craft-based processes, most of the construction quality literature reflects this product focus: either providing a guide to compliance with the ISO 9001 quality system standards or providing pragmatic advice on tools for the control of quality. However, lead organizations in every area of the building and construction industry have recognized that the broad focus that *lean quality* brings to all aspects of organizing and managing is as relevant to building and construction as it is to the manufacturing and service sectors. Furthermore, teachers and researchers in building and construction have recognized that a traditional product-centred paradigm does not provide a sufficiently broad and robust basis for performance improvement within the sector.

This book is designed to address this critical need. As well as providing a broad and robust conceptual platform on which organizations can build their overall process improvement endeavours, the book integrates and places the many seemingly disparate management innovations of the past 20 years into a unified perspective – *Lean quality*.

Increasing the satisfaction of customers and stakeholders through effective goal deployment, cost reduction, productivity and process improvement through lean systems has proven to be essential for organizations to stay in operation. It is now widely recognized, particularly in the case study organizations in this new book, that a lean quality approach provides an important competitive weapon and way of managing for the future. Lean quality is far wider in its application than assuring product or service quality – it is a way of managing organizations and their supply chains so that every aspect of performance, both internally and externally, is improved.

Our second book, therefore, provides guidance on how to manage building and construction sector businesses in a lean quality way. It is structured in the main around four parts of a new model, but based on John's well-known TQM framework: improving *Performance* through better *Planning* and management of *People* and the *Processes* in which they work. The core of the model will always be performance in the eyes of the customer, but must be extended to include performance measures for all the stakeholders. This new core still needs to be surrounded by *commitment* to quality and meeting customer requirements, *communication* of the quality message and recognition of the need to change the *culture* of most organizations to create lean quality. These are the *soft* foundations that must encase the hard management necessities of planning, people and processes. To this we have added continuous improvement as this is now such a key aspect of every successful organization's operations.

Under these headings the essential steps for the successful implementation of lean quality are set out in what we believe is a meaningful and practical way. The book guides the reader through the language of lean quality and all associated recent developments and sets down a clear way for an organization to proceed.

At the end of the book, there are 14 case studies to support the text. Each of these presents the approach and achievements of a different organization within the building and construction sector. They include private and public sector organizations in a wide range of business areas: quarrying and concrete supply, design, construction, MEP services contractors, structural systems production, private owners, public infrastructure providers and a major hospital project. Each case study is linked back to the specific areas of the book that it illustrates.

As stated, many of the new approaches to improving performance appear to present different theories. However, in reality, they are talking the same language, using different dialects. The basic principles of defining quality and process efficiency, and taking these into account throughout all the activities of the business, are common. Lean quality has to be managed – it does not just happen. Understanding and commitment by senior management, effective leadership, teamwork and good process management are fundamental parts of the recipe for success. We have tried to use our extensive research and consultancy experience to take what is, to many, a jigsaw puzzle and to assemble a comprehensive, practical working model for lean quality – the rewards of which are greater efficiencies, lower costs, improved reputation and customer loyalty. Moreover, we have tried to show how holistic this approach now is: embracing the most recent models of Excellence, Lean, Six Sigma, Kaizen and a host of other management methods and teachings.

The book should meet the requirements of both students and practitioners who have or are planning careers within the built environment industry including engineers, architects, building and construction contractors, real estate and facility management professionals. In the operations of any organization within the building and construction sector there is a need to understand the broad implications that the lean quality approach holds for its entire supply chain and internal and external customers.

We hope that this is not seen as a specialist text for specialist practitioners in Lean or Quality. We see *lean quality* as providing the fundamental building blocks for the management of any organization and, hence, people working in every part of each organization need to understand this broad perspective. This book documents a comprehensive approach to the management of any business enterprise – one that has been used successfully by many design and construction-based organizations throughout the world.

We would like to thank our colleagues in Oakland Consulting plc for the sharing of ideas and help in their development. This book is the result of many years of collaboration in assisting organizations to introduce good methods of management and embrace the concepts of lean quality. We are most grateful, yet again, to Marti Marosszeky, who worked with us throughout this project and helped convert our ideas, hopefully, into an error-free, readable, even fluent, text!

Marton Marosszeky and John S. Oakland

Part I

The foundations of lean quality

Understanding lean construction

1

LEAN QUALITY IN CONSTRUCTION

Two major management movements have swept over the construction sector in the past 25 years: first of all, the Quality movement initiated by public sector clients mandating compliance with ISO 9000 in the early 1990s and more recently the Lean Construction movement. Currently leading clients for construction services are beginning to require their suppliers and their supply chains to demonstrate a lean approach to their businesses. Some, like Highways England and Crossrail, are even supporting the transformation with a maturity model, performance measures, metrics and resources/training. Both of these significant transformational philosophies, quality and lean, have taken root in other industry sectors well before early adopters in construction have recognized their value. Today we have mature examples of successful lean quality implementation in the construction sector, though we do have to look around the world to find our examples. This book presents the lead adopters in the form of a series of case studies of lean quality management. The book also postulates that lean and quality are two slightly different lenses through which we view management excellence and argues that these perspectives are merging. For example, the 2015 version of the International Standard on Quality Management System Requirements ISO 9001 introduced continuous improvement into the quality framework in a formal sense for the first time.

A case study

We start this book by exploring the unique characteristics of construction and developing an understanding of the specific challenges faced by design and construction organizations in implementing lean quality. Let us begin with an interesting vignette. In 1913, the Commonwealth Bank of Australia, one of Australia's leading banks at the time, commissioned a new building at the corner of Martin Place and Pitt Street in Sydney. The building was six stories high, had a massive sandstone façade, its footprint was approximately 100 metres square with a light well in the centre for ventilation and natural light.

The drawings for this building, which would cost some $200 million in today's terms, consisted of a single A1 sheet with three images on it, a dimensioned plan, an elevation and a dimensioned cross section, all neatly coloured (Figure 1.1).

The entire contract documentation for the project consisted of:

- these three drawings on one sheet;
- a handwritten statement on the back of this single sheet stating: I hereby undertake to build the building described overleaf for $700;
- 53 pages of scope of work/specification; and
- 4 pages of contract conditions.

FIGURE 1.1
Photograph of the
Commonwealth Bank of
Australia building at its
opening in 1916

Today such limited documentation is unimaginable (of course, the services in our buildings are much more complex), we would have many hundreds of drawings and thousands of pages of contract documentation. What has changed?

When this building was commissioned and built, the architect would have had the quantity surveyor and design engineers in-house as a part of his team. Similarly, the contractor would have had the major trades in-house, with construction teams led by master tradesmen. This story tells us a number of things:

- There was one way of building and master tradesmen had the skill and knowledge to build major buildings from simple depictions, and quality was ensured by the high standard and integrity of the master tradesmen.
- The brief contract (4 pages of contract conditions) demonstrates that there was a high level of trust between the client and contractor for this project. A simple contract such as this for a project of this size is unimaginable today.
- The organizations designing and building were integrated and had control over their resources. They could plan the workflow in detail as they had control of all of the resources required to undertake the work.

DISTINGUISHING FEATURES OF CONSTRUCTION PROJECTS

The construction industry does have a number of unique features which create challenges for the sector:

- projects are uniquely designed or modified to fit a specific site, the needs of a specific client or client group;
- constructed products are large and fixed in location; hence, in contrast to manufacturing, here it is the work teams that move past the product progressively adding value rather than the product moving through a number of work stations;
- generally, the work teams are from different subcontracting organizations rather than from a single organization, creating challenges for coordination and integration of the product design and the production and assembly process;
- the time frame for larger projects is measured in years, and, increasingly, the design phase and construction are, to some extent, concurrent;
- because of the long time frame of projects, the people involved will invariably change over time;
- on larger projects, in particular, the relationship between the parties is mediated by varying commercial terms and organizational arrangements; this diversity creates further significant variables which alter organizational structures, risk allocation, and the responsibilities of the parties from project to project;
- the client is often involved both in the design and construction phases, defining needs and choosing between alternative solutions, hence, influencing function, cost, and risk while the project design is under development and during construction;
- importantly, when the client has an ongoing interaction with the designers and constructers, the quality of the services will influence relationships and can distort customer perceptions of product quality: either positively or negatively;
- the corollary to this is that clients' approach to their role and function can significantly shift the culture and cost of the project: once again, in either a positive or negative direction; and finally,
- quality is considered almost entirely in product terms, yet the creation of a quality design and construction process, an effective collaboration between numerous suppliers from the early design stage to completion of construction is the greatest challenge – this is not owned by any one party; rather, it is achieved through the negotiations and collaboration of the many players; and it is the industry's processes which determine key outcomes.

Recent decades have seen the rapid globalization in all sections of the construction sector. It started with the emergence of global materials businesses in areas such as cement production, concrete production, brick, plasterboard, construction chemicals and timber processing. At the same time, the specialist requirements of major energy and infrastructure projects led to the development of global EPC (engineering, procurement and construction) businesses such as Bechtel and Fluor Daniel. Also during this period, global specialist design, fabricate and install businesses emerged to supply, install and service elevators, air conditioning, fire systems, electrical and control systems. Most recently, global architectural and engineering consulting businesses have been created.

In today's global construction sector, even mid-sized businesses can procure quite sophisticated products and services globally. Local design teams collaborate with others around the world creating the opportunity for 24/7 services. Often, back-office services such as accounting and call centres are relocated to more economical destinations within the country or internationally. Naturally, geographic diversification brings with it special challenges in terms of the seamless, defect-free delivery of complex products and services.

TODAY'S INDUSTRY

Among owners today there is a widespread belief that the lowest cost tendered on a project represents best value. On large complex projects nothing can be further from the truth. In fact, by adopting this method of procurement, owners have traded-off flexibility and the ability to collaborate and innovate across the supply chain for the illusion of cost certainty. We say illusion because large complex projects are almost inevitably time critical; they are unique; and they run over several years. Therefore, the final scope is rarely the same as the one priced at the outset.

Nowadays, clients are reluctant to invest in solution generation upfront; they are generally in a rush to start, and they tend to push decisions and risk down the supply chain. Under these circumstances, all risks are priced conservatively to cover the suppliers' risk in the absence of information; and client changes during the project are priced as variations and usually command a price premium.

Today's industry is extremely fragmented and specialized; numerous alternative technologies exist for each part of a building; and, for reasons of managing industrial relations risk and increased technical specialization, most of today's large projects are built by a team of 100 or more (and on extremely large and complex projects, more than 1,000) different design, fabrication and construction specialist organizations.

This extraordinary fragmentation has brought with it significant challenges to coordination, integration and innovation. Each subcontract organization works on many projects concurrently; and there are competing demands for resources between different projects, making it difficult for subcontractors to make reliable commitments to each project. Subcontractors attempt to manage the widely fluctuating demand on their resources by adopting a pyramid subcontracting strategy. However, this further challenges the overall project team's ability to effectively coordinate work and control critical outcomes such as safety, quality and productivity.

Coupled with the extreme fragmentation of their supply chains, many contractors in the sector favour short-term, cost driven relationships with their suppliers. Most head contractors are more interested in the speed of the job than smoothing out the resource requirements of their subcontractors because speed is the factor that has the greatest impact on costs and profits.

While in the past few decades major contractors have narrowed their supply chains, it is fair to say that their relationships with their suppliers are quite shallow. Almost all contractors collaborate with their suppliers only when working together on projects. Examples of deep, long-term commitment to collaboration with the supply chain are rare.

Within this fragmented delivery model, with low levels of trust between owners, suppliers and within the supply chain, open collaboration is further limited by the commercial terms of the contracts which tend to push as much of the risk and responsibility as possible down the supply chain. The outcome is that each party fiercely protects its margin and ensures that information is not shared openly. The tension between the self-interest of the parties and the common good of the project has never been higher.

During the tender stage, each subcontractor develops their own price in isolation. The head contractor chooses the lowest subcontract prices and submits a tender based on the aggregation of those prices plus a margin and an allowance for contingency. In this process, there is limited opportunity for collaborative innovation between the parties as they are all working in isolation. Once contracts are signed, collaboration between subcontractors working on the same project is limited as each fights to protect his/her own margin.

It is this broad scenario that has led the construction sector to stagnant productivity growth in the last half century while other sectors of the economy have improved their productivity by up to 200 per cent. Figure 1.2, from the US Department of Commerce, Bureau of Labor Statistics, plots the trend of construction sector productivity against the non-farm productivity index between 1964 and 2004.

Productivity improvement in all industries is driven by innovation in technology and through business process change. In today's construction industry, with limited collaboration between the parties in the supply chain, innovation is by and large limited to technical developments in individual product and service areas, and through advances in logistics and global procurement.

Given the size and complexity of major projects, it is essential to tap into the creative opportunities afforded by deep and effective collaboration across the very wide supply chain. To develop the creative solutions needed to drive improvements in quality, safety, productivity and client satisfaction, close collaboration between front-end designers and downstream fabricators and erectors is essential. Such relationships are difficult to achieve when trust is low, risk is simply pushed down the supply chain and there is little willingness to invest in solution generation early in the process.

COMMON CONSTRUCTION CHALLENGES

The complexities that are described in the previous section have led to performance challenges which are remarkably similar the world over.

- **Safety**
  Construction is one of the most hazardous industries – its accident record is third behind mining and forestry. New legislation has now made head contractors, designers and project managers more responsible and potentially liable for injuries incurred by workers on a site, regardless of who had employed them.

- **Quality**
  Research has shown that the cost of rectifying quality errors during and after the contract is of the same order as the profitability of organizations in the sector. Product quality problems are reflected in leaking buildings and premature deterioration of external finishes.[1]

- **Reliability**
  On most project sites only about half of the tasks planned one week out are actually completed according to plan. This means that while overall progress is being made in terms of a measure such as Value Earned, almost half of the work completed on the job each week is not what had been planned. Hence, individual contractors within the supply chain struggle to maintain efficient workflow and resource utilization, driving them to lose confidence in the planning process and to focus inwards on their own profitability.

- **Decision-making**
  Design often takes longer than anticipated and early budgets are rarely accurate. This reflects the fact that the gap between design and construction is too wide and the design decision-making process is out of touch with the real costs and opportunities of construction. It cannot effectively consider alternate construction methods and

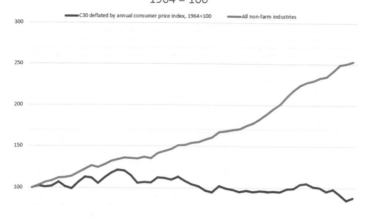

Construction Labor Productivity, 1964–2012
based on various deflators
1964 = 100

— C30 deflated by annual consumer price index, 1964=100   — All non-farm industries

FIGURE 1.2
US Construction vs non-farm productivity 1964–2004

Paul Teicholz, Professor (Emeritus), Department of CEE, Stanford University

materials. Under normal processes, clients are often not afforded the opportunity to make effective choices.

• **Value for money**

Most clients for constructed products the world over clutch to the belief that the lowest price in a tender represents the best value. While this may hold true for simple fast projects, nothing can be further from the truth for complex and uncertain projects. Too often clients call for tenders on a project providing scant documentation and expecting suppliers to invest in the problem solution at their own cost and, often, to accept unrealistic risk. This has created a very real dilemma for the construction sector. Public sector clients face probity issues, as should any client organization; however, under-investing in the generation of the request (design and planning), and insisting that the lowest price represents best value have led to more poor outcomes than good ones. Under such circumstances clients pay an upfront premium for the risk taken and, generally with time, almost always pay a premium for the inevitable variations to the scope of works.

Against this backdrop of complexity and poor performance, contractors and consultants both claim that their profit margins are very low.

The emergence of very large global organizations has led to rapid restructuring in some sectors of the industry. As these organizations have grown in size, the major challenge for them is to remain flexible, adaptive and creative – these are the skills required for the outstanding collaboration needed to optimize the design, fabrication and assembly of large complex projects.

Whatever the type of organization – whether it is an engineering or architectural design practice, a developer, a contractor, a building product manufacturer or a project manager – competition is intense: for end customers, for employees, for projects and for funds. This book is about the strategies that world leaders in the construction sector have deployed to remain at the leading edge of competitiveness.

LEAN PRODUCTION

The ideas of lean production were first introduced to the American automobile industry in 1980 by Fujio Cho, later to become the president of the Toyota Motor Corporation. At that time, these ideas were called the *Toyota Production System* (TPS) or *Just-in-Time* (JIT) manufacturing.

The term 'lean production' was coined by Womack and Jones and their research team a decade later when they were comparing production in the Japanese automobile industry to that of their Western competitors. They introduced the term in their ground-breaking book, *The Machine that Changed the World*. In every measure, they saw that *more was being done with less*: production costs were lower, inventory costs were lower and product development was faster with fewer resources. While in Western automobile factories, a third of all cars produced were sitting in a re-work pool at the end of the production line, waiting for quality errors to be rectified, at the end of Toyota's production line, there was not a single car in the re-work pool. Most importantly, customer perceptions were that the Japanese cars were more reliable.

The traditional definition of lean production is 'creating value for the customer with no waste'; waste being anything that does not add value to the customer. The concept of 'no waste' highlights the need for efficiency. A basic precept of a lean organization is the continuous drive to eliminate non-value-adding processes from all of its activities. Equally inherent in lean is the focus on quality, in the broadest sense. Giving the customer what he/she wants explicitly draws attention to customer needs and satisfaction and, as with the quality movement, this focus is on all customers: the initial customer (owner, developer), all intermediate customers (all the parties in the supply chain) and the end user. It is the dual focus on quality and efficiency that underpins competitiveness; hence, the focus of this book being lean quality (LQ).

JIT is often misunderstood within construction as simply delivering materials and products to site just as they are required. However, the terms JIT Manufacturing or JIT/TPS were used to describe the entire system of management that is considered to form a part of lean production. After the publication of *The Machine that Changed the World*, the term JIT/TPS has been gradually replaced by the term lean production.

However, as with many management terms, lean production is often used loosely. To redefine the term, Jim Womack sent a message to the Lean Enterprise Institute email list entitled 'Deconstructing the Tower of Babel'. He described how in 1987, working with a group of colleagues, they listed the performance attributes of a Toyota-style production system compared with traditional mass production. The Toyota-style production system:

- needed less human effort to design, make and service products;
- required less investment for a given amount of production capacity;
- created products with fewer delivered defects and fewer in-process turn-backs;
- utilized fewer suppliers with higher skills;
- went from concept to launch, order to delivery and problem to repair in less time with less human effort;
- could cost-effectively produce products in lower volume with wider variety to sustain pricing in the market while growing share;
- needed less inventory at every step from order to delivery and in the service system; and
- caused fewer employee injuries, etc.

The group very quickly ascertained that this system needed less of everything to create a given amount of value, so they called it 'lean', hence the term was born. In the intervening period, the term has become loosely applied to a great variety of improvement activity and so to set the record straight, Jim Womack wrote 'here's what lean means to me':

- it always begins with the customer;
- the customer wants value: the right goods or service at the right time, place and price with perfect quality;
- value in any activity – goods, services or some combination – is always the end result of a process (design, manufacture and service for external customers, and business processes for internal customers);
- every process consists of a series of steps that need be taken properly in the proper sequence at the proper time;
- to maximize customer value, these steps must be taken with zero waste;
- to achieve zero waste, every step in a value-creating process must be valuable, capable, available, adequate and flexible, and the steps must flow smoothly and quickly from one to the next at the pull of the downstream customer;
- a truly lean process is a perfect process; perfectly satisfying the customer's desire for value with zero waste; and
- none of us has ever seen a perfect process nor will most of us ever see one; but lean thinkers still believe in perfection, the never-ending journey towards the truly lean process.

Note that identifying the steps in the process, getting them to flow, letting the customer pull, etc. are not the objectives of lean practitioners; these are simply necessary steps to reach the goal of perfect value with zero waste.[2]

Essentially what Womack defined was simply *a focus on the customer, on creating value and on eliminating waste* – the ideal of any production process. The wastes he referred to were those defined by Engineer Ohno of Toyota:

- overproduction;
- waiting;
- excess conveyance;
- extra processing;
- excessive inventory;
- unnecessary motion; and
- defects requiring rework or scrap.

The elimination of these will:

- reduce the proportion of non-value-adding activities;
- reduce lead time;
- reduce variation;
- simplify processes;
- increase flexibility; and
- increase transparency.

The lean production agenda has an increasing amount in common with TQM, the main difference being that the lean focus tends to be on production/delivery processes whereas the TQM focus is on organizations as a whole.

Writing a definition of lean quality for all types of organization is extremely difficult, because the range of products, services and organization structures lead to different impressions of the nature and scope of lean quality. Basically lean quality is a programme directed towards ensuring that the right work is produced at the right quality at the right time throughout the entire supply chain without waste. Anyone who perceives it purely as a material control system, a short-term planning system or a way to reduce overheads and costs is bound to fail.

Lean quality can also be described as a disciplined series of operating concepts which allow the smooth and efficient flow of materials and services as they are required with the minimum amount of resources (facilities, equipment, materials, capital and people). Furthermore, it also creates a basis for the systematic identification of operational problems and proposes a set of tools and strategies for correcting them.

In some organizations JIT was introduced as 'continuous flow production'. This describes the objective of achieving the efficient conversion of purchased material and services to delivery very well (from suppliers to the customer). If this extends into the supplier and customer chains, all operating with lean quality, a perfectly continuous flow of materials, information and services will be achieved. In the VNGC case study there is a section on Takt time planning; this is essentially the application of this concept to the construction planning process.

Lean quality concepts can be used in all parts of all businesses, in administration to manage payments and invoicing, in design to manage the decision-making process as well as the design process, to manage off-site fabrication, logistics and the construction assembly process on site.

There is a well-established range of lean quality and operations management techniques; these include tools and strategies to monitor and analyse processes and outcomes. They include:

- flowcharting to better understand processes;
- process study and analysis to identify potential for improvement;
- preventive maintenance to avoid unplanned disruptions;
- equipment and materials layout to optimize material flow;
- standardized design to reduce process risk;
- statistical process control, applicable to the analysis of all data; and
- value analysis and value engineering to ensure that the focus is on achieving client needs in the most efficient manner.

The following ideas which are directly associated with the operation of JIT are also an important part of the implementation of lean quality:

1 batch or lot size reduction to produce smoother flow of materials and services;
2 flexible workforce to maintain smooth flow and to cope with unanticipated requirements;
3 visible cards that signal material requirements (Kanban);
4 mistake-proofing to ensure errors cannot happen;
5 pull-scheduling: one completed task pulling the other behind it;
6 set-up time reduction such as minimum crane time for assembly operations on site; rapid assembly and stripping of formwork; easy alteration to forms at changes in core configurations; and
7 standardized containers/trolleys for the transport and handling of materials.

In addition to these, joint development programmes with suppliers and customers, establishing long-term relationships, are beneficial. The closer the collaboration between customer and supplier and within the supply chain, the greater the capacity for innovation and the more capable the team is to manage risks in safety, quality and production. These benefits can only be achieved through close collaboration between capable, complementary partners.

There is clear evidence that companies in all industry sectors, all around the world, who, together with their supply chains, have successfully leveraged the benefits of lean quality have made spectacular improvements in their performance. This can be seen in:

- increased flexibility;
- more reliable quality resulting in less rework and disruption;
- more reliable process flow leading to less inventory on site, less damage and less double handling;
- better product and process integration; and
- standardization and simplification of products and processes.

UNDERSTANDING LEAN QUALITY IN CONSTRUCTION

Translated into the construction process, lean quality offers similar benefits to those achieved in manufacturing and service industries. In construction, some of the key areas to target are:

- reliable commitments: all parties (owners, designers, suppliers and contractors) must deliver on their promises for timely decisions and for task completion;
- work handover: all parties must satisfy the needs of the following trade (or designer) in terms of task completeness and quality, to allow everyone's work to proceed smoothly and safely; and
- workflow: all parties need to be well coordinated and progress through the project work (design or construction) at an even completion rate.

The processes of lean construction focus on the planning and management of the production system as a whole. Lean thinking seeks to shift the focus of individual parties in the supply chain from their own efficiency to the efficiency of the entire end-to-end process. For this shift to occur, commercial terms which encourage the seamless collaboration within the supply chain are required.

In construction it is not uncommon for the general contractor to bring a 'cheap' subcontractor on to a project on the basis that they will manage the risks. Generally, this costs everyone on the job, as the disruption to the flow of work caused by a less capable team member has flow-on effects for everyone. Naturally, as far as possible, everyone prices in the increased risk into their tenders as contingency so this is a zero sum game.

Going on to the field with the best team every time creates confidence and trust among the team members, increases the team's capacity for innovation and efficiency and allows all team members to reduce contingency.

THE CONCEPT OF WASTE IN CONSTRUCTION

Toyota introduced the concept of Just in Time (JIT): the idea that production should respond to consumer demand rather than mass production edict, thus improving overall efficiency by minimizing work in progress and inventory. However, a system with minimal inventory and work in progress cannot tolerate defective work, and this led to a focus on waste elimination.

The classical seven wastes in lean production were listed earlier in this chapter; however because construction is one-of-a-kind, project-based production and, therefore, significantly different to mass production, the conceptualization of waste has attracted significant attention. Bolviken *et al.* (2013) propose the following detailed taxonomy of waste. They conclude that some of these categories may be less useful in practice than others; however, we include it here to illustrate the broader philosophical thinking about the nature of waste.

Waste in the transformation of materials

1 Material waste: all materials not used in the final product, including materials that are damaged and need to be replaced, material waste through handling and re-handling, and material ordered surplus to requirements;
2 Sub-optimal use of materials: this can be a result of over-design or overly conservative construction, a relative waste, which will be optimized as technology improves design and construction safety margins;
3 Sub-optimal use of machinery, energy or labour.

Waste in the flow of work

4 Unnecessary movements by people: looking for, reaching for, stacking parts or tools, moving too far to get to facilities, location of site office;
5 Unnecessary work: doing things unnecessarily; for example, in design, preparing 2D documents as well as a 3D model when the latter is sufficient;
6 Inefficient work: doing necessary work in an inefficient manner; for example, working at height unnecessarily, when by redesigning the work it could be done on the ground, ergonomic aspects of work;
7 Waiting: lost time because preceding work is incomplete or because information, approvals, materials or equipment are not available.

Waste in the flow of product

8 Space not being worked in: this waste is specific to construction and can be seen as an equivalent of idle equipment, it is the unrealized opportunity for work;
9 Excess inventory (materials not being worked on): raw material, work in progress (WIP) or finished goods are all at risk of damage, they all accrue storage costs as well as mask production and delivery problems;
10 Unnecessary transport: moving materials or products unnecessarily on site, sending physical documents rather than transmitting them electronically or returning to the site office with information rather than transmitting it from the field.

Loss of value in the main product

11 Lack of quality: production of defective trade work on site or delivery of defective components to site creates rework; the organization of rework; scrap; and inspection, errors in documentation and design;
12 Lack of intended use or overproduction.

Value loss as a by-product of construction

13 Harmful emissions; and
14 Injuries and work-related sickness.

An additional waste, *Making-do, h*as been suggested by Koskela (2004). This refers to starting a task before all preconditions are ready. At times, the team on the ground does not have control over all the preconditions for work to proceed but have to choose between *making do* or delaying the project. When a design discipline gets ahead of overall design resolution resulting in rework, or on fast-tracked projects if foundation construction commences before the design is fully resolved, some foundations may be incorrect.

Greg Howell has suggested that the contingency that estimators include in the pricing of new projects is a waste peculiar to construction. The sector has poor data on productivity and comparative efficiency, and it does not capture the cost of even the simplest forms of waste which is rework arising from defective work. Cost estimates include contingency based on previous business as usual, costs which include rework and inefficiency.

THE LEAN QUALITY TRIANGLE FOR CONSTRUCTION

While the language of each movement (lean and quality) is somewhat different, and hence, the emphasis seems different, the conceptual frameworks of quality and lean at the core level of philosophy and principles have a common goal of meeting or seeking to exceed client needs and expectations as efficiently as possible.

Projects vary along a spectrum between simple and certain at one end, and complex and uncertain on the other. As complexity and uncertainty increase, it becomes more important to build flexibility into the commercial arrangements and organizational structures and to align the commercial interests of the client and the delivery team. Teams constructing complex and uncertain projects will inevitably have to deal with variability in scope and changes in risk profile during the delivery of their project.

As complexity increases, it is also important to effectively link downstream fabricators and erectors with the design team to ensure that solutions generated during design can be efficiently constructed. This type of flexibility is best provided through the use of relational contracts under which risk and reward are shared across the key members of the delivery team.

Lean construction involves all the parties in the design, fabrication and construction processes working together to best understand the needs of the customers (end and internal customers alike) and to jointly strive to eliminate non-value activities (waste) from their operations. This can only be achieved through the end-to-end collaboration of all parties, from the initial client to the end user, and everyone in between.

In traditional construction, the dominant approach to managing the delivery of constructed projects has been through project management, as against production management; however, it is increasingly recognized that the two approaches are fundamentally different, though complementary.

Koskela and Howell argue in their 2010 paper that there are three basic weaknesses in traditional construction project management:

- Planning as supported by sophisticated software products is not closely linked to the execution of the work but is based on an abstract representation of the project.
- There is no systematic approach to managing the execution of the work in the real world.
- Control is limited to taking corrective action after the event rather than as a process of continuous learning and improvement.

The unique aspects of the delivery of constructed products fundamentally alter the production process when compared to other industry sectors. The lean construction community worldwide has over the past two decades developed a three-part framework; which includes commercial terms, organization and a lean operating system to address the challenges described above. This is described in the Lean Construction Institute triangle shown in Figure 1.3.

Often people incorrectly equate lean production with the tools and management processes; while these are invaluable, lean production embraces all three sides of the triangle. The tools without appropriate commercial terms and an integrated organization are limited in what they can accomplish.

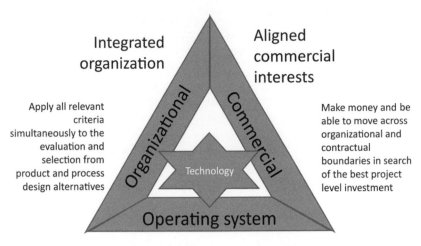

FIGURE 1.3
The Lean Construction Institute triangle

Integrated organization

Aligned commercial interests

Apply all relevant criteria simultaneously to the evaluation and selection from product and process design alternatives

Make money and be able to move across organizational and contractual boundaries in search of the best project level investment

Organizational

Commercial

Technology

Operating system

Lean management methods

Target Value Delivery   Value Stream Mapping   Last Planner System   Built-in Quality and Safety

Integrated organization

Over the years, construction industry clients have made many attempts to improve project outcomes, Design Build, Design Build Operate, Alliances and most recently Early Constrictor Involvement (ECI) and PPPs have sought to better align the interests of the clients and the supply chain and to encourage end-to-end collaboration in the search for better design solutions. While each of these strategies has had benefits, they also have had their limitations.

Northern Californian healthcare provider Sutter Health, in the US, and BA, in the UK, were among the first major building sector owners to adopt a combination of lean construction thinking and methods, and Integrated Project Delivery agreements to drive better performance on its projects. Several major owners, including Intel, Disney and the University of California in San Francisco have followed in their footsteps.

Sutter with support from its consultants developed the five *Big Ideas* to encompass the changes they sought from their project teams, shown in Figure 1.4.

These ideas define the principal objectives required to drive the kind of open, collaborative problem-solving and learning environment that is necessary to deliver large, complex projects effectively and efficiently. They are the guiding principles for how large, multidisciplinary teams engaged in all necessary activities from feasibility design through to fabrication, assembly and operations can work together to identify and manage risk, develop optimum multi-party solutions to complex problems and continuously improve performance through a creative and productive relationship.

True collaboration

This calls for end-to-end collaboration between front-end designers and tail-end fabricators and erectors, and everyone in between. Real collaboration maximizes positive iteration in the design process and reduces wasted effort and rework.

Increase relatedness between all the parties

For the best outcomes, participants must work closely together, be open to each other, jointly learn from mistakes and continue to improve and innovate in an open, collaborative manner.

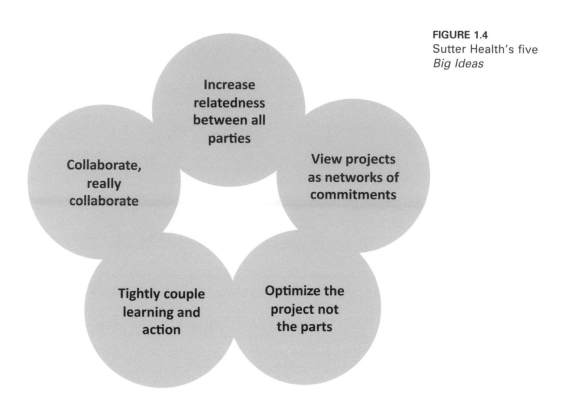

FIGURE 1.4
Sutter Health's five
Big Ideas

Increase
relatedness
between all
parties

View projects
as networks of
commitments

Collaborate,
really
collaborate

Tightly couple
learning and
action

Optimize the
project not
the parts

Projects are a network of commitments

Projects are built as a result of sound promises made between the actors on a project. The reliability of these commitments correlates with the reliability of workflow. Collective striving to improve the reliability of commitments brings personal accountability to the flow of work. This is in stark contrast to the traditional view that work is completed through contract and control.

Optimize the project – not the parts

Projects become more uncertain when the parties focus on self-interest at the expense of the project outcomes. However, the commercial terms have to support the shared focus on what is best for the project.

Tightly couple action with learning

The traditional construction industry tolerates error repetition to a remarkable degree. Errors in safety, quality and productivity are repeated on projects daily. Error repetition can only be eliminated through a focused learning environment.

Aligned commercial interests

To encourage behaviours consistent with the owner's interests and priorities, a commercial agreement is required that binds the interests of the parties together in alignment with the owner's interests, and rewards the achievement of clearly identified, exceptional outcomes that benefit the owner and the key stakeholders. This class of agreements is known as Integrated Project Delivery (IPD) or Integrated Form of Agreements (IFOA).

While infrastructure alliance projects in Australia sought to do this between 1997 and 2012, they did not go far enough to identify benefits that were commercially bankable for the client. Too often the Key Performance Indicators (KPI) and Key Result Areas (KRA) frameworks that mediated rewards were not sufficiently commercially focused, and though they may have been demanding of project teams initially, they became too easy to achieve with time. With the exception of programme alliances, once the KPI and KRA frameworks were set, they were not ratcheted up to drive and demonstrate continuous improvement throughout a project.

Commercially expedient thinking that does not engage with the risks and opportunities of a project and simply pushes risk and responsibility down the supply chain generally costs the client more than any alternative strategy. This is in spite of the fact that the client may tender and accept the lowest price in an open and transparent tender process. Large constructed projects are complex and the client inevitably benefits from engagement with the supply chain in a full exploration of the risks and opportunities of the project.

IPDs, at their simplest, seek to bind the designers and constructors to the objectives of the owner by creating alignment of the interests of all parties in the project with the interests of the owner. The selection of project participants is based on technical and cultural fit rather than price; and, it is accepted that the core members of the supply chain should be appointed early in the life of a project, creating the possibility for fabricators and constructors to participate in design decision-making.

Some clients prefer to tie the whole supply chain into the agreement, seeking to work with a co-operative of equals who collaborate to achieve the owner's objectives without the opportunity for any party to bias processes or decisions towards its self-interest. In these arrangements, clients are involved in weekly discussions about risk and opportunity; and the leadership team of the supply chain (including the owner) meet weekly to resolve problems and leverage opportunities for innovation. A key aspect of such arrangements is that money can move across organization boundaries within the project team in search of the best project level investments, and efficiency gains made by any single party are shared by the team.

Under these agreements, there is a commitment to learning and continuous improvement. All of the partners commit to being effective members of the project executive, putting forward able and empowered leaders to contribute to the team decision-making processes. Such agreements use terms such as the following to express the intent of the owner:

> By forming an Integrated Team, the parties intend to gain the benefit of an open and creative learning environment, where team members are encouraged to share ideas freely in an atmosphere of mutual respect and tolerance. Team Members shall work together and individually to achieve transparent and co-operative exchange of information in all matters relating to the Project, and to share ideas for improving Project Delivery as contemplated in the Project Evaluation Criteria. Team members shall actively promote harmony, collaboration and co-operation among all entities performing on the Project.
>
> The parties recognize that each of their opportunities to succeed on the Project is directly tied to the performance of other Project participants. The parties shall therefore work together in the spirit of co-operation, collaboration and mutual respect for the benefit of the Project, and within the limits of their professional expertise and abilities. Throughout the Project, the parties shall use their best efforts to perform the work in an expeditious and economical manner consistent with the interests of the Project.
>
> (Lichtig, 2005)

Lean quality management system

The third side of the Lean Construction Institute triangle refers to the management systems and tools used to drive excellent outcomes and continuous improvement. These apply to all phases and areas of the work from design to commissioning and, if relevant, operations as well. An underlying principle is the use of data to drive processes and improvement.

During the design phase, the key elements of lean management implementation are to:

* understand clearly the owner's need and priorities;
* use Target Value Design (TVD) to drive the design towards achievement of the owner's budget;
* create genuine collaboration between all the relevant team members to shape design resolution, linking fabricators, constructors and users with the initial designers to develop an optimum, holistic design solution (often supported by the use of 3D modelling);
* develop a Built-In Safety and Quality plan to ensure that safety and quality are considered from the outset and readily achievable; and
* manage the flow of decision and design work during the design phase using the Last Planner® System (LPS) and Reliable Promising, based on the Linguistic Action model.

It has been found, around the world, that typically up to 50 per cent of tasks assigned to a week's plan on the preceding Friday were not completed as planned. Using lean thinking and the LPS® the percentage of tasks completed can be increased to 85 per cent.

Other tools and approaches including Built-In Quality and Safety (BIQS) and Value Stream Mapping (VSM) are commonly used to mistake proof and improve assembly in construction processes. Traditional tools, such as visual management, 5S, standard work, BIM and virtual prototyping and value engineering, from lean production and elsewhere are also used to eliminate waste and improve value, primarily, for the end customer, but also with a view to the needs of intermediate customers. In fact, lean is not a set of tools; rather, it is a mindset to use all appropriate tools and techniques combined with the use of metrics to continuously improve key outcomes.

NOTES

1 Perea, S., Davis, S., Karin, K. and Marosszeky, M. *Enhancing Project Completion* (Research Project reports 1 & 2), ACCI, UNSW, 2003.
2 Email from Jim Womack to the Lean Enterprise Institute list titled 'Deconstructing the Tower of Babel', 7 October 2004.

BIBLIOGRAPHY

Abdelhamid, T. *Lean Construction Principles*, Graduate class offering at Michigan State University, 2007, www.slideshare.net/tabdelhamid/lean-construction-introduction

Bolviken, T., Rooke, J., and Koskela, L. The wastes of production in construction – a TVF based taxonomy, *Proceedings of the 22nd Annual Conference of the International Group for Lean Construction (IGLC)*, Oslo, Norway, 2014.

Koskela, L. Making do: the eighth category of waste, *Proceedings of the 12th Annual Conference of the International Group for Lean Construction (IGLC)*, Copenhagen, Denmark, 3–5 August 2004.

Koskela, L. and Howell, G. Reforming project management: the role of planning, execution and controlling, *IGLC Proceedings*, 2010, womackwww.iglc.net/Papers

Lichtig, W. *Ten Key Decisions to a Successful Construction Project (Choosing Something New: The Integrated Agreement for Lean Project Delivery)*, American Bar Association Forum on the Construction Industry, Sacramento, California, 2005.

Thomas, R., Marosszeky, M., Karim, K., Davis, S. and McGeorge, D. *Enhancing Project Completion*, Australian Centre for Construction Innovation, 2003.

Womack, J., Jones, D. and Roos, D. *The Machine that Changed the World*, Rawson MacMillan, 1990.

CHAPTER HIGHLIGHTS

Lean quality in construction

- Two major management movements have swept over the construction sector in the past 25 years: the Quality movement in the early 1990s and more recently the Lean Construction movement.
- Leading clients for construction services are beginning to require their suppliers and their supply chains to demonstrate a lean quality approach to their businesses.
- Lean and quality are two slightly different lenses through which we view management excellence and these perspectives are merging.

Distinguishing features of construction

- Constructed projects are uniquely designed, fitted to a site, and projects are fixed in location, requiring the resources necessary for construction – materials, products, people and equipment – to be brought to the site.
- Work teams from a myriad of different organizations construct the project in situ, and while they physically work around each other, the workers on site are employed by many different companies, each with different values and performance and business drivers.
- Large projects are undertaken over many years and the members of the design and construction team change significantly over this period; this requires special attention to the handover of work between individuals and teams to ensure that the work continues as smoothly as possible.
- When clients are involved in the design and construction process over an extended period, the quality of the service they receive from their providers influences their perceptions regarding the quality of products.
- Clients have the ability to influence the culture on a project as well as the cost; this influence can be either positive or negative.
- In the last decade the ability of even medium-sized firms to procure both services and products globally has fundamentally changed the capability of organizations; it has also posed a new set of risks and hence changed the demands on their management skills.

Today's industry

- There is a confusion among owners between cost and value; many owners drive towards the lowest cost, without understanding where the value is, nor do they understand their impact on cost.
- Owners are reluctant to invest in defining their needs upfront, they seek to get the supply chain to invest in generating the solution to their needs, paying as little as possible for it.
- Between World War II and the end of the twentieth century contractors moved from a detailed involvement with construction to a position where they simply bought services from subcontractors for a fixed price and passed on the risk.

- Today's industry is extremely fragmented and specialized and in general terms, trust is low. Productivity has been in decline for at least 50 years.
- In the last 15 years leading constructors have narrowed their supply chains and some are starting to deepen their relationships with their suppliers, though they are still wedded to the idea of simply buying services.
- Leading lean contractors have moved back to engaging more closely with the processes of construction, developing strategies to better manage risk jointly with their suppliers and are working more closely with them.

Common construction challenges

- Construction processes are error prone; construction is one of the most dangerous industries, the cost of rework due to quality errors during construction costs as much as the profit margin generated by most businesses in the sector.
- The flow of work on construction sites is chronically unreliable, with as much as half of work, planned one week out, not being completed as planned. This outcome undermines the confidence of all construction workers and managers in the planning process.
- While clients seek value from money above all else, they are rarely in a position to make informed decisions that influence value. There is insufficient information about the cost impact of alternative decisions for them to be practically involved.
- The last three decades have seen the rapid emergence of global organizations in every part of the construction supply chain from design through fabrication and construction. During the same period the size and complexity of projects has increased rapidly in every sector of the market.

Lean production

- The central ideal of lean production is to give the customer what he/she wants (value) with no waste. This provides a focus on understanding what customers want and on understanding what is waste and where it is to be found.
- Lean production and TQM have a lot in common; they are different and complementary lenses through which organizational excellence can be viewed. Lean production has tended to have a greater focus at the process level while TQM has had more of an overall organization focus. Today both frameworks are rapidly converging.
- Both lean and TQM are process focused, both are customer focused and both are focused on continuously improving processes and outcomes. The lean production community worldwide has developed a set of tools for focusing the attention of workers and their management on the improvement of processes. Though it is important to recognize that lean is not just tools-based, it is a holistic philosophy similar to TQM.

Understanding lean quality in construction

- In construction areas of focus for lean improvement are the generation of reliable commitments, satisfying the needs of the immediate customer in all stages of work handover and the creation of smooth and reliable workflow so resources can be applied efficiently throughout the supply chain.
- The processes of lean construction focus on improving the reliability and quality of planning, so that work can be executed as planned. There is also a rediscovery of the need to plan the overall workflow so that all parties can rely on a smooth level of demand for their resources.

The concept of waste in construction

- The lean manufacturing community has developed a taxonomy of waste; it has been found that there are several additional wastes that are unique to construction.
- Waste can occur in several different stages of the construction process: in the transformation of materials; in the flow of work; in the flow of product; and we can see a loss of value in the main product; and there can be value loss to external stakeholders as a by-product of construction.

The lean quality triangle for construction

- Projects vary along a spectrum between simple and certain on one end, and complex and uncertain at the other. As complexity and uncertainty increase, it is important to build flexibility both into the organization and commercial arrangements to allow for the inevitability of change.
- It is important to connect early stage designers with downstream fabricators and erectors in order to arrive at optimum construction solutions.
- Computer-based planning tools have created an illusion of certainty in relation to the planning of work, while in fact it is only the leaders of work teams who are in a position to plan work in detail.
- For lean construction to be successful, an integrated organization needs to be created, the commercial interests of the parties need to align around the efficiency of the project as a whole, and lean tools and management methods should be deployed using the latest available technologies.

Further concepts of lean quality

2

PRODUCT QUALITY THROUGH THE EYES OF THE CUSTOMER

'Is this a quality watch?' If you point to your wrist and ask this question of any group of people, be they undergraduates/postgraduates, experienced managers, engineers, contractors, doctors/nurses, chemists, it matters not who – the answers will vary:

- 'No, it's made in Japan.'
- 'No, it's cheap.'
- 'No, the face is scratched.'
- 'How reliable is it?'
- 'I wouldn't wear it.'

Clearly, the quality of a watch depends on what the wearer requires from a watch: perhaps a piece of jewellery to give an impression of wealth; a timepiece that gives the required data, including the date, in digital form; or one with the ability to perform at 50 metres under the sea? These requirements determine the quality.

Quality is often used to signify 'excellence' of a product or service – people talk about 'Rolls-Royce quality' and 'top quality'; in building, the different standards of finish specified for plasterwork depend on the location and lighting of the surface. Concrete can be manufactured to have low shrinkage and highly impervious characteristics to ensure that it performs well in a harsh environment. In some manufacturing companies, the word quality may be used to indicate that a piece of material or equipment conforms to certain physical dimensional characteristics often set down in the form of a particularly 'tight' specification. In a hospital it might be used to indicate some sort of 'professionalism'. If we are to define quality in a way that is useful in its management, then we must recognize the need to include in the assessment of quality the true requirements of the 'customer' – their needs and expectations.

Quality, then, is simply *meeting the customer requirements*, and this has been expressed in many ways by other authors:

- 'Fitness for purpose or use' – Juran, an early doyen of quality management;
- 'The totality of features and characteristics of a product or service that bear on its ability to satisfy stated or implied needs' – BS 4778. 1987 (ISO 8402, 1986) *Quality Vocabulary; Part 1, International Terms;*
- 'Quality should be aimed at the needs of the consumer, present and future' – Deming, another early doyen of quality management;
- 'The total composite product and service characteristics of marketing, engineering, manufacture and maintenance through which the product and service in use will meet

the expectation by the customer' – Feigenbaum, the first man to publish a book with 'Total Quality' in the title;

- 'Conformance to requirements' – Crosby, an American consultant famous in the 1980s;
- 'The quality of an organization's products and services is determined by the ability to satisfy customers and the intended and unintended impact on relevant interested parties' – ISO (BS EN) 9001:2015 *Quality Management Systems; fundamentals and vocabulary.*

In the case of constructed products, the work of the architect, engineers, product manufacturers, general and specialist contractors all play a critical role in meeting the expectations of the immediate client and the end users who are the final customers for the product. A failure in the work of any one of these can jeopardize the work of the entire team.

Furthermore, because constructed products are generally large and complex, and the user group is often quite diverse, the question of quality is even more vexed. While one group within the client organization may be focused on whole-of-life cost efficiency, another group may be almost entirely driven by the image that their new head office will create in the market. It is an important task for the construction suppliers to make sure that the client is clear regarding the brief and the definition of expectations: refer to the DPR case study (CS9) which describes the processes developed by a leading construction company addresses this complex task.

Every organization basically competes on its *reputation* (for quality, reliability, price and delivery) and most people now recognize that quality and efficiency are the most important of these competitive weapons. If you doubt that just look at the way some organizations, even whole industries in certain countries, have used lean quality to steal a march on their competitors. American, British, French, German, Italian, Japanese, Spanish, Swiss, Swedish organizations and organizations from other countries have used lean quality strategically to win customers, steal business resources or funding and be competitive. This sort of attention to efficiency and quality reduces price and improves reliability, customer satisfaction and long-term customer relations.

For every organization, there are several aspects of reputation that are important:

1 Reputation is built upon the competitive elements of being 'On-Quality, On-Time, On-Cost'.
2 Once you get a poor reputation for product or service quality, reliability or price, it takes a very long time to recover.
3 Reputations, good or bad, earned locally can quickly become recognized nationally and even internationally.

The management of competitive weapons such as lean quality can be learned like any other skill and, in time, they can be used to turn round a poor reputation. However, before anyone will buy the idea that lean quality is critical to success, they have to know what is meant by it.

Performance is currently a very competitive issue in construction and two recent conversations the authors have had with major repeat clients for construction services, one a manufacturer and the other a healthcare provider, suggest that the business of construction is about to change forever. One party said that in future they would prefer to contract with a single co-operative entity which comprises all key members of the construction supply chain (designers, fabricators and erectors) to ensure that they all act with a common purpose. The other said that the last party they would appoint on their next project would be the general contractor, because they add the least value in the process.

Three important things are changing in construction:

- some demanding, repeat clients for construction services are deeply dissatisfied with what the industry offers them and are committed to improving the outcomes they get from their suppliers;
- the maturing of BIM (Building Information Modelling) has created a platform for far better collaboration and communication between all parties in the process enabling solution optimization; and
- the introduction of IPD (Integrated Project Delivery) contracts in which all the key members of the supply chain are bound together in a painshare/gainshare agreement lays the foundation for open collaboration within the supply chain and between the suppliers and the owner.

These three factors have created circumstances in which clients are beginning to embrace radically different procurement strategies for major projects. These changes promise to unlock the potential for significant productivity improvements which have eluded the industry for decades. However, at the same time as they offer a new platform for collaboration, these factors fundamentally challenge the traditional roles and responsibilities of the parties within the supply chain.

While the traditional roles of the general contractor in project delivery are being challenged, these changes also create new opportunities for radically different approaches to service provision by lead suppliers to deliver better value for owners.

Understanding what your customers want

Reliability is used in two ways. Though product reliability and process reliability are two different concepts, they are key elements in building. Product reliability is the ability of the product to *continue* to meet customer requirements. Service reliability is the degree to which a service meets customer expectations, such as delivering the right goods to the right place at the right time, or arriving on time and cleaning up after the work is done.

Reliability ranks with quality and efficiency in importance since it is a key factor in many purchasing decisions where alternatives are being considered. Many lead service suppliers, whether they are contractors or designers, keep a metric indicating repeat business as a percentage of total business. Such a metric reflects a number of attributes including cost and quality. Reliability also ranks high in the composite group of values that inform perceptions of quality.

Many of the general management issues related to achieving product or service quality are also applicable to reliability: why do you buy a Japanese car? quality, reliability and price are the usual answers. Quality and reliability are used synonymously, often in a totally confused way. Clearly, part of the acceptability of a product or service will depend on its ability to function satisfactorily over a period of time, and it is this aspect of performance that is given the name reliability.

It is important to realize that 'meeting the customer requirements' definition of quality does not restrict the functional characteristics of products or services. First of all, cost and quality are always interlinked. Each year, innovation delivers higher levels of functional quality for a lower price, and customers are used to this. Furthermore, anyone with children knows that the quality of some of the products they purchase is more associated with satisfaction in ownership than some functional property. Of course, this is also true of many items purchased by adults, from antiques to clothing. The requirements for status symbols account for the sale of some executive cars, certain bank accounts and charge cards, and even hospital beds! Requirements are of paramount importance in the assessment of the quality of any product or service.

By consistently meeting customer requirements, we can move to a different plane of satisfaction – delighting the customer. There is no doubt that many organizations have so well ordered their capability to meet their customers' requirements time and time again that they have created a reputation for 'excellence'. A development of this thinking regarding customers and their satisfaction is customer loyalty, an important variable in an organization's success. Research shows that focus on customer loyalty can provide several commercial advantages:

- customers cost less to retain than acquire;
- the longer the relationship with the customer, the higher the profitability;
- loyal customers will commit more spend to their chosen supplier; and
- about half of all new customers come through referrals from existing clients (indirectly reducing acquisition costs).

Understanding and building lean quality chains

Typically, general contractors are reluctant to invest in the development of the capability of their supply chain partners believing that by doing so they are supporting their competitors. Because as much as 85 per cent of all work is undertaken by people employed by members of the supply chain, the close collaboration of the entire team is critical to the effective management of safety, quality, risk and innovation, and to the optimization of the construction process on every project.

An example from a building site:

> What the hell is Quality Control doing? We will have to demolish the entire wall. Not only do we lose the tiles, the waterproof membrane and the sheeting but the electrician and plumber will also have to come back to relocate the electrical and hydraulic services so that we can move the wall frame into its correct location. Where is the idiot surveyor who set the thing out?

This was accompanied by a barrage of verbal abuse (which will not be repeated here) aimed at the shrinking quality control manager who tried to slink into his office as the red-faced project manager advanced menacingly.

Was it the surveyor? Or was it the design detail of the architect, or was it the tradesman who built the wall? What about the three subsequent trades who all could have looked at the relationship between the position of the wall and the doorframe and seen that the finished wall would not line up. Each did his or her work well, but in isolation and a fundamental flaw at the outset compromised the entire job.

Do you recognize this situation? Does it not happen every day of the week – possibly every few minutes somewhere on a building site? Is it any different in banking, insurance and the health service? The inquisition of checkers and testers is the last bastion of desperate systems trying in vain to catch mistakes, stop defectives and hold up lousy materials before they reach the external customer and woe betide the idiot who lets them pass through!

Why are events like these so common? The answer to this is that we accept one thing – failure. Not doing it right the first time at every stage of the process. Why do we accept failure in the production of artefacts, the provision of a service or even the transfer of information? On a construction site we appear to accept that the same tradesmen will create the same safety hazards again and again! We close them out each time but they are repeated over and over. And some organizations prefer to litigate over quality issues rather than get the job done right. Yet in many walks of life we do not accept it. We do not say, 'Well, the nurse is bound to drop the odd baby in a thousand – it's just going to happen'. We do not accept that!

Customer – Outside organization

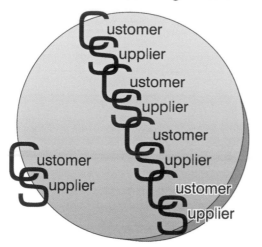

FIGURE 2.1
Lean quality chains

Supplier – Outside organization

In each department, each office, even each household there are a series of suppliers and customers. The PA is a supplier to the boss. Does the PA meet the requirements requested by the boss? Does the boss receive error-free information, set out as it is wanted, when it is wanted? If so, then the PA is delivering a quality service. Does the air steward receive from the supplier to the airline the correct food trays in the right quantity, at the right time; and so on?

Throughout, and beyond, all organizations, whether they be designers, constructors, fabricators, manufacturers, banks, retail stores, universities, hospitals or hotels, there is a series of quality chains of customers and suppliers (Figure 2.1) that may be broken at any point by one person or one piece of equipment not meeting the requirements of the customer, internal or external. The interesting point is that this failure usually finds its way to the interface between the organization and its outside customers and it is the people who operate at that interface – like the air steward – who usually experience the ramifications. The concept of internal and external customers–suppliers forms the core of the so-called 'lean quality' approach.

A great deal is written and spoken about employee motivation as a separate issue. In fact, the key to motivation and lean quality is for everyone to understand who their customers are, and what each customer needs. This takes the concept of customer satisfaction beyond the traditional end customer who buys a product or service, and includes every customer in every process transaction throughout the supply chain.

In the design and construction process these relationships are often implicit rather than explicit. There is a very long chain of events starting with planning and design, where information is built on information, while in construction, material is built upon material. At every step, each person has several customers, including immediate and end-user customers. Often though, because the contractual relationship is with a third party (the client or the head contractor), the process relationship is not recognized as being critical. The simple fact is that the plasterer is the customer of the bricklayer, and the plasterboard sheet-fixer is the customer of the carpenter who has framed the wall; even though the head contractor pays all of them. Making these interdependent relationships explicit is the key to underpinning a change in productive behaviour and an increase in efficiency.

Lean quality has to be managed; it will not just happen. To achieve transformational results, it must involve everyone in the process and be applied throughout all the organizations in the supply chain. Many people in the support functions of organizations never see, experience or touch the products or services that their organizations buy or provide; but they do handle or produce things like purchase orders or invoices. If every fourth invoice carries but one error, what image of quality is transmitted?

Failure to meet the requirements in any part of a lean quality chain has a way of multiplying; and a failure in one part of the system creates problems elsewhere, leading to yet more inefficiency and failure, more problems and so on. The prerequisite of lean quality is the continual examination of all the requirements and our ability to meet them, both in terms of product quality and reliability, and process reliability and efficiency. This will lead to a 'continual improvement' philosophy in both product and process. The benefits of making sure that requirements are met at every stage and every time are truly enormous in terms of increased competitiveness and market share, reduced costs, improved productivity and delivery performance, and the elimination of waste.

Meeting the requirements

When lean quality satisfies customer requirements at every step of the process the outcomes are vast. The requirements may include a full understanding of needs, availability, capability, delivery, reliability, maintainability and cost-effectiveness among many other features. On construction sites availability of resources, accuracy of work, cleanliness of the work areas and the completion of the work in a manner that will allow the follow-on trade to continue in an optimal way are all critical issues. The first item on *the list of things to do* is: find out what the requirements are. If we are dealing with a customer–supplier relationship crossing two organizations then the supplier must establish a 'market research' activity or process charged with this task. A story included in the Southland Industries (SI) case study (CS6) is very pertinent here; on a project the SI supervisor was walking around the site with the client representative, when the client asked the SI supervisor 'what can you see wrong with that ductwork?', 'not much' was the reply because in fact the SI supervisor was very proud of the job they had done. The owner then went on to complain that the standard of finish on the ductwork was far too good; he did not want to pay for the unnecessarily high standard of work that had been delivered on this particular project: a lesson about the need for better market research.

The marketing process must of course understand not only the needs of the customer but also the ability of their own organization to meet them. If your customer places a requirement on you to run 1,500 metres in four minutes, and you know you are unable to meet this demand, then something has to be done to improve your running performance. Of course you may never be able to achieve this requirement and if you believe that to be the case, you need to let your client know this as soon as possible.

Real customer needs are rarely discussed on site. The formworker, who is paid by the square metre, is busy laying as much flat slab sheeting as possible so that his payment is maximized. He rarely stops to think that the steel-fixer, coming right behind him, is slowed down by the absence of edge boards. In the design process the architect and client often continue to make decisions well after the design engineers have started their design work in the belief that the design decisions are locked in. Throughout the construction supply chain we see examples of parties optimizing their own positions at the expense of others, or simply acting independently rather than interdependently.

To achieve quality throughout a project, each party in the quality chain must interrogate every interface as follows:

Customers
- Who are my immediate and final customers?
- What are their true requirements? Do they clash?
- Do the requirements pose unreasonable risk?
- Do the requirements add unnecessary cost?
- How do or can I find out what the requirements are?
- How can I measure my ability to meet the requirements?
- Do I have the necessary capability to meet the requirements?
- (If not, then what must change to improve the capability?)
- Do I consistently meet the requirements?
- (If not, then what prevents this from happening, when the capability exists?)
- How can I get feedback on my performance?
- How do I monitor changes in the requirements?

Suppliers
- Who are my immediate suppliers?
- What are my true requirements?
- Do my requirements create unnecessary risk?
- Are my requirements creating unnecessary cost?
- Are my processes and demands undermining productive collaboration?
- How do I communicate my requirements?
- How do I or they measure their ability to meet the requirements?
- Do my suppliers have the capability to meet the requirements?
- Do my suppliers continually meet the requirements?
- How do I inform them of their performance?
- How do I inform them of changes in the requirements?

The measurement of capability is extremely important if quality chains are to be formed within or outside of an organization. Each person in the organization must also realize that the supplier's needs and expectations must be respected if the requirements are to be fully satisfied. Furthermore, their needs and expectations may change with time.

To understand how lean quality may be built into a product or service, at any stage, it is necessary to examine the two distinct but interrelated aspects of quality:

- quality of design; and
- quality of conformance to the design.

Lean quality in design

We are all familiar with the old story of the tree swing (Figure 2.2); but in how many places, in how many organizations is this chain of activities taking place? To discuss the quality of, say, a chair, it is necessary to describe its purpose. What is it to be used for? If it is to be used for watching TV for three hours at a stretch then the typical office chair will not meet this requirement. The *difference* between the quality of the TV chair and the office chair is not a function of *how* it was manufactured, but a function of its *design.*

Quality of design is a measure of how well a product or service is designed to achieve the agreed requirements. The beautifully presented gourmet meal will not necessarily please the recipient if he or she is travelling on the motorway and has stopped for a quick bite to eat. The most important feature of the design with regard to achieving quality is the specification. Specifications must also exist at the internal supplier–customer interfaces if one is to achieve a total quality performance. For example, the company lawyer asked to

1 What marketing
   suggested

2 What management
   approved

3 As designed by
   engineering

4 What was
   manufactured

5 As maintenance
   installed it

6 What the customer
   wanted

FIGURE 2.2 Quality of design

draw up a purchasing contract for a major item from an overseas supplier by the project manager requires a specification as to its content:

1   supply only or design and fabricate;
2   the type of contract to be used: fixed price or schedule of rates;
3   who are the contracting parties;
4   in which countries are the parties located;
5   what are the products involved;
6   what are the performance requirements;
7   what is the volume;
8   what are the shipping details: timing of deliveries; and
9   what are the financial aspects: price escalation, currency risk?

The financial controller must issue a specification of the information he or she needs, and when it is needed, to ensure that foreign exchange fluctuations do not put the project finances at risk. The business of sitting down and agreeing on a specification at every interface will clarify the true requirements and capabilities. It is the vital first stage for a successful total quality effort.

There must be a corporate understanding of the organization's quality position in the marketplace. It is not sufficient that marketing specifies a product or service 'because that is what the customer wants'. There must be an agreement that the operating departments can achieve that requirement. Should they be incapable of doing so then one of two things must happen: either the organization finds a different position in the marketplace or substantially changes the operational facilities.

A specific challenge in construction is that every design has to meet multiple requirements. It is not enough that the client likes the look of the building or structure; it also must be able to be built safely to the required standard of quality and price. The design process is an exercise in team-based optimization against multiple criteria – there is almost always conflict between some of the objectives. Design may be a part of the supply contract.

Quality of conformance to design

The quality of conformance to design is the extent to which the product or service achieves the quality intended in the design. What the customer actually receives should conform to the design, and operating costs should be tied firmly to the level of conformance required. Quality cannot be inspected into products or services; customer satisfaction must be designed into the whole system at the outset. The conformance check makes sure that the design is achieved as planned, in every aspect of process, service and product.

An important issue on construction sites is the avoidance of design-generated risks. This requires experienced management. Designers are often unaware of the construction implications of their detailing; architects routinely alter details unnecessarily without thinking through the construction consequences; similarly, structural engineering details are at times unnecessarily complex and variable. Especially in high-risk areas, it makes sense to standardize construction details so that tradesmen on site get to know their work and instinctively do it right.

A high level of inspection or checking at the end is generally indicative of an attempt to inspect in quality. This may well result in spiralling costs and decreasing viability. The area of conformance to design is concerned largely with the quality performance of the actual operations. It may be salutary for organizations to use the simple matrix of Figure 2.3

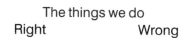

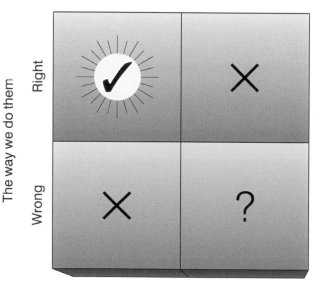

FIGURE 2.3
How much time is spent doing
the right things, right?

to assess how much time they spend doing the right things, right. A lot of people, often through no fault of their own, spend a good proportion of the available time doing the right things, wrong. There are people (and organizations) who spend time doing the wrong things very well, and even those who occupy themselves doing the wrong things, wrong – this can be very confusing!

MANAGING LEAN QUALITY

The quality of finishes in many areas of construction are hard to judge, standards are loosely defined and assessments of quality are subjective. These areas cause more conflict among the architect, subcontractor and head contractor than almost any other issue, often leading to serious disputes between the parties.

On many construction sites we may often see an argument between the quality manager and the construction supervisor. The construction supervisor wants to see progress and wants to see one trade follow on immediately behind the other while the quality manager wants the preceding work to be 100 per cent correct. They argue and debate the evidence before them, the rights and wrongs of the specification, and each tries to convince the other of the validity of their argument. Sometimes things can get quite heated.

This ritual is associated with trying to answer the question, *'Have we done the job correctly?'* Or put another way, *'What is the risk associated with proceeding?'* The words *correctly* and *risk* can depend on the interpretation given to the specification on that particular day. This is not quality *control*; it is *detection* – wasteful detection of questionable product when it is already too late. There is still a belief in some quarters that to achieve quality we must check, test, inspect or measure the ritual pouring on of quality at the end of the process when, in fact, the checks have to be built into the process; they must be an integral part of it. In construction, this is called Built-In Quality.

In the office, there are staff to check people's work before it goes out, validating computer data, checking invoices, word processing and so on. There is also quite a lot of looking for things, chasing why things are late, apologizing to customers for lateness: waste, waste, waste!

To get away from the natural tendency to rush into the detection mode, it is necessary to ask different questions in the first place. We should not ask whether the job has been done correctly; we should ask first, *'Are we capable of doing the job correctly?'* This question has wide implications, and this book is devoted largely to the various activities necessary to ensure that the answer is yes. However, we should realize straight away that such an answer will only be obtained by means of satisfactory methods right through the supply chain; appropriate design and documentation, materials and equipment; suitable skills, instruction and leadership; and a satisfactory overall 'process'.

Lean quality and processes

As we have seen, quality chains can be traced right through the business or service processes used by any organization. A process is the transformation of a set of inputs into outputs that satisfy customer needs and expectations in the form of products, information or services. Everything we do is a process so in each area or function of an organization there will be many processes taking place. For example, a finance department may be engaged in budgeting processes, accounting processes, salary and wage processes, costing processes, etc. Each process in each department or area can be analysed by an examination of the inputs and outputs. This will determine some of the actions necessary to improve quality and remove waste. There are also numerous cross-functional processes.

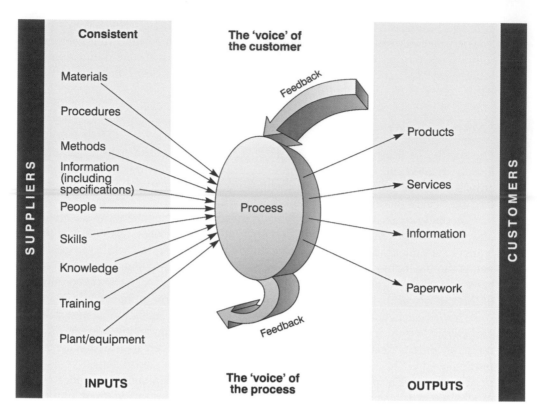

FIGURE 2.4 A process

The output from a process is that which is transferred to somewhere or to someone – the *customer*. Clearly, to produce an output that meets the requirements of the customer, it is necessary to define, monitor and control the inputs to the process which in turn may be supplied as output from an earlier process. At every supplier–customer interface there resides a transformation process (Figure 2.4), and every single task throughout an organization must be viewed as a process in this way.

Once we have established that our process is capable of meeting the requirements, we can address the final question, *'Do we continue to do the job correctly?'* This, of course, brings a requirement to monitor the process and the controls on it.

If we now re-examine the initial question *'Have we done the job correctly?'*, we can see that if we monitor and control our processes, and we can say with confidence that *'We are capable of doing the job correctly'* and *'We continue to do the job correctly'*, we have removed the need to ask the 'inspection' question, and replaced a strategy of *detection* with one of *prevention*. This concentrates all our attention on the front-end of any process – the inputs – and changes the emphasis to making sure that the inputs are capable of meeting the requirements of the process. This is a managerial responsibility and is discharged by efficiently organizing the inputs and resources, and by controlling the processes.

These ideas apply to every transformation process; they all must be subject to the same scrutiny of the methods, the people, skills, equipment and so on to make sure they are correct for the job. A person giving a lecture whose audio/visual equipment will not focus correctly, or whose teaching materials are not appropriate, will soon discover how difficult it is to provide a lecture that meets the requirements of the audience.

In every organization there are some very large processes comprising groups of smaller processes often called *core business processes.* These are activities the organization must carry out especially well if its mission and objectives are to be achieved. In a design organization this includes design progress, documentation quality and documentation progress; in a construction organization this includes procurement, logistics, work planning, constructed quality and reliability of workflow. This area will be dealt with in some detail later on in the book. It is crucial for the management of quality that it be integrated into the strategy of the organization.

The *control* of quality can take place only at the point of operation or production: where the letter is word-processed, the design is documented, the physical work is done or the building product manufactured. The act of *inspection is not quality control.* When the answer to '*Have we done the job correctly*' is given indirectly by answering the questions of capability and control then we have *assured* quality; and the activity of checking becomes one of *quality assurance* – making sure that the product or service represents the output from an effective *system* to ensure capability and control. It is frequently found that organizational barriers between functional or departmental 'empires' encourage the development of testing and checking of services or products in a vacuum, without interaction with other departments.

Quality control then is essentially the activities and techniques employed to achieve and maintain the quality of a product, process or service. It includes a monitoring activity but is also concerned with finding and eliminating causes of quality problems so that the requirements of the customer are continually met.

Quality assurance is broadly the prevention of quality problems through planned and systematic activities (including quality documentation). These will include the establishment of a good quality management system and the assessment of its adequacy, the audit of the operation of the system and the review of the system itself.

In the Heathrow case study (CS10) the client described how the partner contractors are responsible for quality control while the owner's project manager is responsible for quality assurance.

LEAN QUALITY STARTS WITH UNDERSTANDING THE NEEDS

The marketing processes of an organization must be designed to establish the true requirements for the product or service. Having determined the need, the organization should define the market sector and demand to determine such product or service features as grade, price, quality, timing, etc. For example, a major hotel chain thinking of opening a new hotel or refurbishing an old one will need to consider its location and accessibility before deciding whether it will be predominantly a budget, first-class, business or family hotel. This will determine the functions planned and the capital budget for building or refurbishment.

The organization will also need to establish customer requirements by reviewing the market needs, particularly in terms of unclear or unstated expectations or preconceived ideas held by immediate and end-user customers. It is essential to identify the key characteristics that determine the suitability of the product or service in the eyes of the customer. Depending on the situation, this may, of course, call for the use of market research techniques, data gathering and analysis of customer complaints. If necessary, quasi-quantitative methods may be employed, giving proxy variables that can be used to grade the characteristics in importance and decide in which areas superiority over competitors exists. It is often useful to compare these findings with internal perceptions. However, in many instances within the construction supply chain, the immediate customer is the following trade contractor or designer and therefore meetings to tease out perceptions, competing issues and needs have to take place face-to-face, at the very beginning of a new project. Often at this point, support from second-party research to better understand needs and satisfaction of end customers may be of assistance.

Excellent communication between customers and suppliers is the key to achieving a lean quality performance; it will eradicate the 'demanding nuisance/idiot' view of customers, which still pervades some organizations. Poor communications often occur in the supply chain between organizations when the parties are unaware of just how poor they are. Where dissatisfied customers and suppliers do not communicate their dissatisfaction, feedback from both customers and suppliers needs to be improved. Creating a running conversation between suppliers and customers about satisfaction with products and services whenever work is handed over from one party to another throughout the project helps suppliers to better understand how well they are satisfying their customers' needs. In the absence of such conversations, non-conformance of products or services may often be due to customers' inability to communicate their requirements and satisfaction clearly. If these ideas are also used within an organization, then the internal supplier/customer interfaces will operate much more smoothly.

All the efforts devoted to finding the nature and timing of the demand will be pointless if there are failures in communicating the requirements throughout the organization promptly, clearly and accurately. The marketing processes should be capable of producing a formal statement or outline of the requirements for each product or service. This constitutes a preliminary set of *specifications*, which can be used as the basis for service or product design. The information requirements include:

1 characteristics of performance and reliability – these must make reference to the conditions of use and any environmental factors that may be important;
2 aesthetic characteristics such as style, colour, smell, task, feel, etc.;
3 any obligatory regulations or standards governing the nature of the product or service; and
4 response time to a request and reliability of completion required.

The organization must also establish systems for feedback of customer information and reaction; and these systems should be designed on a continuous monitoring basis. To improve the response to customer experience and expectations, any information pertinent to the product or service should be collected and collated, interpreted, analysed and communicated. These same principles must also be applied inside the organization if continuous improvement at every process interface is to be achieved. If one function or department in a company has problems recruiting the correct sort of staff, for example, and HR has not established mechanisms for gathering, analysing and responding to information on new employees then frustration and conflict will replace communication and co-operation.

One aspect of the analysis of market demand that extends back into the organization is the review of market readiness of a new product or service. For example, a product manufacturer introducing a new product into the market would have to pay attention to:

1 suitability of the distribution and customer-service processes;
2 training of personnel in the 'field';
3 availability of 'spare parts' or support staff; and
4 evidence that the organization is capable of meeting customer requirements.

All organizations receive a wide range of information from customers through invoices, payments, requests for information, letters of complaint, responses to advertisements and promotion, etc. An essential component of a system for the analysis of demand is that this data is channelled quickly into the appropriate areas for action and, if necessary, response.

There are various techniques of research which are outside the scope of this book but have been well documented elsewhere. It is worth listing some of the most common and useful general methods that should be considered for use both externally and internally:

- surveys – questionnaires, etc.;
- panel or focus group techniques;
- in-depth interviews;
- brainstorming and discussions;
- role rehearsal and reversal; and
- interrogation of trade associations.

The number of methods and techniques for researching the market is limited only by imagination and funds. The important point to stress is that the supplier, whether an internal individual or an external organization, must keep very close to the customer. Good research coupled with analysis of complaints data is an essential part of identifying requirements and ending the obsession with inward scrutiny, the scourge of quality.

> Have a close look at the Graniterock case study (CS5) and in particular at the way in which the company uses regular survey processes to collect market information to guide the development of strategies for improved performance. Product development and process improvement are both driven by a comprehensive market research process.

LEAN QUALITY IN ALL FUNCTIONS

For an organization to be truly effective, each component of it must work properly together. Each part, each activity, each person in the organization affects and is in turn affected by others. Errors have a way of multiplying and failure to meet the requirements in one part or area creates problems elsewhere, leading to yet more errors, yet more problems, and so on. The benefits of getting it right first time everywhere are enormous. The challenges for achieving this in a construction project setting, where the mix of organizational goals, styles and cultures are very diverse, are considerable and require excellent leadership.

Everyone experiences, almost accepts, problems in working life. This causes people to spend a large part of their time on useless activities: correcting errors, looking for things, finding out why things are late, checking suspect information, rectifying and reworking, apologizing to customers for mistakes, poor quality and lateness. The list is endless, and it is estimated that about one-third of our efforts are still wasted in this way. In the construction sector it can be even higher.

Lean quality, when defined as meeting customer requirements with no waste, provides all the people within any organization a common language for targeting improvement. It enables all the people in the supply chain, with different abilities and priorities, to communicate readily with one another in pursuit of a common goal. As we saw in the first chapter, when business and industry were local, craftsmen could manage more or less on their own. The construction business is now so complex and employs many different specialist skills and specialist organizations, it has forced everyone to rely on the contribution of others to successfully complete their work.

Some of the most exciting applications of lean quality have materialized from groups of people who could see little relevance when first introduced to its concepts. Following training, many different parts of organizations can show the usefulness of the techniques. Using lean quality methods, accounting staff have improved the quality and timeliness of invoicing and payments; cost efficiencies have been designed and built into projects through target cost design; design and construction staff have improved workflow reliability by improving the reliability of their promises; the labour productivity on formwork and scaffolding has been significantly improved through work study and performance measurement; suppliers have improved the timeliness of their material deliveries to site; and the quality of design documentation has improved by using BIM (Building Information Modelling) and measuring the number of deviations between the actual and the BIM model to drive improvement.

It is worthy of mention that the first points of contact for some outside customers are the telephone operator; the security people at the gate; or the person in reception. Equally, the e-business, paperwork and support services associated with the product and services, such as websites, invoices and sales literature, and their handlers must match the needs of the customer. Clearly, lean quality cannot be restricted to the 'production' or 'operations' areas without losing great opportunities to gain maximum benefit.

Managements that rely heavily on exhorting the workforce to 'do the right job, right the first time', or 'accept that quality is your responsibility', will not only fail to achieve quality and eliminate waste but may create division and conflict. These calls for improvement infer that faults are caused only by the workforce and that problems are departmental or functional when, in fact, the opposite is true – most problems are inter-departmental. The commitment of all members of an organization is a requirement of organization-wide LQ improvement. Everyone must work together at every interface to achieve improved performance and this can only happen if the top management is really committed.

BIBLIOGRAPHY

Abdelhamid, T. *Lean Construction Principles*, Graduate class offering at Michigan State University, 2007, www.slideshare.net/tabdelhamid/lean-construction-introduction

Beckford, J., *Quality, a Critical Introduction* (3rd edn), Routledge, London, 2007.

Bolviken, T., Rooke J., and Koskela, L. The wastes of production in construction – a TVF based taxonomy, *Proceedings of the 22nd Annual Conference of the International Group for Lean Construction* (IGLC), Oslo, Norway, 2014.

Crosby, P.B. *Quality is Free*, McGraw-Hill, New York, 1979.

Crosby, P.B. *Quality Without Tears*, McGraw-Hill, New York, 1984.

Deming, W.E. *Out of the Crisis*, MIT, Cambridge, MA, 1982.

Deming, W.E. *The New Economies*, MIT, Cambridge, MA, 1993.

Feigenbaum, A.V. *Total Quality Control* (4th edn), McGraw-Hill, New York, 2004.

Garvin, D.A. *Managing Quality: The Strategic Competitive Edge*, The Free Press (Macmillan), New York, 1988.

Juran, J.M. and DeFeo, J.A. *Juran's Quality Handbook*, McGraw-Hill, New York, 2010.

Koskela, L. Making do: the eighth category of waste, *Proceedings of the 12th Annual Conference of the International Group for Lean Construction* (IGLC), Copenhagen, Denmark, 3–5 August 2004.

Koskela, L. and Howell, G. Reforming project management: the role of planning, execution and controlling, *IGLC Proceedings*, 2010, womackwww.iglc.net/Papers

Lichtig, W. *Ten Key Decisions to a Successful Construction Project (Choosing Something New: The Integrated Agreement for Lean Project Delivery)*, American Bar Association Forum on the Construction Industry, Sacramento, California, 2005.

Murphy, J.A. *Quality in Practice* (3rd edn), Gill and MacMillan, Dublin, 2000.

Thomas, R., Marosszeky, M., Karim, K., Davis, S. and McGeorge, D. *Enhancing Project Completion*, Australian Centre for Construction Innovation, 2003.

Womack, J., Jones, D. and Roos, D. *The Machine that Changed the World*, Rawson MacMillan, 1990.

CHAPTER HIGHLIGHTS

Product quality through the eyes of the customer

- Only when you understand the construction setting and its challenges, can you develop strategies for improvement that will be sound.
- When looking at the problems faced by the sector, it is important to understand the linkages between safety, quality and production; they are all different outcomes from the lack of an process reliability.
- The reputation enjoyed by an organization is built by quality, reliability, delivery and price. Quality is the most important of these competitive weapons.
- Reputations for poor quality last for a long time, and good or bad reputations can become national or international. The management of quality can be learned and used to improve reputation.
- Quality is meeting the customer requirements, and this is not restricted to the functional characteristics of the product or service.
- Reliability is the ability of the product or service to continue to meet the customer requirements over time.
- Organizations 'delight' the customer by consistently meeting customer requirements, and achieve a reputation for excellence and customer loyalty.

Understanding what your customers want

- Product reliability ranks with quality and efficiency in importance in the eyes of customers, and these are key factors in many purchasing decisions where alternatives are being considered.
- Service reliability is a critical aspect of service quality as poor service can influence customers' perception of product quality.
- Research shows that focus on customer loyalty can provide commercial advantage; customers cost less to retain than acquire; the longer the relationship with the customer, the higher the profitability; loyal customers commit more to spend; and about half of all new customers come through referrals.

Understanding and building lean quality chains

- Throughout all organizations there are a series of internal suppliers and customers. These form the so-called 'quality chains', the core of 'company-wide quality improvement'.
- The internal customer/supplier relationships must be managed by interrogation (i.e. using a set of questions at every interface). Measurement of capability is vital.
- Similar quality chains exist across the whole of the construction procurement process between subcontract suppliers who work jointly to produce the end product.
- There are two distinct but interrelated aspects of quality: design and conformance to design. *Quality of design* is a measure of how well the product or service is designed

to achieve the agreed requirements. *Quality of conformance to design* is the extent to which the product or service achieves the design. Organizations should assess how much time they spend doing the right things, right.

Meeting the requirements

- Understanding who the customers are and what they want is the first step to meeting the requirements. Often, this requires market research as the answer is rarely self-evident.
- There is a responsibility on customers to clearly define their needs through a sound request; the more detail that can be given in the request, the more appropriate is the response from the potential service provider.
- If every supplier is focused on satisfying the needs of their customers, and if they receive feedback on their performance at regular intervals, the basis of a virtuous cycle of improvement has been established.

Managing lean quality

- Asking the question 'Have we done the job correctly?' should be replaced by asking, 'Are we capable of doing the job correctly?' and 'Do we continue to do the job correctly?'
- Asking the questions in the right order replaces a strategy of *detection* with one of *prevention*.
- Everything we do is a process that is the transformation of a set of inputs into the desired outputs.
- In every organization there are some core business processes that must be performed especially well if the mission and objectives are to be achieved.
- Inspection is *not* quality control. The latter is the employment of activities and techniques to achieve and maintain the quality of a product, process or service.
- *Quality assurance* is the prevention of quality problems through planned and systematic activities.

Lean quality starts with understanding the needs

- Marketing processes establish the true requirements for the product or service. These must be communicated properly throughout the organization in the form of specifications.
- Excellent communications between customers and suppliers is the key to a total quality performance: the organization must establish feedback systems to gather customer information.
- Appropriate research techniques should be used to understand the 'market' and keep close to customers, and maintain the external perspective.

Lean quality in all functions

- All members of an organization need to work together on organization-wide quality improvement. The co-operation of everyone at every interface is necessary to achieve improvements in performance. This, however, can only happen if the top management is really committed.

Models and frameworks for total lean and quality management

3

EARLY TQM FRAMEWORKS

In the early 1980s manufacturers in the US and Europe found that Japanese industry had become much more competitive. In some instances, Japanese products of similar or better quality were available in the market at less than the cost price of making them in the West. This created an immediate interest in Japanese manufacturing processes; and the most obvious point of difference was the adoption of lean quality management practices. Initially, industry leaders were confused because they saw the practice of quality management as being inextricably linked with Japanese culture. It took some years to untangle the two and develop a clearer understanding of the core management practices that had created the differences in process productivity. At this stage there were many attempts to construct lists and frameworks to help this process.

Famous American 'gurus' of quality management, such as W. Edwards Deming, Joseph M. Juran and Philip B. Crosby, started to try to make sense of the labyrinth of issues involved, including the tremendous competitive performance of Japan's manufacturing industry. Deming and Juran had contributed to building Japan's success in the 1950s and 1960s and it was appropriate that they should set down their ideas for how organizations could achieve success.

Deming's 14 points to help management:

1 Create constancy of purpose towards improvement of product and service.
2 Adopt the new philosophy. We can no longer live with commonly accepted levels of delays, mistakes, defective workmanship.
3 Cease dependence on mass inspection. Require instead statistical evidence that quality is built in.
4 End the practice of awarding business on the basis of price tag.
5 Find problems. It is management's job to work continually on the system.
6 Institute modern methods of training on the job.
7 Institute modern methods of supervision of production workers. The responsibility of foremen must be changed from numbers to quality.
8 Drive out fear, so that everyone may work effectively for the company.
9 Break down barriers between departments.
10 Eliminate numerical goals, posters, and slogans for the workforce asking for new levels of productivity without providing methods.
11 Eliminate work standards that prescribe numerical quotas.
12 Remove barriers that stand between the hourly worker and his right to pride of workmanship.
13 Institute a vigorous programme of education and retraining.
14 Create a structure in top management that will push every day on the above 13 points.

Juran's ten steps to quality improvement:

1 Build awareness of the need and opportunity for improvement.
2 Set goals for improvement.
3 Organize to reach the goals (establish a quality council, identify problems, select projects, appoint teams, designate facilitators).
4 Provide training.
5 Carry out projects to solve problems.
6 Report progress.
7 Give recognition.
8 Communicate results.
9 Keep score.
10 Maintain momentum by making annual improvement part of the regular systems and processes of the company.

Phil Crosby, who spent time as Quality Director of ITT, had:

• Definition – conformance to requirements.
• System – prevention.
• Performance standard – zero defects.
• Measurement – price of non-conformance.

He also offered management 14 steps to improvement:

1 Make it clear that management is committed to quality.
2 Form quality improvement teams with representatives from each department.
3 Determine where current and potential quality problems lie.
4 Evaluate the cost of quality and explain its use as a management tool.
5 Raise the quality awareness and personal concern of all employees.
6 Take actions to correct problems identified through previous steps.
7 Establish a committee for the zero defects programme.
8 Train supervisors to actively carry out their part of the quality improvement programme.
9 Hold a 'zero defects day' to let all employees realize that there has been a change.
10 Encourage individuals to establish improvement goals for themselves and their groups.
11 Encourage employees to communicate to management the obstacles they face in attaining their improvement goals.
12 Recognize and appreciate those who participate.
13 Establish quality councils to communicate on a regular basis.
14 Do it all over again to emphasize that the quality improvement programme never ends.

A comparison

One way to compare directly the various approaches of these three American gurus is in Table 3.1. It shows the differences and similarities clarified under 12 different factors.

Our understanding of 'total quality management' developed through the 1980s and in earlier editions of John Oakland's books on TQM; a broad perspective linking the TQM approach to the direction, policies and strategies of the business or organization was given. These ideas were captured in a basic framework: the TQM Model (Figure 3.1), which was widely promoted in the UK through the activities of the Department of Trade and Industry (DTI) 'Quality Campaign' and 'Managing into the 90s' programmes. These approaches brought together a number of components of the quality approach, including quality circles (*teams*), problem solving and statistical process control (*tools*) and quality systems, such as BS5750 and later ISO 9000 (*systems*). It was recognized that *culture* played an enormous role in whether organizations were successful or not with their TQM approaches. Of course, good *communications* were seen to be vital to success, but the most important of all was *commitment*, not only from the senior management but from everyone in the organization, particularly those operating directly at the customer interface. The customer/supplier or 'quality chains' were the core of this TQM model.

Many companies and organizations in the public sector found this simple framework useful, and it helped groups of senior managers throughout the world get started with TQM. The key was to integrate the TQM activities based on the framework into the business or organization strategy and this has always been a key component of the authors' approach.

QUALITY AWARD MODELS

Starting in Japan with the Deming Prize, companies started to get interested in quality frameworks that could be used essentially in three ways:

1   as the basis for awards;
2   as the basis for a form of 'self-assessment';
3   as a descriptive, 'what-needs-to-be-in-place' model.

The earliest approach to a *total* quality audit process is that established in the Japanese-based Deming Prize, which is based on a highly demanding and intrusive process. The categories of this award were established in 1950 when the Union of Japanese Scientists and Engineers (JUSE) instituted the prize(s) for contributions to quality and dependability of product (*www.juse.or.jp/e/deming*).

This now defines TQM as:

a set of systematic activities carried out by the entire organization to effectively and efficiently achieve the organization's objectives so as to provide products and services with a level of quality that satisfies customers, at the appropriate time and price.

As the Deming Award guidelines say, there is no easy success at this time of constant change and no organization can expect to build excellent quality management systems just by solving problems given by others:

They need to think on their own, set lofty goals and drive themselves to challenge for achieving those goals. For these organizations that introduce and implement TQM in this manner, the Deming Prize aims to be used as a tool for improving and transforming their business management.

The recognition that total quality management is a broad culture change vehicle with internal and external focus embracing behavioural and service issues as well as quality assurance and process control prompted the United States to develop in the late 1980s one of the most famous and now widely used frameworks, the Malcolm Baldrige National Quality Award (MBNQA). The award, composed of two solid crystal prisms 14 inches high, is presented annually in the USA to recognize companies that have excelled in quality management and quality achievement. However, it is not the award or even the fact that it is presented by the President of the USA that attracts the attention of most organizations – it is the excellent framework that it provides for TQM and organizational self-assessments.

TABLE 3.1 The American quality gurus compared

	Crosby	*Deming*	*Juran*
Definition of quality	Conformance to requirements	A predictable degree of uniformity and dependability at low cost and suited to the market	Fitness for use
Degree of senior-management responsibility	Responsible for quality	Responsible for 94% of quality problems	Less than 20% of quality problems are due to workers
Performance standard/motivation	Zero defects	Quality has many scales Use statistics to measure performance in all areas Critical of zero defects	Avoid campaigns to do perfect work
General approach	Prevention, not inspection	Reduce variability by continuous improvement Cease mass inspection	General management approach to quality – especially 'human' elements
Structure	Fourteen steps to quality improvement	Fourteen points for management	Ten steps to quality improvement
Statistical process control (SPC)	Rejects statistically acceptable levels of quality	Statistical methods of quality control must be used	Recommends SPLC but warns that it can lead to too-driven approach
Improvement basis	A 'process', not a program Improvement goals	Continuous to reduce variation. Eliminate goals without methods	Project-by-project team approach Set goals
Teamwork	Quality improvement teams Quality councils	Employee participation in decision-making. Break down barriers between departments	Team and quality circle approach
Costs of quality	Cost of non-conformance Quality is free	No optimum – continuous improvement	Quality is not free – there is an optimum
Purchasing and goods received	State requirements Supplier is extension of business. Most faults due to purchasers themselves	Inspection too late – allows defects to enter system through AQLs. Statistical evidence and control charts required	Problems are complex Carry out formal surveys
Vendor rating	Yes *and* buyers Quality audits useless	No – critical of most systems	Yes, but help supplier improve
Single sources of supply		Yes	No – can neglect to sharpen competitive edge

The Baldrige National Quality Program Criteria for Performance Excellence, as it is now known, aims to:

- help improve organizational performance practices, capabilities and results;
- facilitate communication and sharing of best practices information;
- serve as a working tool for understanding and managing performance, guiding, planning and creating opportunities for learning.

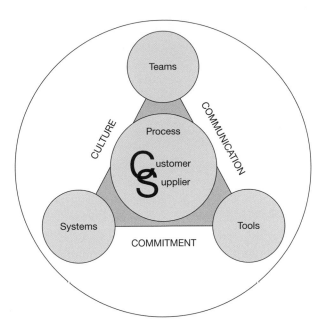

FIGURE 3.1
Total quality management
model – major features

The award criteria are built upon a set of interrelated core values and concepts:

- visionary leadership;
- customer-driven excellence;
- organizational and personal learning;
- valuing employees and partners;
- agility;
- focus on the future;
- managing for innovation;
- management by fact;
- public responsibility and citizenship;
- focus on results and creating value;
- systems developments.

These are embodied in a framework of seven categories which are used to assess organizations:

1. Leadership
   - organizational leadership
   - public responsibility and citizenship
2. Strategic planning
   - strategy development
   - strategy deployment
3. Customer and market focus
   - customer and market knowledge
   - customer relationships and satisfaction
4. Information and analysis
   - measurement and analysis of organizational performance
   - information management

5 Human resource focus
- work systems
- employee education training and development
- employee well-being and satisfaction

6 Process management
- product and service processes
- business processes
- support processes

7 Business results
- customer focused results
- financial and market results
- human resource results
- organizational effectiveness results.

Figure 3.2 shows how the framework's system connects and integrates the categories. This has three basic elements: organizational profile, system, and information and analysis. The main driver is the senior executive leadership which creates the values, goals and systems and guides the sustained pursuit of quality and performance objectives. The system includes a set of well-defined and well-designed processes for meeting the organization's direction and performance requirements. Measures of progress provide a results-oriented basis for channelling actions to deliver ever-improving customer values and organization performance. The overall goal is the delivery of customer satisfaction and market success, leading, in turn, to excellent business results. The seven criteria categories are further divided into items and areas to address. These are described in some detail in the 'Criteria for Performance Excellence' available from the US National Institute of Standards and Technology (NIST), in Gaithesburg USA (*www.nist.gov/baldrige*).

The Baldrige Award led to a huge interest around the world in quality award frameworks that could be used to carry out self-assessment and to build an organization-

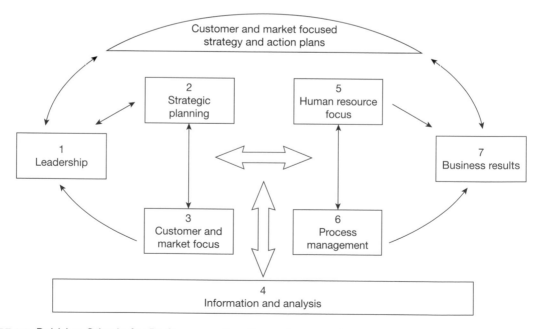

FIGURE 3.2 Baldrige Criteria for Performance Excellence framework

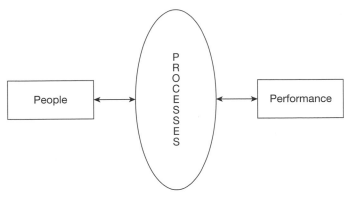

FIGURE 3.3
The simple model for
improved performance

Achieve better performance through involvement of all
employees (people) in continuous improvement
of their processes

wide approach to quality, which was truly integrated into the business strategy. It was followed in Europe in the early 1990s by the launch of the European Quality Award by the European Foundation for Quality Management (EFQM). This framework was the first one to include 'Business Results' and to really represent the whole business model.

Like the Baldrige, the EFQM model recognized that processes are the means by which an organization harnesses and liberates the talents of its people to produce results/performance. Moreover, improvement in performance can be achieved only by improving the processes through involving the people. This simple model is shown in Figure 3.3.

Figure 3.4 displays graphically the 'non-prescriptive' principles of the full Excellence Model. Essentially, customer results, employee results and favourable society results are achieved through leadership driving policy and strategy; people partnerships; resources; and processes which lead ultimately to excellence in key performance results. These enablers deliver the results which, in turn, drive innovation and learning. The EFQM has provided a weighting for each of the criteria which may be used in scoring self-assessments and making awards (see Chapter 10).

Through usage and research, the Baldrige and EFQM Excellence Models continued to grow in stature throughout the 1990s. They were recognized as descriptive, holistic, business models rather than just quality models, and they have mutated into frameworks for (Business) Excellence.

The NIST and EFQM have worked together well over recent years to learn from each other's experience in administering awards and supporting programmes, and from organizations which have used their frameworks 'in anger'.

The EFQM publication for the new millennium of the 'Excellence Model' captures much of this learning and provides a new ten-step framework for organizations to follow:

1 set direction through leadership;
2 establish the results required;
3 establish and drive policy and strategy;
4 set up and manage appropriately the approach to processes, people, partnerships and resources;
5 deploy the approaches to ensure achievement of the policies, strategies and, thereby, the results;
6 assess the 'business' performance in terms of customers, their own people and society results;

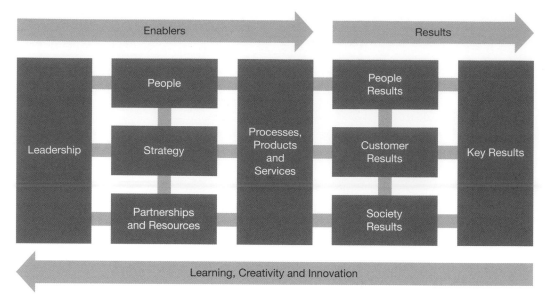

FIGURE 3.4 The EFQM Excellence Model

7 assess the achievements of key performance results;
8 review performance for strengths and areas for improvement;
9 innovate to deliver performance improvements;
10 learn more about the effects of the enablers on the results.

The 2015 version of ISO 9000 introduced, for the first time, the idea of continuous process improvement as an essential part of its structure. This shifts the quality framework described in this international standard towards what we would describe as a lean quality framework.

At this stage, the only award that is based on lean thinking is the Shingo Award, an excellence award developed by the Shingo Institute at the John M. Huntsman School of business at the Utah State University. The Lean Construction Institute of the US is considering the development of a framework for assessing lean excellence; however this has not yet been undertaken.

THE FOUR PS AND THREE CS – THE BASIS FOR A NEW MODEL FOR LEAN QUALITY

We have seen in Chapter 2 how *processes* are the key to delivering quality of products and services to customers. It is clear from Figure 3.4 that processes are a key linkage between the enablers of *planning* (leadership driving policy and strategy partnership and resources), through *people* into the *performance* of people, society, customers and key outcomes.

These 'four Ps' form the basis of a simple model for TQM which was developed by John Oakland in the 1990s. To complete the model these were complemented by the all-important three Cs: *Culture, Communication and Commitment*. This basic TQM model has been further adapted to characterize all aspects of lean quality by wrapping the two key lean ideals of *maximize value for all customers* and *continuous improvement in all processes and outcomes* around it. This new model (Figure 3.5) of lean quality provides both the 'hard'

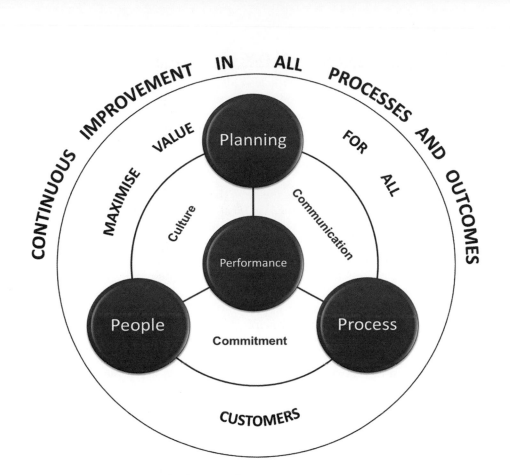

FIGURE 3.5 The new framework for lean quality management

and 'soft' management necessities required to take organizations successfully into the twenty-first century. This model forms the structure of the remainder of this book.

This new lean quality model, based on all the excellent work done during the last century, provides a simple framework for excellent performance, covering all angles and aspects of an organization and its operation.

Performance is achieved using a business excellence approach and by planning the involvement of people in the improvement of processes, by focusing on value creation for customers and by driving continuous improvement in all processes and outcomes.

- *Planning*: the development and deployment of policies and strategies; setting up appropriate partnerships and resources; and designing in quality;
- *Performance*: establishing a performance measurement framework – a 'balanced score-card' for the organization; carrying out self-assessment, audits, reviews and benchmarking;
- *Processes*: understanding, management, design and redesign; quality management systems; continuous improvement;
- *People*: managing the human resources; culture change; teamwork; communications; innovation and learning.

Driving all of this to ensure successful implementation is, of course, effective leadership and commitment, the subject of the next chapter.

Lean can be characterized in terms of the lean ideal, principles, and methods or tools. The lean ideal is to provide a custom product or service, exactly fit for purpose delivered as required with no waste. It describes the ideal outcome from a service or product.

Lean principles on the other hand are the beliefs or rules that guide the actions that support the achievement of the ideal. These are not a prescriptive set of rules; rather they are guidelines to inform thinking about the way forward, either at the individual or organizational level.

The methods and tools of lean are the how; these are practices that have been proven to be productive in moving towards the lean ideal. However, it is important to understand that a fundamental tenet of lean thinking is that this is not a rule-bound approach. Every organization and every situation has its unique characteristics; the lean approach is to use the lean principles and practices (tools or methods) to guide action, lean is a path and the milestones are progress towards the achievement of the ideal.

In this section we propose a set of principles specifically designed to guide enterprises in the construction sector towards the implementation of a lean business framework. The importance of adopting a clear set of principles should not be underestimated; this defines the values which will guide an organization in its development. These principles are in large part very similar to the lean frameworks that have been successfully adopted in manufacturing and service industries. However, there are some principles that are particularly relevant to organizations in the construction sector.

Because much of the construction sector is faced with the challenge of designing, fabricating and erecting unique facilities, every project offers the prospect for design and construction optimization. The lean quality approach builds on the process approach and brings into focus the continuous search for opportunities for improvements in customer service; efficiency; and the elimination of waste. This is as applicable to design processes as it is to fabrication and construction.

Jeffrey Liker, a leading researcher and teacher within the lean movement, in his groundbreaking book *The Toyota Way* (2004) structured his approach around the 14 principles of the Toyota Way arranged in four groupings:

1   Long-term philosophy.
2   The right process will produce the right results.
3   Add value to the organization by developing your people and partners.
4   Continuously solving root problems drives organizational learning.

Liker's principles address the broad organizational issues as well as the detailed processes of production in more detail.

1   Adopt a long-term philosophy.
2   Strive for continuous flow.
3   Use pull systems.
4   Level out workload.
5   Stop to fix problems.
6   Standardize tasks.
7   Use visual control.
8   Only use reliable technology.
9   Grow leaders who understand work.
10   Develop exceptional teams and people.
11   Respect external partners.

12 Go see for yourself.
13 Slow decisions by consensus, implement rapidly.
14 Become a learning organization.

We have built on Liker's approach and expanded it, referring to the contributions of other leading thinkers in the lean construction community of researchers and practitioners.

In contrast to Liker, Lauri Koskela (1992, 2000) focused his attention on the principles of production systems. Early on, Koskela (1992) proposed the following principles for the improvement of production processes:

1 Systematically focus on customer value.
2 Focus control on the whole process.
3 Balance improvement in flow and conversion.
4 Simplify processes.
5 Reduce the share of non-value-adding work.
6 Reduce variability.
7 Increase output flexibility.
8 Reduce cycle time.
9 Benchmark.

Later Koskela (2000) analysed in detail the wide-ranging principles that underpin theories of production and he developed an integrated Transformation-Value-Flow (TVF) view of production which is underpinned by the following three high level principles:

1 Transformation – getting production realized efficiently.
2 Flow – elimination of waste (non-value-adding activities).
3 Value – elimination of value loss (achieved value relative to best possible value).

While this broad set of principles was based on the manufacturing literature, and demonstrated that the principles of lean in manufacturing are applicable in the construction environment, it did not reflect much of the construction-specific thinking developed by the community of researchers and practitioners in the lean construction movement.

Ballard (2016) in a chapter on lean construction turned his attention to the unique characteristics of construction, and he introduced four new principles which reflect the particular characteristics of the modern construction process. They are:

1 allow money and resources to move across contractual and organizational boundaries in search of the best project level investment;
2 improve the predictability of near term workload to drive efficiency and reliability of operations;
3 drive the design to realize an optimum, fit for purpose, product design within the cost constraints of the customer; and
4 involve upstream players in downstream processes to realize innovative and efficient design and construction solutions.

We present a synthesis of these ideas in the context of their application to lean quality in the delivery of capital projects. Our framework is set out in the mind map, which we describe in Figure 3.6.

In this model, the three Ps of *Planning, People and Processes* are clearly articulated, while *Performance* and the three Cs of *Communication, Culture* and *Commitment* are all included in the foundation category of Values and Long-Term Philosophy.

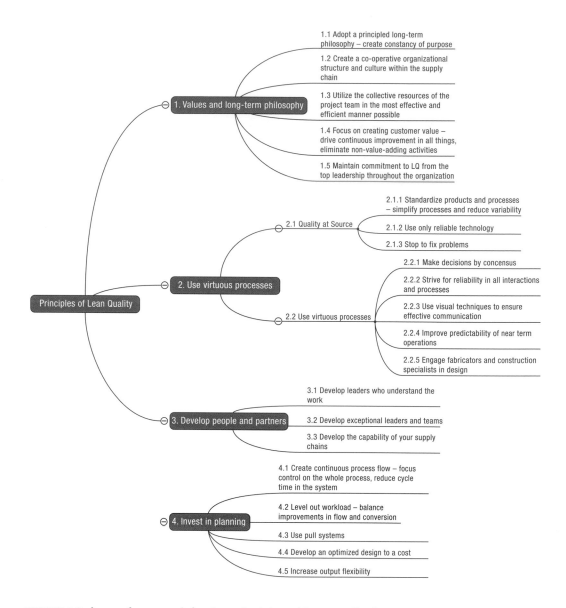

FIGURE 3.6 A new framework for the principles of lean quality in construction

Section 1: Values and long-term philosophy

Basing management decisions on a long-term philosophy will ensure stability and sustainability. It is essential to operate in a co-operative culture with long-term supply chain relationships and to structure agreements to allow for flexibility in response to changes that may occur during project delivery. Visible leadership commitment to lean quality values and processes will ensure compliance by the rest of the workforce, including the supply chain, and lead to the development of a skilled and well-trained workforce that seeks continuous improvement based on learning and reflection.

The first step is the creation of stable production; once this has been achieved, encourage workers at all levels to identify potential improvements in safety, efficiency or quality. If there is a perceived need to improve a process, a team-based process improvement (*Kaizen*) workshop is often the most effective and expeditious way to review and improve the status quo. Essential to continuous improvement is stability in corporate values and leadership.

Principle 1.1: Adopt a principled long-term philosophy – create constancy of purpose

To ensure stability and sustainability, determine a long-term vision and mission based on your values and aspirations. Companies that practise lean quality are values-based and ethically driven enterprises, which seek to maximize their contribution to their customers, to their employees, supply chain partners and the communities in which they operate. Chapter 5 addresses these and related issues.

Principle 1.2: Create a co-operative organizational culture and structure within the supply chain

Large, complex constructed projects require close collaboration across all disciplines and companies in the supply chain to achieve optimum outcomes. The lead companies, designers, contractors and clients must build a culture of open communication and collaboration to ensure that downstream fabricators and constructors are involved in the design development and optimization process. Innovation across disciplinary boundaries requires an open collaborative culture. Commercial frameworks should be designed to encourage such collaboration to occur.

Principle 1.3: Utilize the collective resources of the project team in the most effective and efficient manner possible

Organizational and commercial structures should be designed to permit the most effective and efficient use of collective resources on a project. This should enable resources to be shared across contractual and organizational boundaries to achieve the most efficient deployment of collective resources. As projects increase in complexity and uncertainty, the upfront relationship between work scope and compensation becomes more tenuous. Changes in scope during the project become more likely and with them the need to renegotiate work scope and cost. This principle addresses the need for flexibility in the use of all resources, ensuring that they are used in the most efficient manner possible.

Principle 1.4: Focus on creating customer value – Drive continuous improvement in all things and eliminate non-value-adding activities

A key aspect of lean quality is that targets are set for key inputs and outcomes, and performance is measured as a means of driving continuous improvement. Furthermore, everyone in the organization is tasked with looking for waste, non-value-adding activities that can be modified or eliminated as they do not add value to the end customer.

Principle 1.5: Maintain commitment to lean quality from the top leadership throughout the organization

Lean quality is a committed, long-term approach to doing business. It is not a short- term fix. Lean quality drives incremental improvement in all the key aspects of a business and focuses the organization on continuously improving customer value. It is essential that leadership at all levels in an organization is visibly committed to the long-term values and processes of lean quality.

Section 2: The right processes lead to the best results

There are two key aspects to using the right processes. The first is to ensure that whenever work is done it is done correctly. The second is to ensure that the right decisions are made and communicated effectively. To ensure the achievement of this, communications must be impeccable.

It also requires well-trained workers using standard work processes, and reliable technology which has been tried and tested. Critically, whenever a problem is encountered, immediately stop: identify the root cause, fix the problem and avoid its recurrence.

To ensure that the right decisions are made, decisions should be made by consensus and effort has to be invested to ensure that communications are reliable and effective. Experience has shown that the most effective way to communicate targets and process performance is when information is displayed on a dashboard in an information centre; this is the most effective way to focus people's attention on performance, priorities and problems.

Subsection 2.1 Continuously drive to achieve quality at source

Any rework is wasteful; getting quality at source requires a skilled and trained workforce deploying robust, standard work procedures, and reliable and proven technologies. If any of these elements are missing, the risk of rework is increased. Furthermore, whenever an error occurs, every effort must be made to analyse and understand the cause of the error and prevent its reoccurrence.

Principle 2.1.1: Standardize products and processes to improve quality – simplify processes and reduce variability

Standardizing work lays the foundation for quality achievement as well as production improvement. In construction work, standardization and optimization requires front-end designers to collaborate with fabricators and erectors to develop optimum, standard solutions. Often, the contractual arrangements and the reluctance of project owners/ developers to invest in upfront solution development prevent this kind of collaboration.

Best practice teams are using virtual and physical prototyping to develop improved solutions. These are then documented and shared, and workers doing the work are encouraged to innovate and improve on them. Constructability is optimized through the creation of mock-ups; detailing is standardized; construction processes are error proofed; and improvements in efficiency are designed into the fabrication and assembly process. A process known as DFMA (Design for Manufacture and Assembly), long used in manufacturing, is now being implemented by lead companies at the project level in the construction sector to drive constructability improvement.

Many lead companies are encouraging their workers to create two-minute videos of work improvement ideas. These are posted on the company intranet, productive ideas are adopted and the best ideas recognized and rewarded.

Principle 2.1.2: Use only reliable technology to reduce the risk of failure

Only use materials, products and techniques that are proven. Ensure that you are able to successfully deploy new technologies successfully before you commit to them. Make sure that your people are trained in the use of a new technology and undertake a pilot application to evaluate the processes and identify the risks involved, before you adopt it.

Project developers and general contractors are always looking for faster and cheaper solutions; however, when these are supplied by subcontractors who do not have the capacity to warrant their products and processes, the general contractor ends up being liable.

An unacceptable number of failures arising from the poor application of new systems or materials end up being paid for by insurers and long-term owners. Expensive and often unsuccessful recovery processes benefit only the lawyers and the remedial industry. While caution is recommended in the adoption of new technologies and solutions, this should not discourage enterprises from searching for different and better solutions which improve quality and productivity.

Principle 2.1.3: Stop to fix problems to avoid error repetition

Error repetition undermines efficiency and creates stress. The construction industry is infamous for its acceptance of error and its inability to avoid error repetition. Quality, safety and production errors are repeated again and again, and the 'closing out' of a single instance of an error is lauded as success. For example, the site safety walk, conducted weekly on most projects around the world, generally finds the same workers creating the same hazards, week on week. Site and company management is satisfied by the simple 'closing out' of these hazards within a short time frame, while the avoidance of repetition is rarely a goal that is given sufficient attention.

Best practice teams deploy Built-in Quality and Safety (BIQS) from the outset through the design process, using virtual or physical prototyping if advantageous. BIQS reduces the risk of error and promotes quality, safety and efficiency. Identify errors as close as possible to the source and avoid the repetition of errors once they have been identified. Train operators to certify their own quality at source. Train them to solve problems as they occur and to innovate and develop improved solutions if required.

Use modern quality control and assurance processes to ensure customer quality is achieved every time. Some industry leaders are broadening the role of QA auditors, training them to be problem-solving coaches and lean process improvement leaders.

Subsection 2.2: Use virtuous processes

Use sound principles to ensure that robust and stable decisions drive your organization's processes. This requires that the supply chain is underpinned by stable relationships based on shared values and goals, and excellent communication among all project participants and stakeholders.

Principle 2.2.1: Make decisions by consensus, implement decisions quickly

Constructed products are extremely complex and often there are many solutions to a given need or problem. It is the essence of teamwork to develop solutions through a consensus process. This allows the many different perspectives brought to the process by different experts and stakeholders to be weighed and balanced.

For example, the process of Choosing by Advantage allows teams to develop multiple solutions in parallel, choosing the best from among them at the appropriate time. Similarly, techniques such as physical and virtual prototyping allow users and operators to more effectively engage with the decision-making process and evaluate alternatives and develop optimum solutions. Processes such as these, while they take time in the development phase, set the groundwork for faster and more efficient execution.

Principle 2.2.2: Strive for reliability in all interactions and processes

Reliability is also a key concept in achieving process efficiency. It has been identified in project organizations that workflows are generally unreliable when compared to manufacturing and service industries. This reflects the fact that commitments between parties in the supply chain for the delivery of work are loosely made. It is what we call *unreliable promising* in the lean construction community. Unreliable promises lead to unreliable work-flows, a feature that is all too common in construction sector organizations. Improvements in process efficiency rely on achieving more reliable workflows, and this, in turn, has a cultural implication as it requires reliable promising, a change in practice and behaviour.

Principle 2.2.3: Use visual techniques to ensure effective communication
The use of visual cues has been proven to be a most effective way of influencing people's behaviour in the workplace and focusing attention on critical issues that affect safety and performance.

An important aspect of this is the use of information centres to display performance information in the form of graphs and charts about issues such as progress against targets; performance against plans; error rates in safety, quality, logistics and production. Similarly, the use of problem and countermeasure boards to focus workers' attention on issues that need resolution has been found to be beneficial.

It has also been demonstrated that holding group decision-making and planning meetings in front of an information centre is an effective way of engaging people in the meeting processes; it increases worker involvement and group ownership of the outcomes.

In today's world, with word processing technology and the bureaucratization of management, the norm in reporting has seen a move to longwinded reports, where critical information is easily lost amongst non-essential detail. Whenever possible, the publication of plans and reports in a succinct format, on a single A3 sheet of paper, has been found to be a very effective way of focusing people on the essential.

Principle 2.2.4: Improve predictability of near term activity to optimize operations
The stability and efficiency of overall operations is optimized if near term activity is stable and reliable. The use of the Last Planner® System drives a collaborative planning process that ensures what is planned gets done.

Principle 2.2.5: Engage fabricators and construction specialists in design
Design can only be optimized in a holistic sense, melding the needs of the owner with the requirements for reliable service and product quality, on-site safety and constructability if the expertise of the downstream players in the supply chain is utilized in the design process. This requires the early engagement of fabricators and construction specialists in the project design development process.

Section 3: Develop your people and partners

An organization is only as good as its people: take the people away and you have nothing. This includes people at every level in the organization and in the supply chain. On construction projects, it is not uncommon for almost 85 per cent of the work on site to be done by subcontract labour. This means that safety, quality, environmental and production risks are essentially in the hands of subcontract labour. Companies have to recognize that the management and workers of their subcontractors are as important as their own. There-fore, they need to develop strategies to develop and harness the potential of a closely knit team.

Principle 3.1: Develop leaders who understand the work and can teach others
Leaders should be role models for the company's philosophy and processes. They must understand the work and demonstrate the company's way of working. Good leaders are

teachers and coaches; hence, they must be familiar with the daily work in great detail. It is desirable to grow leaders from within, as this way they will have a deep appreciation of the company, its philosophy and its practices.

In construction, a significant challenge has been the reluctance of leaders in construction companies to recognize that they need to lead and develop their entire supply chain. The issues of leadership are addressed in Chapter 4.

Principle 3.2: Develop exceptional people and teams

Every company's greatest asset is its people. Build a strong and cohesive team-based culture and invest in developing and training your people. This process has to be guided by a clear and stable culture.

Successful construction organizations utilize multidisciplinary teams drawing on their supply chains to develop technical and business opportunities, and to improve business practices and processes. Effective team-based collaborative working requires strong and consistent leadership at the team level and the development of problem-solving and collaboration skills by all team members. These issues are addressed in Chapter 17.

Principle 3.3: Develop the capability of your supply chains

Typically, general contractors are reluctant to invest in the development of the capability of their supply chain partners, believing that by doing so they are supporting their competitors. Because as much as 85 per cent of all work is undertaken by people employed by other companies in the supply chain, the close collaboration of the entire team is critical to the effective management of safety, quality, risk, innovation and to the optimization of the construction process on every project.

The view that by developing their subcontractors, general contractors are supporting their competitors is misguided. The reality is that by learning to work closely with its supply chain, a general contractor is developing, through a unique set of relationships, capability that is exclusive to its own supply chain. While all the subcontractors become more capable, they cannot replicate the same level of collaboration with other general contractors who have not invested in developing a unique, close working relationship.

Section 4: Invest in planning

Encourage investment in planning to create a reliable flow of work

Unreliable planning is the Achilles' heel of the construction sector the world over. Reliable workflow relies on careful and thorough planning, good communication and a genuine collaboration within the supply chain. On construction projects the world over it is quite common for only half of what is planned at the beginning of the week to be completed as planned. Put another other way, about half of what is planned each week is not done as planned.

Such a situation can only be described as highly unreliable and the activities on such construction sites are inevitably somewhat chaotic. There is a confusion within this industry between progress and reliable work; this prevents many from seeing the importance of reliability.

Construction has embraced the concept of Value Earned with enthusiasm, yet while an unreliable flow of work can create the same value as a reliable flow of work, the former is highly inefficient while the latter is highly efficient. Lagging indicators such as value earned, budget achievement and major milestones met give no insight into the efficiency work. Blind faith in lagging indicators allows the industry and its clients to be lulled into a sense of false security, thinking it is doing well when in fact workflows are inefficient and work on sites is somewhat chaotic. The key to predictable quality efficiency and cost is reliable workflow, and the foundation of this is an investment in planning, both schedule planning and cost planning.

Use short-range planning tools such as the Last Planner® System to create reliable workflow and even out demand. This benefits the entire supply chain, improving collaboration and improving the team's capacity to innovate. A more capable team can better manage all outcomes including productivity, quality and safety. A corollary is that poor workflow reliability discourages investment in planning as at the individual level the plans are deemed to be of limited value.

Principle 4.1: Create continuous process flow to bring problems to the surface

Design work processes to achieve reliable, high-value, continuous flow; this includes the flow of materials, parts, information and decisions. Most construction projects experience anything but uniform flow; subcontractors complain about the wildly varying requirement for their resources, only to be told that their bosses priced the contract and they must have known what was coming.

Give visibility to the flow of your processes, continuous process flow brings to the surface problems, exposes systemic weaknesses and drives process improvement and reliability. Continuously strive to reduce cycle time in the system.

Principle 4.2: Level out workload to reduce waste

This is a corollary to Principle 4.1, when the workload is not smooth, it leads to stop–start work. Move from batch processing to smooth, even workflow and resource uniform utilization. Stress increases at times of heightened demand and with higher stress comes an increased risk of error occurrence in safety, quality and production.

On construction projects, unreliable workflow causes work to be brought forward to fill the gap created by work that is not ready to be done. Doing work out of sequence increases the risk of quality, safety and production errors. During lulls in demand, waste due to underutilization is hard to avoid.

Principle 4.3: Use pull systems to avoid overproduction

Provide your downstream customers what they need, when they need it and where they need it. A corollary of this is to minimize 'Work in Progress' (WIP). In the project construction environment, going ahead with design or production can expose a project to unanticipated risk.

In the design phase, going ahead forces you to make assumptions, and assumptions represent risk, incorrect assumptions lead to rework. In construction, materials and products delivered to site early are susceptible to damage and create additional logistics and management problems on site.

Inner city high-rise sites force a just-in-time (JIT) approach to the delivery of materials and products as there is simply nowhere to store them. Sites surrounded by space encourage an undisciplined approach to materials storage and handling, which often leads to wasteful processes, lost and damaged materials and products, and duplicate ordering as material cannot be found.

Principle 4.4: Develop your design to a cost to achieve the client's budget

Rather than allowing cost to be an outcome of design, use knowledge about customer value and about cost to drive the design process. Involve the owner in an ongoing review of targets for value including attributes such as function, capacity and sustainability against cost throughout the design definition process.

Target value design begins with determining customer needs, constraints and preferences and then considers alternate design options to choose the one that best meets the outcomes the client is seeking. Constraints limit the solution space to be explored.

Principle 4.5: Increase output flexibility

As organizations seek to improve production system balance and flow, moving towards uniform production through strategies such as Takt time planning, effective strategies to increase flexibility include adopting increased prefabrication, multiskilling the workforce and minimizing lot sizes to closely match demand.

Conceptual challenges

Today, the construction sector worldwide is faced with a similar dilemma to that confronted by manufacturers in the West in the 1980s. Leaders in the manufacturing sector in the West looked at lean production and saw something which they believed was intrinsically Japanese. It took some years for them to realize that the fundamental conceptual framework of lean production applies to all industries around the globe and has nothing to do with any country's culture.

Construction executives can see the benefits that other industry sectors have won through the adoption of lean quality management practices and they would like to obtain similar benefits for themselves. However, they cannot see how those ideas can be applied in construction because they see construction as being essentially different to manufacturing. In a sense they have the same disconnect as Western industry leaders had in trying to untangle management practices from Japanese culture.

Construction is a one-of-a-kind production: its products are highly complex; its supply chain is extremely fragmented; it has a complex engagement with its clients over a long period of time and, as a result, its processes are significantly different. The challenge for the sector is to untangle the differences between industries and their management practices.

There are two significant impediments to the uptake of lean thinking in the construction sector. First of all large construction projects are complex undertakings, they are always built in a hurry in an attempt to minimize the construction duration and thereby reduce overheads, and increasingly as the capital spend and project size has grown, time to market is becoming a critical factor.

Secondly, this is an industry in which project managers are given almost absolute control over their projects, each project being set up as a quasi-independent business unit. The reason for this is that company management seeks to concentrate responsibility for the achievement of the project business goals with one person, the project director or manager. Head office does not want to interfere with the practices of a project in case they give a project manager an excuse for blaming head office intervention for a failure to achieve critical cost and schedule goals. This approach to organizational structure and governance is itself an impediment to the implementation of lean construction, as the deployment of lean thinking is ultimately left up to each and every project manager.

This book presents the theoretical underpinnings for that analysis and, through the case studies of companies that have bridged the gap, demonstrates how the concepts of lean quality management can be leveraged to benefit construction organizations.

BIBLIOGRAPHY

Ballard, G. 'Lean Construction' in Netland, T. and Powell, D. *The Routledge Companion to Lean Management*, in press , 2016.

BQF (British Quality Foundation). *The Model in Practice* and *The Model in Practice 2*, London, 2000 and 2002.

Crosby, P.B. *Quality is Free*, McGraw-Hill, New York, 1979.

Crosby, P.B. *Quality Without Tears*, McGraw-Hill, New York, 1984.

Deming, W.E. *Out of the Crisis*, MIT, Cambridge, MA, 1982.

Deming, W.E. *The New Economies*, MIT, Cambridge, MA, 1993.

EFQM (European Foundation for Quality Management). *The EFQM Excellence Model*, Brussels, 2015.

Koskela, L. 'Application of the new production philosophy to construction', CIFE technical report #72 Stanford University, 1992.

Koskela, L. 'An exploration towards a production theory and its application to construction', PhD dissertation, VTT Publication 408, 2002.

Liker, J. *The Toyota Way*, McGraw Hill, New York and London, 2004.

National Institute of Standard and Technology. *USA Malcolm Baldrige National Quality Award, Criteria for Performance Excellence*, NIST, Gaithesburg, 2015.

Oakland, J.S. *Total Quality Management – The Route to Improved Performance*, Butterworth-Heinemann, Oxford, 1993.

Pyzdek, T. and Keller, P. *The Handbook for Quality Management; A Complete Guide to Operational Excellence* (2nd edn), ASQ, Milwaukee, 2013.

Chapter highlights

Early TQM frameworks

- In the early 1980s when US and European business leaders first tried to understand how to apply quality business practices that they saw gave their Japanese competitors an advantage, they had a challenge to untangle the business ideas from the cultural differences.
- There have been many attempts to construct lists and frameworks to help organizations understand how to implement good quality management.
- The 'quality gurus' in America, Deming, Juran and Crosby, offered management 14 points, ten steps and four absolutes (plus 14 steps) respectively. These similar but different approaches may be compared using a number of factors, including definition of quality, degree of senior management responsibility and general approach.
- The understanding of quality developed and, in Europe and other parts of the world, the author's early TQM model, based on a customer/supplier chain core surrounded by systems, tools and teams, linked through culture, communications and commitment, gained wide usage.
- Today managers in the construction sector have a similar challenge to that faced by general business leaders in the 1980s; they have to untangle the differences between the business setting of manufacturing and construction and identify those practices that are essential for the achievement of quality outcomes in their operations.

Quality award models

- Quality frameworks may be used as the basis for awards for a form of 'self-assessment' or as a description of what should be in place.
- The Deming Prize in Japan was the first formal quality award framework, established by JUSE in 1950. The examination viewpoints include: top management leadership and strategies; TQM frameworks, concepts and values; QA and management systems; human resources; utilization of information, scientific methods; organizational powers; realization of corporate objectives.
- The US Baldrige Award aims to promote performance excellence and improvement in competitiveness through a framework of seven categories which are used to assess

organizations: leadership; strategic planning; customer and market focus; information and analysis; human resource focus; process management; business results.

- The European (EFQM) Excellence Model operates through a simple framework of performance improvement through involvement of people in improving processes.
- The full Excellence Model is a non-prescriptive framework for achieving good results – customers, people, society, key performance – through the enablers – leadership, policy and strategy, people, processes, partnerships and resources. The framework includes proposed weightings for assessment.

The four Ps and three Cs – a new model for TQM

- *Planning*, *People* and *Processes* are the keys to delivering quality products and services to customers and generally improving overall *Performance*. These four Ps form a structure of 'hard management necessities' for a new simple TQM model which forms the structure of this book.
- The three Cs of culture, communication and commitment provide the glue or 'soft outcomes' of the model which will take organizations successfully into the twenty-first century.
- The encircling lean ideals of 'continuous improvement in all processes and outcomes' embodies the elimination of waste, and 'maximizing value for all customers' focuses everyone's attention on the needs of both internal and external customers.

The principles of lean quality in construction

- Basing management decisions on a long-term philosophy will ensure stability and sustainability. Operate in a co-operative culture with long-term supply chain relationships, structure agreements to allow for flexibility in response to changes, demonstrate visible leadership commitment to lean quality values.
- Any rework is wasteful; getting quality at source requires a skilled and trained workforce deploying robust, standard work procedures, and reliable and proven technologies. If any of these elements are missing, the risk of rework is increased. Furthermore, whenever an error occurs, every effort must be made to analyse and understand the cause of the error and prevent its reoccurrence.
- Use sound principles to ensure that robust and stable decisions drive your organization's processes. This requires that the supply chain is underpinned by stable relationships based on shared values and goals, and excellent communication among all project participants and stakeholders.
- Safety, quality, environmental and production risks are essentially in the hands of sub-contract labour. Recognize that the management and workers of your subcontractors are as important as your own. Develop strategies to develop and harness the potential of a closely knit team which includes your supply chain.
- Reliable workflow relies on careful and thorough planning, good communication and a genuine collaboration within the supply chain. There is a confusion within this industry between progress and reliability; this prevents many from seeing the importance of reliability. Lagging indicators such as value earned, budget achievement and major milestones met give no insight into the efficiency work. The key to predictable quality efficiency and cost is reliable workflow, and the foundation of this is an investment in planning, both schedule planning and cost planning.
- Use short-range planning tools such as the Last Planner® System to create reliable workflow and even out demand. This benefits the entire supply chain, improving collaboration and improving to the team's capacity to innovate. A more capable team can better manage all outcomes including productivity, quality and safety.

Leadership and commitment

4

THE LEAN QUALITY MANAGEMENT APPROACH

'What is lean quality management?' Something that is best left to the experts is often the answer to this question. But this is avoiding the issue because it allows executives and managers to opt out. Lean quality is too important to leave to the so-called 'lean quality professionals'; it cannot be achieved on a company-wide basis if it is left to the experts. Equally dangerous, however, are the uninformed who try to follow their natural instincts because they 'know what lean quality is when they see it'. Many of them say 'we really do this', without understanding the nuances of what they are talking about. This type of intuitive approach can lead to serious attitude problems, which do no more than reflect the limited understanding and knowledge of lean and quality that are present in an organization.

The organization which believes that the traditional quality and production control techniques and the way they have always been used will resolve their lean quality problems may be misguided. Employing more inspectors, tightening up standards, developing correction, repair and rework teams do not improve lean quality. Traditionally lean quality has been regarded as the responsibility of the QA, QC or, in the construction sector, even the planning department. In some organizations, it still has not been recognized that many quality problems originate in the commercial, service or administrative areas and to drive change, leadership must come from the top of the organization.

Leadership excellence is the key requirement for organizational excellence in any enterprise. In terms of the principles of lean quality set out in Chapter 3, Figure 3.6, responsibility for developing the right values and long-term philosophy of the organization rests with the leadership. From a lean quality perspective the leadership sets the direction and standards in all the areas embraced by the principles; however, in particular leadership is crucial in the key areas of:

- creating a co-operative culture within the organization and its supply chains;
- demonstrating visible commitment to lean quality in all areas; and
- creating productive teamwork in all areas of the organization.

This high standard of leadership is required at all levels, at the head of every team and every department.

Lean quality management is far more than shifting the responsibility of *detection* of problems from the customer to the producer, or simply cutting costs. First of all, leadership must understand the core reasons behind problems. They must then develop and implement a comprehensive strategy to move the business forward. Today's business environment is such that managers must plan strategically to maintain a hold on market share, let alone increase it. In construction, waste is everywhere; even though quality problems have been found to seriously erode margins due to the cost of rectifying defective work both during the contract period and after handover, very, very few organizations measure the cost of rework. We have known for years that customers place a higher value on quality than on loyalty to suppliers and price is often not the major determining factor in customer choice. Price has been replaced by quality in industrial, service, hospitality and many other markets.

This perception is somewhat distorted in building construction because of the substantial capital growth of property values in major cities the world over. This has created markets in which there is almost always a speculator prepared to buy even a substandard product. However, many of these markets have matured, real estate values have slumped in some markets as a result of the global financial crisis and consumer protection laws and class legal action against poor quality builders have become more common. As a result of these events many customers are becoming more demanding of quality.

Some companies have been able to obtain a market advantage from the provision of high-quality products and services in the construction sector. In Australia, the medium density housing developer and builder Mirvac has managed to differentiate itself on the basis of product quality and reliability. It is common knowledge in the marketplace that their dwelling units hold their value better than those of other suppliers. In the US, quarry and concrete supplier Graniterock (CS5) for a period was able to command a premium for its higher quality ready-mixed concrete, demonstrating that even purchasers of concrete are prepared to pay a premium for higher standards of service and a better quality product. More recently, other suppliers have matched their service and product quality, thus removing this differentiator from the price equation and challenging the company to find other differentiators to make them stand out in the market.

Lean quality is an approach to improving customer value by focusing on what customers want and improving the competitiveness, effectiveness and flexibility of a whole organization by continuously improving processes. It is essentially a way of planning, organizing and understanding each activity. Its success depends on the combined effort of every individual at each level in the supply chain. For an organization to be truly effective, each part of it must work properly together towards the same goals, recognizing that each person and each activity affects and, in turn, is affected by others.

The DPR case study (CS9) is particularly interesting in this regard. DPR has developed an approach to defining and communicating the customer's quality requirements. They use the term 'key distinguishing feature' to describe the customer's requirements in language that the supply chain understands in a very nuanced way, interpreting the customer requirements into sets of 'key distinguishing features' for each of the supply chain partners and work teams.

Lean quality is also a way of ridding people's lives of wasted effort by eliminating all non-value-adding processes and bringing everyone into the processes of improvement; this way, results are achieved in less time. The methods and techniques used in lean quality can be applied throughout any organization. They are equally useful in the design, construction, manufacturing, public service, healthcare, education and hospitality industries.

The impact of lean quality on an organization is, firstly, to ensure that the management adopts a strategic overview of all aspects of customer service delivery and customer satisfaction. The approach must focus on developing a problem-prevention mentality. But it is easy to underestimate the effort that is required to change attitudes and approaches. Many people will need to undergo a complete change of 'mindset' to unscramble their intuition which often rushes into production without sufficient preparation and believes that the detection/inspection approach to quality will solve quality problems. 'We have a quality problem, we had better double check every single item', irrespective of whether it is service, waterproofing or finishes-related. Managers who worked on the new Australian Parliament House in the late 1980s, where there were three tiers of quality inspection to make sure that everything was right, reported that no one was responsible. The worker relied on the first inspector, the first on the second and so on, no one accepted responsibility.

The correct mindset may be achieved by looking at the sort of barriers that exist in key areas. Staff may need to invest more time in training and learning how to reallocate their time and energy to become more reliable in their work processes; they must review their processes in teams, search for the causes of problems and correct the causes, not merely the symptoms, once and for all.

The JB Henderson case study (CS8) illustrates how this company significantly increased the time work crews spend planning and reviewing work each day and yet increased their labour productivity by some 50 per cent within an 18-month period.

This requires management to take the lead and to be both supportive and empowering, promoting a 'continuous improvement' and 'right-first-time' approach to work. Managers at every level need to become coaches and teachers, helping their people to become problem solvers and innovators. For this to occur, managers also need to develop new skills. Through *process performance improvement teams*, the workers will identify the opportunities for improvements to both products and processes, and they will develop solutions which have Built-in-Quality and Safety (BIQS). This will result in a reduction of the inspection-rejection syndrome. If things are done correctly first time round, the usual problems that create the need for inspection for failure should disappear and a commitment to continuous improvement will create a vitalized organization.

The management of many firms may think that their scale of operation is not sufficiently large; their resources are too slim; or the need for action is not important enough to justify implementing lean quality. However, before arriving at such a conclusion, they should examine their existing performance by asking:

1   Is there a need or an opportunity to improve customer satisfaction?
2   Are the costs of significant wastes such as the costs arising from errors, defects, damaged goods, customer complaints, lost sales, etc. being measured? If so, are these costs minimal or insignificant?
3   Is the standard of management adequate and are attempts being made to ensure that quality and safety are being given proper consideration at the design and planning stages?
4   Are errors common in the organization's production processes or management systems; or, are the management, documentation, operations, etc. in good order?
5   Have people been trained in how to prevent errors and problems? Have they been taught to solve problems and innovate? Do they anticipate and correct potential causes of problems, or do they find and reject?
6   Are subcontract suppliers being selected on the basis of the quality and reliability of their people and services as well as price?
7   Do job instructions contain the necessary quality elements, are they kept up-to-date and are employees doing their work in accordance with them?
8   What is being done to motivate and train employees to do work right the first time? What is being done to motivate employees to seek continuous improvements in their work?
9   How many errors and defects, and how much wastage occurred last year? Is this more or less than the previous year?

If satisfactory answers can be given to most of these questions, an organization can be reassured that it is already well on the way to using the principles of lean quality management. Even so, it may find that the introduction of lean quality causes it to reappraise activities throughout. If answers to the above questions indicate problem areas, it will be beneficial to review the top management's attitude to lean quality. Time and money spent on lean quality-related activities are *not* limitations of profitability; they make significant contributions towards greater efficiency and enhanced profits.

COMMITMENT AND POLICY

To be successful in promoting business efficiency and effectiveness, lean quality must be truly organization-wide. It must include the supply chain, and it must start at the top with the chief executive or equivalent. The most senior directors and management must all demonstrate that they are serious about lean quality. At every level in the organization, it is up to the managers to lead, mentor and teach others in the thinking and methods of lean quality. The middle management have a particularly important role to play since they must not only grasp the principles of lean quality, they must go on to explain and demonstrate them to the people for whom they are responsible and ensure that their own commitment is communicated. Only then will lean quality spread effectively throughout the organization. Middle management also needs to ensure that the efforts and achievements of their subordinates obtain the recognition, attention and the reward that they deserve. Project managers in the construction sector have a critical role as they often have responsibility for selecting the subcontractors and have the challenge of creating a cohesive lean quality focused team on the project. They have to explain the lean quality strategy to all suppliers, their on-site supervisors and workers and ensure that all parts of the team are committed to the shared values and goals.

The Chief Executive of an organization should accept the responsibility for and commitment to a lean quality policy in which he/she must really believe. This commitment is part of a broad approach, extending well beyond the accepted formalities of a QA function. It creates responsibilities for a chain of lean quality interactions between the marketing, design, production/operations, purchasing, distribution and service functions. Within each and every department of the organization at all levels, starting at the top, basic changes of attitude may be required to implement lean quality approaches. If the owners or directors of the organization do not recognize and accept their responsibilities for the initiation and operation of lean quality, then these changes will not happen. Controls, systems and techniques are very important in lean quality, but they are not the primary requirement. It is more an attitude of mind, based on pride in the job and teamwork, and it requires from the management total commitment, which must then be extended to all employees at all levels and in all departments.

Senior management commitment should be obsessional, and not just pay lip service. It *is* possible to detect real commitment. Where lean quality is implemented, it *shows* on the construction sites; in the head office; and in the design office. In organizations sporting poster-campaigns of lean quality without commitment, one is quickly able to detect the falseness. The people are told not to worry if problems arise: 'Just do the best you can. The customer will never notice,' quite the opposite to an organization where lean quality means something. There, one can see, feel, and hear the enthusiasm and commitment. Things happen at this operating interface as a result of *real* commitment. Material problems are corrected with suppliers; equipment difficulties are put right by improved mainten-ance programmes or replacement; people are trained and empowered to contribute to the improvement of their work processes; change takes place; partnerships are built; and continuous improvement means just that – at every level of the supply chain.

At the project level, similar leadership is called for. In the building construction sector in Australia, the UK and USA subcontractors, who both prefabricate and assemble materials and products on site, do more than 85 per cent of the actual work. This means that the bulk of the operational workforce on site is from the subcontract supply chain. One of the major challenges on such projects is the need to influence the values and behaviours of the entire delivery team. To be successful, the project manager needs to select suppliers on the basis of the fit in values as well as on their technical competence and price. It is the effectiveness of the inter-organizational team working side by side that will determine the success or failure of any project. In such a setting, the challenge of leadership is to create cohesion in values and behaviours across the entire project around a lean quality framework.

The lean quality policy

A sound lean quality policy, supported by the organization and facilities to put it into effect, is a fundamental requirement for effective implementation. The content of the policy together with the arrangements for its implementation should be made known to all employees. This together with continuous monitoring will result in smoother production or service operation, few errors and reduced waste.

Management should be dedicated to the regular improvement of lean quality, not simply a one-step improvement to an acceptable plateau. These ideas can be set out in a *lean quality policy* that requires top management to:

1 identify the end customer's needs (including perceptions);
2 assess the ability of the organization to meet these needs economically;
3 actively look for opportunities to eliminate non-value-adding tasks and continuously improve processes and outcomes;
4 ensure that bought-in materials' reliability meet the required standards of performance and efficiency;
5 ensure that subcontract suppliers working on site share your values and process goals;
6 concentrate on the prevention rather than detection philosophy;
7 educate and train for lean quality improvement and ensure that your subcontractors do so as well;
8 measure customer satisfaction at all levels: the end customer as well as customer satisfaction between the links of the supply chain;
9 measure and report key performance metrics and trends at every level of the organization, and report performance and challenges through visible information centres throughout the organization; and
10 review the lean quality management systems to maintain progress.

The lean quality policy should be the concern of all employees. The principles and objectives should be communicated as widely as possible so that they are understood at all levels of the organization and within the subcontract supply chain on construction projects. Practical assistance and training should be given, where necessary, to ensure the relevant knowledge and experience are acquired for the successful implementation of the policy throughout the supply chain.

Examples of two company quality policies are given below.

Quality policy example 1 (a process industry company supplying the automotive sector)

- The company will concentrate on its customers and suppliers, both external and internal.
- The performance of our competitors will be communicated to all relevant units.
- Important suppliers and partners will be closely involved in our quality policy. This relates to both external and internal suppliers of goods, resources and services.
- Lean quality management systems will be designed, implemented, audited and reviewed to drive continuous improvement. They will be integrated into the overall management system.
- Lean quality improvement is primarily the responsibility of management and will be tackled and followed up in a systematic and planned manner. This applies to every part of our organization.
- In order to involve everyone in the organization of quality improvement, management will enable all employees to participate in the preparation, implementation and evaluation of improvement activities.
- Lean quality improvement will be a continuous process and widespread and attention will be given to education, training and skills development activities, which will be assessed with regard to their contribution to the quality policy.
- Publicity will be given to the quality policy in every part of the organization so that everyone may understand it. All available methods and media will be used for its internal and external promotion and communication.
- Reporting on the progress of the implementation of the policy will be a permanent agenda item in management meetings.

Quality policy example 2

See Figure 4.1, p. 66.

As an exercise, look at the way in which the case study companies have enunciated their quality policies. These demonstrate the strategies of some of the best companies in the sector, worldwide.

CREATING OR CHANGING THE CULTURE

The culture within an organization is formed by a number of components:

1. behaviours based on people interactions;
2. norms resulting from working groups;
3. dominant values adopted by the organization;
4. rules of the game for 'getting on'; and
5. the climate.

Culture in any business may be defined by how the business is conducted, and how employees behave and are treated. Any organization needs a *vision framework* that includes its *guiding philosophy, core values and beliefs* and a *purpose*. These should be combined into a *mission*, which provides a vivid description of what things will be like when it has been achieved (Figure 4.2).

FIGURE 4.1 Quality policy example 2

The *guiding philosophy* drives the organization and is shaped by the leaders through their thoughts and actions. It should reflect the vision of an organization rather than the vision of a single leader, and should evolve with time, although organizations must hold on to *core* elements.

The *core values and beliefs* represent the organization's basic principles about what is important in its business: its conduct, social responsibility and response to changes in its environment. The *core values and beliefs* should act as a guiding force with clear and authentic values which are focused on employees, suppliers, customers, society at large, safety, shareholders and generally stakeholders.

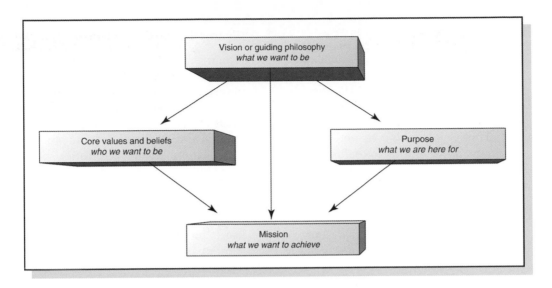

FIGURE 4.2 Vision framework for an organization

The *purpose* of the organization should be a development from the vision, core values and beliefs and should succinctly convey how the organization is to fulfil its role.

The *mission* will translate the abstractness of philosophy into tangible goals that will move the organization forward and make it perform to its optimum. It should not be limited by the constraints of strategic analysis and should be proactive not reactive. Strategy is subservient to mission: the strategic analysis should be done after, not during, the mission setting process.

Two examples of how leaders of organizations, one in the private sector and one in the public sector, develop their vision, mission and values and are role models of a culture of lean quality excellence are given in the boxes below (p. 68).

Some of the case study companies address the question of business strategy, while others are more focused at the operational level. It is interesting to look at the very different ways that the case study companies set about defining their vision, mission and core values. Look at Graniterock (CS5), Crossrail (CS7), the DPR (CS9), Heathrow (CS10) and Costain (CS12). They are very different businesses with widely differing goals, but all are striving for excellence.

Control

The effectiveness of an organization and its people depends on the extent to which individuals and departments perform their role and move towards agreed common goals and objectives. Control is maintained by using information or feedback to track all functions, including improvement initiatives, to make sure they are kept on track. It is the sum total of these activities that increase the probability of the planned results being achieved. Depending upon their position in the managerial process, control mechanisms fall into three categories, as shown in Table 4.1.

Many organizations use after-the-fact controls, causing managers to take a reactive rather than a proactive position. Such 'crisis-orientation' needs to be replaced by a more anticipative one in which the focus is on preventive or before-the-fact controls.

Example 1: Private sector

To enable the company to set direction and achieve its vision, the senior management team address priorities for improvement. These are driven by a business improvement process which consists of articulating a vision, determining the actions to realize the vision, defining measures and setting targets, then implementing a rigorous review mechanism.

Each member of the team takes responsibility for one of the Excellence Model criteria. They develop improvement plans and personally ensure that these are properly resourced and implemented, and that progress is monitored. Improvements identified at local level are prioritized and resourced by local management against the organization's annual business plan.

Example 2: Public sector

The mission of the organization (the purpose and direction) is developed by a task team. Senior, middle and junior managers review and update the mission, vision and values annually to ensure it supports policy and strategy.

Leaders invite input from stakeholders via the Employee Involvement initiative, monthly update meetings and customer service seminars. The values have been placed on help-cards for every employee and are continually re-emphasized at monthly update meetings.

Leaders act as role models and have their own list of Role Model Standards to follow and are measured against these in their Performance Management System. All managers include lean quality objectives in their Performance Agreements and Personal Development Plan, which are reviewed through the Review.

TABLE 4.1 The three categories of control mechanisms

Before the fact	Operational	After the fact
Strategic plan	Observation	Annual reports
Action plans	Inspection and correction	Variance reports
Budgets	Progress review	Audits
Job descriptions	Staff meetings	Surveys
Individual performance objectives	Internal information and data systems	Performance review
Training and development	Training programmes	Evaluation of training

There is a brief but interesting discussion on the DPR case study (CS9) about this very issue, one of the things that DPR's Rodney Spencely was seeking to achieve in developing the 'key distinguishing features' of client quality requirements in a very detailed way was to create structured activities that could be used as a leading indicator of the quality-related efforts of DPR and its supply chains.

Attempting to control performance through systems, procedures or techniques *external* to the individual is not an effective approach since it relies on 'controlling' others; individuals should be responsible for their own actions. An externally based control system can result in a high degree of concentrated effort in a specific area if the system is overly structured, but it can also cause negative consequences to surface:

1 Since all rewards are based on external measures, which are imposed, the 'team members' often focus all their effort on the measure itself: they may have it set lower or higher than possible, manipulate the information which serves to monitor it, or dismiss it as someone else's goal not theirs. In the budgeting process, for example, distorted figures are often submitted by those who have learned that their 'honest projections' will be automatically altered anyway.

2 When the rewards are dependent on only one or two limited targets, all efforts are directed at these, even at the expense of others. If short-term profitability is the sole criterion for bonus distribution or promotion, it is likely that investment for longer-term growth areas will be substantially reduced. Similarly, strong emphasis and reward for output or production may result in lowered quality.

3 The fear of not being rewarded or even being criticized for performance that is less than desirable may cause some to withhold information that is unfavourable but nevertheless should be flowing into the system.

4 When reward and punishment are used to motivate performance, the degree of risk-taking may lessen and be replaced by a more cautious and conservative approach. In essence, the fear of failure replaces the desire to achieve.

The following problem situations have been observed by the authors and their colleagues within companies that have taken part in research and consultancy:

• The goals imposed are seen or known to be unrealistic. If the goals perceived by the subordinate are in fact accomplished, then the subordinate has been proven to be wrong. This clearly has a negative effect on the effort expended, since few people are motivated to prove themselves wrong!

• Where individuals are stimulated to commit themselves to a goal, and where their personal pride and self-esteem are at stake, then the level of motivation is at a peak. For most people, the toughest critic and the hardest taskmaster they confront is not their immediate boss, but themselves.

• Directors and managers are often afraid of allowing subordinates to set the goals for fear of them being set too low or loss of control over subordinate behaviour. It is also true that many workers do not wish to set their own targets, but prefer to be told what is to be accomplished.

• Where external project managers are recruited to run projects and a reward is negotiated on the basis of a bonus package reflecting time and cost performance, all too often the company is left with the legacy of quality defects long after the project manager has finished the assignment and pocketed his/her bonuses.

• Some public sector client organizations in Australia, when they adopted alliance arrangements to deliver infrastructure projects, developed very complex performance frameworks to incentivize project outcomes in non-cost areas such as safety, quality, community and legacy. In some cases the complexity of these were such that the performance measures became an end in themselves and got in the way of manage-ment initiative and continuous improvement.

Lean quality is concerned with moving the focus of control from outside the individual to within. The objective is to make everyone accountable for their own performance, and to get them committed to attaining lean quality in a highly motivated fashion. Managers have to recognize that people do not need to be coerced to perform well: they want to achieve, accomplish, influence activity and challenge their abilities of their own volition. If there is belief in this, then only the techniques remain to be discussed.

Lean quality is user-driven: it cannot be imposed from outside the organization, as perhaps a quality management standard or statistical process controls may be. This means that the ideas for improvement must come from those with knowledge and experience of the processes, activities and tasks; this has massive implications for training and follow-up. Lean quality is not a cost-cutting or productivity improvement device in the traditional sense, and it must not be used as such. Although the effects of a successful programme will certainly reduce costs and improve productivity, lean quality is concerned chiefly with changing attitudes and skills so that the culture of the organization becomes one of preventing failure: doing the right things, right first time, every time and always seeking ways for improvement and waste elimination.

Most construction organizations rely on external control, and it is commonplace to believe that workers don't care about anything other than their pay and their beer, that they are relatively unskilled and do not have the potential to change. It is important to realize that management's attitudes can easily become self-fulfilling, and the task of changing workers' behaviour and attitudes is purely a management challenge. Without a change in the attitudes and strategies of leadership, workers' behaviours will not change.

EFFECTIVE LEADERSHIP

Some management teams have broken away from the traditional style of management; they have made a 'managerial breakthrough'. Their approach puts their organization head and shoulders above others in the fight for sales, profits, resources, funding and jobs. Many public service organizations are beginning to move in the same way, and the successful lean quality-based strategy they are adopting depends very much on effective leadership.

Effective leadership begins with the Chief Executive and their top team's vision, their ability to capitalize on market or service opportunities, their innovative development of a strategy that will give the organization competitive or other advantage and results in business or service success. It goes on to embrace all the beliefs and values held, the decisions taken and the plans made by anyone, anywhere in the organization and turns them into effective, value-adding action.

Together, effective leadership and lean quality result in the company or organization doing the right thing, right, the first time and every time.

The five requirements for effective leadership are:

1 Develop and publish clearly documented corporate beliefs and purpose – a vision

Executives should express values and beliefs through a clear vision of what they want their company to be and its purpose – what they specifically want to achieve in line with their basic beliefs. The senior management team will need to spend some time away from the 'coal face' to do this and to develop their programme and strategy for implementation.

Clearly defined and properly communicated beliefs and objectives which can be summarized in the form of vision and mission statements are essential if the directors, managers and other employees are to work together as a winning team. The beliefs and objectives should address:

- business definition (e.g. the needs that are satisfied or the benefits provided);
- commitment to effective leadership and lean quality;
- target sectors, relationships with customers, and market or service position;

- role or contribution of the company, organization or unit (e.g. example, profit-generator, service department, opportunity-seeker);
- the distinctive competence – a brief statement which applies only to that organization, company or unit;
- indications for future direction – a brief statement of the principal plans which would be considered;
- commitment to monitoring performance against customers' needs and expectations, and to continuous improvement in all areas.

These together with broad beliefs and objectives may then be used to communicate an inspiring vision of the organization's future. The top management must then show *TOTAL COMMITMENT* to it.

2 Develop clear and effective strategies and supporting plans for achieving the vision

To achieve the company or service vision and mission requires the development of business or service strategies, including the strategic positioning in the 'marketplace'. Plans can then be developed for implementing the strategies. While such strategies and plans can be developed by senior managers alone, there will be more commitment to them if employee participation in their development and implementation is invited and encouraged.

3 Identify the critical success factors and critical processes (Figure 4.3)

For the mission to be realized, the *critical success factors* (CSFs) which define the most important sub-goals of the business or organization must be identified. For the realization of the CSFs, the organization's *core business processes* must be done particularly well. This process is described in more detail in later chapters. It is also critical for each of these to develop some performance measures that indicate progress towards each goal that has been set. An interesting example is found in the Crossrail case study (CS7). Crossrail, the rail client, set out to make quality as important as safety on this project. Crossrail have identified five CSFs that are key to delivering its quality vision.

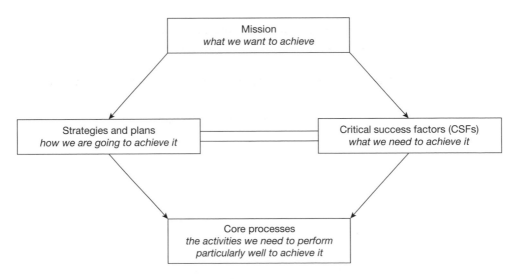

FIGURE 4.3 Mission into action through CSFs and core processes

4 Review the management structure

Defining the corporate vision, mission, strategies, CSFs and core processes might make it necessary to review the organizational structure. Directors, managers and other employees can be fully effective only if an effective structure based on process management exists. This includes both the definition of responsibilities for the organization's management and the operational procedures they will use. These must be the agreed best ways of carrying out the core processes.

The review of the management structure should also include the establishment of a process improvement team structure throughout the organization. Improvement activities must be properly resourced and progress monitored against plans.

5 Empowerment – encouraging effective employee participation

For effective leadership, it is necessary for management to get very close to the employees. They must develop effective communications – up, down and across the organization – and take action on what is communicated; and they should encourage good communications between all suppliers and customers.

Particular attention must be paid to the following:

Attitudes

The key attitude for managing any winning company or organization may be expressed as follows: 'I will personally understand who my customers are and what their needs and expectations are. And I will take whatever action is necessary to understand and satisfy them fully. I will also communicate my requirements to my suppliers, inform them of changes and provide feedback on their performance.' This attitude should start at the top, with the chairman or chief executive. It must then percolate down, to be adopted by each and every employee. This will happen, only, if managers lead by example. Words are cheap and will be meaningless if employees see from managers' actions that they do not actually believe or intend what they say.

Abilities

Every employee should be able to do what is needed and expected of him or her, but first, it is necessary to determine and communicate what is actually needed and expected. If it is not clear what employees or subcontractors are required to do and what standards of performance are expected of them, how can managers expect these requirements to be fulfilled? Over the past few decades, examples of such confusion are evident on many construction projects. Management repeatedly says that 'we want the job done fast!' but rarely does it stress that it 'we want it done correctly the first time'. This has resulted in a culture of 'we can fix it later so let's just get on with the job'. This attitude results in numerous defects being incorporated into buildings, defects that are more expensive to rectify later.

The essence of getting this right is to train, train, train and train again. Training is very important, but it can be expensive if the money is not spent wisely. The training should be related to needs, expectations and process improvement. It must be planned and *always* its effectiveness reviewed.

Participation

If all employees are to participate in making the company or organization successful, directors and managers included, then the management must also be trained in the basics of disciplined management.

They must be trained to:

> **Evaluate**: the situation and define their *objectives*.
> **Plan**: to achieve those objectives fully.
> **Do**: implement the plans.
> **Check**: that the objectives are being achieved.
> **Amend**: take corrective action if they are not.

The word 'disciplined' applied to people at all levels means that they will do what they say they will do. It also means that in whatever they do, they will go through the full process of Evaluate, Plan, Do, Check and Amend, rather than the more traditional and easier option of starting by doing. This will lead to a never-ending improvement helix (Figure 4.4).

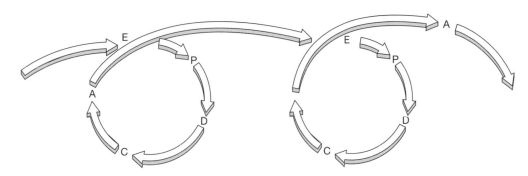

FIGURE 4.4 The helix of never-ending improvement

This basic approach needs to be backed up with good project management, planning and problem-solving techniques, which can be taught to anyone in a relatively short period of time. Project management enables changes to be made successfully and empowers people to remove the obstacles in their way. Directors and managers need this training as much as other employees.

In construction projects, the project manager has a very important and difficult leadership challenge. The project manager has to create a cohesive team at the project level, often with more than 85 per cent of the operational staff and their supervisors employed by subcontractors. It is up to the project manager to galvanize the entire project team towards a set of shared goals. It is generally true that project managers who do this well, do so instinctively; it is not an area that has been identified and taught with any focus. Furthermore, because on construction projects most problems occur at the interfaces between suppliers, the challenge is that process innovation and problem solving have to be achieved by teams of workers from the collaborating organizations. Hence, while the lean quality principles in construction are the similar to other enterprises, there is this added challenge of realizing goals and creating concerted action across a broad supply chain which is working in a shared workspace.

EXCELLENCE IN LEADERSHIP

The vehicle for achieving excellence in leadership is the lean quality framework. We have seen that its framework covers the entire organization, all the people and all the functions, including external organizations and suppliers. In the first three chapters, several facets of lean quality have been reviewed, including:

- Recognizing customers and discovering their needs. This refers to immediate and end-user customers equally.
- Setting standards that are consistent with internal and end-user customer requirements.
- Controlling processes, including systems, and improving the capability of teams.
- Management's responsibility for setting the guiding lean quality philosophy, lean quality policy, etc., and providing motivation and empowerment to equip people to achieve lean quality.
- Selecting the right employees and supply chain partners, and empowering people at all levels in the organization and across the supply chain to act for lean quality improvement.

The task of implementing lean quality can be daunting, and the Chief Executive and directors faced with it may become confused and irritated by the proliferation of theories and packages. A simplification is required. The *core* of lean quality is the customer–supplier interfaces, both internally and externally, and the fact that at each interface there are processes to convert inputs to outputs. Clearly, there must be commitment to building-in quality through management of the inputs and processes.

How can senior managers and directors be helped in their understanding of what needs to be done to become committed to lean quality and implement the vision? The American and Japanese quality 'gurus' each set down a number of points or absolutes – words of wisdom in management and leadership – and many organizations have used these to establish a policy based on quality.

Similarly, the EFQM have defined the criterion of leadership and its sub-criteria as part of their Excellence Model. The fundamental principle behind all these approaches is that the behaviours of the leaders in an organization need to create clarity and constancy of purpose. The first steps are through the development of the vision, values, purpose and mission needed for longer-term performance success. However, to be effective, leaders have to accept the roles of mentor, teacher and advocate for lean quality.

In this next section, we turn briefly to each of the elements of the new 'Oakland–Marosszeky model for lean quality' to define the main issues for attention of leadership in delivering organizational excellence.

Planning
- Develop the vision and mission needed for constancy of purpose and for long-term success.
- Develop, deploy and update policy and strategy.
- Align organizational structure to support delivery of policy and strategy.

Performance – Continuous improvement in all processes and outcomes
- Identify critical areas of performance in processes and outcomes.
- Develop measures to indicate levels of current performance.
- Set goals, identify gaps and benchmark progress.
- Provide feedback to people, at all levels, regarding their performance against agreed goals.

Processes
- Ensure a system for managing processes is developed and implemented.
- Ensure through personal involvement that the management system is developed, implemented and continuously improved.

- Prioritize and properly resource improvement activities and ensure they are planned on an organization-wide basis.

People

- Train managers and team leaders, at all levels, in leadership skills and problem solving, including mentoring their direct reports.
- Empower people and stimulate teamwork to encourage creativity and innovation.
- Encourage, support and act on results of training, education and learning activities.
- Motivate, support and recognize the organization's people – both individually and in teams.
- Help and support people to achieve plans, goals, objectives and targets.
- Respond to people and encourage them to participate in improvement activities.

Customers – Maximize value for all customers

- Be involved with customers and other stakeholders, ensure that external customer end needs and expectations are fully understood.
- Ensure that every team fully understands the needs and expectations of their immediate customers (external and internal) and respond to them.
- Establish and participate in partnerships – as a customer demand continuous improvement in everything.
- Ensure that near real-time-evaluation of the performance of all teams is obtained from the viewpoint of their customers and fed back to them.

Commitment

- Be personally and actively involved in quality and improvement activities.
- Review and improve effectiveness of own leadership.

Culture

- Develop the values and ethics to support the creation of a lean quality culture across the entire supply chain.
- Implement values and ethics through actions and behaviours.
- Ensure creativity, innovation and learning activities are developed and implemented.

Communications

- Stimulate and encourage communication and collaboration.
- Personally communicate the vision, values, mission, policies and strategies.
- Be accessible and actively listen.

Lean quality should not be regarded as a woolly minded approach to running an organization. It requires strong leadership with clear direction and a carefully planned and fully integrated strategy derived from the vision. One of the greatest tangible benefits of excellence in leadership is the improved overall performance of the organization. The evidence for this can be seen in some of the major consumer and industrial markets of the world. Moreover, effective leadership leads to improvements and superior quality which can be converted into premium prices. Research now shows that leadership and lean quality clearly correlate with profit, but the less tangible benefit of greater employee participation is equally, if not more, important in the longer term. The pursuit of continual improvement must become a way of life for everyone in an organization if it is to succeed in today's competitive environment.

BIBLIOGRAPHY

Davidson, H. *The Committed Enterprise,* Butterworth-Heinemann, Oxford, 2002.
Juran, M. *Juran on Leadership for Quality: An Executive Handbook,* The Free Press (Macmillan), New York, 1989.
Townsend, P.L. and Gebhardt, J.E. *Quality in Action – 93 Lessons in Leadership, Participation and Measurement,* John Wiley Press, New York, 1992.

CHAPTER HIGHLIGHTS

The lean quality management approach

- Lean quality is a comprehensive approach to improving competitiveness, effectiveness and flexibility through planning, organizing and understanding each activity, and involving each individual at every level. It is useful in all types of organization.
- Lean quality ensures that management adopts a strategic overview of performance, and focuses on prevention, not detection, of problems.
- Invest in understanding customer needs and areas of dissatisfaction to provide a focus for improvement, measure performance, set goals and benchmark your rate of improvement.
- It often requires a mindset change to break down existing barriers. Managements that doubt the applicability of lean quality should ask questions about the operation's costs, errors, wastes, standards, systems, training and job instructions.

Commitment and policy

- Lean quality starts at the top, where serious obsession and commitment to quality and leadership must be demonstrated. Middle management also has a key role to play in communicating the message.
- Every chief executive must demonstrate real commitment to a lean quality policy that deals with the resources and organizational structure for lean quality, the customer needs, the ability of the organization, supplied materials and services, education and training, and review of the management systems for never-ending improvement.

Creating or changing the culture

- The culture of an organization is formed by the beliefs, behaviours, norms, dominant values, rules and climate within the organization.
- Any organization needs a vision framework, comprising its guiding philosophy, core values and beliefs, purpose and mission.
- The effectiveness of an organization depends on the extent to which people perform their roles and move towards the common goals and objectives.
- Lean quality is concerned with moving the focus of control from the outside to the inside of individuals so that everyone is accountable for his/her own performance.

Effective leadership

- Effective leadership starts with the chief executive's vision, and develops into a strategy for implementation.
- Top management should develop the following for effective leadership: clear beliefs and objectives in the form of a mission statement; clear and effective strategies and

supporting plans; the critical success factors and core processes; the appropriate management structure; employee participation through empowerment, and the EPDCA helix.

Excellence in leadership

- The vehicle for achieving excellence in leadership is lean quality. Using the construct of the new Oakland–Marosszeky model for lean quality, the four Ps and four Cs provide a framework for this: Planning, Performance, Processes, People, Customers, Commitment, Culture, Communications together with the all-embracing focus on maximizing value and continuous improvement in processes and outcomes ensures operational excellence.

Part I Discussion questions

1 You are planning to start a medium density housing design and construction business; this is a crowded market. Your aim is to attract customers who are seeking a better quality product and service. Discuss the key implications of this for the management of the business.

2 Explain the difference between product quality and product reliability; and between quality of design and quality of conformance, illustrating your answer with examples taken from the construction context.

3 Discuss how service quality impacts customer perceptions of product quality and describe how this relationship plays out in construction in relation to the end customer and intermediate customers.

4 Discuss the various facets of lean in construction, paying particular attention to its interfaces with the functional areas within the organization.

5 Explain what you understand by the term 'lean quality management', paying particular attention to the following terms: quality, supplier/customer interfaces, and process.

6 Present a 'model' for lean quality management, describing briefly the various elements of the model.

7 Select one of the so-called 'Gurus' of quality management, such as Juran, Deming, Crosby, Ishikawa, and explain their approach, with respect to the 'Oakland–Marosszeky Model'. Discuss the strengths and weaknesses of their approach using this framework.

8 Compare the three models for total quality described by the Deming Prize in Japan, the Baldrige Award in the USA and the European Excellence Award in the light of the principles of lean quality and identify key differences.

9 In your new role as quality manager of the high-tech unit of a large national company, you identify a problem which is typified by the two internal memos shown below. Discuss in some detail the problems illustrated by this conflict, explaining how you would set about trying to make improvements.

From: Marketing Director
To: Managing Director

c.c.
Production Director
Works Manager

Date: 4 August

We have recently carried out a customer survey to examine how well we are doing in the market. With regard to our product range, the reactions were generally good, but the solid concrete plaster sections are a problem. Without exception everyone we interviewed said that its quality is not good enough. Although it is not yet apparent, we will inevitably lose our market share.

As a matter of urgency, therefore, will you please authorize a complete redesign of this product?

From: Works Manager
To: Production Director

Date: 6 August

This really is ridiculous!

I have all the QC records for the past three years on this product. How can there be anything seriously wrong with the quality when we only get 0.1 per cent rejects at final inspection and less than 0.01 per cent returns from contractors?

---

10 Explain how the culture in an organization develops over time and describe the main components. How would you go about addressing negative cultural and behavioural aspects in a construction business which are clearly leading to quality problems in the marketplace?

11 What are the aspects of leadership which are key to a successful lean quality approach? You have been appointed to take leadership for strengthening the quality and efficiency of the business processes and outputs in a construction company. Describe how you would go about driving change through the business, both internally and throughout your supply chain.

Part II

Planning

Policy, strategy and goal deployment

5

INTEGRATING LEAN QUALITY INTO POLICY AND STRATEGY

One thing that all writers on strategy agree is that the leaders of any organization need a clear sense of direction and purpose, which they must communicate effectively throughout the organization – a clear message from the previous chapter on leadership. This typically involves the development of the vision, values and mission which define the fundamental nature of the organization, and the strategic plan which determines where resources will be invested to greatest net benefit.

For this to happen, the vision and mission and their deployment must be based on the needs and expectations of the organization's stakeholders – present and future – and a thorough examination of the environment in which the organization exists. In today's dynamic environment, this requires information from research and learning activities and, even more importantly, accurate and timely measures of key performance criteria. Without this information, regularly reviewed and updated, it is difficult for managers to make the right strategic choices.

Included in the EFQM Excellence Model is a 'strategy' criterion that is concerned with how the organization implements its vision and mission via a clear stakeholder-focused strategy, supported by relevant policies, plans, objectives, targets and processes.

The challenge for management is how to ensure that a gap doesn't appear between the chosen direction for the organization and the day-to-day operations. Any misalignment wastes resources, squanders opportunities and increases risk.

The Crossrail case study (CS7) provides a very interesting example of policy development around a very clear vision. The board of Crossrail decided that it wanted to see quality elevated to the same importance in terms of outcomes as safety. The case study describes how Crossrail went about developing its strategy to achieve this objective.

The Highways England (HE) case study (CS14) shows how this client through clear-minded vision and policy has driven process and capacity improvement in its supply chains. HE determined that a key goal for the organization was to improve the value it derived from its construction supply chains, and to support this it implemented a procurement strategy which included the development of a maturity model, of training materials, provision of coaching and benchmarking supplier performance to drive value creation.

Many companies have difficulty starting a continuous improvement programme. They do not know how to state their vision and mission simply because management is not used to thinking in these terms. A good place to start is through research into external customer satisfaction. Feedback can then provide the foundation definition of corporate vision, and as the process matures it will involve feedback from all stakeholders.

In construction organizations, there are two significant obstacles to effective strategy and goal deployment. The first is that larger projects are treated as independent business units and the project director is given almost complete independence. Head Office (HO) does not like to interfere as full responsibility for achieving critical cost and schedule targets is in the hands of the project director and HO does not want to dilute that responsibility. This high degree of independence often becomes an obstacle to the achievement of broader corporate strategic goals. The second challenge is that relatively few construction organizations would recognize that their supply chain partners are stakeholders in their organization; yet, subcontractors supply most of the labour, materials and equipment on construction sites. Construction safety, quality and production risks are largely in the hands of their subcontractor workers and supervisors.

Because *design and construction* as a method of procurement has become more popular, the supply chain now includes the designers as well. Subcontract employees constitute upwards of 85 per cent of workers on most building construction sites, fewer on civil projects. Hence, a very important part of a construction organization's strategy must consider such issues as the values, skills, knowledge and attitudes of subcontractor management and employees. It is at the interfaces between the subcontractors that most of the opportunities for innovation and increased value creation lie; it is also at these interfaces that a great deal of process-related risk is to be found.

A close examination of the case studies at the end of this book illustrates that there is no one answer for how to proceed with lean quality implementation. Many of the construction companies started off recognizing the need to improve the quality of assignments in short-range plans on projects, realizing that the instability of work on a daily basis created risks that could not be effectively managed. From this point of departure, these organizations have matured as lean quality thinking has spread from the operational weekly work plan to all aspects of strategy and operations. The JB Henderson case study (CS8) and the Crossrail case study (CS7) illustrate perhaps two very different approaches to driving change, reflecting their genesis and the size of operations being changed. The JB Henderson case study shows a rapid change and improvement in labour productivity at the operational level on site and in the fabrication shop led by worker ideas for improvement within a management framework which includes empowerment, a search for waste, collective learning and collaborative planning within a relational contract. The Crossrail case study shows a client-led strategy to elevate the achievement of 'right first time' quality to the same status as safety in a major construction undertaking. In this case the board of the client organization led by the CEO has developed a carefully crafted policy framework and strategy to create an effective focus on quality achievement throughout its supply chains. The lesson that stands out for the authors in these case studies is that lean quality must be clearly and visibly led from the top, but the realization of lean quality comes from the engagement of individuals and work teams with the principles of maximizing customer satisfaction whilst continuously striving to improve the efficiency of processes and the quality of outcomes at every level throughout the organization.

ALIGNING STRATEGY AND ACTION

There are six basic steps for achieving the alignment between the strategic choices, critical success factors, processes and people, and providing a foundation for the implementation of effective improvement. What is more, senior lean quality professionals can play an active role in facilitating this process.

Step 1. Develop a shared vision and mission for the business/organization

Once the top team is reasonably clear about the direction the organization should be taking it can develop vision and mission statements that will help to define process alignment,

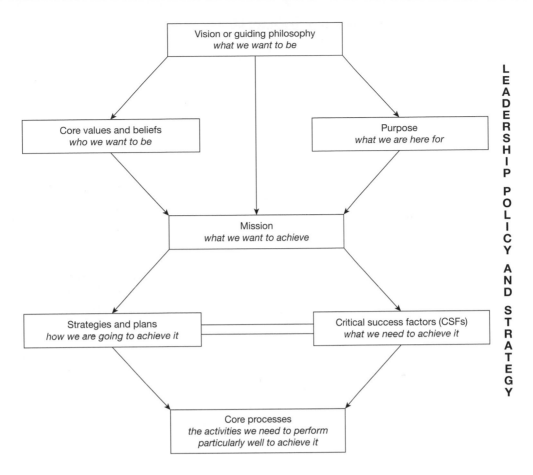

FIGURE 5.1 Vision framework for an organization

roles and responsibilities. This will lead to a coordinated flow of analysis of processes that cross the traditional functional areas at all levels of the organization without changing formal structures, titles and systems which can create resistance. The vision framework was introduced in Chapter 4 (Figure 5.1).

The mission statement gives a purpose to the organization or unit. It should answer the questions: What are we here for? What is our basic purpose? What have we got to achieve? It therefore, defines the boundaries of the business in which the organization operates. This will help to focus on the 'distinctive competence' of the organization and to orient everyone in the (same) direction of what has to be done. The mission must be documented; agreed to by the top management team; sufficiently explicit, so that its eventual accomplishment may be credible; and, ideally, it should be no more than four sentences. The statement must be understandable, communicable, believable and usable.

The mission statement is:

- an expression of the aspiration of the organization;
- the touchstone against which all actions or proposed actions can be judged; and
- long term (usually), though short term if the mission is for survival.

In framing the mission, leadership should have an eye to the principles of lean quality (as set out in Chapter 3).

Typical content may include a statement of:

- the role or contribution of the business or unit: profit generator, service department, opportunity seeker;
- the definition of the business: the needs you satisfy or the benefits you provide. Do not be too specific or too general;
- your distinctive competence: a brief statement that only applies to your specific unit (a statement which could apply equally to any organization is unsatisfactory);
- indications for future direction: a brief statement of the principal things to which you would give serious consideration.

Some questions that may be asked of a mission statement. Does it:

- define the organization's role;
- fulfil the organization's aims
    - is it worthwhile/admirable,
    - will employees identify with it,
    - how will it be viewed by key stakeholders and external customers;
- take a long-term view based on stable long-term lean quality values, including a commitment to continuous improvement;
- indicate plans which might lead to commitment to a new product or service development, or training of personnel;
- recognize the need for collaboration among all the 'stakeholders' of the organization, especially the outsourced subcontract supply chain;
- ensure the purpose remains constant despite changes in top management?

It is important to establish in some organizations whether or not the mission is survival. This does not preclude a longer-term mission, but the short-term survival mission must be expressed, if it is relevant. The management team can then decide how to balance short-term survival with long-term values. If survival is a real issue it is inadvisable to concentrate on long-term planning initially.

There must be open and spontaneous discussion during the generation of the mission, but, in the end, there must be convergence on one statement. If the mission statement is wrong, everything that follows will be wrong too, so a clear understanding is vital.

Step 2. Develop the 'mission' into its critical success factors (CSFs) to coerce and move it forward

The development of the mission is clearly not enough to ensure its implementation. This is the 'danger gap' into which many organizations fall because they do not foster the skills needed to translate the mission through its CSFs into the core processes. Hence, they have 'goals without methods', change is not integrated properly into the business and the mission is rarely realized.

Once the top managers begin to list the CSFs, they will gain some understanding of what the mission or the change requires. The first step in going from mission to CSFs is to brainstorm all the possible factors that might impact on the mission. In this way, 30 to 50 factors ranging from politics to costs, from national cultures to regional market peculiarities will be identified.

The CSFs (what the organization must accomplish to achieve the mission) may now be defined through the examination and categorization of the impacts. This should lead to a balanced set of deliverables for the organization in terms of:

- financial and non-financial performance;
- customer/market satisfaction;
- people/internal organization satisfaction;
- supply chain satisfaction; and
- environmental/societal satisfaction.

There should be no more than eight CSFs, and no more than four if the mission is survival. They are the building blocks of the mission, the minimum key factors or sub-goals that the organization **must have** or **needs** and which together will achieve the mission. They are the *whats* not the *hows*, and are not directly manageable. They may be, in some case, statements of hope or fear. But they provide direction and the success criteria and are the end product of applying the processes. In CSF determination, a management team should follow the rule that *each* CSF is **necessary** and *together* they are **sufficient** for the mission to be achieved.

Some examples of CSFs may clarify their understanding:

- We must have right-first-time suppliers.
- We must have motivated, skilled workers.
- Suppliers working on our sites must have motivated, skilled workers.
- Our outsourced subcontract supply chain must function as a seamless team.
- We need new products that satisfy market needs.
- Our designs must be easily constructible and well documented.
- We need new business opportunities.
- We must have best-in-the-field product quality.

The list of CSFs should be an agreed balance of strategic and tactical issues, each of which deals with a 'pure' factor, the use of 'and' being forbidden. It will be important to

Divisional or functional CSFs	CSF No.	1	2	3	4	5	6	7	8	KPIs
	1									
	2									
	3									
	4									
	5									
	6									
	7									
	8									

FIGURE 5.2 Interaction of corporate and divisional CSFs

know when the CSFs have been achieved, but an equally important step is to use the CSFs to enable the identification of the processes.

Senior managers in large complex organizations may find it necessary or useful to show the interaction of divisional CSFs with the corporate CSFs in an impact matrix (see Figure 5.2 and discussion under Step 6).

Step 3. Define the key performance indicators (KPIs) as being the quantifiable indicators of success in terms of the mission and CSFs

The mission and CSFs provide the 'what' of the organization, but they must be supported by measurable key performance indicators (KPIs) that are tightly and inarguably linked. These will help to translate the directional and sometimes 'loose' statements of the mission into clear *targets* and, in turn, to simplify management's thinking. The KPIs will be used to monitor progress and as evidence of success for the organization, in every direction, internally and externally. They are the measures that will be reported on management dashboards and which will be displayed in information centres at all levels of reporting on the company.

Each CSF should have an 'owner' who is a member of the management team that agreed to the mission and CSFs. The task of an owner is to:

- define and agree the KPIs and associated targets;
- ensure that appropriate data is collected and recorded;
- monitor and report progress towards achieving the CSF (KPIs and targets) on a regular basis;
- review and modify the KPIs and targets where appropriate.

A typical CSF data sheet for completion by owners is shown in Figure 5.3.

The derivation of KPIs may follow the 'balanced scorecard' model, proposed by Kaplan, which divides measures into financial, customer, internal business and innovation, and learning perspectives (see Chapter 8). In large complex organizations, the skills and resources needed to achieve such demanding targets are rarely contained within one discrete 'business unit'. Therefore, it is vital that leaders are able to delegate responsibility for achieving contributory elements of the CSFs and KPIs to other members of the organization. It is through simple and effective performance management, allied with clarity of the processes, that this can be achieved.

A company practising lean quality will be biased towards the use of measurable lead indicators of reliability and quality in those operations that can help steer the company towards the realization of its mission day-to-day and week-by-week.

Some CSFs may involve gathering feedback from supply chain partners. It is important in these areas to bring the supply chain partners along in the process of developing and implementing those CSFs.

Step 4. Understand the core processes and gain process sponsorship

This is a critical step where many companies falter; the ideas of a mission are easier to define than the means to accomplish them. This is the point when the top management team has to consider how to institutionalize the mission in the form of processes that will continue to be in place throughout the organization, until major changes are required.

The core business processes describe what actually is or needs to be done so that the organization meets its CSFs. As with the CSFs and the mission, each process which is **necessary** for a given CSF must be identified and, together, all the processes listed for a CSF must be **sufficient** for that CSF to be accomplished. To ensure that **processes** are listed, they should be in the form of verb plus object, as in *research the market, recruit competent*

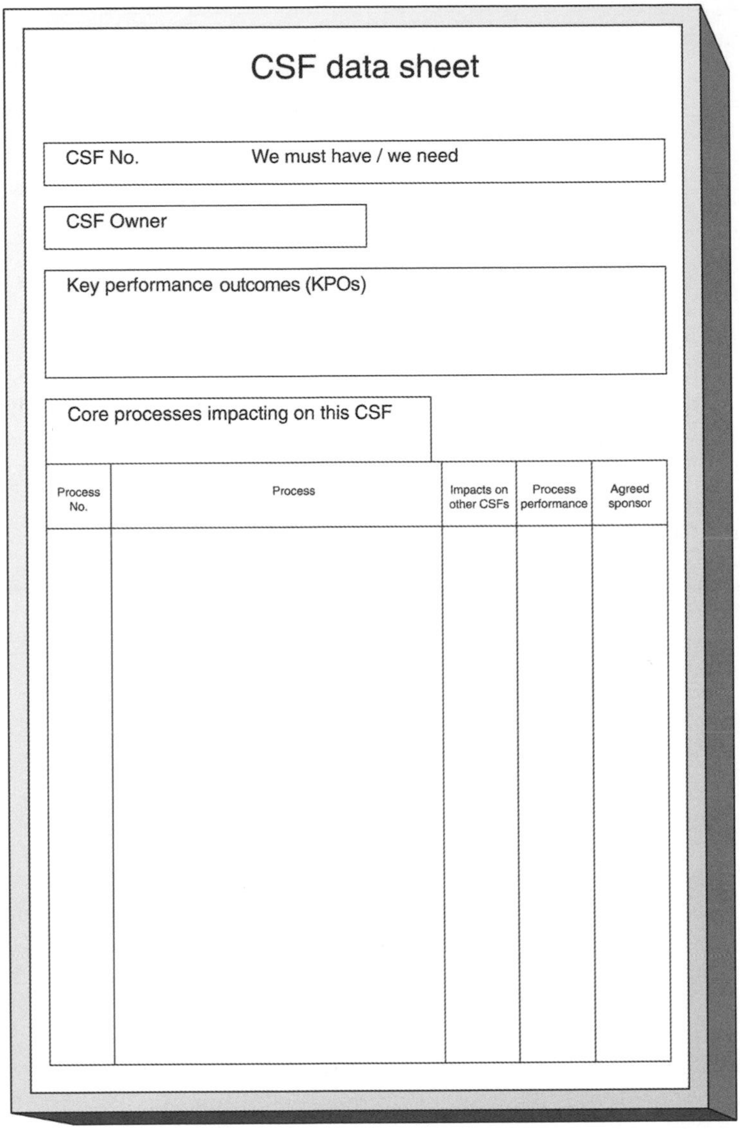

CSF data sheet

CSF No.	We must have / we need

CSF Owner

Key performance outcomes (KPOs)

Core processes impacting on this CSF

Process No.	Process	Impacts on other CSFs	Process performance	Agreed sponsor

FIGURE 5.3 CSF data sheet

staff or *manage supplier performance*. The core processes identified frequently run across 'departments' or functions, yet they must be measurable.

Each core process should have a sponsor, preferably a member of the management team that agreed to the CSFs.

The task of a sponsor is to:

- ensure that appropriate resources are made available to map, investigate, report and improve the process;
- assist in selecting the process improvement team leader and members;
- remove blocks to the team's progress;
- ensure that supply chain collaborations are in place, where necessary; and
- report progress to the senior management team.

No.	Process	CSF No.									Number of CSF impacts	A–E ranking

A–E process ranking: A–Excellent; B–Good; C–Average; D–Poor; E–Embryonic

FIGURE 5.4 Process/CSF matrix

The first stage in understanding the core processes is to produce a set of processes of a common order of magnitude. Some smaller processes identified may combine into core processes; others may be already at the appropriate level. This will ensure that the change becomes entrenched, the core processes are identified and that the right people are in place to sponsor or take responsibility for them. This will be the start of getting the process team organization up and running.

The questions will now come thick and fast. Is the process currently carried out? By whom? When? How frequently? With what performance and how well compared with competitors? Responding to these will force process ownership into the business. The process sponsor may form a process team which will take quality improvement into the next steps. Deciding where to focus is somewhat of a balancing act. Some form of prioritization using process performance measures is necessary at this stage to enable effort to be focused on the key areas for improvement. However, trying to do too much will lead to failure, while not doing enough will lead to a loss of momentum and the improvement initiative may falter. Determining priorities may be carried out by a form of impact matrix analysis (see Figure 5.4). The outcome should be a set of 'most critical processes' (MCPs) which receive priority attention for improvement, based on the number of CSFs impacted by each process and its performance on a scale A to E.

This high-level picture as to how 'the business is wired up' provides a valuable forum for informed debate about where and how value is designed, created, delivered and communicated. Without these informed debates, executives can take a very functional approach to delivering the strategy, missing the huge opportunities created by taking an end-to-end view of the value chain. If they don't, their customers certainly will.

Step 5. Break down the core processes into sub-processes, activities and tasks and form improvement teams around these

Once an organization has defined and mapped out the core processes, people need to develop the skills to understand how the new process structure will be analysed and made to work. The very existence of new process teams with new goals and responsibilities will force the organization into a learning phase. The changes should foster new attitudes and behaviours.

In the procurement of construction projects, whether in the design stage or in the construction stage, many of these teams will necessarily involve supply chain partners. Often more than 85 per cent of the people working side by side in design or on the construction site are employed by subcontractors. Hence, in this industry creating shared values among the supply chain partners is a prerequisite for success. Inter-organizational process teams are essential if the potential of the design and construction team for innovation and process improvement is to be realized.

Current thinking in relation to procurement practice is an obstacle to collaboration, as the mindset of minimizing cost and outsourcing all risk is not consistent with close collaboration. It is important to identify and sell the benefits of the new process structures to all parties involved, including commercial management as well as the leadership of outsourced suppliers.

An illustration of the breakdown from mission through CSFs and core processes, to individual tasks may assist in understanding the process required (Figure 5.5).

Mission

Two of the statements in a well-known management consultancy's mission statement are:

Gain and maintain a position as Europe's foremost management consultancy in the development of organizations through management of change.

Provide the consultancy, training and facilitation necessary to assist with making continuous improvement an integral part of our customers' business strategy.

Critical success factor

One of the CSFs which clearly relates to this is:

<u>We need a high level of awareness</u>

↓

Core process

One of the core processes which clearly must be done particularly well to achieve this CSF is to:

<u>Promote, advertise and communicate the company's business capability.</u>

↓

Sub-process

One of the sub-processes which results from a breakdown of this core process is:

↓

<u>Prepare the company's information pack.</u>

Activity

One of the activities which contributes to this sub-process is:

<u>Prepare **one** of the subject booklets,</u>
<u>i.e. 'Business Excellence and Self-Assessment'.</u>

↓

Task

One of the tasks which contributes to this is:

<u>Write the detailed leaflet for a particular seminar</u>
<u>e.g. a one or three-day seminar on self-assessment.</u>

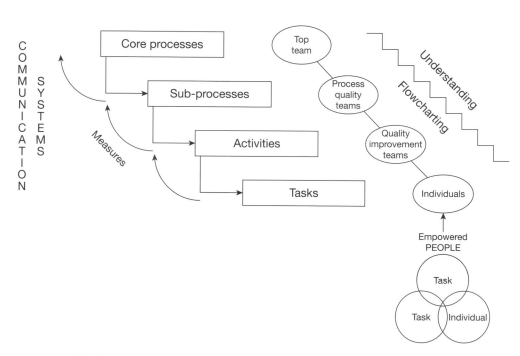

FIGURE 5.5 Breakdown of core processes into sub-processes, activities and tasks

Individuals, tasks and teams

Having broken down the processes into sub-processes, activities and tasks in this way, it is now possible to link this with the Adair model of action-centred leadership and team-work (see Chapter 17).

Clearly the tasks are performed, at least initially, by individuals. For example, somebody has to sit down and draft out the first version of a seminar leaflet mentioned in the previous paragraph. There has to be an understanding by individuals of the task and its position in the hierarchy of processes. Once the initial task has been performed, the results must be checked against the activity of coordinating the promotional booklet, say, for lean quality. This clearly brings in the team, and there must be interfaces between the needs of the *tasks*, the *individuals* who performed them and the *team* concerned with the *activities*.

Performance measurement and metrics

Once the processes have been analysed in this way, it should be possible to develop metrics for measuring the performance of the processes, sub-processes, activities and tasks. These must be meaningful in terms of the *inputs* and *outputs* of the processes and in terms of the *customers* and of *suppliers* to the processes (Figure 5.5).

At first thought, this form of measurement can seem difficult for processes such as preparing a sales brochure or writing leaflets advertising seminars. However, if we think carefully about the *customers* for the leaflet-writing tasks, they will include the *internal* ones, the consultants, and we can ask whether the output meets their requirements. Does it really say what the seminar is about, what its objectives are and what the programme will be? Clearly, one of the 'measures' of the seminar leaflet-writing task could be the number of typing errors in it. But is this a *key* measure of the performance of the process? In the context of office management this is essential, but it is not an important performance measure. Metrics must relate to key processes and outcomes, or they will devalue the improvement process.

The same goes for the *activity* of preparing the subject booklet. Does it tell the 'customer' what lean quality or SPC is and how the consultancy can help? For the *sub-process* of preparing the company brochure, does it inform people about the company, and does it bring in enquiries from which customers can be developed? Clearly, some of these measures require *external market research,* and some of them *internal research.* The main point is that metrics must be developed and used to reflect the *true performance* of the processes, sub-processes, activities and tasks. These must involve good contact with external and internal customers of the processes. The metrics may be quoted as *ratios*: numbers of customers derived per number of brochures mailed out. Good data collection, record-keeping and analysis are essential.

It is hoped that this illustration will help the reader to:

- understand the breakdown of processes into sub-processes, activities and tasks;
- understand the links between the process breakdowns and the task (individual and team concepts);
- link the hierarchy of processes with the hierarchy of quality teams;
- begin to assemble a cascade of flowcharts representing the process breakdowns, which can form the basis of the lean quality management system, and communicate what is going on throughout the business; and
- understand the way in which metrics must be developed to measure the true performance of the process, and their links with the customers, suppliers, inputs and outputs of the processes.

The changed patterns of coordination, driven by the process maps, should increase collaboration and information sharing. Clearly the senior and middle managers need to provide the right leadership, mentoring and support. Once employees, at all levels, identify what kinds of new skills are needed, they will ask for formal training programmes in order to develop those skills further. This is a key area because teamwork around the processes will ask more of employees, so they will need increasing support from their managers.

This has been called 'just-in-time' training, which describes very well the nature of the training process required. This contrasts with the blanket or carpet bombing training associated with many unsuccessful change programmes which target competencies or skills but do not change the organization's patterns of collaboration and coordination.

Step 6. Ensure process and people alignment through a policy deployment or goal translation process

One of the keys to integrating excellence into the business strategy is a formal 'goal translation' or 'policy deployment' process. If the mission and measurable goals have been analysed in terms of critical success factors and core processes, then the organization has begun to understand how to achieve the mission. Goal translation ensures that the 'whats' are converted into 'hows', passing this right down through the organization using a quality function deployment (QFD) type process (Figure 5.6 – Chapter 8). The method is best described by an example.

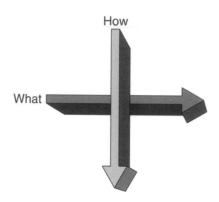

FIGURE 5.6
The goal translation process

At the top of an organization in the chemical process industries, five measurable goals have been identified. These are listed under the heading 'What' in Figure 5.7. The top team listens to the 'voice of the customer' and tries to understand *how* these business goals will be achieved. They realize that product consistency, on-time delivery and speed or quality of response are the keys. These CSFs are placed along the first row of the matrix and the relationships between the *what* and the *how* are estimated as strong, medium or weak. A measurement target for the *hows* is then specified.

The *how* becomes the *what* for the next layer of management. The top team shares their goals with their immediate reports and asks them to determine their *hows*, indicate the relationship and set measurement targets. This continues down the organization through a 'catch-ball' process until the senior management goals have been translated through the *what/how; what/how; what/how* matrices to the individual tasks within the organization. This provides a good discipline to support the breakdown and understanding of the business process mapping described in Chapter 10.

A successful approach to policy/goal deployment and strategic planning in an organization with several business units or division is that mission, CSFs with KPIs and targets, and core processes are determined at the corporate level, typically by the board and the senior executive. Whilst there needs to be some flexibility about exactly how this is translated into the business units, typically it would be expected that the process is repeated with the senior team in each business unit or division. Each business unit head should be part of the top team that did the work at the corporate level, and each of them would develop a version of the same process with which they feel comfortable.

Each business unit would then follow a similar series of steps to develop their own mission (perhaps) and certainly their own CSFs and KPIs with targets. A matrix for each business unit showing the impact of achieving the business unit CSFs on the corporate CSFs would be developed. In other words, the first deployment of the corporate 'whats' CSFs is into the 'hows' – the business unit CSFs (Figure 5.2).

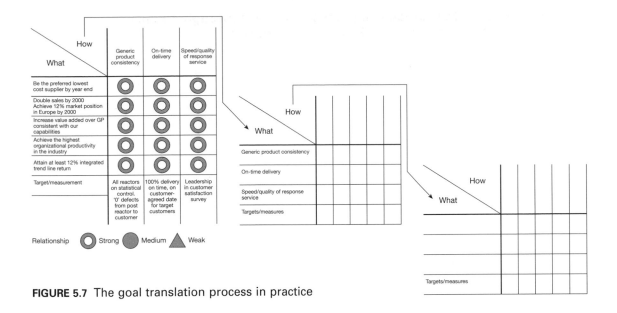

FIGURE 5.7 The goal translation process in practice

If each business unit follows the same pattern, then each unit team will identify CSFs, KPIs with targets and core processes which are interlinked with the ones at corporate level. Indeed the core processes at corporate and business unit level may be the same, with any specific additional processes identified at business unit level to catch the flavour and business needs of the unit. It cannot be over-emphasized how much ownership there needs to be at the business unit management level for this to work properly.

The point of weakness in the organizational structure of construction companies is that major projects are regarded as semiautonomous business units. With the exception of safety practices, project leadership is allowed unqualified freedom to run its business unit as it sees fit in order to achieve the cost and schedule goals of the project. These are seen as the primary risk factors in any construction business, and head office does not want to interfere in any way that might allow excuses for failure in these critical areas of performance.

This means that critical, broader performance improvement goals, such as productivity improvement of subcontractors through better project level planning, innovation driven by closer collaboration among key supply chain partners, are difficult to realize. Such relationships are left to the whim and skill of each project director, and corporate policy is often left languishing for lack of effective support at the project level of the organization.

With regard to core processes, each business unit or function will begin to map these at the top level. This will lead to an understanding of the purpose, scope, inputs, control and resources for each process and provide an understanding of how the sub-processes are linked together. Flowcharting showing connections with procedures will then allow specific areas for improvement to be identified so that the continuous improvement, 'bottom-up' activities can be deployed, and benefit derived from the process improvement training to be provided (Figure 5.8).

In construction organizations, it is critical that key, long-term subcontract service suppliers are involved, providing suggestions into both the top-down and bottom-up processes of idea generation. Their collaboration and support are essential to getting the best out of this typically fragmented supply chain.

It is important to get clarity at the corporate and business unit management levels about the 'whats/hows' relationships; but the ethos of the whole process is one of

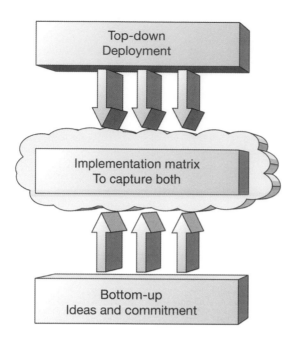

FIGURE 5.8
Implementation top-down
and bottom-up

involvement and participation in goal/target setting, based on good understanding of processes, so that what can be achieved and what needs measuring and targeting at the business unit level is known and agreed to.

Senior management may find it useful to monitor performance against the CSFs, KPIs and targets, and to keep track of process using a reporting matrix, perhaps at their monthly meetings. A simplified version of this developed for use in a small company is shown in Figure 5.9. The frequency of reporting for each CSF, KPI and process can be determined in a business planning calendar.

As previously described, in a larger organization, this approach may be used to deploy the goals from the corporate level through divisions to site/departmental level (see Figure 5.10). This form of implementation should ensure the *top-down and bottom-up* approach to the deployment of policies and goals.

Deliverables

The deliverables after one planning cycle of this process in a business will be:

1 an agreed framework for policy/goal deployment through the business;
2 agreed mission statement for the business and, if required, for the business units/ divisions;
3 agreed critical success factors (CSFs) with ownership at top team level for the business and business units/divisions;
4 agreed key performance indicators (KPIs) with targets throughout the business;
5 agreed core business processes, with sponsorship at top team level;
6 a corporate CSF/business unit CSF matrix showing the impacts and the first whats/ hows deployment;
7 a what/how (CSF/process) matrix approach for deploying the goals into the organization through process definition, understanding and measured improvement at the business unit level;

Conduct research	Manage int. systems	Manage financials	Manage our accounts	Develop new business	Develop products	Manage people	Core processes			
							CSFs: We must have	**Measures**	**Year targets**	**Target CSF owner**
	×	×	×	×		×	Satisfactory financial and non-financial performance	Sales volume. Profit. Costs vs plan. Shareholder return Associate/employee utilization figures	Turnover £2m. Profit £200k. Return for shareholders. Days/month per person	
×	×		×	×	×	×	A growing base of satisfied customers	Sales/customer Complaints/recommendations Customer satisfaction	>£200k = 1 client. £100k–£200k = 5 clients. £50k–£100k = 6 clients <£50k = 12 clients	
	×	×				×	A sufficient number of committed and competent people	No. of employed staff/associates Gaps in competency matrix Appraisal results Perceptions of associates and staff	15 employed staff 10 associates including 6 new by end of year	
×			×			×	Research projects properly completed and published	Proportion completed on time, in budget with customers satisfied. Number of publications per project	3 completed on time, in budget with satisfied customers	
			* *	* *		* *	** = Priority for improvement			
							Process owner			
							Process performance			
							Measures and targets			

FIGURE 5.9 CSF/core process matrix

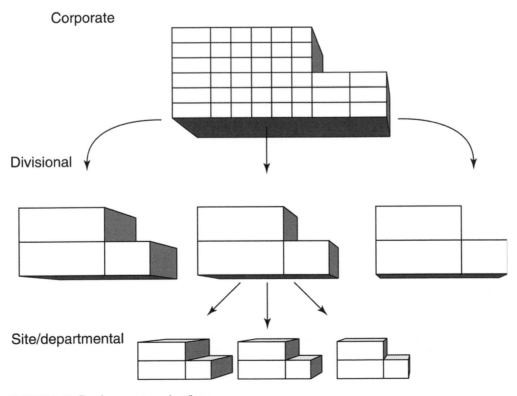

Corporate

Divisional

Site/departmental

FIGURE 5.10 Deployment – what/how

8   focused business improvement at the project level, linked back to the CSFs, with prioritized action plans and involvement of employees and key subcontract service providers.

Strategic and operational planning

Changing the culture of an organization to incorporate a sustainable ethos of continuous improvement and responsive business planning will only come about as the result of a carefully planned and managed process. In any organization, this must involve key long-term supply chain partners. In construction, this is particularly critical as on many building construction projects almost all the workers, working side by side, work for different companies. Clearly many factors are involved, including:

- identifying strategic issues to be considered by the senior management team;
- balancing the present needs of the business against the vital needs of the future;
- concentrating finite resources on important things;
- providing awareness of impending changes in the business environment in order to adapt more rapidly, and more appropriately.

Strategic planning is the continuous process by which any organization will describe its destination, assess barriers standing in the way of reaching that destination and select approaches for dealing with those barriers and moving forward. Of course, the real contributors to a successful strategic plan are the participants.

The strategic and operational planning process described in this chapter will:

- provide the senior management team with the means to manage the organization and take account of its strengths and weaknesses through the change process;
- allow the senior management team members to have a clear understanding of and achieve agreement on the strategic direction, including vision and mission;
- identify and document those factors critical to success (CSFs) in achieving the strategic direction and the means by which success will be measured (KPIs) and targeted;
- identify, document and encourage ownership of the core processes that drive the business;
- reach agreement on the priority processes for action by the process improvement teams through incorporating current initiatives into an overall, cohesive framework;
- provide a framework for successfully deploying all goals and objectives at all organizational levels through a two-way 'catch-ball' process;
- provide a mechanism by which goals and objectives are monitored, reviewed and appropriate actions taken at required frequencies throughout the operational year;
- transfer the skills and knowledge necessary to sustain the process.

The components outlined above will provide a means of effectively deploying a common vision and strategy throughout the organization. They will also allow for the incorporation of all change projects as well as 'business as usual' activities into a common framework which will form the basis of detailed operating plans.

THE DEVELOPMENT OF POLICIES AND STRATEGIES

Let us assume that a management team is to develop the policies and strategies based on stakeholder needs and the organization's capabilities, and that it wants to ensure these are communicated, implemented, reviewed and updated. Clearly a detailed review is required of the major stakeholders' needs, the performance of competitors, customer satisfaction, the state of the market and industry/sector conditions. This can then form the basis of top-level goals, planning activities and setting of objectives and targets.

How individual organizations do this varies greatly, of course, and some of this variation can be seen in the case studies at the end of the book. However, some common themes emerge under six headings.

Customer/market

- Data collected, analysed and understood in terms of where the organization will operate.
- Customers' needs and expectations understood, now and in the future.
- Developments anticipated and understood, including those of competitors and their performance.
- The organization's performance in the marketplace known.
- Benchmarking against best in class organizations.

Shareholders/major stakeholders

- Shareholders/major stakeholders needs and ideas understood.
- Appropriate economic trends/indicators and their impact analysed and understood.
- Policies and strategies appropriate to shareholder/stakeholder needs and expectations developed.
- Needs and expectations balanced.
- Various scenarios and plans to manage risks developed.

People

- The needs and expectations of the employees understood.
- The needs and expectations of the subcontractors' employees understood.
- Data collected, analysed and understood in terms of the internal performance of the organization.
- Output from learning activities understood.
- Everyone appropriately informed about the policies and strategies.

Processes

- A key process framework to deliver the policies and strategies designed, understood and implemented.
- Key process owners identified.
- Each key process and its major stakeholders defined.
- Key process framework, including criteria and metrics, reviewed periodically in terms of its suitability to deliver to the organization's requirements.

Partners/resources

- Appropriate technology understood.
- Impact of new technologies analysed.
- Needs and expectations of supply chain and business partners understood.
- Policies and strategies aligned with those of partners.
- Financial strategies developed.
- Appropriate buildings, equipment and materials identified/sourced.

Society

- Social, legal and environmental issues understood.
- Environment and corporate responsibility policies developed.

The whole field of business policy, strategy development and planning is huge and there are many excellent texts on the subject. It is outside the scope of this book to cover this area in detail, of course, but one of the most widely used and comprehensive texts is *Exploring Corporate Strategy – Text and Cases* (Johnson *et al.*, 2014). This covers strategic positioning and choices, and strategy implementation at all levels.

BIBLIOGRAPHY

Collins, J.C. *Good to Great*, Random House, New York, 2001.

Collins, J.C. and Hansen, M.T. *Great by Choice*, Random House, New York, 2011.

Collins, J.C. and Porras, J.I. *Built to Last: Successful Habits of Visionary Companies*, Random House, New York, 2005.

Hardaker, M. and Ward, B.K. 'Getting things done – how to make a team work', *Harvard Business Review:* 112–119, 1987.

Hutchins, D. *Hoshin Kanri – The Strategic Approach to Continuous Improvement*, Gower Publishing, Aldershot, 2008.

Johnson, G., Scholes, K. and Whittington, R. *Exploring Corporate Strategy, Text & Cases* (10th edn), Prentice Hall, London, 2014.

CHAPTER HIGHLIGHTS

Integrating lean quality into policy and strategy

- Policy and strategy is concerned with how the organization implements its mission and vision in a clear stakeholder-focused strategy supported by relevant policies, plans, objectives, targets and processes.

Aligning strategy and action

- Senior management may begin the task of alignment through six steps:
  1. develop a shared vision and mission;
  2. develop the critical success actors;
  3. define the key performance indicators (balanced scorecard);
  4. understanding the core process and gain ownership;
  5. break down the core processes into sub-processes, activities and tasks;
  6. ensure process and people alignment through a policy deployment or goal translation process.
- The deliverables after one planning cycle will include: an agreed policy/goal deployment framework; agreed mission statements; agreed CSFs and owners; agreed KPIs and targets; agreed core processes and sponsors; whats/hows deployment matrices; and focused business improvement plans.

The development of policies and strategies

- The development of policies and a detailed review of the major stakeholders' needs, the performance of competitors, the market/industry/sector conditions to form the basis of top-level goals, planning activities and setting of objectives and targets.
- The common themes for planning strategies may be considered under the headings of customers/market, shareholders/major stakeholders, people, processes, partners, resources and society.
- The field of policy and strategy development is huge and the text by Johnson *et al.* (2014) is recommended reading.

Partnerships and resources

6

PARTNERING AND COLLABORATION

Business, technologies and economies have developed in such a way that organizations recognize the increasing needs to establish mutually beneficial relationships with other organizations, often called 'partners'. The philosophies behind the various TQM, lean quality and Excellence models support the establishment of partnerships and lay down principles and guidelines for them. This is particularly important in construction where because of the cost-focused mindset, organizations within the supply chain are often out of alignment. This creates situations in which parties that rely on each other for services have inconsistent objectives.

Because in the late 1980s and early 1990s, *partnering* was promoted and adopted as a formal procurement process in construction, many people in the construction sector use the word *partnering* to describe a form of contracting. We, however, are *not* using the word in this sense. In this book when we use the term *partnering* we simply mean an effective and close collaboration between two or more parties working together towards shared objectives.

How companies in the private sector plan and manage their partnerships can mean the difference between success and failure; for in today's economic climate, it is extremely rare to find companies which can sustain a credible business operation without a network of co-operation between individuals and organizations. This extends the internal customer–supplier relationship ideas into the supply chain of a company, making sure that all the necessary materials, services, equipment, information, skills and experience are available in totality to deliver the right products or services to the end customer. Gone are the days, hopefully, of conflict and dispute between customers and their suppliers. An efficient supply chain process built on strong confident partnerships will create high levels of people satisfaction, customer satisfaction and support and, in turn, good business results.

This is especially true on construction projects where much of the work is outsourced and more than 85 per cent of the workforce is in the employ of subcontractors. These subcontractors design, fabricate, supply and fix more basic materials or design, supply and fix complex subsystems. Construction contractors are increasingly finding that their risks, whether they are safety, quality, production or environment related, lie in the hands of their subcontractors' designers, employees and supervisors. They are the people at the workface who must get it *right-first-time* if the process is to be truly efficient. It is a critical challenge for general contractors to work out how they can build long-term relationships with their subcontractors, even when their joint work is discontinuous and project-based. They need to work out how to create viable and stable relationships which will create advantages they can leverage to their mutual benefit when working together.

Many general contractors who practise lean quality have further narrowed their supply chains and deepened their relationships with the key subcontract suppliers. Leading lean clients have gone further by creating integrated teams on their projects working under relational contracts with painshare/gainshare commercial terms and project processes in which the entire team, including the owner, work together to manage risks and identify opportunities for innovation and improvement (refer to the case studies about the Sutter Van Ness and Geary Campus (CS2) project and of the Multinational High-Tech Manufacturer lean journey (CS4)).

How an organization plans and manages the external partnerships must be in line with its overall policies and strategies, designed and developed to support the effective operation of its processes. A key part of this, of course, is identifying with whom those key strategic partnerships will be formed. Whether it is working with key suppliers to deliver materials or components to the required quality, plan (lead-times) and costs or the supply of information technology, transport, broadcasting or design consultancy services. The quality of these partnerships has been recognized throughout the world as a key success criterion.

Understanding and applying a sound approach to improving strategic partnerships and supplier relationships will have a significant return on investment by:

- significantly reducing the management overhead through increasing the level of collaboration on a number of key dimensions that research shows to be critical in effective partnerships;
- establishing the right partnerships from the outset through understanding what makes partnerships work and building that into the selection, review, and 're-contracting' process.

Once the goals for the partnership are established, partners may carry out a 'health check' on the current operations to assess strengths and areas for improvement.

The approach recommended looks at five key dimensions of good collaboration (see Figure 6.1):

- *Strategic alignment*: how well aligned are the partners and how do they achieve this alignment at all key levels of management?
- *Customer focus*: to what extent do both parties develop and deliver the desired standards of service and experience across the whole service chain?
- *Decision-making and governance*: how is the partnership managed to achieve best effect and efficiency?
- *Communications and transparency*: how well is data and knowledge captured, shared and disseminated in a way that builds value and not cost?
- *Investment and improvement*: how well does the partnership jointly invest in and improve the partnership operations and outcome measures?

There are various ways of ensuring the partnership processes work well for an organization. These range from the use of quality management system audits and reviews through certificates of competence to performance reviews and joint action plans. A key aspect of successful partnerships is good communications and exchange of information which supports learning between two organizations and often leads to innovative solutions to problems that have remained unsolved in the separate organizations, prior to their close collaboration.

Although in construction the focus for some years has been on fixed price contracts, some examples of genuine partnerships do exist. For example, in recent years, long-term, stable and beneficial partnering arrangements have been created in the maintenance of

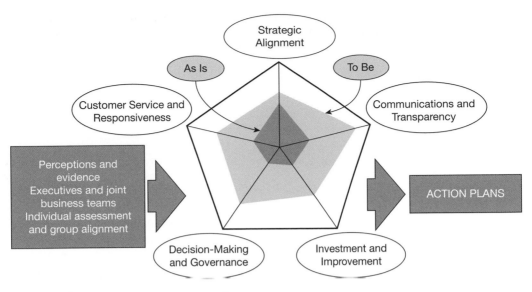

FIGURE 6.1 Partnering collaboration radar

buildings, roads and rail infrastructure. Many of these have yielded very significant benefits to the service providers and clients alike.

In the construction process, examples of such partnerships are relatively rare. An interesting example of a partnership which paid good dividends was on a contract to widen a section of the M25 around London. The road easement was tight and to accommodate the widened road it was necessary to drive piles in both cut and filled embankments to stabilize them. Information about the need for piling was scant hence all the tenderers had to make similar conservative allowances regarding the number of piles required. The company that won the work under a fixed price contract went to its piling subcontractor and agreed that, once they were on site, they would jointly invest in detailed investigation and redesign the piling based on real data from the field. They also agreed on a split of the savings.

At the end of the contract they found that they had been able to save a very significant 40 per cent of the originally allowed piling. It is interesting to speculate that some of this saving could have accrued to the client, the Highways Agency, had the commercial arrangements been set up in such a manner as to encourage the further investigation and a three way split of the savings. As it was, the client traded off its potential saving for cost certainty at contract award.

When establishing partnerships, attention should be given to:

- maximizing the understanding of what is to be delivered by the partnership – the needs of the customer and the capability of the supplier must match perfectly if satisfaction and loyalty are to be the result;
- understanding what represents value for money – getting the commercial relationship right, includes the clear understanding of scope, quality expectations, and challenges;
- understanding the respective roles and ensuring an appropriate allocation of responsibilities to the party best able to manage them;
- working in a supportive, constructive and a team-based relationship;
- having solid programme of work, comprising agreed plans, timetables, targets, key milestones and decision points;

- identifying areas where the achievement of agreed goals requires training – and implementing joint training programmes between the partners;
- structuring the resolution of complaints, concerns or disputes rapidly, and at the lowest practical level;
- enabling the incorporation of knowledge transfer – and making sure it adds value;
- developing very close working relationships and gearing the supply chain to deliver better and better value products and/or services to the end customer, based on continuous improvement principles.

GLOBAL OUTSOURCING

Ongoing unpublished research by The Oakland Institute shows the major reasons for selection of 'global suppliers' are the pursuit of new markets and reducing costs. Product/ service quality and customer service improvements did feature, but they were down the list of priorities. Yet, a mountain of evidence now points to the fact that poor quality can quickly, quietly and devastatingly demolish the benefits of any new market gains or cost reductions. Moreover, the difficulty of managing quality increases in direct relation to the distance between an organization and its partners, hence, the need for a good 'assurance' model.

Managing performance and exposure to risk

As with other industries, the last decade has seen a period of rapid expansion of global sourcing by the construction sector: materials, products and equipment including windows and doors, building facades, fabricated steelwork, elevator systems and mechanical equipment to name but a few. These products fall into four groups:

- purchase of simple stock materials such as tiles, cavity wall ties and sheet cladding;
- purchase of more sophisticated standard products such as AC units, elevator systems, pumps, etc.;
- design, fabricate and install relatively simple bespoke products such as joinery for shop fittings, made to measure windows and reinforcement fabricated to size; and
- sophisticated equipment and building systems such as entire glazed facade systems, equipment for gas trains for natural gas treatment, construction of offshore oil platforms and prefabricated structural steel structures.

We know that quality cannot be inspected into a product or service at the end; it must be built in. Therefore, it is surprising that many companies in developed economies that rule out inspection as a quality control methodology on the grounds of prohibitive cost and ineffectiveness decide that it is appropriate when importing from developing economies. Inspection and trying to control at the end do not prevent poor quality happening, nor are they effective in ensuring that customers get good quality.

The real problem, all too often, is that there is no clear view of risk across a global manufacturing or operations footprint. There may be a lot of data and 'noise', but little in the way of real knowledge in a form that executives can digest and act upon – we can only manage what we can measure.

Getting organized for success

Building a global quality management capability needs leadership and a clear quality management policy which is implemented on a global scale. The approach must integrate quality assurance, quality control and quality improvement techniques such as Lean Six-

Sigma. Simple reliance on inspection and quality control will not eliminate the risk of delivering poor quality products or services.

Without the right organization in place, Chief Operating Officers of global operations can find themselves exposed to a cost and quality 'killer' – VARIATION. Research indicates that variation can add up to 30 per cent to manufacturing and support costs. The challenge is to move from inspection to 'assurance'. Relatively small improvements here can multiply into big benefits.

The authors are aware of numerous problems that have been experienced by the construction sector worldwide with global sourcing in recent years; these include issues such as:

- poor dimensional control on products such as imported tiles and windows;
- quality levels that are widely different for seemingly identical products;
- protective finishes such as galvanizing thinner than standards require;
- cladding materials that are combustible;
- products that include asbestos fibres; and
- delivery delays due either to the provider or the logistics.

A recently built gas project sourced its major equipment from an overseas fabricator; when the plant arrived on shore, the remedial cost of getting the plant to a satisfactory condition was as much as the original purchase price.

Each of these problems ends up costing corporations both financially and in terms of reputation – a high price to pay for ineffective quality systems.

Delivering better value from the global supply chain

The ongoing management of quality on a global scale requires an appropriate balance of stewarding global quality performance, facilitating global improvement and managing risk. It is also vital to ensure that quality is not solely seen as the responsibility of the lean quality department. As lean quality is everyone's responsibility, it must rank with equal importance with other critical areas. Part of the solution is to create a central team that will genuinely be regarded as a value-centre rather than a cost-centre. It is all a matter of getting value, not just cutting costs.

SUPPLY CHAIN EFFECTIVENESS IN THE GLOBAL ECONOMY

The urgent pursuit of greater security of supply or improved efficiencies can necessitate the transfer of manufacturing or re-sourcing to alternative supply routes. These strategies represent great opportunities but also carry big risks. In turn, supply chains can represent the source of real savings if better managed.

Reliance upon a single source of supply can be risky at any time, but in unpredictable economic conditions it can be downright dangerous. An organization must establish whether it is at risk. Below is a brief, high-level overview of a process to strengthen sourcing security in terms of quality, timeliness and cost, by sourcing a second supplier.

1. Decide the selection criteria that are right for the business

Consider sourcing security needs by reference to a set of sourcing criteria. The list will be unique to a business or organization and its needs, but an example to show the idea is given in Figure 6.2.

The list might include sourcing and supply chain capability, technical capability, product cost profile and so on. The important thing is to create a tailored list – time spent on this exercise will repay itself many times.

2. Make an initial selection

Using the selection criteria make an initial selection of candidates.

3. Toughen up the selection criteria and make a shortlist

Using a more rigorous set of selection criteria reduce the list to, say, five candidates.

4. Conduct site visits

At this stage, really put the candidates under the microscope by scrupulously examining all the factors around the potential deal, including:

1 approach to managing a large production and assembly account;
2 level of interest in the commercial opportunity; and
3 ability to quickly accommodate site visits and assessments.

Factor	Spectrum		
Business size and scale	Scale		Upper tier
Geography	Single plant locations		Global footprint
Business relationship	Simple contract		Business partnership
Business maturity and capability	Capability		Capability + maturity

FIGURE 6.2 Simple set of sourcing criteria

5. Make a final evaluation . . . and choice

When the site visits are complete, re-evaluate the key business criteria scores, by moving them up and down as appropriate. Also, determine the candidates' key strengths and weaknesses together with associated risks, and draw conclusions.

In one company, a potential list of 14 new suppliers was reduced to a shortlist of five, with one finally being selected as an additional supplier to the original single source supplier. The benefits included manufacturing costs being halved compared to the existing source, and expectations built of a step change in quality with indications that the defect rate would drop over the first twelve months from the initial level of 2 per cent to less than 0.5 per cent. This is a typical return from a more secure supply chain.

THE ROLE OF PROCUREMENT/PURCHASING IN PARTNERSHIPS

As we have seen, very few organizations are self-contained to the extent that their products and services are all generated at one location, from basic materials. Some materials or services are usually purchased from outside organizations, and the primary objective of a 'purchasing' or 'procurement' function is to obtain the correct equipment, materials and services in the right quantity, of the right quality, from the right origin, at the right time and cost. Procurement or purchasing can also play a vital role as the organization's 'window-on-the-world', providing information on any new products, processes, materials and services that become available. It can also advise on probable prices, deliveries and performance of products under consideration by the research, design and development functions. In other words, it should support any partnership in the supply chain.

The purchasing or procurement system should be documented and include:

- assigning responsibilities for and within the purchasing/procurement function;
- defining the manner in which suppliers are selected – to ensure that they are continually capable of supplying the requirements; and
- specifying the purchasing documentation – written orders, specifications, etc. required in any modern procurement activity.

Principle 2.1.2 in Chapter 3 states: *use only reliable technology to reduce the risk of failure.* This maxim pays for itself very quickly once the cost of rework is considered.

Most leading organizations now employ supply chain or procurement specialists. Their role is to work across the business functions exploring ways to optimize the supply process through strategies such as outsourcing, early supplier involvement and offshoring. In large organizations, the purchase of goods and services tends to be negotiated centrally with the detailed management of their requirements organized locally. Purchasing/procurement should bring value to organizations by improving contract management and fostering supplier compliance across the entire lifecycle of contracts, resulting in continued overall cost reduction. Nevertheless, according to the UK Chartered Institute of Purchasing and Supply (CIPS), much work remains to be done in understanding how to capture the breadth and depth of the role of supplier management in achieving cost reductions and other benefits – the traditional measures of cost saving are no longer appropriate. Clearly, the link between procurement strategy and business performance and the resulting shareholder value need to be considered.

Procurement needs to expand its remit into managing risk and vulnerability within the supply chain, particularly in the context of geographically dispersed and distant suppliers. Additional complexity also arises as a direct consequence of the volatility of commodities, currencies and interest rates. Senior management should demand that procurement avoids or at least reduces the vulnerability of supply chains to disruptions, thereby increasing their resilience should problems occur. Risk-related issues include managing exposure to commercial risks and loss of reputation as well as providing improved supply market intelligence, such as forecasting future shortages. Collaborative relationships are a prerequisite for tapping into innovations available from suppliers globally. Suppliers are becoming increasingly integrated into new product development efforts. In some organizations, this requires the development of new remuneration and risk sharing models to share the benefits of technological development with suppliers. Most large organizations and governments worldwide are increasingly focusing on the issue of sustainability, both from the societal and the environmental perspectives. This has the potential to be the single most significant influence on organizational strategy, with procurement playing an important role in this effort.

On construction projects it is not uncommon for the project director and their team to award one or two subcontracts on price alone in the belief that their team will be able to manage the shortcomings of the low bid subcontractor. In a recent example, one of the authors observed that a low-bid HVAC subcontractor not only fell behind the rest of the construction team in delivering and installing fabricated duct to the site, but when the duct did arrive, it was often delivered to the wrong floor. The result was an extremely chaotic construction process; workflow went out the window and all the services subcontractors worked on top of each other on every floor in the building. What general contractors do not realize is that the short-sighted benefits they gain in such an instance, while they pocket a cash saving, is far outweighed by the cost imposed on all the other subcontractors on the site. Everyone is penalized by the chaos; no one can work efficiently. Yet the costs and benefits of this kind of thinking are never measured.

Another example relates to workmanship in the waterproofing of wet areas in high-rise residential buildings in the Sydney market. In recent years, it has become all too common that either the waterproofing subcontractor or the tiler leaves poorly finished or damaged waterproofing membrane in shower recesses. The result is that relatively shortly after the occupants move in and start using the shower, water damage appears in the adjoining room. The cost of rectifying this problem is in excess of $5,000 for every occurrence, while the cost of avoiding it in the first instance would have been between $20 and $100. This situation is the result of using cheap, poorly trained and unmotivated workmen to save money; however, any surpluses on the project quickly evaporate in rectification once occupants have moved in. Worse still, the company can never recover the lost goodwill caused as a result of the nuisance of both the initial leak and the subsequent rectification.

In construction, there is a history of dividing projects into their subsystems and writing contracts for each of the separate parts as if they were to be built in isolation. Similarly, risk is generally parcelled up and passed on to the subcontractors who supply and install materials and equipment. To top it off, many contractors then 'buy' some of the packages on price to the detriment of one or more of the other critical performance areas. Principle 1.3 in Chapter 3 states: *utilize the collective resources of the project team in the most efficient manner.* To achieve this objective the commercial terms and the relationships should all be aligned to maximize collaboration; yet, the above procurement strategies often get in the way of genuine collaboration.

A final critical area is timely delivery. In the construction sector this can be problematic both for the supply of products and services. The authors are aware of numerous instances where suppliers have not met their commitments with severe consequences for the process on site.

Historically, many organizations in the construction sector have operated an inspection-oriented quality system for bought-in parts, materials and on-site workmanship. Such an approach has many disadvantages. It is expensive, imprecise and impossible to apply evenly across all materials and parts, all of which lead to variability in the extent and reliability of appraisal. Many organizations have found that survival and future growth in both volume and variety demand that changes be made to this approach.

A reliance on inspection (detection) to achieve quality is unproductive in two ways: it reduces any sense of the responsibility at the workface (inspectors become responsible) and adds wasteful cost. The inspector cannot be ever-present; therefore, a responsible, careful, skilful and innovative workforce is essential for successful operations.

The limited space for materials on most construction sites has meant that holding large stocks of components and raw materials is impractical; this has led to the creation of the 'just-in-time' (JIT) concept. This requires that suppliers make frequent, on-time deliveries of smaller quantity of materials, parts, components, etc. Often deliveries are made straight to the point of use so that materials on site can be kept to a minimum. This approach, however, requires an effective supplier network: one that produces goods and services that can be trusted to conform to the real requirements with a high degree of confidence.

Commitment and involvement

The process of improving supplier performance is complex and clearly relies very heavily on securing real commitment to a partnership from the senior management of both organizations. Activities such as joint training of head contractor and subcontractor managers, meetings with the directors of supply chain partners to review performance and to develop improvement strategies can only help to improve the capacity of supply chains. If this can be achieved within the constraints of business and technical confidentiality, it is always a better approach than the arms-length approach to purchasing still used by many companies (refer to the VNGC case study (CS2) to see close collaboration).

The authors recall the benefits that accrued from a partnership between a structural engineer and his client, a builder in the Singapore high-rise residential sector. The builder approached the structural engineer, who had worked for him for some years, with a view to revising their agreement for work. At first the engineer was fearful that the builder was looking for an excuse to cut his fees – a request that many readers would not find unusual. In fact, the builder's approach was along quite different lines: he had been monitoring the reinforcing steel and concrete quantities in his buildings. His offer was to pay a bonus, a share of the savings in materials in any new building. The structural engineer quickly found that his bonuses exceeded his original fees. It is interesting to speculate on what had changed. The structural engineer can work to three different value propositions:

- make sure that the building stands up;
- minimize the materials in the building; and
- minimize the overall construction cost of the building.

In fact, when fees are being squeezed and scope is not clearly defined, any organization will look for ways to minimize costs to stay profitable. In this case, the client managed to align his interests with the second proposition. The third is the highest value proposition; however, it is harder to judge and perhaps harder to manage.

Policy

One of the first things to communicate to any external supplier is the purchasing organization's policy on quality of incoming goods and services. This can include such statements as:

- It is the policy of this company to ensure that the quality of all purchased materials and services meets its requirements.
- Suppliers who incorporate a quality management system into their operations will be selected. This system should be designed, implemented and operated accordingly to the International Standards Organization (ISO) 9000 series (see Chapter 14).
- Suppliers who incorporate statistical process control (SPC) and continuous improvement methods into their operations (see Chapter 15) will be preferred.
- Suppliers willing and able to collaborate with all relevant supply chain partners to optimize project processes and outcomes will be preferred.
- Routine inspection, checking, measurement and testing of incoming goods and services will not be carried out by this company on receipt.

- Suppliers will be audited and their operating procedures, systems and SPC methods will be reviewed periodically to ensure a never-ending improvement approach.
- It is the policy of this company to pursue uniformity of supply and to encourage suppliers to strive for continual reduction in variability. (This may well lead to the narrowing of specification ranges.)
- It is the policy of this company to work with suppliers who are committed to safe work practices and to achieving right-first-time outcomes.

Lean quality management system assessment certification

Many customers examine their suppliers' lean quality management systems themselves, operating a second-party assessment scheme (see Chapters 10 and 14). Inevitably this leads to high costs and duplication of activity, for both the customer and supplier. If a qualified, independent third party is used instead to carry out the assessment, attention may be focused by the customer on any special needs and in developing closer partnerships with suppliers. Visits and dialogue across the customer/supplier interface are a necessity for the true requirements to be met, and for future growth of the whole business chain. Visits should be concentrated, however, on improving understanding and capability, rather than on close scrutiny of operating procedures, which is best left to experts, including those within the supplier organizations charged with carrying out internal system audits and reviews. Highways England (HE) has developed a lean maturity model for its suppliers, with associated training, incentivization and support as a strategy for improving the outcomes of the work of its supply chains (refer to the lean quality maturity model in Chapter 10).

LEAN QUALITY IN PARTNERSHIPS AND THE SUPPLY CHAIN

The development of long-term partnerships with a few suppliers rather than short-term ones with many leads to the concept of *co-producers* in networks of trust, providing dependable quality and delivery of goods and services. Each organization in the chain of supply is often encouraged to extend lean quality methods to its suppliers. The requirements of lean quality mean that suppliers are usually local to the project and deliver smaller quantities regularly to match the usage rate. Administration is kept to a minimum and standard quantities in standard containers are usual. The requirement for suppliers to be located near the project (which places those at some distance at a competitive disadvantage) makes lead times shorter and deliveries to be more reliable.

Marton Marosszeky witnessed the structures subcontractor on a major high-rise site in a city purchase reinforcement for a large part of the building at a discounted price in one job lot. A few days later several semitrailers of reinforcing steel arrived on site at the same time. You can imagine the shock and dismay of the project manager. The commercial agreement between the parties gave the subcontractor the freedom to purchase and bring materials to site in any way that best suited the subcontractor. The 30 tonnes of reinforcement was moved around the site for some weeks before it was returned to storage. This is the antithesis of lean quality; the commercial terms allowed the parties to act in their own interests at the expense of the project.

Some argue that lean quality in purchasing and delivery are suitable mainly for assembly line operations, and less so for certain process and service industries. However, the reduction in the inventory, the smooth flow of materials required and the associated reduction in transport costs should encourage innovations that will lead to the widespread adoption of lean quality. The development of closer relationships and a more productive dialogue – the sharing of information and problems – will lead to the delivery of the right product quality, in the right quantity, at the right time.

The Rosendin case study (CS11) describes how the company won an award for the development of a mobile app that works with the Auodesk®BIM360 Field Platform to streamline material and tool management for field personnel.

RESOURCES

All organizations assemble resources (financial resources, buildings, equipment, materials, technology, information and knowledge), other than human, to support the effective operation of the processes that hopefully will deliver their strategy. How these resources are managed will have a serious effect on the effectiveness and efficiency of any establishment, whether it be in manufacturing, service provision or the public sector.

Financial resources

The ability to attract investment often determines the strategic direction of commercial enterprises. The development and implementation of appropriate financial strategies and processes will therefore be driven by the financial goals and performance of the business. For example, in a private company, focus on improving earnings before interest and tax (EBIT) – the measure of profitability – and economic value add (EVA) – the measure of the degree to which the returns generated exceed the costs of financing the assets used – can be the drivers for linking the strategy to action. Once policies and strategy are set, and plans for action agreed in principle, financial resources must be allocated to ensure that significant appropriate activities can be carried out to deliver the strategy. The corollary of this is that process improvement activities will fail if sufficient resources are not allocated. Unfortunately, in construction organizations almost all investment in improvement is expected to be funded at the project level and at the final discretion of the project manager; this often leads to a failure to follow through on intent. This is a failure of leadership.

Consolidation of these plans, coupled with an iterative review and approval process, provides a mechanism for the best possible chance of success. Use of a 'balanced score-card' approach (see Chapter 9) can help to ensure that the long-term impact of financial decisions on processes, innovation and customer satisfaction are understood and taken into account. The extent to which financial resources are being used to support strategy needs to be subject to continuous appraisal – this will include evaluating investment in the tangible and non-tangible assets such as knowledge.

Compared to other business sectors, in construction it is more difficult to assess the benefits of investing in change and improvement strategies. This is primarily due to the project basis of production: each project being different, with a different production team, makes it difficult to compare one to the other. Hence, investment in change strategies is often resisted. One way of overcoming this resistance is to assess the cost of current inefficiencies. This may produce stronger arguments for investments in strategies and technologies that will most likely result in improvement.

In small and medium-sized enterprises it is even more important that the financial strategy forms a key part of the strategic planning system and that key financial goals are identified, deployed and regularly scrutinized.

Other resources

Many different types of resources are deployed by organizations: most have built assets, use equipment and consume materials; therefore, directors and managers must pay attention to:

- utilization of these resources;
- security of the assets;
- maintenance of building and equipment;
- managing material inventories and consumption;
- reduction of waste and recycling; and
- environmental impact – conservation of non-renewable resources and adverse impact of products and processes.

Principle 1.3: *Utilize the collective resources of the project team in the most efficient manner possible* refers to the benefits that can be achieved at the project level if resources are shared between the supply chain partners. This principle is very well illustrated by a consortium in Florida that operates as an integrated supply chain. On site, the parties utilize equipment in the most efficient way possible. If the plumber needs a core-hole drilled and the electrical contractor has a core-drill on the job, they collaborate to get the hole drilled as quickly as possible and work out the allocation of costs afterwards. Refer to the VNGC case study (CS2) where under an IFOA contract, the entire supply chain works in an open collaborative environment sharing resources on a best for project basis, both in design and construction.

Technology has become a vital resource in modern business. Existing alternative and emerging technologies need to be identified, evaluated and appropriately deployed in the drive towards achieving organizational goals. This will include managing the replacement of 'old technologies' and using innovations that will lead to the adoption of new ones. There are clear links here, of course, with process redesign and re-engineering (see Chapter 13 and in particular refer to the DPR case study (CS9)). It is not yet possible to create the 'paper-less' construction site, though lead practice is certainly approaching this. A traditional major hospital project, for example, generates well over a million pieces of paper, and these are shared by the many members of the delivery supply chain. To replace this with computer systems and back-up files requires more than just a flick of a switch. The whole end-to-end process of design and construction is undergoing a revolution and soon documentation will be near paperless. Over the past 15 years, documentation and correspondence have been increasingly kept on shared servers, giving access to a common body of documents for everyone. The DPR case study showcases the Chinese Hospital project in San Francisco where the entire design documentation is digital. Every day, the files are brought up to date, and every morning each foreman and engineer docks his/her iPad and the full digital documentation for the entire hospital is updated on everyone's iPad. For the first time, it is easy for everyone on the site to have the same information at their fingertips.

The VNGC project case study (CS2) showcases how the Sutter Health team helped the government authority OSHPD to accelerate the review and approval process by going on-line with the documents in a shared server environment. As with other industries, the construction sector is rapidly adopting IT technologies to enhance communication, underpin problem solving and accelerate decision processes.

Construction projects rely on vast quantities of information, hundreds of thousands of documents prepared by numerous independent and, yet, interdependent parties. Some estimate the cost of generating and managing information in the entire construction process can be as much as 33 per cent of the total moneys expended to procure a complex building. To convert the entire system to computer systems takes time, training, development and investment. It also requires smaller operators in the sector to become more computer literate and to increase their investment in technology. The involvement of everyone, from one end of the process to the other, is essential if IT-based systems are to achieve their potential in this sector.

Most organizations' strategies these days have some, if not considerable focus, on technology and information systems since they play significant roles in how they supply products and services to and communicate with customers. They need to identify technology requirements through business planning processes and work with technology partners and IT system providers to exploit technology to best advantage, improve processes and meet business objectives. Whether this requires a dedicated IT team to develop the strategy will depend on the size and nature of the business, but it will always be necessary to assess information resource requirements, provide the right balance and ensure value for money is provided. This is often a tall order, it seems, in the provision of IT services! Close effective partnerships that deliver in this area are often essential.

In the piloting and evaluation of new technology, the impact on customers and the business itself should be determined. The rollout of any new systems involves people across the organization, and communication cycles need to be used to identify any IT issues and feedback to partners (see Chapter 18). IT support should be designed in collaboration with users to confirm business processes, functionality, and the expected utilization and availability. Responsibilities and accountabilities are important here, of course, and in smaller organizations this usually falls on line management.

Like any other resource, knowledge and information need managing, and this requires careful consideration in its own right. Chapter 18 on communication, innovation and learning covers this in some detail.

In the design of lean quality management systems, resource management is an important consideration and is covered by the detail to be found in the ISO 9000:2015 family of standards (see Chapter 14).

COLLABORATIVE BUSINESS RELATIONSHIPS (BS 11000)

The British Standard BS 11000, first published in 2010, provides a strategic framework to establish and improve collaborative relationships in organizations of all sizes, with a view to ensuring that they are effective, optimized and deliver enhanced benefits to the stakeholders.

BS 11000 outlines different approaches to collaborative working that have proven to be successful in businesses of all sizes and sectors. It shows how to eliminate the known pitfalls of poor communication by defining roles and responsibilities, and creating partnerships that do nothing but add value to a business. The benefits of BS 11000 Collaborative Business Relationships are claimed to be:

- collaborating successfully with chosen partners;
- creating a neutral platform for mutual benefit with business partners;
- defining roles and responsibilities to improve decision-making processes;
- sharing cost, risks, resources and responsibilities;
- providing staff with wider training opportunities;
- building better relationships that lead to quicker results.

The standard gives 'ten top tips' for implementing BS 11000

1 Get commitment and support from senior management.
2 Engage the whole business with good internal communication.
3 Compare existing business relationships with BS 11000 requirements.
4 Get partner and stakeholder feedback on current collaborative working.
5 Establish an implementation team to get the best results.

6 Map out and share roles, responsibilities and timescales.
7 Adapt the basic principles of the BS 11000 standard to the business.
8 Motivate staff involvement with training and incentives.
9 Share BS 11000 knowledge and encourage staff to train as internal auditors.
10 Regularly review the BS 11000 system to make sure it remains effective and is being continually improved.

BS 11000-based collaborative business relationship management systems can be integrated with other management systems in place. This allows organizations to combine and streamline the way they manage the processes that apply to more than one system.

BIBLIOGRAPHY

Ansari, A. and Modarress, B. *Just-in-time Purchasing,* The Free Press (Macmillan), New York, 1990.

Bineno, J. *Implementing JIT,* IFS, Bedford, 1991.

British Standards Institution. *BS 11000 Partnering,* BSI, Milton Keynes, 2011.

Harrison, A. *Just-in-Time Manufacturing in Perspective,* Prentice Hall, Englewood Cliffs, NJ, 1992.

Lysons, K. *Purchasing and Supply Chain Management,* Prentice Hall, New York, 2000.

Muhlemann, A.P., Oakland, J.S. and Lockyer, K.G. *Production and Operations Management* (6th edn), Pitman, London, 1992.

Voss, C.A. (ed.). *Just-in-Time Manufacture,* IFS, Bedford, 1989.

CHAPTER HIGHLIGHTS

Partnering and collaboration

- Partnering is particularly important in construction where so many organizations collaborate to produce a completed project, and the ability to innovate, work efficiently and reliably depends on the effectiveness of the collaboration.
- Organizations increasingly recognize the need to establish mutually beneficial relationships in partnerships. The philosophies behind lean quality and business excellence define principles and guidelines to achieve this.
- How partnerships are planned and managed must be in line with overall business policies and strategies, and supply chain partners must be able to support operational excellence at the process level.
- Establishing key strategic partnerships requires attention to the development of the relationship, cultural fit and mutual development, shared knowledge and learning, joint training and development, joint process innovation, measured performance and feedback.

Global outsourcing

- The top two reasons organizations select global suppliers are the pursuit of new markets and reducing cost, yet poor quality quickly, quietly and devastatingly demolishes the benefits of any new market gains or cost reductions.
- The difficulty of managing quality increases in direct relation to the distance between an organization and its partners, so a good 'assurance' model is needed.
- Three key areas of opportunity and risk that are common across both global in-sourcing and outsourcing are: managing performance and exposure to risk, getting organized for success and delivering better value from the supply chain.

Supply chain effectiveness in global economies – the secret of safe sourcing

- Reliance upon a single source of supply can be risky so an organization must establish whether it is at risk.
- To strengthen sourcing security in terms of quality, timeliness and cost, by sourcing a second supplier: decide the selection criteria; make an initial selection; toughen up the criteria and make a shortlist; conduct site visits, make a final selection and choice.

Role of procurement/purchasing in partnerships

- The prime objective of purchasing is to obtain the correct equipment, materials and services in the right quantity, of the right quality, from the right origin, at the right time and cost. Purchasing also acts as a 'window-on-the-world'.
- The documented purchasing system should specify the documentation to be used; it should assign responsibilities and define the means of selecting suppliers.
- It is critical that the purchasing policy is driven by the achievement of value rather than simply by minimizing price. The challenge is to identify the value to be created and to ensure that the partners are aligned towards its achievement.
- Improving supplier performance requires from the suppliers' senior management commitment, education, a policy, an assessed quality system and supplier approval.

Lean quality in partnerships and the supply chain

- Single sourcing – the close relationship with one supplier for each item or service – depends on technical, conditional and full stages of approval.

Resources

- All organizations assemble resources to support the operation of the processes and delivery of the strategy. These include finance, buildings, equipment, materials, technology, information and knowledge.
- Investment and/or funding is key for future development of all organizations and often determines strategic direction. Financial goals and performance will therefore drive strategies and processes. Use of a 'balanced scorecard' approach with continuous appraisal helps in understanding the long-term impact of financial decisions.
- In the management of buildings, equipment and materials attention must be given to utilization, security, maintenance, inventory, consumption, waste and environmental aspects.
- Technology plays a key role in most organizations; therefore, the management of existing, alternative and emerging technologies need to be identified, evaluated and deployed to achieve organizational goals.
- There are clear links between the introduction of new or the replacement of old technologies and process redesign/engineering. The rollout of any new systems always involves people across the whole organization and therefore good communication is vital.

Collaborative business relationships (BS 11000)

- BS 11000 provides a strategic framework to establish and improve collaborative relationships in organizations of all sizes to ensure they are effective, optimized and deliver enhanced benefits to the stakeholders.

- BS 11000 outlines different approaches to collaborative working that can be successful in businesses of all sizes and sectors. It shows how to eliminate the known pitfalls of poor communication by defining roles and responsibilities, and creating partnerships that add value to a business.
- The standard lists the benefits of using the approach, including defining roles and responsibilities to improve decision-making processes, sharing cost, risks, resources and responsibilities, providing staff with wider training opportunities and building better relationships that lead to quicker results. BS 11000 also gives 'ten top tips' for implementation.

The planning and execution of work

7

THE UNRELIABILITY OF WORK

In this chapter we primarily address the issue of the unreliability of workflow in the construction of complex, major projects and recent techniques for improving the stability, reliability, throughput and cost of work. The building and construction industry is extraordinarily diverse, so practices, processes and products vary widely across different sectors of the industry. They include the construction of:

- modern hospitals, sophisticated laboratory building, manufacturing and processing plants, power plants and petrochemical plants;
- high-rise office buildings and medium density housing;
- civil infrastructure projects including road, rail and water; and
- industrial sheds and warehouses, and the project home building sector, which builds large numbers of standardized free-standing homes.

One of the earliest insights of the co-founders of the lean construction movement, Glenn Ballard and Greg Howell, was that work executed on construction sites was chronically unreliable. Early analyses showed that little more than half of what was planned to be done on site within a one-week planning horizon was actually done as planned. In other words, half of what was planned didn't get done as planned. This research work has been extended, in the intervening period, with larger studies all around the world, and the rough measure of some 50 per cent of planned work not being completed as planned has been confirmed, worldwide, as being typical.

In Chapter 1 we discuss the structural challenges that face the construction sector; in summary, the modern industry can be characterized as follows:

- It is highly fragmented with some 85 per cent of site work being delivered by sub-contractors.
- Many primary subcontractors further subcontract work to smaller and more specialized enterprises in what can only be described as pyramid subcontracting, further challenging coordination and control.
- The industry is transaction-cost driven and many of the traditional general contractors have broad supply chains and relatively shallow cost-based relationships with their suppliers.
- In many trade areas there are few barriers to entry and hence there is constant downwards pressure on costs, perhaps, in part, driven by efficiency gains – but primarily driven by new entrants willing to take risks.
- General contractors have changed from being builders who plan and craft buildings to contractors who integrate and buy products and services from others.

- Relationships between general contractors and their subcontractors only exist during projects; hence they do not have long-term relationships with their supply chains.
- General contractors are more interested in driving project speed, in order to minimize the cost of preliminaries, the area of primary cost savings and risk for the general contractor, rather than the production efficiency of the whole enterprise.
- Subcontractors are focused on optimizing their piece of the work, not the whole.
- Designers have become more and more specialized and removed from the construction process, dealing in abstractions rather than the real processes of construction.
- Clients and owners have become increasingly risk averse and less willing to invest in early solution generation and needs definition, seeking to push all risk down to the supply chain.
- General contractors reflect the attitudes of their clients to risk by seeking to push all risk further down the supply chain to their suppliers.

As a result of this fragmented industry structure, the reliability of workflow defined in weekly plans is low. It seems that, on most projects, the parties sitting around the table each week at the weekly planning meeting are focused on the productive efficiency of *their* own workforce, rather than the efficiency of the project as a whole.

The design process is even more variable and the flow of work is much less visible. In construction, one party builds on the physical work of another and progress can be easily gauged; in design, however, the building blocks are information flows and decisions and therefore progress and commitment reliability are more difficult to gauge. Designers are chronically unreliable in meeting their commitments, making design processes appear to be even less reliable than construction processes. When asked what they would like to see improved, designers working in multidisciplinary design teams invariably give the same answer: 'To receive information when it's promised'. Individual designers are often working on several projects concurrently and find it difficult to create a reliable and predictable workflow within their own practice. The case study of Boulder Associates (CS1) illustrates how this architectural practice has developed a framework for planning work and making commitments that has markedly increased the reliability of individual work commitments. The result has been less overtime, less stress and better productivity.

The exceptions to this description of construction are projects such as the VNGC project (CS2) and the Multinational High-Technology Manufacturer (CS4). Both of these owners have changed to using relational IPD agreements because they found that their traditional procurement arrangements were not giving them the value they needed in their projects. Such IPD agreements bind the entire supply chain to a common purpose through painshare/gainshare provisions. The client is also a part of the team, managing risk and exploiting opportunities for efficiencies throughout the projects. Both of the above clients have found that these agreements are a much better way to drive innovation, efficiency and reliability on their large, complex projects.

The unreliability of promises for the completion of work has very significant repercussions on efficiency throughout the supply chain. Everyone who is operating under a fixed price agreement, and that is most parties in normal contractual situations, loses margin if their workers cannot work in a smooth, predictable and reliable manner. Whether they are waiting on other workers to get out of their way, waiting on workers to properly complete prior work, waiting on materials to arrive or waiting on information, they are similarly impacted. They cannot get a clear run at their own work and hence everyone is operating in a sub-optimal manner, most of the time.

TRADITIONAL STRATEGIES FOR STABILIZING WORK ON SITE

Construction is a risky activity in many ways: building contractors are particularly exposed to risk arising from poor weather. In addition, it is also an industry that has a history of poor safety and quality outcomes, and historically high levels of industrial unrest. Each of these risks is greatest during the execution of work on site. Hence, the focus of contractors has been to reduce the number of workers on site and build as quickly as possible – to minimize risk. There have been many strategies for accelerating construction and reducing the amount of work and the number of workers on site.

The period since the Second World War has seen the emergence of large-scale prefabrication in some building forms. In Europe, large-scale precast concrete building systems were developed for the construction of schools, industrial buildings and even entire towns. In the Netherlands, large-scale, site-built, steam-cured concrete construction systems were developed for the construction of row houses. Industrial building frames were also developed in steel.

In the US, some builders specialized in the construction of stacked modular buildings. Hotel buildings were built of reinforced concrete units manufactured in a factory on site, while condominiums were built of timber and transported from factories nearby. More recently, we have seen stacked units made of modified steel containers, manufactured in China, shipped to mature markets in the US, Australia and elsewhere.

Apart from these larger scale industrialized developments, the last half century in particular has seen an enormous development in building materials and equipment, all of which is designed to improve labour productivity on site. Equipment has replaced labour in numerous ways, making work safer and faster on site. There has been a move away from the labour-intensive wet trades wherever possible. Bricks and blocks are being replaced by drywall construction and in situ concrete by precast and curtain walling wherever possible.

A recent innovation in the US construction sector is described in the ConXtech case study (CS3). ConXtech has developed a very sophisticated, precision made 3D steel framing system that is suitable for a wide range of multi-storey building applications, including medium density housing, office buildings and hospitals. The frame clips together easily and quickly and is then overclamped with a bolted connection for permanency.

Given these myriad changes, it is surprising that productivity in the sector has failed to increase. It would appear that the fragmentation and poor coordination of work at an overall level has offset the gains achieved by individual innovations.

HISTORIC CHANGES IN THE PLANNING AND MANAGEMENT OF CONSTRUCTION

Until relatively recently the architect, as the head of the design team, had the role of interpreting the client's brief, overseeing and integrating the design of the other professionals and overseeing the contract works procured by tender. This was a role that the profession had fulfilled for many hundreds of years. However, as the twentieth century unfolded, the development of technology exploded and projects became much larger, more sophisticated and complex, architects were unable to keep pace with the increasing demands of technical complexity and project management. Increasingly, the profession saw its role narrowed to focus on the aesthetic and functional aspects of projects. This transition is ongoing and has been underway for half a century.

As architects withdrew from the broader management role on major projects, the new profession of project management emerged to fill the critical niche of project coordination and management on behalf of the owner. The new profession codified its practices in the PMBOK (the project management body of knowledge) and ranked its members on the basis of their competencies. However, critically, the profession of project management did not recognize the underlying causes of the capacity gap that was emerging between clients' expectations and industry capability. Professional project management limited its purview to better administering the status quo. As a result, to the present time, the new profession has made only a modest impact on the critical issues of better meeting clients' needs and increasing the efficiency and reliability of the construction process.

During this period, computer technology enabled the creation of a range of powerful tools to assist project and construction managers on large projects. The Gantt chart, developed in the early twentieth century by William Gantt in the US, once available on computers gained widespread use from the 1980s for creating ever more detailed schedules. The application of spreadsheet and database-based software to the challenges of cost planning and control did the same for the area of cost management. These tools armed both project management businesses and construction companies with more powerful ways of creating abstractions of the project in the planning stage and controlling the administration of the contract during the project delivery phase. However, in the planning and delivery stages, these tools only create abstractions and are unable to predict reality in any detail. During construction they certainly offer more effective ways of tracking the project and dealing with variations and claims.

A problem for the industry is that these powerful tools created an illusion of sophisticated planning and control but, in doing so, they held back the development of a more pragmatic engagement with the complex reality of work on the ground. The challenges of the practical integration of work in design offices and on construction sites went largely unaddressed. This growing gap was not addressed until the advent of the lean construction movement in the mid-1990s. The realization that only half of all work occurred as planned turned the attention of a small vanguard of academic and industry thinkers to the structural problems inherent in the traditional approaches to procurement and delivery of constructed products.

The lean construction movement

During the last 20 years the lean construction movement has spearheaded a re-engagement with the myriad practical production challenges that the industry faces, and has capitalized on the potential that new technologies offer to the delivery of capital projects.

- The development and implementation of the Last Planner® System (LPS) has created a basis for better involving the leaders of work teams in the creation of reliable short and medium-range plans.
- The more detailed evaluation and planning of construction activities using work structuring, work study and Value Stream Analysis are driving greater efficiencies at the process level.
- The co-location of the design teams on major projects creates the basis for much improved communication and collaboration within the design team. When this is extended to co-location with the construction team, further benefits accrue from the bringing together of downstream fabricators and erectors with the initial designers of the project.
- Conversations between front-end designers and downstream fabricators and erectors are enabled within IPD style agreements and, through the use of powerful 3D modelling tools, they are able to develop more constructible design solutions.

- The use of 3D modelling tools has created the basis for front-end designers to engage in a productive dialogue with owners, as well as end users and external stakeholders.
- Powerful 4D and 5D modelling tools support target value design and enable teams to evaluate design alternatives and assess the impacts of different construction scenarios on schedule and cost.
- The adoption of the language action theory has provided a basis for exploring sound agreements between customers and service providers in an industry in which commitments are chronically unreliable.
- The realization that in the case of large, complex projects, a collaborative enterprise that engages with the management of risk is much more effective at optimizing outcomes and managing inevitable changes on projects than an enterprise driven by risk avoidance and led by principles of command and control.
- The use of location-based planning and Takt time planning – as tools for smoothing the workflow – provide a basis for the reliable prediction of stable resource requirements on construction projects.

THE LAST PLANNER® SYSTEM

The Last Planner system of production control was developed in the early 1990s and was more widely published in the 1998 paper titled 'Shielding Production' (Ballard and Howell, 1998). Since that time, it has been used in short-range planning meetings on projects around the world as a method for improving the reliability of work assignment execution, in both design and construction activities. There is also a very significant literature around the theory and practice of LPS® on the IGLC conference website (www.IGLC.net).

The LPS® differentiates between five levels of planning, each getting into more and more detail regarding the work to be done:

1   Initial
2   Phase
3   Lookahead
4   Commitment planning
5   Work structuring.

Initial planning is the creation of the master schedule at the outset of a project to specify the milestones between phases. It is used for producing the project budget and contract schedule and is the basis for the definition of major project milestones. It is also used to identify timings for the ordering of long-lead time items. The lean construction community believes that this level of planning should not go beyond milestone planning, as the greater the level of detail, the greater the level of abstraction and risk that is locked into the plan. At this stage there are too many unknowns and assumptions in the programme for extensive detail to be of value.

Phase planning or scheduling is for the period between significant milestones and is generally undertaken between three and six-monthly. This level of planning is called pull planning as it pulls activities to the next milestone. It is created by the production team leaders, either design team leaders or construction team leaders, who understand critical interdependencies and constraints and are in a position to define critical handoffs between production teams. These secondary milestones set the goals that will drive production control through the more detailed Lookahead and Commitment planning stages.

The purpose of Lookahead planning, which looks ahead for a three to six week period, is to make work ready. This involves the team of production team leaders identifying any potential constraints to work, and specifying and allocating necessary actions to ensure that when activities come into the 1-week Commitment planning stage, they are ready to be done.

Commitment planning is what is done at the weekly level by the last planners: defining the assignments that will be undertaken in the forthcoming week in as much detail as possible.

Work structuring is the detailed planning and analysis of activities at the task level to optimize production in detail. It involves considering logistics, materials, workplace design, equipment and skills at the task level.

Since 1998 the LPS® has been increasingly widely used among those design and construction companies that practise lean construction worldwide. Among the case study companies, the major contractors and subcontractors all practise LPS® on their projects. They do so to improve the productivity of their on-site workforce and also to create the basis for reliable requests for prefabricated products for site assembly from their fabrication shops and materials and equipment from their stores. Clients such as the Multinational High-Technology Manufacturer (CS4) and Sutter Health (CS2) have invested in providing training in the practice of the LPS® to their supply chains with a view to better integrating their workflows. The basis of Boulder Associates' management system for improving the reliability of work plan commitments by individual designers is also based on LPS®.

It has been found by work teams who use LPS® in their weekly planning processes that the rate of reliable commitments on a weekly basis can be raised from near 50 per cent up to 85 per cent. It has been shown that this increases the productivity of work teams and the profitability of the businesses working on these projects. More recently it has become obvious that stabilizing work plans using LPS® not only improves productivity but provides the basis for improvements in safety and quality. Work that is done as planned has a greater chance of being 'right first time' than work that is brought into play prematurely.

Process notes for phase planning

This is a participative planning process in which the production (design or construction) team leaders work together to create a plan of the workflow within a phase.

1 Team leaders have relevant schedules and plans with them. If client reviews and approvals are critical within the phase, a client representative should participate. The meeting coordinator sets the scene, defining the works to be included in the planning session and issues relevant support information to the participants.

2 The meeting coordinator sets the completion date for the phase, defines any intermediate dates that are fixed and indicates any major interim releases of work from prior phases. These should be posted on sticky notes at the top of the planning wall for all to see. These dates create a time frame for the planning session.

3 Using different colours for each discipline, develop the network of activities required to complete all work in the phase, working back from the completion date and working around all fixed intermediate dates. Working back from the completion date defines this process as pull-planning.

4 Apply durations to all activities with no allowance for contingency.

5 Re-examine the logic of the workflow logic and tighten the schedule wherever possible.

6 Determine the earliest practical start date for the phase.

7 If there is time available for buffering activities, allocate buffers to the most fragile activities.

8 Reserve any unallocated buffer as a general contingency for the phase.

Process notes for weekly work planning

This is a process for creating a reliable weekly work plan; it should be led by the design manager in design planning meetings and by the general foreman in construction meetings. Attendees are design leaders and trade supervisors/foremen. Project managers and site engineers should be attending as advisers and coaches. A draft plan for the next week and the Lookahead period (three to six weeks) should be circulated prior to the meeting so people will have been thinking about the plan before coming into the room. Some discussions will have been held one-on-one prior to the meeting to solve problems and tease out issues. However, you must come to the meeting with an open mind; this is the meeting of last planners and at this meeting, this group will agree the best plan going forward. If the general contractor's supervisor comes in with a fixed plan in their mind and their hand, an opportunity for collaboratively arriving at the best plan will be squandered.

Before starting, decide on a process that will engage the last planners. This could be:

- a schedule on a spreadsheet projected onto a screen or wall;
- the use of sticky notes on the wall, with a different colour for each discipline;
- a large-scale coloured and annotated plan on the wall; or
- a proprietary Last Planner software run off tablets and projected on to a shared screen.

It is very important that whatever means are used, it should be at a large scale so that everyone in the room can see it. The meeting chair should preferably be standing by the plan on the wall to hold everyone's attention. The last thing you want is everyone looking at their own piece of paper or, even worse, at their messages on their phones or tablets.

Attention should be paid to some important meeting discipline issues: one conversation at a time in the room; avoiding the use of ambiguous language (hope, think); and not accepting phone calls or responding to messages as far as absolutely possible.

1   Review the last week's plan, you will have assessed Plan Percent Complete (PPC%), the ratio of completed tasks over the total number planned prior to the meeting. In addition, you will have spoken to people about the reasons for specific plan failures. You should spend the first, say, ten minutes reviewing PPC, trends in PPC, conducting 5 Whys analyses to determine the root causes of last week's planning failures and reflecting on what can be done better to improve planning reliability.

2   If the planning session is to be projected onto a wall, the meeting chair should stand at the front by the plan, leading the conversation, encouraging the participants to identify any impediments and either changing or validating the plan as it stands. At the end of the meeting, the plan confirming the next week's work and the three-week Lookahead can be circulated to each of the participants and others on the distribution list. Everybody immediately has the same plan and they can fix it to the wall in a visible location for all to see.

3   If the planning process relies on individuals negotiating their workflow commitments and posting them on sticky notes (each discipline having a different colour), this process will involve the participants talking more informally in small groups by the planning wall. On a road project, the planning background could be a plan of the road alignment fixed to the wall, and work commitments can be posted on the road alignment at the appropriate locations. In other situations, the plan might be a more conventional workflow chart. In either case, the stickies become the plan for the forthcoming week, and this can be photographed or transcribed and circulated. Commitments not met at the end of the week could be highlighted with a red dot to draw attention to them.

4   In such a meeting it is not unusual to identify problem areas where work cannot be easily allocated because of some identified constraints; these should be transferred to a Problem and Countermeasure board nearby, where the problem is set out so that everyone on the project is aware of it. This practice invites ideas from everyone for the solution.

5   A part of this process should always include the identification of workable backlog (i.e. work that can be brought into next week's window, should it be required) in case of bad weather or other unexpected event.

6   Work that is identified as being not ready should not be included, but should be carefully considered and actions identified to make sure it is made ready so it can be picked up and completed as soon as practicable.

There are some well accepted rules for the assessment of every activity in relation to:

- clarity of task, precisely what it includes;
- everything that needs to be known is known;
- does the work flow; and
- for ongoing work this should specify the quantity.

Process notes for Lookahead planning

Lookahead planning is a part of the weekly work planning meeting; however, it must be a separate agenda item. Once next week's plan is set, the less detailed plan in the Lookahead window should be briefly reviewed. The question to be asked is: 'Can anyone see any problems, impediments to work or issues that could pose a risk to workflow within the Lookahead period?'

The purpose of this process is to identify any risks that might disrupt workflow far enough in advance so that they can be managed and necessary actions taken so that the work can be made certain within the planning time frame. One of the authors was working on a project where the implications of a change in the thickness of a core wall was not identified early enough as a risk to the work plan. The core wall was to reduce from 300 mm to 200 mm at level 15. Some steelwork had to be fabricated to accommodate the external change in the dimension of the shaft. Workshop drawings had to be prepared. These had to be reviewed and approved; steelwork had to be fabricated and galvanized. This process took six weeks, yet the issue was only identified three-weeks out and the oversight created disruptions to the flow of work on the job.

The main challenge in this process is to train last planners to look farther ahead than the next week, and this may be something that some people at the weekly meeting are not used to doing.

PROCESS EFFICIENCY THROUGH WORK STRUCTURING AND WORK STUDY

As builders have moved from engagement with the processes of building to become purchasers of products and services, generally for a fixed price, and as subcontractors have increased pyramid subcontracted work to labour-only subcontractors, innovation is largely left to the lowest level of subcontractor, a level at which the capacity for innovation is at a minimum.

The lean construction community has re-engaged with work efficiency through two mechanisms: work structuring and work study. The Lean Construction Institute defines work structuring as the fundamental level of production system design, which involves designing the construction process in its entirety, including technology selection, construction sequencing, organizing resources and suppliers. Work study is the detailed analysis of a task on the field to identify improvements in method, tools and/or work sequence that can improve throughput, quality or safety.

Techniques for work structuring include first run studies and computer simulations. First run studies are either undertaken on prototypes or on the first of a run or series of similar tasks, and may involve trials of alternate techniques, alternate materials and different degrees of off-suite preparation. The aim is to simplify tasks, minimize waste, remove risk and improve throughput. For example, on Terminal 5 Heathrow, where several kilometres of pedestrian tunnel were constructed under the runways, computer simulation was used to detail reinforcement and to optimize the sequence of reinforcement placement.

A paper titled 'Work Structuring of Construction Crews in the Installation of Light Fittings' (Nerwal and Abdelhamid, 2010) summarizes the results of experimentation in crew and equipment configuration on the time and cost of a light fitting installation. The batch of lights to be fitted was 300. This full-scale, live study demonstrates through the evaluation of three different scenarios that the duration, cost and labour utilization varied widely between the three alternative scenarios.

The base case cost $9.00 per installation, and the other two cases cost $7.61 and the lean case $5.69 respectively; the lean solution was 37.19 per cent more cost-effective than the base case. In terms of duration of operations, the base case was 130 hours and the other two cases were 45 hours and 60 hours respectively, the most cost-effective taking 33 per cent longer that the 2nd alternative. The lean case that was the most cost-effective had the least wasted time and the lowest work in progress (WIP), but was not the fastest.

The findings of the work study team described in the Rosendin case study (CS11) found that after detailed trials and a value stream analysis, they were able to install a light fitting in exactly half the time. The authors are not aware of the team configuration and cost savings in this case.

These cases illustrate that, at the detailed level, there is significant opportunity to increased efficiency in construction work. It is also evident that optimizing time or cost may lead to a different crew or equipment configuration so it is important to identify the objective before starting.

LOCATION-BASED PLANNING AND TAKT TIME PLANNING

Some companies who have practised lean quality and have been using the LPS® for some time and have experienced the benefits of improved reliability in workflow on their projects are now looking to further improve efficiency. This has, over the past ten years, led to a growing interest in location-based planning and more recently in Takt time planning.

As we describe earlier in this chapter, the supply side of this sector has become increasingly fragmented and, hence, it has become more difficult to integrate and control the industry's processes. This has led to an even greater interest in the creation of uniform workflows and an increase in the levelling of resource in the construction process using balanced workflow planning. It is not until all stakeholders in the value stream are aligned around the same production rhythm that the true potential lean principles can be realized.

In many lean implementations, the focus is set on eliminating waste, *muda*, with less attention to workflow levelling, *mura*. The waste approach is straightforward and can give some results quickly; however, it does little in a structured way to optimize the system as a whole. A critical step in implementing lean quality is to first of all create stability in a process so that it can then be assessed and improved. Standardized work cannot be established if there is variability within a system.

In manufacturing, to create balanced workflow, the lean community has spent half a century developing tools around the metric of Takt time. The first American employee, and later manager at Toyota, John Shook explains the purpose of Takt time as 'first and foremost, to serve as a management tool to indicate at a glance whether production is ahead or behind', thus providing instant feedback to any discipline that is overproducing or causing delays so that they can alter their production to maintain flow. John Shook continues to stress that Takt planning further 'serves as an alignment tool: aligning proceeding with following processes, aligning resource requirements with demand, aligning corporate functions with real-time production needs' so that all parts of the value stream align around the same rhythm.

The question that many ask is how can this concept be used in construction? After all, work on a construction sites is anything but similar to an assembly line. In manufacturing, the overall flow of production can be seen through the process cycle time (Takt), as the product passes through sequential process steps. The speed of the assembly line provides a tangible and transparent indicator of pace. Consequently, all sub-assemblies or works stations can efficiently design their production processes to stay on pace with the overall speed of the line (Figure 7.1).

Since the construction industry does not have a physical assembly line, to get a better appreciation of the true potential of Takt time planning, we need to picture a construction situation where the ideas of Takt time are self-evident. This is provided by high-rise building construction where on a regular cycle of, say, four days, work crews move up the building behind each other in a continuous process. In such an example, it would be common for the floor plate to be divided into four quarters, and each crew would need to finish all its work on a quarter of the floor plate, each day, before moving on to the next quarter. Apart from the interference of the weather, such a system can work on a daily cycle for weeks on end.

In such a situation, the principles of flow and pull are made just as tangible as in the assembly line. However, it is the crews that move at a set pace rather than the product. Each crew has a set pace, materials and equipment are supplied to it according to plan and a set amount of work is finished daily. The work can be broken down into tasks of differing durations and the progress of each task and of the whole work crew can be tracked. With time, the crews solve problems, work runs more smoothly and the hand-offs between crews will become more regular and systematic. The workers and the crews soon find the rhythm for what needs to be done each hour of the day for the work to be completed reliably. In turn, the production manager on the floor gets to know the work tasks, material and equipment requirements, work allocation and the rhythm of the work. The only metric the production manager needs to monitor is that each crew is on plan with its tasks, hour by hour, so that at the end of the day, the work crew can move on to the next quarter of the slab.

Since both expectations and tracking are simplified, management resources are freed up to focus on more valuable tasks such as optimization of individual tasks, employee satisfaction, problem solving and the efficiency of the process as a whole. At first glance, the indicator of flow is not as tangible as in the assembly line example; however, Takt time planning successfully aligns the production efforts to optimize flow. The expectations are clearly communicated and all resources align to a common pace, reducing both over-production and delays.

Once the work involves building a hospital or shopping centre, with a large slab area and high services intensity in some areas and low intensity in others, this rhythm is lost. Work crews move around the building on the basis of the project schedule and while that schedule is designed to keep everyone engaged, the reality is that workers are often waiting

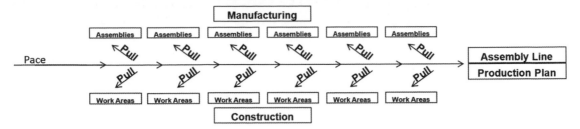

FIGURE 7.1 The sub-assemblies on the assembly line are equivalent to the areas of the high-rise building and the speed of production is regulated by the cycle time

for work; and work is often waiting for workers. Crew sizes fluctuate almost daily. Often crews are working in several areas simultaneously, waiting on others in some places and running ahead in others. Management effort is expended on keeping everything moving and there is little time if any for process optimization (Figure 7.2).

Interestingly, it is possible to interpret the principles of lean in a similar fashion as we have illustrated with the high-rise building example above, and develop a production system that shares the benefit of flow through completely balanced and aligned resources, material and equipment; however, on such projects this is rarely achieved.

In manufacturing, and in the case of the high-rise building, the pace of production is set by the production team and based on available time per needed unit of output, and also through the definition of cycle time or Takt. The work crews in the supply chain draw materials and equipment as they are needed, and in accordance with the plan.

In normal construction, by consciously rightsizing weekly/daily handoff areas to contain equal amount of work, the production planners can produce as transparent and tangible display of flow as the pace of the high-rise building or assembly line. This can release a consistent amount of work daily and set a clear, steady and predictable pace that all trade disciplines can design their production capacity to (Figure 7.3). The defined hand-offs are designed to break up the planning efforts into manageable portions for the last planners. As the experience level and predictability of the team increases, the frequency of handoffs can decrease to daily or even hourly handoffs.

Once the scope of work is levelled out between the handoff areas, each individual crew's production rates need to be aligned with the overall cycle time or Takt. In this exercise,

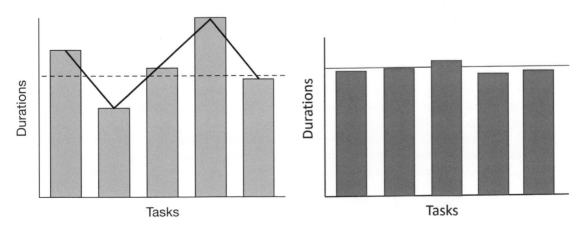

FIGURE 7.2 CPM schedule durations

FIGURE 7.3 Takt planning durations

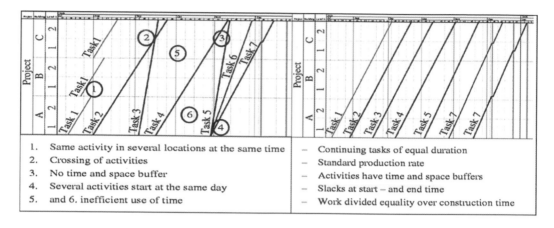

FIGURE 7.4 Flowline diagram, traditional vs Takt

the faster trades might have to reduce resources to avoid overproducing, while the slower trades, or bottlenecks, need to be carefully analysed to discover opportunities to increase their throughput. It is important to explore prefabrication opportunities, waste elimination and production process improvements before resorting to short-term variations in manpower.

To understand how the suggested process can be made to work in a construction setting, we recommend the following planning approach (Figure 7.4):

• Each floor is broken up into areas with similar amount of scope.
• Each discipline has one cycle (X days) to complete each area.
• Only one trade can occupy an area at once.
• Material is delivered directly to each crew's work area.
• Crews plan and monitor progress towards completion of their tasks daily.
• All disciplines complete and move to the next area at the end of each cycle.

The focus is to make sure that each crew is moving at the same speed, following each other through the building in a coordinated fashion. The trade crews should optimize their resources to fit the plan and benefit from a stable workflow and resource demand. The predictability can allow disciplines to plan their work better and achieve the needed throughput with fewer resources. When schedule optimization is the prime objective, and all trades can increase throughput without major disruptions to the flow of work, the Takt can be reduced to accelerate the project and to further to stress the system. Always bear in mind that *the slower but consistent tortoise causes less waste and is much more desirable than the speedy hare that races ahead and then stops to doze occasionally.* Engineer Ohno from Toyota wrote 'The Toyota Production System can be realized only when all the workers become tortoises' (Ohno, 1988).

When schedule optimization is the prime goal, and all trades are able to increase their throughput without major disruption, the Takt time can be reduced further; however, this will stress the system. When doing this, it is essential that all trades are able to accelerate at the same pace. It should be kept in mind that a slower and more reliable system is much more desirable than one that is faster but stops and starts. The knowledge that the pace of construction is predictable enables the entire supply chain to plan its resources and logistics in a reliable matter.

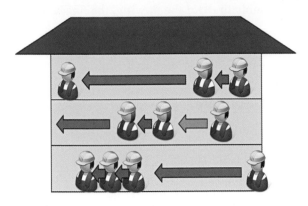

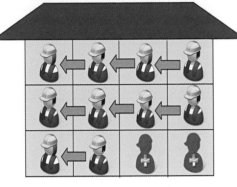

FIGURE 7.5 Traditional scheduling   **FIGURE 7.6** Takt time planning

To illustrate the above, Figure 7.5 shows a typical construction project that has multiple trades on each floor and there are some areas where little work is being done. In the crowded areas there is very little time buffer between trades, forcing the trades to continuously re-plan their work to keep everyone busy. In the areas that are vacant, a trade has completed its work earlier than anticipated and the next trade has yet to catch up. Overall, there is no clear indicator of progress to plan. In this scenario, work is waiting for workers; and workers are waiting for work.

In Figure 7.6 the team has replaced the regular schedule with a Takt time-based schedule. Handoffs are defined in areas, and the planned durations are the same for all disciplines. In this scenario, because of the predictability of the pace of work, it is easier for the trades to manage their resources and rightsize their buffers to hit the completion date. While the production flow will not be free from disruptions, each foreman is able to make an educated guess of how much capacity he needs to be able to stay on track.

If visible targets for weekly and daily handoffs are created, the transparency of progress promotes competition and drives the last planners to keep up with their commitments. If the incremental targets are aligned with the Takt time of the job, it not only provides targets and a metric to assess progress, it also provides a mechanism that makes sure all parties are on the same page and optimizes the flow of work. The key is that all areas of the building are being worked on without trade stacking. No areas are being left vacant due to unrealized buffers.

- Within each area, capacity is managed at an incremental level to eliminate excess resources, buffers and waste.
- The scheduling efforts are clear, simple and predictable, shifting resources to plan production execution, process optimization and waste elimination.

This creates a structure that provides manageable planning areas for the Last Planner process, and provides incremental feedback on alignment and progress to manage production flow.

To focus on creating a balanced system is not new to the construction industry. The construction worker Richmond Shreve, for example, had this to say about his experiences at the Empire State Building project:

> We always thought of it as a parade in which each marcher kept pace and the parade marched out of the top of the building, still in perfect step . . . Sometimes we thought

of it as a great assembly line, only the assembly line did the moving, the finished product stayed in place.

(quoted in Tauranac, 2014, p. 214)

This exemplifies the power of controlling the entire value stream to a common rhythm.

BIM (Building Information Modelling)

In today's construction industry the potential and limitations of BIM are not clearly understood. While some companies are still sitting on the sideline, thinking that they can do without it for now, others think that the technology is a silver bullet and of itself will improve efficiency and productivity. Meanwhile the early adopters are already reaping the significant benefits that the technology offers. In fact, BIM has changed the processes of design forever; it has the potential to radically improve communication within the design process and on site, and it will disrupt the traditional processes of design and construction, changing roles, relationships and creating opportunities for new business models.

There are many clients and stakeholders who have difficulty in visualizing the three-dimensional object being designed on the basis of two-dimensional drawings. Furthermore, in the most complex areas of design, especially in plant rooms, even the designers have great difficulty in understanding how all the subsystems fit together. BIM has the potential to improve communication at every level, it supports design development, problem resolution between designers, and it provides a platform which enables designers and constructors to collaborate and develop more constructible designs. It also supports prefabrication by increasing the accuracy of design and construction and most recently it provides detailed documentation on site in digital form that can be refreshed regularly (world's best practice is daily), ensuring that everyone has the same information on site (refer to CS9).

However, the main challenge for organizations, and even more particularly for entire supply chains, is to use BIM effectively otherwise these potential benefits cannot be realized. The effective uptake of BIM is slow in many organizations for several reasons, which include:

- The company does not recognize the potential of BIM to improve its operations.
- BIM, a powerful collaboration tool, is deployed in a contractual environment that does not encourage collaboration.
- Insufficient planning goes into the deployment and management of BIM. Implementing BIM is not simple, it requires leadership, careful planning, training and coordination across the supply chain.

A recognized approach to managing the implementation of BIM on a project is through what is known as the BIM Execution Plan or BEP. In this plan details are defined in relation to issues such as:

- roles, responsibilities and authorities;
- project milestones to monitor progress;
- deliverable strategy, scope in terms of the whole of asset life;
- existing legacy data use;
- approval of information;
- authorization process.

The BEP needs to be a contract document as it sets out critical relationships and responsibilities. However, there are a number of critical issues to ensure that BIM is effectively deployed; these include the need to:

- ensure that all consultants provide accurate and consistent data quality;
- ensure that all parties provide information according to an agreed timetable;
- ensure that the legacy information creates maximum value for the long-term owner and operator.

It is important to be clear about who will lead and manage the implementation of the BEP; this could be the client, the lead designer or the lead contractor. Each party will bring to the task different skills and biases, and each arrangement has advantages and disadvantages. Table 7.1 sets out the main issues for consideration.

TABLE 7.1 Biases that different parties bring to BIM management

	Advantages	*Disadvantages*
Client-driven policy	• End requirements are more clearly defined • Client is in a position of authority to make decisions	• Poor understanding of operational design issues • Client can be too removed from the issues requiring resolution
Designer-driven policy	• Likely to encourage early implementation • Would encourage a holistic approach • This would be a sensible additional service for the lead designer	• Will have a focus on geometry and dimensions • Is likely to be unaware of construction and building operational issues • Designers have limited contractual power, and therefore may not have enough authority
General contractor-driven	• Financial incentive to get value from the BIM • Contractual power and effective decision-maker • Focus on creating value in constructability	• May have a construction focus at the expense of design • A greater focus on cost may neglect important aspects of value

Traditionally, the processes of design development and documentation have been both undertaken and managed by designers and draftsmen. More recently however, under many contractual arrangements, general contractors are taking more responsibility for design management; this represents a structural change that has not been without its difficulties, as contractors have had to develop skills in this challenging area which is very different to construction management. However, this also offers great potential benefits, as contractors are the best placed to drive overall efficiencies in the prefabrication, transport, assembly and on-site construction processes by integrating the inputs of all the relevant parties.

The effective adoption of BIM represents further changes which bring both opportunities and challenges to all parties. A critical insight is, that to take full advantage of the opportunities BIM offers, organizations need to adopt a data centric business model. The fundamental processes of design creation, documentation, review and approval are changing forever. The VNGC example (CS2) illustrates how a novel approach to data sharing and web-based collaboration accelerated the review and approval process beyond expectations. The DPR example (CS9) illustrates how the general contractor has been able to integrate the entire design and associated data processes to deliver integrated information simultaneously to all parties in the design development, documentation and production process on site updated daily. Because of the vast amount of information on construction projects, never before has everyone been looking at the same information every day of a project.

ACKNOWLEDGEMENT

The authors are grateful to Klas Berghede of Boldt Constructions for his assistance with the section on Takt time planning.

BIBLIOGRAPHY

Ballard, G. and Howell, G. Shielding production: essential step in production control, *Journal of Construction Engineering and Management*, 124(1), 1998.

Ballard, G. and Howell, G.A. An update on Last Planner, *Proceedings of the 11th IGLC Conference*, Blacksburg, VA, 2003.

Nerwal, N. and Abdelhamid, T.S. Work structuring of construction crews in the installation of light fittings, *Proceedings of the 18th IGLC Conference*, Haifa, Israel, 2010.

Ohno, T. *Toyota Production System, Beyond Large Scale Production,* Productivity Press, New York, 1988.

Shook, J. *Misunderstandings About Value-Stream Mapping, Flow Analysis, and Takt Time,* Lean Enterprise Institute, n.d., www.lean.org/Library/Shook_on_VSM_Misunderstandings.pdf

Tauranac, J. *The Empire State Building: The Making of a Landmark,* Cornell University Press, Ithaca, NY, 2014.

CHAPTER HIGHLIGHTS

The unreliability of work

- Only some 50 per cent of what is planned one week out is completed as planned on most construction sites. A similar level of unreliability exists in the design process.
- The reasons for that are summarized, essentially the commercial arrangements do not encourage collaboration and the level of fragmentation in the industry makes it difficult to organize work to be efficient.
- Contractors focus on project speed and do not appreciate the cost impact of unreliable workflows. These costs are borne by the subcontractors and the owner, not by the general contractor.

Traditional strategies for stabilizing work on site

- In Europe in the early post World War II era, many prefabricated systems and on-site equipment intense systems were developed to increase speed and to build with less labour.
- Project size has been increasing and this has put more pressure on the planning and control of cost and time schedules. Computer-based systems have come to the fore, increasing the industry's capability for dealing with large projects; however, these programmes do not deal with the reality of daily construction on site.
- Construction management has become disengaged from issues of construction efficiency as, increasingly, contractors purchase products and services from their supply chains and, generally, do not engage with the detail of construction.

The Last Planner® System

- LPS® is a planning system to support the last planners who will execute the work; it helps in the creation of reliability in short-range work plans.

- LPS®-based thinking has helped case study company Boulder Associates develop a management support system to aid their design staff to improve the soundness of their work commitments, and hence the efficiency of their work.
- Lead owners are using LPS® training to help their supply chains to improve workflow reliability on their projects, and hence to drive efficiency.

Process efficiency through work structuring and work study

- As builders have become more accustomed to purchasing services and products than planning construction, and as pyramid subcontracting has become more prevalent, the industry's capacity for innovation has been stifled and efficiency has suffered.
- Work structuring is a process for planning construction work in a detailed manner to examine opportunities for improving efficiency and speed.
- Work studies have shown that at the task level, in some areas, there is the potential to double productivity and speed.

Location-based planning and Takt time planning

- Location-based planning and Takt time planning are planning approaches for smoothing and balancing workflow across an entire production system.
- In high-rise construction, all efficient builders unconsciously use location-based planning and Takt time planning to some degree, since this is the natural way to build such structures – floor by floor, trade following trade.
- The same principles of balancing work team capacity across the supply chain can be applied to all types of construction.
- In a Takt time-balanced workflow construction project, work teams follow each other, close coupled like carriages on a train, to progressively complete a building, area by area.
- As projects teams become more efficient, work can gradually be accelerated; but it is important that all work teams speed up together.
- Once a project is being run in this way, stability is created and management has more time to focus on optimization of the entire process.

Opportunities and challenges raised by BIM

- BIM offers great opportunities for enhanced communication and collaboration; logically then, to maximize the benefits of this technology, commercial terms also need to support collaboration.
- Effective BIM implementation requires a BIM Execution Plan (BEP) to set out the management framework for the deployment of BIM. This includes roles, responsibilities and milestones.
- The BEP must be a part of the contract.

Design for quality

DESIGN, INNOVATION AND IMPROVEMENT

Products, services and processes are designed both to add value to customers and to become more profitable. But leadership and management style are also designed and reflected through the use of symbols and processes and in internal communication methods and materials. Every area of an organization's operations is influenced by design decisions or the lack of them.

Design can be used to:

- gain and hold on to competitive edge;
- save time and effort;
- deliver innovation;
- stimulate and motivate staff;
- take the drudgery out of the mundane and turn it into something inspiring;
- simplify complex tasks;
- delight clients and stakeholders;
- dishearten competitors;
- achieve impact in a crowded market;
- justify a premium price; or
- simply to make money.

Design is important to every organization as it influences internal culture, as well as perceptions of the organization by others. Design is practised by many members of the supply chain, from architecture to marketing, on-site construction planning to the design of stationery. At its essence, it is a problem-solving process.

In the Collins *Cobuild English Language Dictionary*, design is defined as: 'the way in which something has been planned and made, including what it looks like and how well it works'. Using this definition, there is very little of an organization's activities that is not covered by 'planning' or 'making'. Clearly the consideration of what it looks like and how well it works in the eyes of the customer determines the success of products or services in the marketplace.

Innovation entails both the invention and design of radically new products and services, embodying novel ideas, discoveries and advanced technologies, *and* the continuous development and improvement of existing products, services and processes to enhance their performance and quality. It may also be directed at reducing costs of production or operations throughout the lifecycle of the product or service system.

All organizations need to update their products, processes and services periodically. In markets such as electronics, audio and visual goods, and office automation, new variants of products are offered frequently, almost like fashion goods. While constructed products changed relatively slowly in the past, recent decades have seen the rapid development of new materials, equipment and building systems, new technologies, a new aesthetic and a myriad of new services.

While in some markets the pace of innovation may not be as fast and furious, there is no doubt that the rate of change for products, services, technology and process design has accelerated on a broad front.

Each of the case studies at the end of this book gives examples of how the companies use design innovation to improve aspects of their services or products. The following are examples to illustrate the diversity of design innovation in the sector.

- The Graniterock case study illustrates innovation in the design and production of concrete, a traditional material designed with improved properties, innovation in delivery logistics and the use of IT to enable customers to self-serve in the purchase of aggregates outside normal business hours.
- The DPR case study illustrates innovation in the use of BIM data in the field, and innovation in the conceptualization and communication of client quality require-ments.
- The Rosendin case study illustrates the use of IT to streamline logistics and VSM to streamline the installation of light fittings.
- The Southland Industries case study illustrates the use of BIM to improve the sophistication of composite prefabricated components, thereby reducing work in the field.
- The Crossrail case study illustrates innovation in the development of the vision and implementation strategies to elevate quality performance to a similar standard as safety.
- The ConXtech case study is a classical product design innovation, where an entire building system has been developed to improve quality, safety and to accelerate construction on site.

Innovation entails both the invention and design of radically *new* products and services; it embodies novel ideas, discoveries and advanced technologies, as well as the continuous development and improvement of *existing* products, services and processes to enhance performance and quality. It may also be directed at reducing costs of production or operations throughout the lifecycle of the product or service system.

Within all industries, rapid innovation is changing every aspect of business, including the products and the services offered. These include an increase in the use of IT-based technologies in design, communication, management, manufacturing and service delivery. In addition, there are numerous examples of new technologies such as new equipment for materials handling and assembly, new materials, new financing arrangements and new procurement processes which involve risk sharing.

In many organizations innovation is predominantly either technology-led (e.g. in the design phase and the manufacture of engineered-to-order products, 3D technology and virtual reality are driving change in processes) or marketing-led (e.g. in the design of the actual end product – dictated by user preferences). What is always striking about leading product or service innovators is that their developments are *market*-led, which isdifferent from *marketing*-led. The latter means that the *marketing* function takes the lead in product and service developments. But most leading innovators identify and set out to meet existing and potential demands profitably and, therefore, are market-led and constantly striving to meet the requirements ever more effectively through appropriate experimentation.

Everything we experience in or from an organization is the result of a design decision, or lack of one. This applies not just to the tangible things like products and services, but the intangibles too: the systems and processes which affect the generation of products and delivery of services. Design is about combining function and form to achieve fitness for purpose: be it an improvement to a supersonic aircraft, a new structural type, the development of a new building material or product, a new management process, a staff incentive scheme or this book.

Of course the goal posts change, once fitness for purpose has been achieved. Events force a re-assessment of needs and expectations and customers are sure to want something different. In such a changing world, design is an ongoing activity; dynamic not static – a verb not a noun – to *design is to be part of an ever evolving process of change.*

THE DESIGN PROCESS

Commitment in the most senior management helps to build quality throughout the design process and to ensure good relationships and communication between various groups and functional areas both within the organization and across the supply chain. Designing customer satisfaction and loyalty into products and services contributes greatly to competitive success. Clearly, it does not guarantee it, because the conformance aspect of quality must be present and the operational processes must be capable of producing to the design. As in the marketing/operations interfaces, it is never acceptable to design a product, service, system or process that the customer wants, but the organization is incapable of delivering.

In the construction industry, different sectors have entirely different design processes. For example, a large company specializing in project homes as a production builder will normally have a number of standard models which it refines in response to consumer market research, changes in materials, products, equipment and on-site construction processes. In contrast, a company specializing in the design and construction of one-off, large complex projects such as hospitals, industrial facilities or laboratory buildings is faced with a one-of-a-kind production where each project is designed and built by a different supply chain team. A collaborative team culture has to be built anew for each project.

The design process often concerns technological innovation in response to or in anticipation of changing market requirements and trends in technology. Those companies with impressive records of product or service-led growth have demonstrated a state-of-the-art approach to innovation based on three principles:

1 *Strategic balance* between product and process development to ensure that product and service innovation maintains market position, while process innovation ensures that production risks in safety and quality are effectively controlled and reduced, and productivity is enhanced.
2 *Top management approach* to design to set the tone and, by visibly supporting the design effort, ensure that commitment to the process is universal. Direct control should be concentrated on critical decision points, since over-meddling by very senior people in day-to-day project management may delay and demotivate staff.
3 *Teamwork* to ensure that once projects are underway, specialist inputs (e.g. from marketing and technical experts) are fused and problems are tackled simultaneously. The teamwork should be urgent yet informal – too much formality will stifle initiative, flair and the fun of design.

The extent of the design process should not be underestimated, though it often is. Many people associate design with *styling* of products, and while this is certainly an important aspect, for certain products and for many service operations the *secondary design* considerations are vital. For example, anyone who has bought an assemble-it-yourself kitchen unit will know the importance of the design of the assembly instructions. Other aspects of design that affect quality in this way are packaging, customer-service arrangements, maintenance routines, warranty details and their fulfilment, spare-part availability, etc. In construction the quality and completeness of documentation is critical for people fabricating in the shop and constructing in the field.

Personal computers are an excellent example of an industry that has learned much about the secondary design features of its products. Much of the customer dissatisfaction experienced in this market has not been product design features but problems with user manuals, availability and loading of software and the software applications themselves.

Built infrastructure certainly falls into this category of technically complex products and service systems. The design and marketing of after-sales arrangements are an essential component of the design activity. The ease of maintenance, cleaning and servicing of built infrastructure are critical aspects of design. Of course it is equally important that the durability of structures must satisfy the client's requirements. It is critical to ensure that the design team addresses whole of life issues as well as the initial appearance and constructability of the item. This widens the management of design quality into suppliers, subcontractors and general contractors and requires everyone's total commitment.

In one-of-a-kind major projects, traditional design processes are put under pressure by the inherent complexity of large-scale projects and the time pressure under which they typically operate. On such projects, the design is inevitably undertaken by a large multidisciplinary team of design professionals working concurrently rather than in series. Generally, the team meets weekly or fortnightly to review progress, exchange documents and discuss broad-based problems. Individual problems are addressed off-line in small groups in between these regular meetings.

The flow of design work on such projects is generally highly unreliable, with individual designers struggling to keep their commitments on a regular basis. This appears to be widely accepted and, hence, this tends to be the norm. Design teams work under pressure, with members often working on several projects concurrently, and designers have to juggle their availability as best they can. Meanwhile design businesses struggle to remain profitable, and project designs are almost always behind schedule.

There is a tension between price-based purchasing, which has become the most common commercial arrangement for design in many countries, and the delivery of innovative designs, together with the timely delivery of accurate, high-quality documentation for the downstream fabricators and constructors. The service quality aspects of design are often jeopardized by the commercial arrangements. Recent research has suggested that the early, more creative phase of design development is best managed through a cost reimbursable mechanism, because a fixed cost agreement may limit the design creativity necessary to explore design options. In contrast, it suggests that in the later phases of detailing and documentation, the commercial arrangements may be managed either on a fixed price basis or a cost reimbursable basis.

However, it is interesting to note in the VNGC case study (CS2) that during the construction detailing stage, close collaboration between the constructor and the engineering designer, working through 3D modelling, was able to significantly improve constructability, and hence improve efficiency in the field. Such collaboration can only be achieved through a cost reimbursable mechanism.

Furthermore, designing constructability into the design is critical and hard to achieve. In construction, design has a much larger role than is generally recognized; for example, on site, in the process of assembling a building, the selection and positioning of equipment during construction is a design task. Similarly, the selection of technology – the decision between using cast in place or prefabricated components – is also essentially a design problem.

Designers are used to working with abstractions and, too often, construction issues are not well enough understood by them. These issues are almost entirely left up to fabricators and constructors. Generally, these latter parties are engaged well after a design has been completed, or is at a stage when design decisions have been fixed, and it is too late for construction input to be influential. To ensure that designs are efficient to construct is a critical challenge for procurement, and it is essential that the input of downstream fabricators and constructors is available early in the design process to help influence the design solution, so that it is efficient to build.

The proper design detailing of buildings and structures plays a major role in the elimination of errors, defects and waste. Correct initial design also obviates the need for costly and wasteful modifications to be carried out once construction has commenced. It is at the design stage that such important matters as variability of details, reproducibility, technical risk of failure due to workmanship, ease of use in operation, maintainability, etc. should receive detailed consideration. As most constructed projects are one-of-a-kind, the design phase is even more difficult to manage than it is in process and service industries: as on most projects the design team will not have previously worked together.

Designing

If design quality is taking care of all aspects of the customer's requirements, including cost, production, safe and easy use, and maintainability of products and services, then *designing* must take place in all aspects of:

• identifying the need (including need for change);
• developing that which satisfies the need;
• checking the conformance to the need;
• ensuring that the need is satisfied.

Designing covers every aspect, from the identification of a problem to be solved, usually a market need, through the development of design concepts and prototypes to the generation of the detailed specifications or instructions required to produce the artefact or provide the service. It is the process whereby needs are satisfied through action or in some physical form, initially as a solution, and then as a specific configuration or arrangement of materials resources, equipment and people. Design influences many operational areas of an organization as well as many strategic solutions, and although design professionals may control detailed product styling, decisions about many aspects of design involve people from many other functions within an organization. Lean quality management supports such a cross-functional interpretation of design.

In the construction environment, this broad conceptualization of the design function is essential as design impacts on every stage of the production process: safety during construction, constructability, the cost and ease of prefabrication of engineered products, and the reliable achievement of product quality on site. Design also impacts the ongoing performance of the artefact, and can shape the ongoing relationship between the provider and the end user.

Design, like any other activity, must be carefully managed. A flowchart of the various stages and activities involved in the design and development process appears in Figure 8.1.

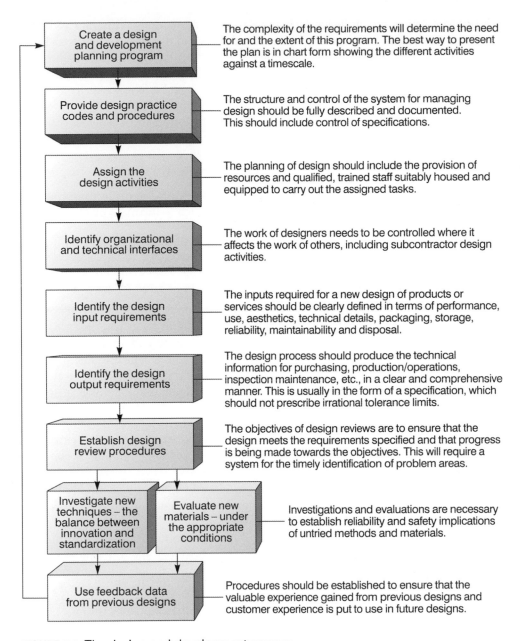

FIGURE 8.1 The design and development process

By structuring the design process in this way, it is possible to:

- control the various stages;
- check that they have been completed;
- decide which management functions need to be brought in and at what stage; and
- estimate the level of resources needed.

The control of the design process must be carefully handled to avoid stifling the creativity of the designer(s). It is clear that the design process requires a range of specialized skills. The way in which these skills are managed, the way they interact and the amount of effort devoted to the different stages of the design and development process all influence the quality, constructability and price of the service or final product. A team approach to the management of design is critical to the success of a project. The input of manufacturers, engineering fabricators and site assemblers are as crucial as the input from the end-user market.

It is never possible to exert the same tight control on the design effort as on other operational phases; yet, the cost and the time used are often substantial and both must appear somewhere within the organization's budget.

Certain features make control of the design process difficult:

1   The creation of constructed products is more design intensive than manufacturing because, generally, unless you are in a business such as the mass production of housing or a product system, each project is designed anew.
2   The design, in a holistic sense, of an artefact will never be 'complete' in the sense that, with changes in materials, manufacturing technology and equipment, some modification or improvements are inevitable.
3   Few designs are entirely novel. An examination of most 'new' products, services or processes will show that they employ existing techniques, components or systems to which novel elements have been added.
4   The more time spent on a design, the less will be the increase in the value of the design unless a technological breakthrough has been achieved. This diminishing return from the design effort must be carefully managed. However, this has to be balanced with the need for adequate design resolution and sound documentation because production risk increases when the design is not properly resolved and effectively communicated.
5   The design process is information intensive and the timing of decision-making, both by the clients and the design team, is critical to the efficiency of the entire process. It is not practical to manage the design process in the same manner as we do the production process – on the basis of tasks. For every task there may be up to ten information flows and the ratio of information flows to tasks is highly variable. Also there are a great number of concurrent and interdependent activities which need skill and experience in their effective resolution.
6   External and/or internal customers will impose limitations on design time and cost. It is as difficult to imagine a design project whose completion date is not implicitly fixed either by a promise to a customer, the opening of a trade show or exhibition, a seasonal 'deadline', a production schedule or some other constraint as it is to imagine an organization whose funds are unlimited, or a product whose price has no ceiling.

Choosing by advantages (CBA)

Because of the complexity of the design process for large built assets, a method known as CBA is gaining some popularity. There are two very real challenges in design: the first is to give everyone in the design process a voice and ensure they are heard, the second is to give credence to alternative solutions so they can be effectively considered. CBA addresses the latter challenge.

Essentially, a design team will develop a range of possible solutions. Then using the CBA method, these are evaluated and reduced in number, and the best are developed further until a decision is made as to which of these creates the greatest value for the customer.

CBA is a decision-making system that was developed through a collaboration between the US Forest Service and Utah State University. The main insight that this method provides is that decisions should be made on the basis of the differences between the key attributes of potential solutions, rather than on the importance of the specific attributes in any solution. It is by focusing on relative differences, especially in the desired attributes, that sound decisions are best made. The method does rely on making decisions regarding the relative value of different attributes and is readily facilitated so that project teams can make sound collective decisions. The method relies on making rationally based comparative judgements rather than relying on a numerical result derived through the use of weighting factors to rank attributes and preferences. An example of CBA to a design problem can be found in the Boulder Associates case study (CS1).

Set-based design

Another approach to design problem solving, which is gaining currency, is set-based design. In the early stages of design, a very large number of options – some technical, some market influenced and others perhaps financial – all need to be considered concurrently in arriving at the initial design schema which will be carried forward to detailed design and construction. During the preliminary design phase, there is a unique opportunity for trade-offs among individual performance criteria.

It has been found in a large number of different industries, where complex products are being routinely designed, that using traditional approaches, the design team has converged to a single solution too early, before all the options have been comprehensively considered. This has led to the development of sub-optimal designs which are then not built. Convergence to a single design relatively early has generally come to be known as point-based design.

In general, point-based design strategies consist of five basic steps (Likert *et al.*, 1996):

1   First the problem is defined.
2   Engineers generate a large number of alternative design concepts, usually through individual or group brainstorming sessions.
3   Engineers conduct preliminary analyses on the alternatives, leading to the selection of a single concept for further development.
4   The selected concept is further analysed and modified until all the product's goals and requirements are met.
5   If the selected concept fails to meet the stated goals, the process begins again either from step 1 or step 2, until a solution is found.

One step beyond a point-based design approach is concurrent engineering (CE). In CE, a single initial design solution is adopted, and then the various designers work in parallel. Often, as a part of CE, the design team is co-located and this increases the opportunity for improved communication and collaboration. Co-location generally shortens the design process, creates improved conditions for innovation and reduces errors because of closer and more effective collaboration. Nowadays, on major complex construction projects, CE and co-location are often implemented.

During the International Automobile Study, the findings of which were published in *The Machine that Changed the World* (Womack *et al.*, 1991), the Toyota design process is characterized as being world-class. In a subsequent study of Toyota's design process at the University of Michigan, Ward and his colleagues (1995) describe the four main features of this design process as follows:

1   Broad sets of design parameters are defined to allow concurrent design to begin.
2   These steps are kept open longer than typical to more fully design trade-off information.

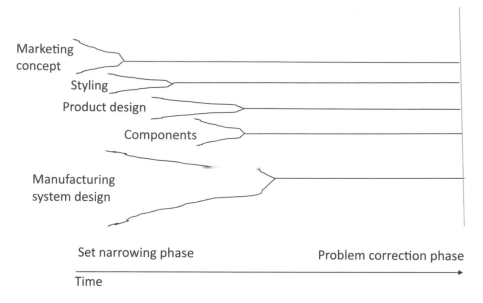

FIGURE 8.2 Parallel set narrowing process as sketched by a Toyota manager
(Ward *et al.*, 1995)

3   The sets are gradually narrowed until a more globally optimum solution is revealed
     and refined.
4   As the sets narrow the level of detail (or design fidelity) increases.

This was illustrated in a sketch produced by a Toyota manager and is reproduced in
Figure 8.2. Alan Ward characterized this design approach as set-based design. It differs
from point-based design in that design options across a broad set of parameters are kept
open for longer, allowing a broader range of solutions and trade-offs to be considered by
the team.

A detailed description of the Toyota process and culture can be found in James Morgan
and Jeffrey Liker's (2006) book, *The Toyota Product Development System: Integrating People,
Process and Technology*. Toyota's approach emphasizes the need to front load the product
development process to allow alternative solutions to be thoroughly explored, while there
is still maximum flexibility in the design process and in the solution generation space. An
approach based on this thinking is increasingly being applied to the design processes on
major, complex construction projects.

Modelling in the design process

In designing built products, one of the greatest challenges is to effectively engage all the
key stakeholders in a productive discussion in the design process. Many people cannot
read two-dimensional plans, and this limits their ability to fully understand the design
and engage with the design process. Up until 30 years ago, young architects were taught
to build models of their designs so that they could more effectively communicate with
their clients and other stakeholders. Thus 3D models from cardboard and balsa wood
helped to facilitate communication.

In recent years, with the development of 3D modelling, realistic representations of the
design could be generated and viewed on a computer screen. Up until ten years ago these

images were virtual reality simulations: however, more recently, they are 3D images of the actual building information model (BIM). This model is collaborative created by the design team as it designs and details the project.

The DPR case study (CS9) shows the federated model of an entire hospital made available on iPads to the entire construction team. The screen viewed on the iPad is location and direction sensitive, so that the image on the screen shows what the iPad is being pointed at. The model is built up in layers so that the general view can be seen, or the structure behind it, or the services behind that. Everything bigger than 1 cm is included in the model and, depending on the view, its position is dimensioned on the screen. The information is sufficient for foremen and trade workers to build from.

The Boulder Associates case study gives an example of a design team, working with its key stakeholders through the medium of a full sized, cardboard cut-out model of the proposed spaces being designed. The Boulder Associates team typically runs a one-week workshop with the key stakeholders for the design of a new health facility. Rooms together with all the fittings are cut out and assembled in full size out of cardboard. Stakeholders are able to experience the layout of the space, the positioning of equipment and the functionality of their use of the space at full scale. Boulder Associates find that this enables them to understand and interpret their clients' and stakeholders' needs much more effectively than any other technique. The time taken in building the full-scale models is more than compensated for by the time saved in the design and documentation process.

Total design processes

Design quality is about far more than the product or service design and its ability to meet the customer requirements. It is also about the activities of design and development. The appropriateness of the actual *design process* has a profound influence on the performance of any organization, and much can be learned by examining successful companies and how their strategies for research, design and development are linked to the efforts of marketing and operations. In some quarters this is referred to as 'total design', and the term 'simultaneous engineering' has been used. This is an integrated approach to a new product or service introduction, similar in many ways to Quality Function Deployment (QFD – see next section) in using multifunction teams or task forces to ensure that research, design, development, manufacturing, purchasing, supply and marketing all work in parallel from concept through to the final launch of the product or service into the marketplace, including servicing and maintenance.

Quality function deployment (QFD) – the house of quality

Quality function deployment (QFD) is a process for translating the requirements of the customer into language that is clear and definite for the technical design teams. Many years ago in the manufacturing sector, it was recognized that consumer terms used to describe the attributes of products were too ambiguous to be of use to the technical design team. QFD was developed as an approach to translating meaning from common language to the technical language of material properties and product attributers. In the DPR case study (CS9) the approach taken in developing the company's quality management system is conceptually similar to QFD. The DPR team developed an approach in which the client's requirements are defined as a set of 'Key Distinguishing Features'. Once these are agreed, these characteristics are translated into 'Key Distinguishing Features' for each trade package. This defines, for each trade area, what it must do to meet the client's requirements.

The 'house of quality' is the framework of an approach to design management known as quality function deployment (QFD). Dr Yoji Akao originally developed QFD in Japan in 1966, combining his work on quality assurance and quality control with function deployment as used in value engineering. It was first deployed at Mitsubishi's Kobe shipyard in 1972, but it has been developed in numerous ways by Toyota and its suppliers, and many other organizations. The house of quality (HoQ) concept, initially referred to as quality tables, has been used successfully by medium density housing providers, manufacturers of integrated circuits, synthetic rubber, construction equipment, engines, home appliances, clothing and electronics. The process has been used by leading companies around the world. In Japan, its design applications include public services and retail outlets. The application of QFD in construction is limited to companies that have specialized in a specific market sector; for example, in Japan, Turkey and Brazil it has been used to design and market apartments.

The activities included in QFD are:

1   market research;
2   basic research;
3   innovation;
4   concept design;
5   prototype testing;
6   final-product or service testing; and
7   after-sales service and trouble-shooting.

These are performed by people with different skills in a team whose composition depends on many factors, including the products or services being developed and the size of the operation. In many industries, such as cars, video equipment, electronics and computers, engineering designers are seen to be heavily into 'designing'. But in other industries and service operations designing is carried out by people who do not carry the word 'designer' in their job title. Particularly in the services sector, the failure to recognize the necessary design inputs of 'non-designers' and the failure to provide them with training and support will limit the success of the design effort.

The QFD team in operation

The first step of a QFD exercise is to form a cross-functional QFD team. Its purpose is to take the needs of the market and translate them into such a form that they can be satisfied within the operating unit and delivered to the customers.

As with all organizational problems, the structure of the QFD team must be decided on the basis of the detailed requirements of each organization. One thing, however, is clear, close liaison must be maintained at all times between the design, marketing and operational functions represented in the team.

The QFD team must answer three questions – **WHO**, **WHAT** and **HOW**.

WHO are the customers?
WHAT does the customer need?
HOW will the needs be satisfied?

WHO may be decided by asking: 'Who will benefit from the successful introduction of this product, service or process?' Once the customers have been identified: 'WHAT can be ascertained through interview/questionnaire/focus group processes, or from the knowledge and judgement of the QFD team members?' HOW is more difficult to determine and will consist of the attributes of the product, service or process under development. This will constitute many of the action steps in a QFD strategic plan.

WHO, WHAT and HOW are entered into a QFD matrix or grid of 'house of quality' (HoQ), which is a simple 'quality table'. The *WHAT*s are recorded in rows and the *HOW*s are placed in the columns.

The house of quality provides structure to the design and development cycle, often likened to the construction of a house, because of the shape of matrices when they are fitted together. The key to building the house is to focus on the customer requirements so that the design and development processes are driven more by what the customer needs than by innovations in technology. This ensures that more effort is used to obtain vital customer information. It may increase the initial planning time in a particular development project, but the overall time, including design and redesign, taken to bring a product of service to the market will be reduced.

This requires that marketing people, design staff (including architects and engineers) and production/operations personnel work closely together from the time the new service, process or product is conceived. It will need to replace, in many organizations, the 'throwing it over the wall' approach, where a solid wall exists between each pair of functions (Figure 8.3).

The HoQ provides an organization with the means for inter-departmental or inter-functional planning and communications, starting with the so-called customer attributes (CAs). These are phrases customers use to describe product, process and service characteristics.

A complete QFD project will lead to the construction of a sequence of house of quality diagrams, which translate the customer requirements into specific operational process steps. For example, the 'feel' that customers like on the steering wheel of a motor car may translate into a specification for 45 standard degrees of synthetic polymer hardness, which in turn translates into specific manufacturing process steps, including the use of certain catalysts, temperatures, processes and additives. Similarly in construction, the acoustic privacy that home-owners want is translated into a measurable decibel transfer rate and specific construction systems to achieve it. The cycle of QFD leads to a consideration of the product as a whole and then subsequent steps will consider the individual components. For example, in the design of an apartment block, the overall building would be the starting point, but subsequent QFD exercises would tackle the kitchen, bathroom, bedrooms and lounge-dining room in detail. Each of the areas would have specific customer

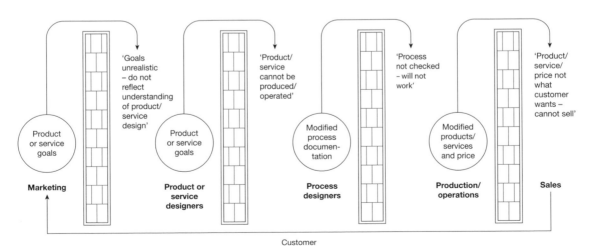

FIGURE 8.3 Throw it over the wall

requirements but these would all need to be compatible with the overall service concept. An example of the application of this approach to apartment design in Brazil has been described in some detail by Alarcon (Alarcon and Mardones, 1998). For further detail about QFD, refer to Chapter 6 of *TQM and Operational Excellence* (Oakland, 2014).

Target value design

Elemental cost planning was first promoted by the UK Ministry of Education in its *Building Bulletin No. 4* in 1951; this introduced the concept of cost planning for the primary elements of school buildings. Elemental cost planning relies on the adoption of a standard format of cost analysis for buildings, and allows costs to be compared on the basis of a common format. This then provides the basis for the benchmarking of costs for similar buildings. The purpose of elemental cost planning is described as:

- creating the basis for a realistic estimate of costs during the design stage;
- ensuring that the money available for a project is allocated consciously and economically to the various components of the building;
- relying on the measurement and pricing of approximate quantities; and
- aiming to help achieve good value at the desired price level.

Elemental cost planning is often referred to as 'designing to a cost' or 'target cost planning', because a cost plan is determined even before the design commences. Throughout the design process, the design team is focused on the achievement of the client's needs at the planned cost.

The building cost information service known as BCIS is a part of the Royal Institution of Chartered Surveyors, and it is the leading provider of cost and price information to the UK construction industry. The BCIS carries out statistical analysis of the UK construction industry as a basis for all early cost advice and elemental cost planning. The institution has been building its cost database since the 1950s.

However, in the intervening period, procurement practices in the construction sector have changed significantly. Builders have become more used to buying and selling, rather than designing and making, and hence, the historic data has become less and less relevant. The organizations who have had the greatest interest in maintaining strong cost planning databases are owners of specialist buildings such as schools, companies with a large stock of similar buildings such as Intel, McDonald's and Caltex, and organizations who build and operate hospitals and healthcare facilities.

The application of Target Costing in the United States construction industry was reported within the lean construction community by Ballard and Reiser (2004). They reported on a Boldt construction project, the St Olaf Fieldhouse project. The term Target Value Design (TVD) was coined later, and it appropriately reflects the focus on value in managing the design process to meet the client's needs within its budget. It also reflects the shift from cost management post design to managing costs within the design stage, at a time when the team is most able to influence cost.

The basic idea that underpins TVD is the setting of a well-constructed target cost. This puts artificial pressure on the design team to dig deep and be as creative as possible within their design process while working to a target cost. The important distinction between traditional cost planning and TVD is that cost planning uses historical costs to inform the design team as to the likely cost of the design. This is essentially a feed-back system based on historical costs, whereas TVD is a feed-forward system where high-quality construction cost information is developed by the construction and fabrication teams and is used to refine the design to achieve the cost outcomes sought by the client.

TVD in the construction industry is a team effort; it involves all the design disciplines and all the fabricators and constructors in working together, subsystem by subsystem, to guide the overall construction cost towards the client's target. Refer to the VNGC case study (CS2) where the team worked over a ten-month period to bring the cost down to the client's business case budget. Naturally such an approach is easier to implement when the whole supply chain is working as a cohesive unit, focused on optimizing the project and not their own individual positions within it. The VNGC team was working under an IFOA contract, which saw the entire team co-located and working together as a seamless unit.

There are two primary challenges to the adoption of TVD in the construction industry. The first is that good quality cost information is necessary so that a realistic target cost plan can be created to guide the design process. The second challenge is that in order to work collaboratively, the entire team needs to be working under commercial terms that encourage collaboration and innovation rather than commercial terms which tend to lock the individual service providers into silos, responsible for risk management and profit maintenance to the exclusion of all else.

A cardinal rule in the implementation of TVD is that only the customer can change the target scope, cost, quality or schedule. It is the team's task to work within that rule, making trade-offs between parts of the project with the approval of the customer, but not moving beyond the overall envelope that the customer has defined.

DESIGN MANAGEMENT WITH ADePT

The Analytical Design Process (ADePT) developed by Simon Austin and his colleagues at Loughborough University in the UK is a powerful, complex problem-solving tool that is ideal for the management of the design process. Design is much more complex to manage than construction, yet relatively little research or development has been undertaken in the construction sector on issues of design management. Design is iterative and many decisions and tasks are undertaken concurrently.

Current practice is to consider the design process in terms of the timing of deliverables (drawings and specifications) which are scheduled in a similar manner to construction tasks; yet this does not recognize the fact that the actual design progress is determined by the flow of information and decisions rather than of documents. Hence, while network analysis and CPM are generally used to plan and schedule design work on large to medium-sized projects, these tools do nothing to recognize the ill-defined design process and cannot represent its iterative nature. The ADePT approach consists of four basic steps (the first three being discussed further) (Figure 8.4):

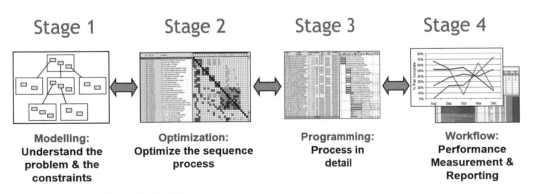

| Stage 1 | Stage 2 | Stage 3 | Stage 4 |

Modelling:
Understand the problem & the constraints

Optimization:
Optimize the sequence process

Programming:
Process in detail

Workflow:
Performance Measurement & Reporting

FIGURE 8.4 Overview of ADePT process

1   a work breakdown structure-based model of the building process is created and information flows between design tasks are modelled;
2   a Dependency Structure Matrix (DSM) prioritizes design decisions;
3   a design schedule is produced and integrated with the construction programme; and
4   the project is managed and controlled against the schedule which is constantly reviewed and progress monitored using the Last Planner concept.

Modelling the design process

First of all, Austin and his colleagues developed a modification of the Integrated Manufacturing (IDEF) Definition technique (refer to Chapter 12), called IDEF-0v to support their Analytical Design Planning Technique (ADePT). In Figures 8.5 and 8.6 controls and mechanisms have been replaced to make explicit the three sources of information input.

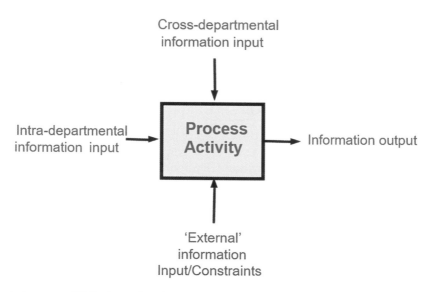

FIGURE 8.5 IDEF view of a design activity

Information flows between activities are modelled and the actual transformation of work within activities is simply assumed. It is noteworthy that the number of information flows between tasks vary widely within any process. Rarely, there is a single information flow in and out. It is more common to have several in and several out with no regular relationship between the numbers of incoming and outgoing information flows. Whilst such graphical representations have their merits they can be time-consuming to create and are not essential – an information dependency table mapping inputs and their source tasks for every design activity is sufficient.

Every information flow is prioritized according to a three-point scale from critical (3) to 'nice to have'. This scale is based on three parameters:

1   its importance or 'strength of dependency';
2   the sensitivity of subsequent activities to changes in the information; and
3   the ability to create reliable estimates for the information.

Expert judgement is required to define these classifications.

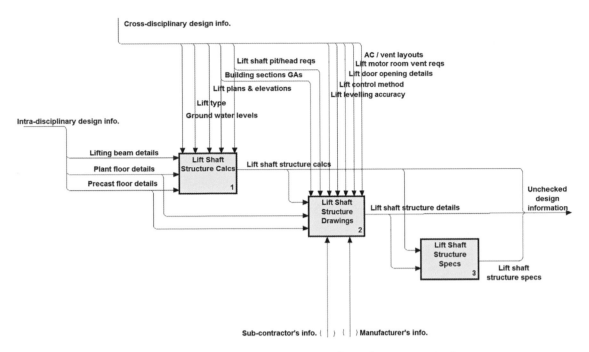

FIGURE 8.6 An example of a design process diagram for lift shaft structure design

Information flows need to be defined for every project; however, generic process models have been developed and it has been found that for buildings, for example, over 90 per cent of the required process was contained in the generic model. The same will apply to other types of projects. The building model has a hierarchical or work breakdown structure as illustrated in Figure 8.7. The first level subdivides the process into design undertaken by the professional disciplines, and then breakdowns into the building, sub-systems and components. In other project management applications the process can be divided by function or system.

The project planning for a particular building will require some modification of the generic process model by the consultant team. Redundant sections are deleted, some added and others altered.

Optimizing the process using the Dependency Structure Matrix

In the 1960s, Steward developed a theory that a complex problem such as design could be solved more efficiently by representing the interrelationships between activities in the form of a Design Structure Matrix (Steward, 1965) or Dependency Structure Matrix (DSM).

Figure 8.8 shows a matrix for a very simple design problem: activities are listed down the left hand side of the matrix in the order they are planned to be undertaken and in the same order across the top. The rows in turn represent each activity and the columns the precedent ones. Information flows (dependencies) are shown by placing a 1, 2 or 3 where the row (activity) intersects the relevant column (precedent activity) to indicate the criticality of each information flow. This then logs all the information flows within the process and notes the criticality of each. Dependencies noted below the diagonal indicate that the particular activity is dependent on a previous activity, whereas dependencies

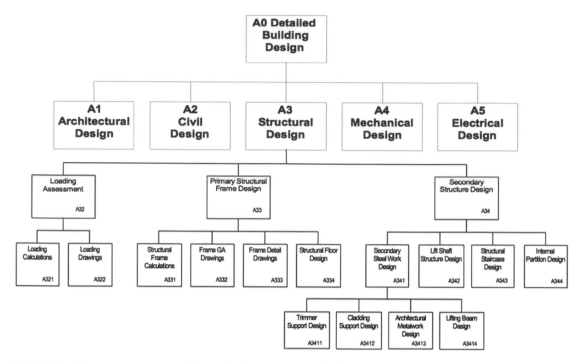

FIGURE 8.7 The design process hierarchy for structural design

above the diagonal indicate dependency on information that has yet to be produced. Critical dependencies above the diagonal will hold up progress, and hence, relevant tasks that provide information for such critical tasks will have to be brought forward or, as a last resort, the information will have to be assumed and subsequently checked. Such assumptions create risk and should be logged in a risk register and reviewed at the earliest time.

Early in his career, Marton Marosszeky was set the task of designing the structure for a 30-storey building, however, there was still a building on the site, and so geo-technical advice was based on the immediately adjoining vacant site. After demolition it was found that the foundation conditions were worse than anticipated and the entire structure had to be redesigned to create a more flexible structure, one that could absorb the predicted differential movements. In fact it would have been possible though expensive and inconvenient to drill from the basement of the existing building so that the design could have been based on known information rather than assumptions.

The need for some assumptions can be eliminated by reordering the activities within the matrix so that all critical dependencies are below the diagonal or as close to it as possible, thus producing the optimum sequence. This maximizes the availability of information, and minimizes the amount of wasteful iteration and rework.

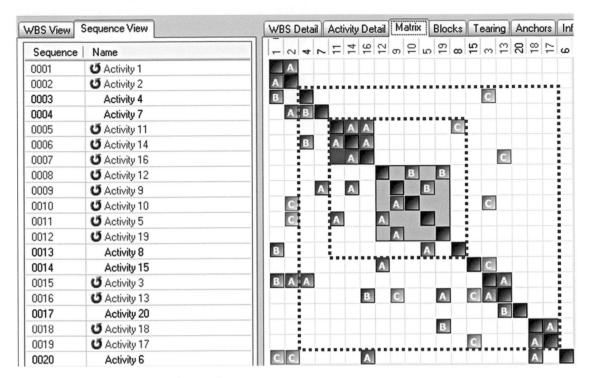

FIGURE 8.8 Optimized example matrix

Optimizing a matrix identifies the interdependent activities that are within an iterative block and the block's location in the overall order. In real projects the iterative blocks can be very large, including hundreds of activities and the challenge of reducing them is considerable; however, the effort is well worthwhile as it creates a significantly more efficient design process.

The other strategy for reducing the amount of necessary iteration to a minimum is to declassify critical information by making decisions early, by building contingency into the design or by finding a way of creating sufficiently reliable estimates. Strategies might include increasing the expertise in the design team by bringing in an additional expert. It might also involve getting all the experts in the team to work together to identify those decisions in the design process that must be made in order to improve the process flow.

Austin and his team have developed software (ADePT Design Builder and Design Manager) to support this process of optimizing the decision-making process. This helps to identify the declassifications that have the greatest effect on reducing the size of the block and, hence, the scale of iteration. Within this modelling process each block of activity can be hierarchically subdivided to show finer detail on another diagram, ensuring that any single diagram does not become too cumbersome. Information flows for each discipline area are distinguished because they may require different management approaches.

Creating the integrated design schedule (programme)

Once the order of design activities is optimized on the basis of information flows and decisions, the design schedule can be represented in the form of a traditional bar chart (Figure 8.9). The software can download the optimized sequence of activities into proprietary planning software to produce a typical programme.

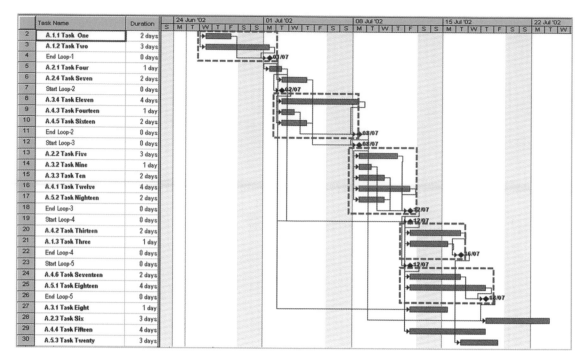

	Task Name	Duration
2	A.1.1 Task One	2 days
3	A.1.2 Task Two	3 days
4	End Loop-1	0 days
5	A.2.1 Task Four	1 day
6	A.2.4 Task Seven	2 days
7	Start Loop-2	0 days
8	A.3.4 Task Eleven	4 days
9	A.4.3 Task Fourteen	1 day
10	A.4.5 Task Sixteen	2 days
11	End Loop-2	0 days
12	Start Loop-3	0 days
13	A.2.2 Task Five	3 days
14	A.3.2 Task Nine	1 day
15	A.3.3 Task Ten	2 days
16	A.4.1 Task Twelve	4 days
17	A.5.2 Task Nighteen	2 days
18	End Loop-3	0 days
19	Start Loop-4	0 days
20	A.4.2 Task Thirteen	2 days
21	A.1.3 Task Three	1 day
22	End Loop-4	0 days
23	Start Loop-5	0 days
24	A.4.6 Task Seventeen	2 days
25	A.5.1 Task Eighteen	4 days
26	End Loop-5	0 days
27	A.3.1 Task Eight	1 day
28	A.2.3 Task Six	3 days
29	A.4.4 Task Fifteen	4 days
30	A.5.3 Task Twenty	3 days

FIGURE 8.9 Bar chart for the example

Conventional project management software represents sequential processes and does not allow elements of work containing iteration to be scheduled. Thus, feedback is not identified, resulting in coordination failures and rework both during design development and in production. The output from the DSM is entered in a way that incorporates the iteration within the process. This is done by grouping tasks that form a block under a 'rolled up' activity and removing interrelationships from within the loop so that they can be programmed to occur in parallel. The group's relationships with previous and subsequent tasks remain. The overall duration of the group of tasks must allow for the necessary information exchanges to achieve coordination.

Whilst the end result, a bar chart, looks the same as that of conventional systems, the way you get there with ADePT is fundamentally different. Full account has been taken of the complex and interdependent nature of the design process.

The overall process delivers a number of clear benefits. The careful analysis of information flows and the classification of decisions require close consultation among the design team and this creates an integrated team approach. The final schedule gives clear guidance as to the organization of the design process: indicating when the whole team must meet to resolve critical interdependent decisions. It also indicates when the logical milestones occur from a decision flow perspective, and provides a basis for cleanly integrating design and construction with transparent decision points and hold points. For more detail about the ADePT process refer to Austin *et al.* (2005) or contact ADePT at *www.adeptmanagement.org*.

STANDARDIZATION VS DEVELOPMENT

In Chapter 3, principle 2.1.1 states: *standardize products and processes to improve quality, reduce stress and drive efficiency*. Standardizing work lays the foundation for quality achievement as well as production improvement.

There is a strong relationship between standardization and work specification. To ensure that a product or service is *standardized* and may be repeated a large number of times in exactly the manner required, *specifications* must be written so that they are open to only one interpretation. The requirements, and therefore the various dimensions of lean quality, must be built into the design specification.

There are many examples in the construction sector of buildings such as McDonald's and Pizza Hut stores, service stations, Formula One hotels and project homes where many features of the design are standardized. However, in the one-of-a-kind construction sector for large projects, the opposite is the case. Architects are given free rein in terms of finishes and details, and structural engineers and services engineers follow suit by designing one-off bespoke solutions throughout the project.

This means, for example, that a structural engineer will not adopt standard reinforcing details at column junctions, even though the construction team would prefer it. Nor will the engineer select one bolt size for structural steel connections, though on site this would make the work easier and less prone to error. There seems to be little attention given to standardization at this level; yet, there is no doubt in any constructor's mind that the standardization of details would reduce errors on site and increase productivity. Small savings in material quantities as a result of design optimization are more than offset by increases in construction cost and a slowing of the project. The same thinking applies to all of the subsystems of the building: the building facade, the electrical services, and the hydraulic and mechanical services, to name but a few.

An outstanding example of standardization occurred in the reinforcement detailing of the concrete pedestrian access tunnels on the Heathrow Terminal 5 project. The reinforcing detail for the access tunnels under the runways, of which there were several kilometres, was developed and refined through virtual prototyping.

Lean quality also addresses the question of standard work for people at all levels of an organization. The idea of standard work is that planned patterns of work create stability and ensure that all issues get dealt with in a regular, planned manner. On many construction projects and in design offices, the opposite holds true. Work tends to be reactive to the many pressures from concurrent projects, and time is rescheduled and rescheduled again, to fit in the unexpected. The approach taken in standard work is that essential tasks and meetings are scheduled and cannot be changed, time is allocated to react to demands that emerge each day so that these issues do not displace planned standard work. Such an approach to work planning has the effect of steadying the habits of the organization. Commitments for meetings are kept, and everyone can rely on them happening, and rely on everyone being present. Stabilizing meetings, site inspections and reviews, and ensuring that the key participants are present has the effect of improving communication and reflects the importance of reliable behaviour to everyone in the organization.

There are national and international standards which, if used, help to ensure that specifications will meet certain accepted criteria of technical or managerial performance, safety, etc. Standardization does not guarantee that the best design or specification is selected. It may be argued that the whole process of standardization slows down the rate and direction of technological development, and it affects what is produced. If standards are used correctly, however, the process of drawing up specifications should provide opportunities to learn more about particular innovations and to change the standards accordingly.

These ideas are well illustrated worldwide by the construction sector's approach to the adoption of performance-based specifications wherever possible. Performance-based standards encourage innovation in keeping with measurable and transparent technical requirements. This allows the opportunity for manufacturers with new products and innovative solutions to have their ideas accredited and gain market entry. In areas like waterproofing, however, a building contractor might prefer to be very prescriptive in specifying the precise technical solution they want. This is a particularly important area of construction where, based on everyday experience, we know that the risk of failure is high and its consequences – water leaking through the roof or out of a bathroom into adjoining rooms – is simply unacceptable. In such areas, correct design and implementation is critical for the general contractor if they are to manage an important area of risk.

It is possible to strike a balance between innovation and standardization; however, a sound approach to innovation clearly recognizes the areas of design innovation that add value for the customer and the areas of standardization that reduce risk in the production process. Clearly, it is desirable for designers to adhere where possible to past-proven materials and methods, in the interests of reliability, maintainability and variety control. Hindering designers from using recent developed materials, components or techniques, however, can cause the design process to stagnate. A balance must be achieved by the analysis of materials, products and processes proposed in the design, against the background of their known reproducibility and reliability. If breakthrough innovations are proposed, then analysis or testing should be indicated and their adoption in preference to the established alternatives justified objectively.

It is useful to define a specification. The International Standards Organization (ISO) defines it in ISO 8402 (1986) as 'The document that prescribes the requirements with which the product or service has to conform'. A document not giving a detailed statement or description of the requirements to which the product, service or process must comply cannot be regarded as a specification, and this is true of much sales literature.

The specification conveys the customer requirements to the supplier to allow the product or service to be designed, engineered, produced or operated by means of conventional or stipulated equipment, techniques and technology. The basic requirements of a specification are that it gives the:

- performance requirements of the product or service in measurable terms;
- parameters – such as dimensions, concentration, turn-round time – which describe the product or service adequately (these should be quantified and include the units of measurement);
- materials to be used by stipulating properties or referring to other specifications;
- method of production or delivery of the service;
- inspection/testing/checking requirements; and
- references to other applicable specifications or documents.

To fulfil its purpose, the specifications must be written in terminology that is readily understood, and in a manner that is unambiguous and so cannot be subject to differing interpretation. This requires expertise and knowledge.

It is in relation to the clear communication of process specifications that the use of 3D and virtual reality (VR) technologies are showing great potential. At many stages of the design and construction process, complex information has to be communicated to the partners in the supply chain or to customers and their design teams. Often, end-clients and other stakeholders are not be able to conceptualize the design elements of a project; however, through the use of visualization tools their ability to interact with the design team is greatly enhanced. In other instances, the process design and detailing of parts of a structure can be very complex and VR simulation can assist both in optimizing the process through virtual prototyping and then in communicating the process to the people executing the work.

Good specifications are usually the product of much discussion, deliberation and sifting of information and data, and may represent tangible output from a QFD team.

QUALITY IN THE SERVICE SECTOR[1]

The emergence of the services sector has been suggested by economists to be part of the natural progression in which economic dominance changes first from agriculture to manufacturing and then to services. It is argued that if income elasticity of demand is higher for services that it is for goods, then as incomes rise, resources will shift towards services. The continuing growth of services verifies this, and is further explained by changes in culture, fitness, safety, demography and life styles.

In considering the design of services it is important to consider the differences between goods and services. Some authors argue that the marketing and design of goods and services should conform to the same fundamental rules, whereas others claim that there is a need for a different approach to services because of the recognizable differences between the goods and services themselves.

Constructed products are interesting in that where a customer is involved with the process from design through to construction, they experience the process as a service and the output as a product. Furthermore, the quality of the service will influence their perception of the quality of the product. Great service will make them accepting of minor quality errors, whereas poor service will expand even tiny quality problems into ones of monumental proportion, creating obstacles to payment and the complete loss of goodwill.

Furthermore, throughout the extremely fragmented construction supply chain, the achievement of successful outcomes relies on excellence in service quality among all the partners in the process. The focus on service to immediate customers (see Chapter 1) is essential, whether we are looking at the relationships within the design process or the relationships among subcontract suppliers on site. An interesting aspect of service within the site construction process is that while the contractual relationships of subcontractors are all with the general contractor, their service relationships are with preceding and sub-sequent trades. This is in marked contrast to manufacturing where subcontract suppliers deliver their fabricated components to the lead manufacturer's facility where they are integrated into the final product by the employees of the lead manufacturer. In construction, the relational inconsistency between contractual links and production process flow may have been an obstacle to managers in subcontracting and in general contracting organizations recognizing the importance of service quality among the subcontractors within the process.

In terms of design, it is possible to recognize three distinct elements in the service package – the physical elements or facilitating goods, the explicit service or sensual benefits, and implicit service or psychological benefits. In addition, the particular characteristics of service delivery systems may be itemized.

- Intangibility.
- Simultaneity.
- Heterogeneity.

It is difficult, if not impossible, to design the intangible aspects of a service, since consumers often must use experience or the reputation of a service organization and its representatives to judge quality.

Simultaneity occurs because the consumer must be present before many services can take place. Hence, services are often formed in small and dispersed units, and it is difficult to take advantage of economies of scale. The rapid developments in computing and communications technologies are changing this in sectors such as banking, but contact continues to be necessary for many service sectors. Design considerations here include the environment and the systems used. Service facilities, procedures and systems should be designed with the customer in mind, as well as the 'product' and the human resources. Managers need a picture of the total span of the operation, so factors which are crucial to success are not neglected. This clearly means that the functions of marketing, design and operations cannot be separated in services, and this must be taken into account in the design of the operational controls, such as the diagnosing of individual customer expectations. A QFD approach here may be helpful to analyse the process in more detail when dealing with a standard product.

Heterogeneity of services occurs in consequence of explicit and implicit service elements relying on individual preferences and perceptions. Differences exist in the outputs of organizations generating the same service, within the same organization, and even the same employee on different occasions. Clearly, unnecessary variation needs to be controlled, but the variation attributed to estimating, and then matching, the consumers' requirements is essential to customer satisfaction and loyalty and must be designed into the systems. This inherent variability does, however, make it difficult to set precise quantifiable standards for all the elements of the service.

In the design of services it is useful to classify them in some way. Several sources from the literature on the subject help us to place services in one of five categories:

- Service factory;
- Service shop;
- Mass service;
- Professional service;
- Personal service.

Several other service attributes have particular significance for the design of service operations:

1 *Labour intensity* – the ratio of labour costs incurred to the value of assets and equipment used (people vs equipment-based services).
2 *Contact* – the proportion of the total time required to provide the service for which the consumer is present in the system.
3 *Interaction* – the extent to which the consumer actively intervenes in the service process to change the content of the service; this includes customer participation to provide information from which needs can be assessed, and customer feedback from which satisfaction levels can be inferred.
4 *Customization* – which includes *choice* (providing one or more selections from a range of options, which can be single or *fixed*) and *adaptation* (the interactions process in which the requirement is decided, designed and delivered to match the need).
5 *Nature of service act* – either tangible, i.e. perceptible to touch and can be owned, or intangible, i.e. insubstantial.
6 *Recipient of service* – either people or things.

Table 8.1 gives a list of some typical construction sector services with their assigned attribute types.

TABLE 8.1 A classification of selected services

Service	Labour Intensity	Contact	Inter- action	Custom- isation	Nature of act	Recipient of service
Architect	High	Low	High	Adapt	Intangible	Things
Cleaning firm	High	Low	Low	Fixed	Tangible	People
Eng'g design	High	Low	High	Adapt	Intangible	Things
Equip. hire	Low	Low	Low	Choice	Tangible	Things
House Manuf.	Low	Low	High	Adapt	Tangible	Things
S/cont Labour	High	Low	Low	Adapt	Tangible	Things
Maintenance	Low	Low	Low	Choice	Tangible	Things
Mgt. consult.	High	High	High	Adapt	Intangible	People
Nursery	High	Low	Low	Fixed	Tangible	People
Product Man.	Low	Low	Low	Fixed	Tangible	Things
Repair firm	Low	Low	Low	Adapt	Tangible	Things
Subcontractor	Low	Low	Low	Adapt	Tangible	Things

It is apparent that services are part of almost all organizations and not confined to the service sector. What is clear is that the service classifications and different attributes must be considered in any service design process.

In the design of services it is useful to classify them in some way. The authors, in their research are working with the SERVQUAL assessment tool (Parasuraman *et al.*, 1988, 1991, 1994). This has been used in a study of the relationship between service quality and customer perceptions of product quality, and is currently being used to research service quality between subcontractors and between the general contractor and subcontractors within the construction process on building sites.

Parasuraman's five dimensions are:

- *Reliability*: Ability to perform the promised service dependably and accurately.
- *Responsiveness*: Willingness to help customers and provide prompt service.
- *Assurance*: Knowledge and courtesy of employees and their ability to inspire trust and confidence.
- *Empathy*: Caring, individualized attention the firm provides to its customers.
- *Tangibles*: Physical facilities, equipment, and appearance of personnel.

As a part of their work Parasuraman and his co-researchers developed a generic survey instrument and this is widely recognized as an excellent tool for measuring *service quality*. SERVQUAL scores *service quality* using 22 standardized statements to canvass customer views on the dimensions of *service quality*. Statements from the instrument are shown in Table 8.2.

Responses to these questions using a 9-point Likert scale are used to enable customer satisfaction to be assessed and benchmarked.

FAILURE MODE, EFFECT AND CRITICALITY ANALYSIS (FMECA)

In the design of products, services and processes it is possible to determine possible modes of failure and their effects on the performance of the product or operation of the process or service system. Failure mode and effect analysis (FMEA) is the study of potential failures to determine their effects. If the results of an FMEA are ranked in order of seriousness, then the word CRITICALITY is added to give FMECA. The primary objective of a FMECA is to determine the features of product design, production or operation and distribution

TABLE 8.2 SERVQUAL survey statements

Reliability	1.	Providing service as promised
	2.	Dependability in handling customers' service problems
	3.	Performing services right the first time
	4.	Providing services at the promised time
	5.	Maintaining error-free records (e.g. financial)
Responsiveness	6.	Keeping customers informed of when services will be performed
	7.	Prompt service to customers
	8.	Willingness to help customers
	9.	Readiness to respond to customers' requests
Assurance	10.	Instilling confidence in customers
	11.	Make customers feel safe in their transactions
	12.	Being consistently courteous
	13.	Having the knowledge to answer questions
Empathy	14.	Giving customers individual attention
	15.	Dealing with customers in a caring fashion
	16.	Having the customers' best interests at heart
	17.	Understanding the needs of their customers
	18.	Convenient business hours
Tangibles	19.	Modern equipment
	20.	Visually appealing facilities
	21.	Having a neat, professional appearance
	22.	Visually appealing materials associated with the service

(Parasuraman *et al.*, 1994)

that are critical to the various modes of failure, in order to reduce failure. It uses all the available experience and expertise, from marketing, design, technology, purchasing, production/operation, distribution, service, etc., to identify the importance levels or criticality of potential problems and stimulate action to reduce these levels. FMECA should be a major consideration at the design stage of a product or service.

The elements of a complete FMECA are:

- *Failure mode* – the anticipated conditions of operation are used as the background to study the most probable failure mode, location and mechanism of the product or system and its components.
- *Failure effect* – the potential failures are studied to determine their probable effects on the performance of the whole product, process or service, and the effects of the various components on each other.
- *Failure criticality* – the potential failures on the various parts of the product or service system are examined to determine the severity of each failure effect in terms of lowering of performance, safety hazard, total loss of function, etc.

FMECA may be applied to any stage of design, development, production/operation or use, but since its main aim is to prevent failure, it is most suitably applied at the design stage to identify and eliminate causes. With more complex product or service systems, it may be appropriate to consider these as smaller units or subsystems, each one being the subject of a separate FMECA.

Special FMECA pro formas are available and they set out the steps of the analysis as follows:

1   Identify the product or system components, or process function.
2   List all possible failure modes of each component.
3   Set down the effects that each mode of failure would have on the function of the product or system.
4   List all the possible causes of each failure mode.
5   Assess numerically the failure modes on a scale from 1 to 10. Experience and reliability data should be used, together with judgement, to determine the values, on a scale 1–10, for:
    P   the probability of each failure mode occurring (1 = low, 10 = high).
    S   the seriousness or criticality of the failure (1 = low, 10 = high).
    D   the difficulty of detecting the failure before the product or service is used by the consumer (1 = easy, 10 = very difficult). See Table 8.3.
6   Calculate the product of the ratings, $C = P \times S \times D$, known as the criticality index or risk priority number (RPN) for each failure mode. This indicates the relative priority of each mode in the failure prevention activities.
7   Indicate briefly the corrective action required and, if possible, which department or person is responsible and the expected completion date.

TABLE 8.3 Probability and seriousness of failure and difficulty of detection

Value	1	2	3	4	5	6	7	8	9	10
P	low chance of occurrence ———————————————————— almost certain to occur									
S	not serious, minor nuisance ——————————————— total failure, safety hazard									
D	easily detected ——————————————————————— unlikely to be detected									

When the criticality index has been calculated, the failures may be ranked accordingly. It is usually advisable, therefore, to determine the value of C for each failure mode before completing the last columns. In this way the action required against each item can be judged in the light of the ranked severity and the resources available.

Moments of truth

Moments of truth (MoT) is a concept that has much in common with FMEA. The idea was created by Jan Carlzon, CEO of Scandinavian Airlines (SAS) and was made popular by Albrecht and Zemke (1985). A MoT is the moment in time when a customer first comes into contact with the people, systems, procedures or products of an organization, which leads to the customer making a judgement about the quality of the organization's services or products.

In MoT analysis the points of potential dissatisfaction are identified proactively, beginning with the assembly of process flowchart type diagrams. Every small step taken by a customer in his/her dealings with the organization's people, products or services is recorded. It may be difficult or impossible to identify all the MoTs, but the systematic approach should lead to a minimization of the number of the number and severity of unexpected failures, and this provides the link with FMEA.

One Australian residential builder has a policy of checking a dwelling immediately before the purchaser is handed the keys; the final inspection is recorded on a check sheet and fixed to the door. If the record is missing, the customer relations officer will not allow the new owners to enter the unit in case it is not entirely ready. The company does not want to risk the new owners having a negative impression on their first visit to their new home. This is an example of MoT being used to shape company policy within the Strand construction sector.

THE LINKS BETWEEN GOOD DESIGN AND MANAGING THE BUSINESS

Research carried out by the European Centre for Business Excellence (1998) has led to a series of specific aspects that should be addressed to integrate design into the business or organization. These are presented under various business criteria below.

Leadership and management style
- Listening is designed into the organization.
- Management communicates the importance of good design in good partnerships and vice versa.
- A management style is adopted that fosters innovation and creativity, and that motivates employees to work together effectively.

Customers, strategy and planning
- The customer is designed into the organization as a focus to shape policy and strategy decisions.
- Designers and customers communicate directly.
- Customers are included in the design process.
- Customers are helped to articulate and participate in the understanding of their own requirements.
- Systems are in place to ensure that the changing needs of the customers inform changes to policy and strategy.
- Design and innovation performance measures are incorporated into policy and strategy reviews.
- The design process responds quickly to customers.

People – their management and satisfaction
- People are encouraged to gain a holistic view of design within the organization.
- There is commitment to design teams and their motivation, particularly in cross-functional teamwork (e.g. Quality Function Deployment teams).
- The training programme is designed, with respect to design, in terms of people skills training (e.g. interpersonal, management teamwork) and technical training (e.g. resources, software).
- Training helps integrate design activities into the business.
- Training impacts on design (e.g. honing creativity and keeping people up to date with design concepts and activity).
- Design activities are communicated (including new product or service concepts).
- Job satisfaction is harnessed to foster good design.
- The results of employee surveys are fed back into the design process.

Resource management
- Knowledge is managed proactively, including investment in technology.
- Information is shared in the organization.

- Past experience and learning is captured from design projects and staff.
- Information resources are available for planning design projects.
- Suppliers contribute to innovation, creativity and design concepts.
- Concurrent engineering and design is integrated through the supply chains.

Process management

- Design is placed at the centre of process planning to integrate different functions within the organization and form partnerships outside the organization.
- 'Process thinking' is used to resolve design problems and foster teamwork within the organization and with external partners.

Impact on society and business performance

- Consideration is given to how the design of a product or service impacts on:
    - the environment;
    - the recyclability and disposal of materials;
    - packaging and wastage of resources;
    - the (local) economy (e.g. reduction of labour requirements).
- There is understanding of the impact of design on the business results, both financial and non-financial.

This same research showed that strong links exist between good design and proactive flexible deployment of business policies and strategies. These can be used to further improve design by encouraging the sharing of best practice within and across industries, by allowing designers and customers to communicate directly, by instigating new product/service introduction policies, project audits and design/innovation measurement policies and by communicating the strategy to employees. The findings of this work may be summarized by thinking in terms of the 'value chain', as shown in Figure 8.10.

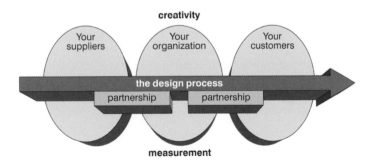

FIGURE 8.10 The value chain and design process

The built environment has a greater impact on the community and the environment than any other sector in the modern economy. Many leading companies have articulated their social responsibilities very clearly and they see this as a central element of their lean quality commitments. Several of the case studies in this book indicate how companies are working to meet their service obligations to the communities within which they operate.

Effective people management skills are essential for good design – these include the ability to listen and communicate, to motivate employees and encourage teamwork, as well as the ability to create an organizational climate which is conducive to creativity and continuous innovation.

The only way to ensure that design actively contributes to business performance is to make sure it happens 'by design', rather than by accident. In short, it needs coordinating and managing right across the organization. The case studies at the end of the book describe companies that have excelled in the design of their management processes and products, and through this broad commitment, they have created viable and sustainable businesses in the construction sector.

NOTES

1 The authors are grateful to the contribution made by John Dotchin and Simon Austin to this section of Chapter 8.

BIBLIOGRAPHY

Alarcon, F.L. and Mardones, D.A. Improving the design-construction interface, *IGLC Proceedings*, Guaruja, Brazil, 1998.

Albrecht, K. and Zemke, R., *Service America – Doing Business in the New Economy*, Dow Jones-Irwin, Homewood, IL, 1985.

Austin, S.A. *Integrated Design Planning*, Loughborough University, UK, 2008.

Austin, S., Baldwin, A., Li, B. and Waskett, P. Analytical Design Planning Technique (ADePT): a dependency structure matrix tool to schedule the building design process, *Construction Management and Economics*, 18: 173–182, 2000.

Austin, S.A., Newton, A.J. and Baldwin, A.N. 'Improving project management with the Analytical Design Planning Technique' in Carayannis, E., Kwak, Y.H. and Anbari, F.T. *The Story of Managing Projects*, Praeger Publishers, Westport, USA, 2005, pp 271–292.

Austin, S.A., Thorpe, A., Root, D.S., Thomson, D.S. and Hammond, J.W. Integrated collaborative design, *Journal of Engineering, Design and Technology*, 5(1): 7–22, 2007.

Ballard, G. and Reiser, P. The St Olaf College fieldhouse project: a case study in designing to target cost, *IGLC* 12, Helsingor, Denmark, 2004.

Carlzon, J. *Moments of Truth*, Harper & Row, London, 1987.

Caulcutt, R. *Statistics in Research and Development* (2nd edn), Chapman and Hall, London, 1991.

Choo, H.J., Hammond, J., Tommelein, I.D., Ballard, G. and Austin, S.A. DePlan: A tool for integrated design management, *Automation in Construction*, 13(3): 313–326, 2004.

European Centre for Business Excellence (the Research and Education Division of Oakland Consulting plc, 33 Park Square, Leeds LS1 2PF)/British Quality Foundation/Design Council. *Designing Business Excellence*, 1998.

Gargione, L.A. Using quality function deployment (QFD) in the design phase of an apartment construction project, *Proceedings IGLC 7*, Berkeley, USA, 1999.

Liker, J., Sobek, D., Ward, A. and Cristiano, J. Involving suppliers in product development in the U.S. and Japan: evidence for set-based concurrent engineering, *IEEE Transactions on Engineering Management*, 43(2): 165–177, 1996.

Lockyer, K.G., Muhlemann, A.P. and Oakland, J.S. *Production and Operations Management* (6th edn), Pitman, London, 1992.

Morgan, J. and Liker, J. *The Toyota Product Development System: Integrating People, Process, and Technology*, Productivity Press, New York, 2006.

Newton, A., Steele, J., Austin, S. and Waskett, P. Benefits derived from use of DSM as part of the ADePT approach to managing engineering projects, *9th International Design Structure Matrix Conference*, DSM'07, Munich, 16–18 October 2007.

Oakland, J.S. *Total Quality Management* (2nd edn), Butterworth-Heinemann, Oxford, 1993.

Oakland, J.S. *Statistical Process Control* (5th edn), Butterworth-Heinemann, Oxford, 2002.

Oakland, J.S. *Total Quality Management and Operational Excellence,* Routledge, London, 2014.

Parasuraman, A., Zeithaml, V.A. and Berry, L.L. SERVQUAL: a multiple-item scale for consumer perceptions of service quality, *Journal of Retailing*, 64(1): 12–40, 1988.

Parasuraman, A., Zeithaml, V.A. and Berry, L.L. Refinement and reassessment of the SERVQUAL scale, *Journal of Retailing*, 67(4): 420–450, 1991.

Parasuraman, A., Zeithaml, V.A. and Berry, L.L. Alternative scales for measuring service quality: a comparative assessment based on psychometric and diagnostic criteria, *Journal of Retailing*, 70(3): 201–230, 1994.

Steward, D.V. Partitioning and tearing systems of equations, *SIAM Journal on Numerical Analysis*, 2(2): 345–365, 1965.

Ward, A., Liker, J.K., Christiano, J.J. and Sobek II, D. The second Toyota paradox: how delaying decisions can make better cars faster, *Sloan Management Review*, 36: 43–61, 1995.

Womack, J.P. Jones, D.T. and Roos, D. *The Machine that Changed the World: the Story of Lean Production,* Harper Perennial, New York, 1991.

CHAPTER HIGHLIGHTS

Design innovation and improvement

- Design is a multifaceted activity that covers many aspects of an organization.
- All businesses need to update their products, processes and services.
- Innovation entails both invention and design, *and* continuous improvement of existing products, services and processes.
- Leading product/service innovations are market-led, not marketing-led.
- Everything in or from an organization results from design decisions.
- Design is an ongoing activity – dynamic not static, a verb not a noun – design is a process.

The design process

- To build in quality throughout the design process, commitment at the top is required. Moreover, the operational processes must be capable of achieving the design.
- State-of-the-art approach to innovation is based on a strategic balance of old and new, top management engagement with design, and teamwork.
- The 'styling' of products must also be matched by such secondary design considerations as operating instructions and software support.
- Designing takes in all aspects of identifying *the need*; developing something to satisfy *the need*; checking conformance to *the need*; and ensuring *the need* is satisfied.
- The design process must be carefully managed and can be flowcharted, like any other process, into: planning; practice codes; procedures; activities; assignments; identification of organizational and technical interfaces and design input requirements; review investigation and evaluation of new techniques and materials; and use of feedback data from previous designs.
- Total design or 'simultaneous engineering' is similar to quality function deployment and uses multifunction teams to provide an integrated approach to product or service introduction.

Quality function deployment (QFD): the house of quality

- The 'house of quality' is the framework of the approach to design management known as quality function deployment (QFD). It provides structure to the design and development cycle based on customer requirements.

- QFD is a system for translating the needs of customers into the language of the industry; all members of the supplier organization must be involved in the process.
- The purpose of a QFD team is to take the requirements of the customers and translate them into a form that can be understood and satisfied within the operating unit.
- The QFD team answers the following questions: **WHO** are the customers? **WHAT** do the customers need? **HOW** will the needs be satisfied?
- The answers to the **WHO**, **WHAT** and **HOW** questions are entered into the QFD matrix or quality table: one of the seven new tools of planning and design.

Design management with the ADePT

- Design processes are complex and many activities are concurrent. The relationship between tasks is better defined on the basis of information than simple task sequencing or outputs such as plans and specifications.
- The ADePT design management tool provides a planning approach that recognizes the dependency between tasks on the basis of information flows.
- The outputs of the ADePT process is in the form of a typical bar chart of activities, with interdependent activities blocked together. This format is ideal for planning the design process in detail.
- The relationship between design and construction can be designed in detail through information flows and hold points, enabling the fast tracking of projects with a far greater degree of control than is normally the case.

Specifications and standards

- There is a strong relation between standardization and specifications.
- If standards are used correctly, the process of drawing up specifications should provide opportunities to learn more about innovations, and change standards accordingly.
- The aim of specifications should be to reflect the true requirements of the product/service that are capable of being achieved.

Quality in the service sector

- In the design of services, three distinct elements may be recognized in the service package: physical (facilitating goods), explicit service (sensual benefits) and implicit service (psychological benefits). Moreover, the characteristics of service delivery may be itemized as intangibility, simultaneity and heterogeneity.
- The five dimensions of service quality (reliability, responsiveness, assurance, empathy and tangibles) are a very useful framework for assessing service quality weaknesses and for benchmarking service quality.
- A standard set of survey questions have been developed and validated; and hence they provide a mature and well-tried performance measurement and benchmarking tool for industry.
- The service attributes that are important in designing services include labour intensity, contact interaction, customization, nature of service act and the direct recipient of the act.
- Use of this framework allows services to be grouped under the five classifications.

Failure mode, effect and criticality analysis (FMECA)

- FMEA is the study of potential product, service or process failures and their effects. When the results are ranked in order of criticality, the approach is called FMECA. Its aim is to reduce the probability of failure.

- The elements of a complete FMECA are to study failure mode, effect and criticality. It may be applied at any stage of design, development, production/operation or use.
- Moments of truth (MoT) is a similar concept to FMEA. It refers to the moments in time when customers first come into contact with an organization, leading them to judgements about quality that can distort their overall satisfaction with the product or service.

The links between good design and managing the business

- Research has identified a series of specific issues which need to be addressed in order to effectively integrate design into an organization.
- The aspects may be summarized under the headings of: leadership and management style; customers, strategy and planning; people – their management and satisfaction; resource management; process management; impact on society and business performance.
- The research shows that strong links exist between good design and proactive flexible deployment of business policies and strategies: design needs coordinating and managing right across the organization.

Part II Discussion questions

1 Describe the key stages of integrating lean quality into the strategy of an organization of your choice in the construction supply chain.

2 Explain the difference between the 'Whats' and the 'Hows' of a company in the rail infrastructure sector. Identify likely critical success factors for such an organization and list possible key performance indicators for each one.

3 A design firm is concerned about the changes taking place in its sector and believes that the market is 'polarizing' into a high-value–high-fees end and a 'quick and dirty'– cheaper service associated with lower quality. Prepare a presentation to the senior management of the firm which provides an alternative view and show them how they could look at developing their business accordingly in the future.

4 You are the manager of a busy remedial contractor specializing in insurance claims work. Last year's abnormal winter gales led to an exceptionally high level of insurance claims for house damage caused by strong winds, and you had considerable problems in coping with the greatly increased workload. The result was excessively long delays in both acknowledging and settling customers' claims. Your area manager has asked you to outline a plan for dealing with such a situation should it arise again. The plan should identify what actions you would take to deal with the work and what, if anything, should be done now to enable you to take those actions should the need arise. What proposals would you make, and why?

5 Discuss the preparations required for the negotiation of a one-year contract with a major material supplier. What are the major factors to consider in partnering with key suppliers?

6 Imagine that you are the chief executive or equivalent in a company which has construction operations across the country, and that you plan to introduce best practice supply chain management into the organization.

   a Prepare a briefing of your senior managers, which should include your assessment of the aims, objectives and benefits to be gained from the approach.

   b Outline the steps you would take to develop an excellent supply chain and explain how you would attempt to ensure its success.

7 You are a management consultant with particular expertise in the area of lean quality in construction design and development. You are at present working on projects for four firms:

   a a hotel chain in its role a client for construction services;

   b an MEP fabrication and construction business;

   c an architectural design practice;

   d a road construction contractor.

   What factors do you consider are important generally in your area of specialization? Compare and contrast how these factors apply to your four current projects.

8 Discuss the application of quality function deployment (QFD) and the 'house of quality' in a residential housing company which designs and builds medium density housing developments.

9 Explain in full how Failure Mode Effect and Criticality Analysis (FMECA) can help a medium density housing design and construction company improve customer satisfaction. Describe in detail the method and consider some of the barriers that may arise.

10 Discuss how the ADePT design planning approach differs from traditional task scheduling as it is practised in construction.

11 Discuss the elements of the Last Planner® System and describe how they change the planning process on a construction project.

12 The techniques of Choosing by Advantages and Set-Based Design are gaining increasing popularity in the design of complex projects; describe the differences between these two approaches and their specific benefits.

13 Consider Takt time planning and its application on construction projects; describe how you would approach using this method for a medium-sized commercial project which includes a food court, retail and offices on the upper floors.

Part III

Performance

Performance measurement frameworks

9

PERFORMANCE MEASUREMENT AND THE IMPROVEMENT CYCLE

Traditionally, performance measures and indicators have been derived from cost-accounting information often based on outdated and arbitrary principles. These provide little motivation to support attempts to introduce lean quality and, in some cases, actually inhibit continuous improvement because they are unable to map process performance. In the organization that is to succeed over the long term, performance must begin to be measured by the improvements seen by the customer.

In the cycle of never-ending improvement, measurement plays an important role in:

- tracking progress against organizational goals;
- identifying opportunities for improvement;
- comparing performance against internal standards; and
- comparing performance against external standards.

Measures are used in *process control*, e.g. control charts (see Chapter 12), and in *performance improvement*, e.g. quality improvement teams (see Chapters 13 and 14) so they should give information about how well processes and people are doing and motivate them to perform better in the future.

The author and his colleagues have seen many examples of so-called performance measurement systems that frustrated improvement efforts. Various problems include systems that:

1. produce irrelevant or misleading information;
2. track performance in single, isolated dimensions;
3. generate financial measures too late, e.g. quarterly, for mid-course corrections or remedial action;
4. do not take account of the customer perspective, both internal and external;
5. distort management's understanding of how effective the organization has been in implementing its strategy; and
6. promote unproductive behaviour and *undermine* the achievement of the strategic objectives.

Typical harmful summary measures of local performance are purchase price, machine or plant efficiencies, direct labour costs and ratios of direct to indirect labour. In the construction sector, typically, the primary measures used are cost against budget, and time against scheduled time. These are incompatible with quality and productivity improvement measures because they cannot provide feedback that will motivate improvement. A popular recent measure is Value Earned; this too is of limited utility from a performance improvement perspective, it registers how much has been earned but says nothing about the efficiency of the process. In contrast, measures such as process and throughput times, supply chain performance, inventory reductions and increases in flexibility can motivate improvement – especially if they are *non-financial*. Financial summaries provide valuable information, of course, but they should not be used for control. Effective decision-making requires direct measures for operational feedback and improvement.

One example of a 'measure' with these shortcomings is return on investment (ROI). ROI can be computed only after profits have been totalled for a given period. It was designed therefore as a single-period, long-term measure but is often used as a short-term one. Perhaps this is because most executive bonus 'packages' in the West are based on short-term measures. ROI tells us what happened, not what is happening or what will happen and, for complex and detailed projects, ROI is inaccurate and irrelevant.

Many managers have a poor or incomplete understanding of their processes and products or services and, looking for an alternative stimulus, become interested in financial indicators. The use of ROI, for example, for evaluating strategic requirements and performance can lead to a discriminatory allocation of resources. In many ways, the financial indicators used in many organizations have remained static while the environment in which they operate has changed dramatically.

Traditionally, the measures used have not been linked to the processes where the value-adding activities take place. What has been missing is a performance measurement framework that provides feedback to people in all areas of business operations. Of course, lean quality stresses the need to start with the process for fulfilling customer needs.

The critical elements of a good performance measurement framework (PMF) are:

- Leadership and commitment;
- Full employee involvement;
- Good planning;
- Sound implementation strategy;
- Measurement and evaluation;
- Control and improvement;
- Achieving and maintaining standards of excellence.

The Deming Cycle of continuous improvement – PLAN DO CHECK ACT – clearly requires measurement to drive it; and yet it is a useful design aid for the measurement system itself:

PLAN: establish performance objective and standards.
DO: measure actual performance.
CHECK: compare actual performance with the objectives and standards – determine the gap.
ACT: take the necessary actions to close the gap, and make the necessary improvements.

Before we use performance measurement in the improvement cycle, however, we should attempt to answer four basic questions:

1   Why measure?
2   What to measure?
3   Where to measure?
4   How to measure?

Why measure?

It has often been said that it is not possible to manage what cannot be measured. Whether this is strictly true or not there are clear arguments for measuring. In a quality-driven, never-ending improvement environment the following are some of the main reasons *why measurement is needed,* and why it plays a key role in quality and productivity improvement. We must measure to:

- ensure customer requirements *have* been met;
- set sensible *objectives* and comply with them;
- provide *standards* for establishing comparisons;
- provide *visibility* and provide a 'score-board' for people to *monitor* their own performance levels;
- highlight *quality problems* and determine which areas require *priority attention*;
- give an indication of the *costs of poor quality*;
- justify the *use of resources*; and
- provide *feedback* for driving the improvement effort;

It is also important to know the impact of lean quality on improvements in business performance, on sustaining current performance and perhaps on reducing any decline in performance. In the construction environment there is a need to develop performance measurement frameworks for projects as well as for enterprises. This is also important at the process level for processes both in design and construction that are to be targeted for improvement.

What to measure?

A good start point for deciding what to measure is to look at: what the key goals of senior management are; what problems need to be solved; what opportunities are there to be taken advantage of; and what customers perceive to be the key ingredients that influence their satisfaction. In the case studies there are numerous examples of performance measurement in different areas of enterprise and project management. These examples reflect the primary business goals of senior management in each case, and they embrace the entire spectrum of issues that are addressed in a company's core values.

In the business of process improvement, process understanding, definition, measurement and management are tied inextricably together. In order to assess and evaluate performance accurately, appropriate measurement must be designed, developed and maintained by people who *own* the processes concerned. They may find it necessary to measure effectiveness, efficiency, reliability, quality, impact and productivity. In these areas there are many types of measurement, including direct output or input figures; the cost of poor quality; economic data; comments and complaints from customers; information from customer or employee surveys, etc., generally continuous variable measures (such as time) or discrete attribute measures (such as absentee defective units).

No one can provide a generic list of what should be measured, but once it has been decided in any one organization what measures are appropriate, they may be converted into indicators. These include ratios, scales, rankings and financial and time-based indicators. Whichever measures and indicators are used by the process owners, they must reflect the true performance of the process in customer/supplier terms and emphasize continuous improvement. Time-related measures and indicators have great value.

Current and recent research by the authors in the area of performance measurement has been at the enterprise, the project and the process level, and these serve to illustrate both the approach to performance measurement and some specific measures that are effective. Performance measures for site safety management, quality management and environmental waste measurement have also been developed and are now used on construction sites in Sydney (Trethewy *et al.*, 2000; Karim *et al.*, 2003).

Each of the case studies refers to one or more areas of performance measurement. The Graniterock (CS5) illustrates how performance in shrinkage reduction, a competitive attribute of concrete performance, drove competitive advantage for the company. The JB Henderson case study (CS8) shows how measurement and benchmarking of value added and wasted labour time provided the initial stimulus for embarking on the lean quality journey. Compliance with the Costain Way (CS12) six weeks into a project, a whole of project performance assessment framework which evaluates compliance with detailed requirements during the initiation phase of projects, has been found to be a sound indicator of project performance. The quality dashboard in the Crossrail case study (CS7) shows a client-led framework development which measures among other things NCR and RFI response times, on time certification of works and causes of NCRs are designed to drive quality performance to the same level as safety performance.

Where to measure?

If true measures of the effectiveness of lean quality are to be obtained, there are three components that must be examined – the human, technical and business components.

The human component is clearly of major importance, and the key tests are that wherever measures are used they must be:

1   *transparent* – understood by all the people being measured;
2   *non-controversial* – accepted by the individuals concerned;
3   *internally consistent* – compatible with the rewards and recognition systems;
4   *objective* – designed to offer minimal opportunity for manipulation; and
5   *motivational* – trigger a response to improve outcomes.

Technically, the measures must be the ones that truly represent the controllable aspects of the processes rather than simple output measures that cannot be related to process management. They must also be readily measurable, correct, precise and accurate.

The business component requires that the measures are objective, timely and result-oriented, and above all they must mean something to those working in and around the process, *including the customers*.

How to measure?

Measurement, as any other management system, must be implemented through the stages of design, analysis, development, evaluation, implementation and review – the PDCA cycle. The system must be designed to measure *progress*, otherwise it will not engage the improvement cycle. Progress is important in five main areas: effectiveness, efficiency, productivity, process reliability (quality and safety) and impact.

Effectiveness

Effectiveness may be defined as the percentage actual output over the expected output:

$$\text{Effectiveness} = \frac{\text{Actual output}}{\text{Expected output}} \times 100 \text{ per cent}$$

Effectiveness focuses on the *output* side of the process and is about the implementation of the objectives – doing what you said you would do. Effectiveness measures should reflect whether the organization, group or process owner(s) are achieving the desired results, accomplishing the right things. Measures of this may include:

- Quality, e.g. a grade of product, or a level of service.
- Quantity, e.g. tonnes, lots, bedrooms cleaned, accounts opened.
- Timeliness, e.g. speed of response, product lead times, cycle time.
- Cost/price, e.g. unit costs.

Efficiency
Efficiency is concerned with the percentage resource actually used over the resources that were planned to be used:

$$\text{Efficiency} = \frac{\text{Resources actually used}}{\text{Resources planned to be used}} \times 100 \text{ per cent}$$

Clearly, this is a process *input* issue and measures the performance of the process system management; however, in construction because of the variable nature of construction work it also reflects on the skill of the initial estimation. It is, of course, possible to use resources 'efficiently' while being *ineffective*, so performance efficiency improvement must be related to certain output objectives.

All process inputs may be subjected to efficiency measurement, so we may use labour/staff efficiency, equipment efficiency (or utilization), materials efficiency, information efficiency, etc. Inventory data and throughput times are often used in efficiency and productivity ratios.

Productivity
Productivity measures should be designed to relate the process outputs to its inputs:

$$\text{Productivity} = \frac{\text{Outputs}}{\text{Inputs}}$$

and this may be quoted as expected or actual productivity.

$$\text{Expected productivity} = \frac{\text{Expected output}}{\text{Resources expected to be consumed}}$$

$$\text{Actual productivity} = \frac{\text{Actual output}}{\text{Resources actually consumed}}$$

There is vast literature on productivity and its measurement, but simple ratios such as tonnes per man-hour (expected and actual), sales output per telephone operator-day and many others like this are in use. Productivity measures may be developed for each combination of inputs, for example, sales/all employee costs.

Process reliability (quality and safety)
This has been defined elsewhere of course (see Chapter 1). The *non-quality*-related measures include the simple counts of defect or error rates (perhaps in numbers per square metre or per thousand dollars spent); percentage outside specification or Cp/Cpk values; resources delivered compared to commitments (people, materials or equipment); deliveries not on time; or, more generally, as the costs of poor quality such as the measure of cost of rectification as a percentage of gross expenditure on site. When the positive costs of

prevention if poor quality are included, these provide a balanced measure of the costs of quality. (See next section.)

In a research collaboration, one of the authors was working in the areas of safety and quality performance on construction sites for over a decade. He essentially explored three areas of process management: management actions compared to those planned; management reaction to problems; and outcome measures that indicate performance improvements in the specific focus area (Marosszeky et al., 2004; Marosszeky, 2005). He was also involved in research on several projects to evaluate the service quality of suppliers within the on-site construction supply chain. This work also involved the assessment of project culture, the former through the use of a SERVQUAL[5]-based questionnaire, and the latter through the use of the Competing Values Framework developed by Quinn (see Parasuraman et al., 1988; Cameraon and Quinn, 2006). In all cases, these measures were being correlated against more traditional outcomes such as the incidence of errors and the cost of rectifying them.

The quality measures should also indicate positively whether we are doing a good job in terms of customer satisfaction, implementing the objectives, and whether the designs, systems and solutions to problems are meeting the requirements. These really are voice-of-the-customer measures.

Impact and value added

Impact measures should lead to key performance indicators for the business or organization, including monitoring improvement over time. Value-added management (VAM) requires the identification and elimination of all non-value-adding wastes, including time. Value added is simply the volume of sales (or other measure of 'turnover') minus the total input costs, and provides a good direct measure of the impact of the improvement process on the performance of the business. A related ratio, the percentage return on value added (ROVA) is another financial indicator that may be used:

$$\text{ROVA} = \frac{\text{Net profits before tax}}{\text{Value added}} \times 100 \text{ per cent}$$

Other measures or indicators of impact on the business are *growth* in sales, assets, numbers of passengers/students, etc., and *asset-utilization* measures such as return on investment (ROI) or capital employed (ROCE), earnings per share, etc.

Some of the impact measures may be converted to people productivity ratios, e.g.:

$$\frac{\text{Value added}}{\text{Number of employees (or employee costs)}}$$

Activity-Based Costing (ABC) is an information system that maintains and processes data on an organization's activities and cost objectives. It is based on the activities performed being identified and the costs being traced to them. ABC uses various 'cost drivers' to trace the cost of activities to the cost of the products or services. The activity and cost-driver concepts are the heart of ABC. Cost drivers reflect the demands placed on activities by products, services or other cost targets. Activities are processes or procedures that cause work and thereby consume resources. This clearly measures impact, both on and by the organization.

Reporting performance

Organizations practising lean quality tend to report performance in highly visible information centres, sometimes known as war rooms (Figure 9.1). These provide a dashboard

FIGURE 9.1 Typical information centre on a construction site

of current performance against key metrics. In the construction environment, they often provide safety and quality performance measures; indicate progress against plans; outline short-range plans in an A3 format; and include Problem and Countermeasure boards to focus people's attention on issues requiring resolution.

Research conducted at the Australian Centre for Construction Innovation showed conclusively that making performance results visible in information centres has a measurably greater influence on outcomes than reporting performance in reports and meeting minutes. The photographs in the VNGC case study (CS2) give the feel of a lean quality project site, in terms of the presentation of policy and performance information on large presentations on the wall, where it captures people attention.

Marton Marosszeky was privileged to visit the war room of a previous CEO of Rio Tinto in Perth, where the walls of the room were covered with an extensive dashboard of performance charts. This is where the CEO held his regular stand up meetings with his direct reports.

COSTS OF QUALITY

Manufacturing a quality product, providing a quality service or doing a quality job – one with a high degree of customer satisfaction – is not enough. The cost of achieving these goals must be carefully managed, so that the long-term effect on the business or organization is a desirable one. These costs are a true measure of the quality effort. A competitive product or service based on a balance between quality and cost factors is the principal goal of responsible management and may be aided by a competent analysis of the costs of quality (COQ).

The analysis of quality-related costs is a significant management tool that provides a:

1 method of assessing the effectiveness of the management of quality;
2 means of determining problem areas, opportunities, savings and action priorities.

The costs of quality are no different from any other costs. Like the costs of maintenance, design, sales, production/operations and other activities, they can be budgeted, measured and analysed. Having said this, a major difficulty in construction is capturing the totality of the costs. The construction process is highly fragmented and many parties including design professionals, general contractor supervisors and workers, and subcontractor supervisors and workers incur costs. Unless costs are to be recovered from another party either under the contract or as a variation, no one is interested in recording the costs. Yet a detailed knowledge of costs is potentially one of the main drivers for improvement and the authors are experimenting with industry partners to develop an accounting framework that effectively captures the significant costs.

Having specified the quality of design, the operating units have the task of matching it. The necessary activities will incur costs that may be separated into prevention costs, appraisal costs and failure costs: the so-called P-A-F model, first presented by Feigenbaum. Failure costs can be further split into those resulting from internal and external failure.

Prevention costs

Prevention costs are associated with the design, implementation and maintenance of the quality management system. They are planned and are incurred before actual operation. Prevention includes the following.

Product or service requirements

The determination of requirements and the setting of corresponding specifications (which also takes account of process capability) for incoming materials, processes, intermediates, finished products and services.

Quality planning

The creation of quality, reliability, operational, production, supervision, process control, inspection and other special plans, e.g. pre-production trials required to achieve the quality objective.

Quality assurance

The creation and maintenance of the quality system.

Inspection equipment

The design, development and/or purchase of equipment for use in inspection work.

Training

The development, preparation and maintenance of training programmes for operators, supervisors, staff and managers both to achieve and maintain capability.

Miscellaneous

Clerical, travel, supply, shipping, communications and other general office management activities associated with quality.

Resources devoted to prevention give rise to the *costs of doing it right the first time*.

Appraisal costs

These costs associated with the supplier's and customer's evaluation of purchased materials, processes, intermediates, products and services to assure conformance with the specified requirements. Appraisal includes:

Verification

Checking of incoming material, process set-up, first-offs, running processes, intermediates and final products, including produce or service performance appraisal against agreed specifications.

Quality audits

To check that the quality system is functioning satisfactorily.

Inspection equipment

The calibration and maintenance of equipment used in all inspection activities.

Supply chain and vendor rating

The assessment and approval of all suppliers, of both products and services.

Appraisal activities result in the 'costs of checking it is right'.

Internal failure costs

These costs occur when the results of work fail to reach designed quality standards and are detected before transfer to the customer takes place. Internal failure includes the following:

Scrap

Defective products or materials that cannot be repaired, used or sold.

Identification and organization of rework

The inspection, listing and organization of workers to return to the correct locations and re-do the work.

Rework or rectification

The correction of defective material or errors to meet the requirements.

Re-inspection

The re-examination of rectified work by independent design professionals and construction supervisors.

Downgrading

A product that is usable but does not meet specifications may be downgraded and sold as 'second quality' at a low price. While it may be argued that this is not relevant in construction, recent experience in the Sydney property market has shown that defective buildings become stigmatized and dwellings in them lose value. The products of recognized 'poor quality' builders are discounted because of the builder's poor reputation even though they may not appear to be defective.

Failure analysis

The activity required to establish the causes of internal product or service failure.

External failure costs

These costs occur when products or services fail to reach design quality standards but are not detected until after transfer to the consumer. External failure includes:

Repair and servicing

The repair of defective construction work has to be done on site and the cost can be prohibitive. One of the authors recently visited a building where the cost of repairing failed waterproofing in bathrooms was more than $20,000 per bathroom, while the cost of avoidance would have been less than $100 per occurrence.

Warranty claims

Failed products that are replaced or services re-performed under some form of guarantee. Once again, the labour involved in replacement is usually far greater than the value of the defective part.

Complaints

All work and costs associated with handling and servicing of customers' complaints.

Returns

The handling and investigation of rejected or recalled products or materials including transport costs. While this has limited relevance in construction, the authors know of cases where the only way a builder was able to pacify an angry owner of a defective building was to purchase the building back at market value.

Liability

The result of product or service liability litigation and other claims, which may include a change of contract. In the case of construction quality failures, these costs can be very significant and include expert reports on top of legal costs.

Loss of goodwill

The impact on reputation and image which will impinge directly on future prospects for sales.

External and internal failure produce the 'costs of getting it wrong'.

Order re-entry, unnecessary travel, telephone calls and conflict are just a few examples of wastage or failure costs that are often excluded. Every organization should be aware of the costs of getting it wrong and management needs to obtain some idea of how much failure is costing each year.

Clearly, this classification of cost elements may be used to interrogate any internal transformation process. Using the internal customer requirements concept as the standard for failure, these cost assessments can be made wherever information, data, materials, service or artefacts are transferred from one person or one department to another. It is the 'internal' costs of lack of quality that lead to the claim that approximately one-third of *all* our efforts are wasted.

The relationship between the quality-related costs of prevention, appraisal and failure against increasing quality awareness and improvement in the organization is shown in Figure 9.2. Where the quality awareness is low, the total quality-related costs are high – the failure costs predominating. As awareness of the cost to the organization of failure gets off the ground, through initial investment in training, an increase in appraisal costs usually results. As the increased appraisal leads to investigations and further awareness, further investment in prevention is made to improve design features, processes and systems. As the preventive action takes effect, the failure *and* appraisal costs fall, and the total costs reduce.

The first presentations of the P-A-F model suggested that there may be an optimum operating level at which the combined costs are at the minimum. The authors, however,

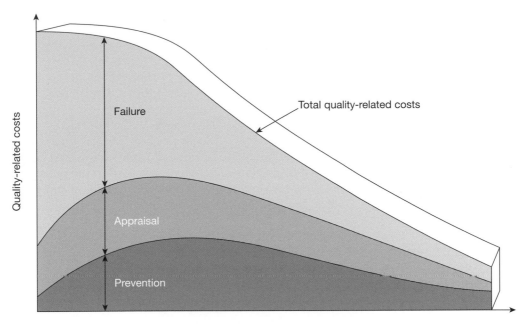

FIGURE 9.2 Increasing quality awareness and improvement activities

have not yet found one organization in which the total costs have risen following investment in prevention.

Levels of quality, cost reported in the construction sector

There is very little research worldwide into failure costs in the construction sector. A number of studies have been undertaken, and failure costs between 1 per cent and 12 per cent. These studies, however, were not taken under a consistent conceptual or measurement framework.

Research at the Australian Centre for Construction Innovation measured the total cost of rework on four projects in Sydney with a value of completed work within the study period of $60 million. The research included in error identification, rectification and re-inspection on these high-rise residential housing developments prior to practical completion. Some 3,500 instances of rework were investigated, assessed and costed. Rework was found to represent 5.5 per cent of the total cost of the completed work. The study did not include costs incurred after the buildings were contractually completed, and these are often substantial.

It was found that the direct cost of rework to the subcontractors was equivalent to the indirect cost of organizing and re-inspecting the work by the general contractor. It is noteworthy that rework as a percentage of total costs assessed was equivalent to the profit margins of most major contractors in the sector.

Costs incurred by contractors post Practical Completion are impossible to track, as project failure costs on one project are simply absorbed into the next project. Team members simply return to their previous projects to sort out problems and these costs are never measured or brought to account.

THE PROCESS MODEL FOR QUALITY COSTING

The P-A-F model for quality costing has a number of drawbacks. In lean quality, prevention of problems, defects, errors, waste, etc., is one of the prime functions, but it can be argued that everything a well-managed organization does is directed at preventing quality problems. This makes separation of *prevention costs* very difficult. There are clearly a range of prevention activities in any organization that are integral to ensuring quality but may never be included in the schedule of quality-related costs.

It may be impossible and unnecessary to categorize costs into the three categories of P-A-F. For example, a design review may be considered a prevention cost, an appraisal cost or even a failure cost, depending on how and where it is used in the process. Another criticism of the P-A-F model is that it focuses attention on cost reduction and plays down, or in some cases even ignores, the positive contribution made to price and sales volume by improved quality.

The most serious criticism of the original P-A-F model presented by Feigenbaum and used in, for example, British Standard 6143 (1981) *'Guide to the determination and use of quality related costs'* is that it implies an acceptable 'optimum' quality level above which there is a trade-off between investment in prevention and failure costs. Clearly, this is not in tune with the never-ending improvement philosophy of lean quality. The key focus of lean

Prevention	Appraisal	Internal failure	External failure
Preventive maintenance costs • People hours • Materials costs • Improvements to processing equipment	Inspection costs • Hours spent on QC inspection and test • Hours spent on operator self/peer group inspections	NCR data • Cost of time and materials to repair or rework products by product group • Confirmed companies	Warranty • Warranty provision • Warranty claims
Training costs • Internal trainer hours • External training costs • Hours spent in training	External inspections • Hours to support customer inspection visits • Factory costs to support visits • Third-party examinations	Scrapped product • Materials and labour to point of scrapping including allowance for added value from previous operations	Customer related – direct • Refunds • Product recalls • Excess carriage to replace products
Quality system • Requirements analysis and development of clear specifications • Development and maintenance of clear internal documentation	Checking of incoming raw materials • Hours spent • Testing materials used	Waste above the expected yield from the process • Materials • Time	Customer related – indirect • Loss of reputation and repeat business • Loss of goodwill
Investment in improvement projects including defect investigation and root-cause analysis • Hours spent • Costs of materials used	Quality system • Product design reviews • Hours spent on internal audits	Costs of rejected incoming materials – including factory downtime if this is an effect	Technical product support in the field (or via call centre) to answer questions and respond to incorrect user perceptions

FIGURE 9.3 The drivers of CoQ

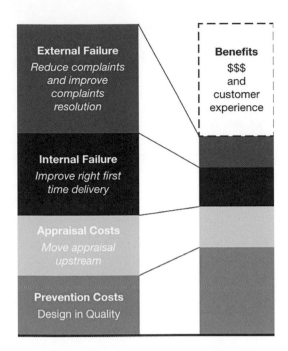

FIGURE 9.4
The Costs of Quality Failure (CoQF)

External Failure
Reduce complaints and improve complaints resolution

Benefits
$$$ and customer experience

Internal Failure
Improve right first time delivery

Appraisal Costs
Move appraisal upstream

Prevention Costs
Design in Quality

Cost of Quality Failure (CoQF) is the difference between the actual cost of a product or service and what the reduced cost would be if there were no possibility of substandard service, failure of products or defects in their manufacture

quality on waste elimination and process improvement includes improvement through product and process redesign, and a cost categorization scheme that does not consider process costs such as the P-A-F model, has limitations. (BS6143–2 was re-published in 2002 as 'Guide to the economies of quality: prevention, appraisal and failure model'.)

In a lean quality-related cost system that focuses on processes rather than products or services, the operating costs of generating customer satisfaction will be of prime importance. The so-called 'process cost model', described in the revised BS6143–1 (2002), sets out a method for applying quality costing to any process or service. It recognizes the importance of process ownership and measurement and uses process modelling to simplify classification. The categories of the cost of quality (COQ) have been rationalized into the cost of conformance (COC) and the cost of non-conformance (CONC):

COQ = COC + CONC

The cost of conformance (COC) is the process cost of providing products or services to the required standards by a given specified process in the most effective manner; for example, the cost of the ideal process where every activity is carried out according to the requirements first time, every time. The cost of non-conformance (CONC) is the failure cost associated with the process not being operated to the requirements, or the cost due to variability in the process. Part 2 of BS6143 (2002) still deals with the P-A-F model, but without the 'optimum'/minimum cost theory (see Figure 9.2).

Process cost models can be used for any process within an organization and can be developed by flowcharting or use of the ICOR methodology (see Chapter 12). This will identify the key steps and the parameters that are monitored in the process. The process cost elements should then be identified and recorded under the categories of product/service (outputs); and people, systems, plant or equipment, materials, environment, information (inputs). The COC and CONC for each stage of the process will comprise a list of all the parameters monitored.

At this stage, the use of detailed modelling for quality costs is not used on construction sites and because of the fragmentation of the supply chain it is unlikely to be effective in that setting. However, the application of this technique in the production and design processes of specialist suppliers shows more promise. There, errors could be assessed in this way and both the conformance and non-conformance costs would be in-house and could be realistically captured.

Steps in process cost modelling

Process cost modelling is a methodology that lends itself to stepwise analysis and while the following example is for the retrieval of medical records, it illustrates the process clearly and could be applied to any routine process in a volume production or service setting within the construction sector. The following are the key stages in building the model.

1 Choose a key process to be analysed; identify and name it, for example, Retrieval of Medical Records (Acute Admissions).
2 Define the process and its boundaries.
3 Construct the process diagram:
   a   identify the outputs and customers (for example see Figure 9.5);
   b   identify the inputs and suppliers (for example see Figure 9.6); and
   c   identify the controls and resources (for example see Figure 9.7).
4 Flowchart the process and identify the process owners (for example see Figure 9.8). Note, the process owners will form the improvement team.
5 Allocate the activities as COC or CONC (see Table 9.1).

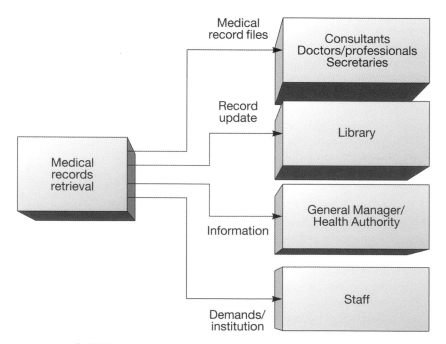

FIGURE 9.5 Building the model: outputs and customers

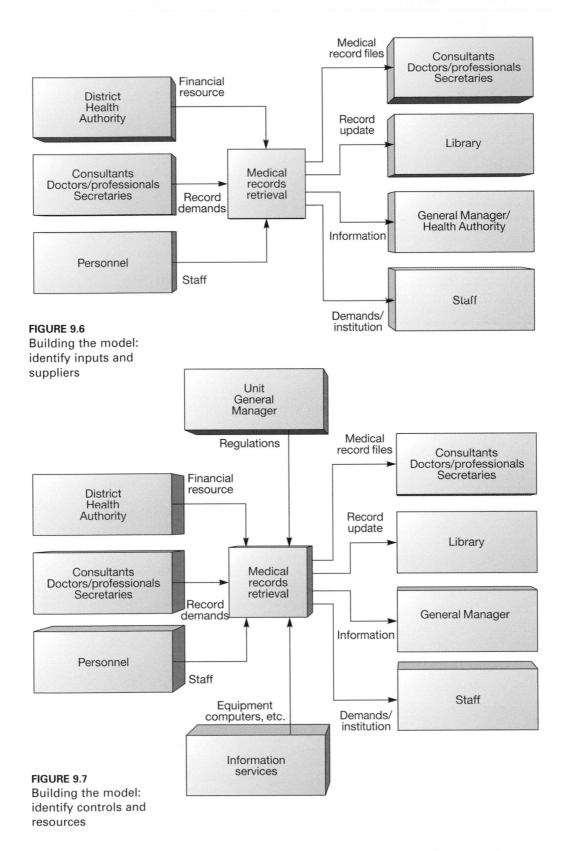

FIGURE 9.6
Building the model:
identify inputs and
suppliers

FIGURE 9.7
Building the model:
identify controls and
resources

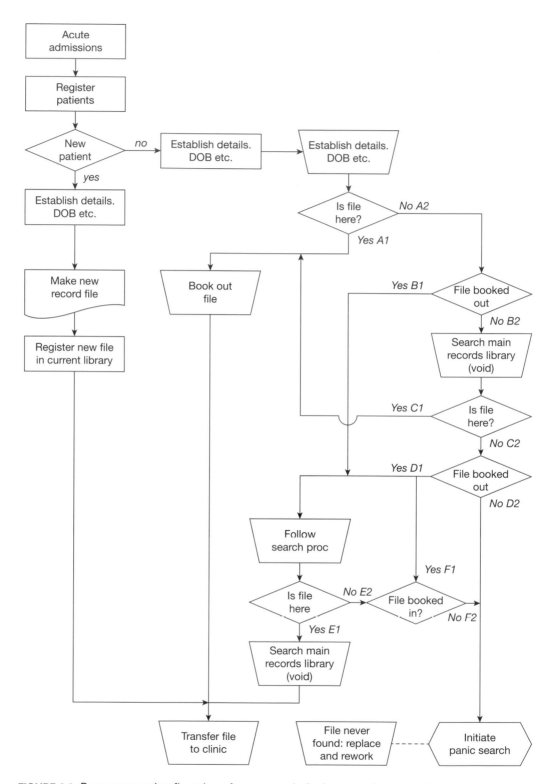

FIGURE 9.8 Present practice flowchart for acute admissions medical records retrieval

6　Calculate or estimate the quality costs (COQ) at each stage (COC + CONC). Estimates may be required where the accounting system is unable to generate the necessary information.

7　Construct a process cost report (see Table 9.2). The report summary and results are given in Table 9.3.

There are three further steps carried out by the process owners – the improvement team – which take the process forward into the improvement stage:

8　Prioritize the failure costs and select the process stages for improvement through reduction in costs of non-conformance (CONC). This should indicate any requirements for investment in prevention activities. An excessive cost of conformance (COC) may suggest the need for process redesign.

9　Review the flowchart to identify the scope for reductions in the cost of conformance. Attempts to reduce COC require a thorough process understanding; and a second flowchart of what the new process should be, may help (see Chapter 12).

10　Monitor conformance and non-conformance costs on a regular basis, using the model, and review for further improvements.

The process cost model approach should be seen as more than a simple tool to measure the financial implications of the gap between the actual and potential performance of a process. The emphasis given to the process, improving the understanding and seeing in detail where the costs occur should be an integral part of quality improvement. In generating the costs data, the authors and his colleagues have found that the involvement of the finance group in the organization is essential.

A PERFORMANCE MEASUREMENT FRAMEWORK (PMF)

A performance measurement framework (PMF) is proposed based on the strategic planning and process management models outlined in Chapters 5 and 12. The framework has four elements: strategy development/goal deployment, process management, individual performance management and review (Figure 9.9). This reflects an amalgamation of the approaches used by a range of organizations in performance measurement.

As we have seen in earlier chapters, the key to strategic planning and goal deployment is the identification of a set of critical success factors (CSFs) and associated key performance indicators (KPIs). These factors should be derived from the organization's mission and should represent a balanced mix of stakeholder issues. Action plans over both the short and medium term should be developed and responsibility clearly assigned for

TABLE 9.1 Building the model: allocate activities as COC or CONC

Key activities	COC	CONC
Search for files	Labour cost incurred finding a record while adhering to standard procedure	Labour cost incurred finding a record while unable to adhere to standard procedure
Make up new files	New patient files	Patients whose original files cannot be located
Rework		Cost of labour and materials for all rework files/records never found as a direct consequence of . . .
Duplication		Cost incurred in duplicating existing files

TABLE 9.2 Building the model: process cost report

Process cost report
Process: medical records retrieval (acute admissions)
Process owner: various
Time allocation: 4 days (96 hrs)

Process COC	Process CONC	Cost details Act	Cost details Synth	Definition	Source	
	Labour cost incurred finding records	# ref. Sample		Cost of time required to find missing records	Medical records	£210
	Cost incurred making up replacement files		#	Labour and material costs multiplied by number of files replaced	Medical records	£108
	Rework		#	Labour and material cost of all rework	Medical records	£80
	Duplication		#		Medical records	£24

TABLE 9.3 Process cost model: report summary

Labour cost
  14 hrs x £12.00/hr = £168
  £168 + overhead and contribution factor 25% = £210

Replacement costs
  No of files unfound 9
  Cost to replace each file £12.00
  Overall cost £108

Rework costs
  2 x Pathology reports to be word processed £80

Duplication costs
  No of files duplicated 2
  Cost per file £12.00
  Overall cost £24

                    TOTAL COST £422

RESULTS
Acute admissions operated 24 hrs/day 365 days/year
This project established a cost of non-conformance of approx. £422
This equates to £422 x 365/4 = £38,507.50
Or two personnel fully employed for 12 months.

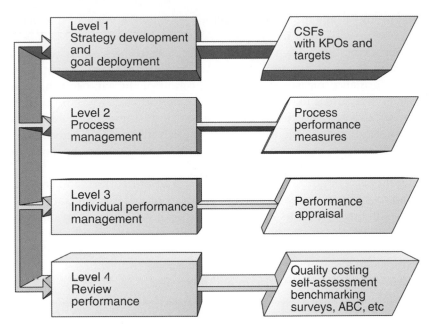

FIGURE 9.9 Performance measurement framework

performance. The strategic goals of the organization should then be clearly communicated to all individuals and translated into measures of performance at the process/functional level.

The key to successful performance measurement at the process level is the identification and translation of customer requirements and strategic objectives into an integrated set of process performance measures. The documentation and management of processes have been found to be vital in this translation process. Even when a functional organization is retained, it is necessary to treat the measurement of performance between departments as the measurement of customer–supplier performance.

Performance measurement at the individual level usually relies on performance appraisal, executed as formal planned performance reviews and performance management or, in other words, the day-to-day management of individuals. A major drawback with some performance appraisal systems, of course, is the lack of their integration with other aspects of performance measurement.

Performance review techniques are used by many world-class organizations to identify improvement opportunities and to motivate performance improvement. These companies typically use a wide range of such techniques and, in their drive for continuous improvement, they are innovative in performance measurement.

The links between performance measurement at the four levels of the framework are based on the need for measurement to be part of a systematic process of continuous improvement rather than for 'control'. The framework provides for the development and use of measurement rather than prescriptive lists of measures that should be used. It is, therefore, applicable in all types of organization.

The elements of performance measurement are distinct from the budgetary control process and also from the informal control systems used within organizations and should not be treated as a separate isolated system. Instead, measurement should be documented as and when it is used at the organizational, process and individual level. In this way it can facilitate the alignment of the goals of all individuals, teams, departments and processes with the strategic aims of the organization and incorporate the voice of the stakeholders in all planning and management activities.

A number of factors have been found to be critical to the success of performance measurement systems. These factors include the level of top management support for non-financial performance measures; the identification of the vital few measures; the involvement of all individuals in the development of performance measurement; the clear communication of strategic objectives; the inclusion of customers and suppliers in the measurement process; and the identification of the key drivers of performance. These factors will need to be taken into account by managers wishing to develop a new performance measurement system, or refine an existing one.

In most world-class organizations there are no separate performance measurement systems. Instead, performance measurement forms part of wider organizational management processes. Although elements of measurement can be identified at many different points within organizations, measurement itself usually forms the 'check' stage of the continuous improvement PDCA cycle. This is important since measurement data that is collected but not acted upon in some way is clearly a waste of resources.

The four elements of the framework in Figure 9.9 are:

Level 1: Strategy development and goal deployment leading to mission/vision, critical success factors and key performance outcomes (KPOs).

Level 2: Process management and process performance measurement through key performance indicators (KPIs) (including input, in-process and output measures, management of internal and external customer–supplier relationships and the use of management control systems).

Level 3: Individual performance management and performance appraisal.

Level 4: Review performance (including internal and external benchmarking, self-assessment against quality award criteria and quality costing).

Level 1 – Strategy development and goal deployment

The first level of the performance measurement framework is the development of organizational strategy and the consequent deployment of goals throughout the organization. Steps in the strategy development and goal deployment measurement process are (see Chapter 5):

1 Develop a mission statement based on recognizing the needs of all organizational stakeholders, customers, employees, shareholders and society. Based on the mission statement, identify those factors critical to the success of the organization achieving its stated mission. The CSFs should represent all the stakeholder groups, customers, employees, shareholders and society.

2 Define performance measures for each CSF – i.e. key performance outcomes (KPOs). There may be one or several KPOs for each CSF. Definition of KPO should include:
   a title of KPO;
   b data used in calculation of KPO;
   c method of calculation of KPO;
   d sources of data used in calculation;
   e proposed measurement frequency;
   f responsibility for the measurement process.

3 Set targets for each KPO. If KPOs are new, targets should be based on customer requirements, competitor performance or known organizational criteria. If no such

data exists, a target should be set based on best guess criteria. If the latter is used, the target should be updated as soon as enough data has been collected.

4 Assign responsibility at the organizational level for achievement of desired performance against KPO targets. Responsibility should rest with directors and very senior managers.

5 Develop plans to achieve the target performance. This includes both action plans for one year, and longer-term strategic plans.

6 Deploy mission, CSFs, KPOs, targets, responsibilities and plans to the core business processes. This includes the communication of goals, objectives, plans and the assignment of responsibility to appropriate individuals.

7 Measure performance against organizational KPOs, and compare them to target performance.

8 Communicate performance and proposed actions throughout the organization.

9 At the end of the planning cycle, compare organizational capability against the target against all KPOs, and begin again at step 2 above.

10 Reward and recognize superior organizational performance.

Strategy development and goal deployment is clearly the responsibility of senior management within the organization, although there should be as much input to the process as possible by employees to achieve 'buy-in' to the process.

The system outlined above is similar to the policy deployment approach known as Hoshin Kanri, developed in Japan and adapted in the West.

Level 2 – Process management and measurement

The second level of the performance measurement framework is process management and measurement, the steps of which are:

1 If not already completed, identify and map processes. This information should include identification of:
   a   process customers and suppliers (internal and external);
   b   customer requirements (internal and external);
   c   core and non-core activities;
   d   measurement points and feedback loops.

2 Translate organizational goals, action plans, and customer requirements into process performance measures (input, in-process and output) – key performance indicators (KPIs). This includes definition of measures, data collection procedures and measurement frequency.

3 Define appropriate performance targets based on known process capability, competitor performance and customer requirements.

4 Assign responsibility and develop plans for achieving process performance targets.

5 Deploy measures, targets, plans and responsibility to all sub-processes.

6 Operate processes.

7 Measure process performance and compare to target performance.

8 Use performance information to:
   a   implement continuous improvement activities;
   b   identify areas for improvement;
   c   update action plans;
   d   update performance targets;
   e   redesign processes, where appropriate;

  f manage the performance of teams, individuals (performance management and appraisal) and external suppliers;

  g provide leading indicators and explain performance against organizational KPIs.

 9 At the end of each planning cycle compare process capability to customer requirements against all measures, and begin again at step 2.

10 Reward and recognize superior process performance, including sub-processes and teams.

The same approach should be deployed to sub-processes and to the activity and task levels.

 The above steps should be managed by the process owners, with inputs wherever possible from the owners of sub-processes. The process outlined should be used whether an organization is organized and managed on a process or functional departmental basis. If functionally organized, the key task is to identify the customer–supplier relationships between functions and for functions to see themselves as part of a customer–supplier chain.

The balanced scorecard

The derivation of KPOs and KPIs may follow the 'balanced scorecard' model, proposed by Kaplan, which divides measures into financial, customer, internal business and innovation, and learning perspectives (Figure 9.10).

 A balanced scorecard derived from the business excellence model described in Chapters 3 and 8 would include key performance results, customer results (measured via the use of customer satisfaction surveys and other measures, including quality and delivery), people results (employee development and satisfaction) and society results (including community perceptions and environmental performance). In the areas of customers, people and society there needs to be a clear distinction between perception measures and other performance measures.

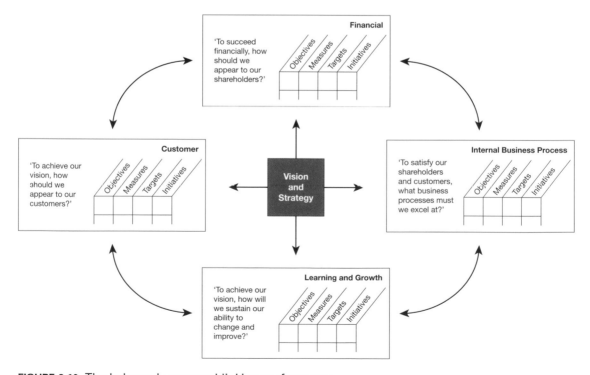

FIGURE 9.10 The balanced scorecard linking performance

Financial performance for external reporting purposes may be seen as a result of performance across the other KPOs, while the non-financial KPOs and KPIs are assumed to be the leading indicators of performance. The only aspect of financial performance that is cascaded throughout the organization is the budgetary process, which acts as a constraint rather than a performance improvement measure.

In summary then, organizational KPOs and KPIs should be derived from the balancing of internal capabilities against the requirements of identified stakeholder groups. This has implications for both the choice of KPOs/KPIs and the setting of appropriate targets. There is a need to develop appropriate action plans and clearly define responsibility for meeting targets if they are to be taken seriously.

Performance measures used at the process level differ widely between different organizations. Some organizations measure process performance using a balanced scorecard approach whilst others monitor performance across different dimensions according to the process. Whichever method is used, measurements should be identified as input (supplier), in-process and output (or results–customers).

It is usually at the process level that the greatest differences can be observed between the measurement used in manufacturing and services organizations. However, all organizations should measure quality, delivery, customer service/satisfaction and cost.

Depending on the process, measurement frequency varies from daily (for example, in the measurement of delivery performance) to annual (for example, in the measurement of employee satisfaction). Naturally, this will have implications for the PDCA cycle time of the particular process(es). Measurement frequency at the process level may, of course, be affected by the use of information technology. Cross-functional process performance measurement is a vital component in the removal of 'functional silos', and the consequent potential for sub-optimization and failure to take account of customer requirements. The success of performance measurement at the process level is dependent on the degree of management of processes and on the clarity of the deployment of strategic organizational objectives.

Measuring and managing the 'whats' and the 'hows'

Busy senior management teams find it useful to distil as many things as possible down to one piece of paper or one spreadsheet. The use of key performance outcomes (KPOs), with targets, as measures for CSFs, and the use of key performance indicators (KPIs) for processes may be combined into one matrix which is used by the senior management team to 'run the business'.

Figure 9.11 (also shown in Chapter 4) is an example of a matrix which is used to show all the useful information and data needed:

- the CSFs and their owner – the *whats*;
- the KPIs and their targets;
- the core business processes and their sponsors – the *hows*;
- the process performance measures – KPIs.

It also shows the impacts of the core processes on the CSFs. This is used in conjunction with a 'business management calendar' which shows when to report/monitor performance and identify process areas for improvement. This slick process offers senior teams a way of:

- gaining clarity about what is important and how it is measured;
- remaining focused on what is important and what the performance is;
- knowing where to look if problems occur.

Conduct research	Manage int. systems	Manage financials	Manage our accounts	Develop new business	Develop products	Manage people	Core processes	CSFs: We must have	Measures	Year targets	Target: CSF owner	
	×	×	×	×		×		Satisfactory financial and non-financial performance	Sales volume. Profit. Costs vs plan. Shareholder return Associate/employee utilisation figures	Turnover £2m. Profit £200k. Return for shareholders. Days/month per person		
×	×		×	×	×	×		A growing base of satisfied customers	Sales/customer Complaints/recommendations Customer satisfaction	>£200k = 1 client £100k–£200k = 5 clients £50k–£100k = 6 clients <£50k = 12 clients		
	×	×				×		A sufficient number of committed and competent people	No. of employed staff/associates Gaps in competency matrix. Appraisal results Perceptions of associates and staff	15 employed staff 10 associates including 6 new by end of year		
×			×			×		Research projects properly completed and published	Proportion completed on time, in budget with customers satisfied. Number of publications per project	3 completed on time, in budget with satisfied customers		
		* *	* *			* *		** = Priority for improvement				
								Process owner				
								Process performance				
								Measures and targets				

FIGURE 9.11 CSF/core process reporting matrix

Level 3 – Individual performance and appraisal management

The third level of the performance measurement framework is the management of individuals. Performance appraisal and management is usually the responsibility of the direct managers of individuals whose performance is to be appraised. At all stages in the process, the individuals concerned must be included to ensure 'buy in'.

Steps in performance and management appraisal are:

1 If not already completed, identify and document job descriptions based on process requirements and personal characteristics. This information should include identification of:
   a   activities to be undertaken in performing the job;
   b   requirements of the individual with respect to the identified activities, in terms of experience, skills, and training; and
   c   requirements for development of the individual, in terms of personal training and development.
2 Translate process goals and action plans, and personal training and development requirements into personal performance measures.
3 Define appropriate performance targets based on known capability and desired characteristics (or desired characteristics alone if there is no prior knowledge of capability).
4 Develop plans towards achievement of personal performance targets.
5 Document 1 to 4 using appropriate forms, which should include space for the results of performance appraisal.
6 Manage performance. This includes:
   a   planning tasks on a daily/weekly basis;

    b   managing performance of the tasks;

    c   monitoring performance against task objectives using both quantitative (process) and qualitative information on a daily and/or weekly basis;

    d   giving feedback to individuals of their performance in carrying out tasks; and

    e   giving recognition to individuals for superior performance.

7 Formally appraise performance against a developed range of measures, and compare to target performance.

8 Use comparison with target to:

    a   identify areas for improvement;

    b   update action plans;

    c   update performance targets;

    d   redesign jobs, where appropriate. (This impacts step 1 of the process.)

9 After a suitable period, ideally more than once a year, compare capability to job requirements and begin again at step 2.

10 Reward and recognize superior performance.

The above activities should be undertaken by the individual whose performance is being managed together with their immediate superior.

The major differences in approaches in the management of individuals lies in the reward of effort as well as achievement and the consequently different measures used, and in the use of information in continuous improvement required to reward and recognize

One of the authors undertook a major study of the Key Result Areas (KRAs) and KPI frameworks used to determine the risk/reward mechanisms on major public sector infrastructure projects in Australia. These projects had been procured through Alliance framework agreements in which the public sector owner joined with its supply chain partners to design and construct public infrastructure.

The reward formulae included painshare/gainshare provisions. While the basis of cost reimbursement was agreed and costs were not at risk, profits were determined on the basis of an agreed performance assessment framework. These KRA/KPI frameworks included such cost and non-cost areas as safety, quality, environment, legacy, alliance health and community. The study was undertaken on the basis of detailed interviews and workshops with groups representing four major infrastructure owners and their supply chains. The main findings included:

- KPI frameworks were generally found to be too complex; they should be simple rather than complex and target value creation for the owner.

- Not all KRA areas needed to be incentivized. Incentives should be seen as an investment by the owner, underpinned by a business case.

- Effective frameworks should not merely have set targets. They should be designed to motivate and reward continuous improvement and on longer projects should have provision for performance target revision.

- By identifying an area for reward, an owner stimulates resource allocation to that area. The cost benefit of that increased investment needs to be considered in the overall modelling.

- Measures should demonstrate the value created through continuous improvement.

performance, including teamwork. Unlike management by objectives (MBO), where the focus is on measurement of results, which are often beyond the control of the individual whose performance is appraised, good performance management systems attempt to measure a combination of process/task performance (effort and achievement) and personal development.

The frequency of formal performance appraisal is defined by the frequency of the appraisal process, usually with a minimum frequency of six months. Between the formal performance appraisal reviews, most organizations rely on the use of other performance management techniques to manage individuals. To improve team performance, the measures of performance or that of participation in various teams should be included in the appraisal systems where possible. In many organizations, the performance appraisal system is probably the least successfully implemented element of the framework. Appraisal systems are often designed to motivate individuals to achieve process and personal development objectives rather than to perform in teams. One of the limitations of appraisal processes is the frequency of measurement, which could be increased, but few organizations would consider doing so.

The Graniterock case study describes how the company uses Individual Professional Development Plans (IPDPs) in conjunction with the written role description to ascertain knowledge and skill development needs as well as long-term career aspirations consistent with the needs of the company. The IPDP process is more a career and educational planning than a review process.

Level 4 – Performance review

The fourth level of the performance measurement framework is the use of Performance Review techniques. Steps in a review are as follows:

1 Identify the need for review, which arise from:
    a poor performance at the organizational or process levels against KPO/KPIs;
    b identified superior performance of competitors;
    c customer inputs;
    d the desire to better direct improvement efforts;
    e the desire to concentrate attention on the need for performance improvement.
2 Identify the method of performance review to be used. This involves determining whether the review should be carried out internally from within the organization or externally, and the method that should be used. Some techniques are mainly internal, e.g. self-assessment and quality costing; whilst others, e.g. benchmarking, involve obtaining information from sources external to the organization. The choice should depend on:
    a how the need for review was identified (see 1);
    b the aim of the review: if, for example, the aim is to improve performance relative to competitors, then external benchmarking may be a better option than internally measuring the cost of quality;
    c the relative costs and expected benefits of each technique.
3 Carry out the review.
4 Feed results into the planning process at the organizational or process level.

5   Determine whether to repeat the exercise. If it is decided to repeat the exercise, then
     the following points should be considered:
     a   frequency of review;
     b   at what levels to carry out future reviews (e.g. organization-wide or process-by-
          process);
     c   decide whether the review technique should be incorporated into regular perform-
          ance measurement processes and, if so, how will this be managed.

Review methods will at times require the use of a level of resources greater than that
normally associated with performance measurement, often due to the need to develop
data collection procedures, train people in their use and the cost of data collection itself.
However, review techniques usually give a broader view of performance than most individual measures.

The use of review techniques is most successful when it is based on a clearly identified
need, perhaps due to perceived poor performance against existing performance measures
or against competitors, and the activity itself is clearly planned and the results used for
performance improvement. This is often the difference between the success and failure of
quality costing and benchmarking, in particular. The use of most of the review techniques
has been widely documented but often without regard to their integration into the wider
processes of measurement and management.

Review techniques

Techniques identified for review include:

1   quality costing using either presentation-appraisal-failure or process costing methods;
2   self-assessment against Baldrige, EFQM Excellence Model or internally developed
     criteria;
3   benchmarking, internal or external;
4   customer satisfaction surveys; and
5   activity-based costing (ABC).

THE IMPLEMENTATION OF PERFORMANCE MEASUREMENT SYSTEMS

It has already been established that a good measurement system must start with the customer and measure the right things. The value of any measure clearly needs to be compared
with the cost of producing it. There will be appropriate measures for different parts of the
organization, but everywhere they must relate process performance to the needs of the
process customer. All critical parts of the process must be measured, and it is often better
to start with simple measures and then improve them.

There must be a recognition of the need to distinguish between different measures for
different purposes. For example, an operator may measure time, various process parameters and amounts, while at the management level measuring costs and delivery
timeliness may be more appropriate.

Participation in the development of measures enhances their understanding and
acceptance. Process-owners can assist in defining the required performance measures,
provided that senior managers have communicated their mission clearly, determined the
critical success factors and identified the critical processes.

If all employees participate and own the measurement processes, there will be lower
resistance to the system, and a positive commitment towards future changes will be
engaged. This will be derived from the 'volunteered accountability', which will in turn

make the individual contribution more visible. Involvement in measurement also strengthens the links in the customer–supplier chains and gives quality improvement teams much clearer objectives. This should lead to greater short-term and long-term productivity gains.

The mnemonic SMART has been associated with designing measures and measurement systems in organizations – they should be Simple, Meaningful, Appropriate, Relevant and Timely. (The same mnemonic is also used to help set Specific, Measurable, Achievable, Realistic and Time-based objectives.)

There are a number of possible reasons why measurement systems fail:

1 They do not define performance operationally.
2 They do not relate performance to the process.
3 The boundaries of the process are not defined.
4 The measures are misunderstood or misused or measure the wrong things.
5 There is no distinction between control and improvement.
6 There is a fear of exposing poor and good performance.
7 It is seen as an extra burden in terms of time and reporting.
8 There is a perception of reduced autonomy.
9 Too many measurements are focused internally and too few are focused externally.
10 There is a fear of the introduction of tighter management controls.

These and other problems are frequently due to poor planning at the implementation stage or a failure to assess current systems of measurement. Before the introduction of a total quality-based performance measurement system, an audit of the existing systems should be carried out. Its purpose is to establish the effectiveness of existing measures, their compatibility with the quality drive, their relationship with the processes concerned and their closeness to the objectives of meeting customer requirements. The audit should also highlight areas where performance has not been measured previously, and indicate the degree of understanding and participation of the employees in the existing systems and the actions that result.

Generic questions that may be asked during the audit include:

- Is there a performance measurement system in use?
- Has it been effectively communicated throughout the organization?
- Is it systematic?
- Is it efficient?
- Is it well understood?
- Is it applied?
- Is it linked to the mission and objectives of the organization?
- Is there a regular review and update?
- Is action taken to improve performance following the measurement?
- Are the people who own the processes engaged in measuring their own performance?
- Have employees been properly trained to conduct the measurement?

Following such an audit, there are twelve basic steps for the introduction of lean quality-based performance measurement. Half of these are planning steps and the other half implementation.

Planning

1 Identify the purpose of conducting measurement. Is it for:
   a   reporting (ROI reported to shareholders);
   b   controlling (using process data on control charts);
   c   improving (monitoring the results of a quality improvement team project).

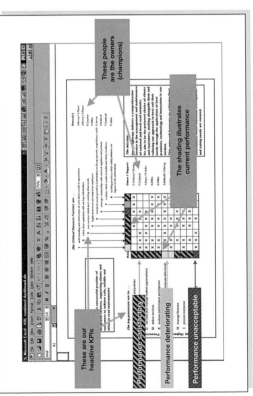
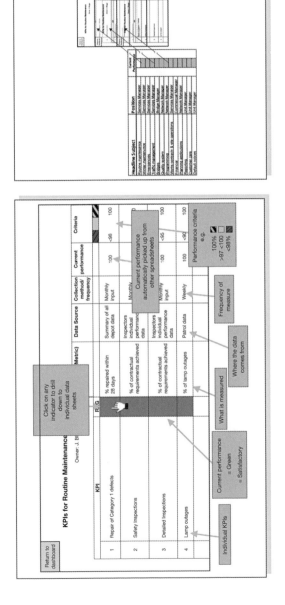
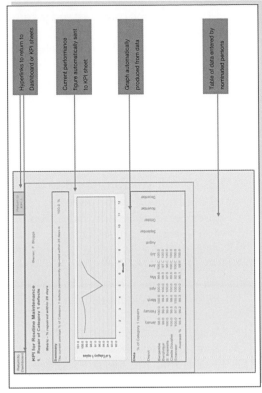

FIGURE 9.12 Performance dashboard and measurement framework

2 Choose the right balance between individual measures (activity or task-related) and group measures (process- and sub-process-related) and make sure they reflect process performance.

3 Plan to measure all the key elements of performance, not just one – time, cost and product quality variables may all be important.

4 Ensure that the measures will reflect the voice of the internal/external customers.

5 Carefully select measures that will be used to establish standards of performance.

6 Allow time for the learning process during the introduction of a new measurement system.

Implementation

7 Ensure full participation during the introductory period and allow the system to mould through participation.

8 Carry out cost/benefit analysis on the data generation, and ensure measures that have high 'leverage' are selected.

9 Make the effort to spread the measurement system as widely as possible since effective decision-making will be based on measures from all areas of the business operation.

10 Use surrogate measures for subjective areas where quantification is difficult (e.g. improvements in morale may be 'measured' by reductions in absenteeism or staff turn-over rates).

11 Design the measurement systems to be as flexible as possible to allow for changes in strategic direction and continual review.

12 Ensure that the measures reflect the quality drive by showing small incremental achievements that match the never-ending improvement approach.

In summary, the measurement system must be designed, planned and implemented to reflect customer requirements; give visibility to the processes and the progress made; communicate the total quality effort; and engage the never-ending improvement cycle. It must itself be periodically reviewed. Figure 9.12 shows a complete practical performance dashboard and measurement framework.

BIBLIOGRAPHY

British Standard. *BS6143–2: 2002, Guide to the Economics of Quality, Prevention, Appraisal and Failure Model*, BSI, London, 2002.

Cameron, K.S. and Quinn, R.E. *Diagnosing and Changing Organizational Culture Based on the Competing Value Framework*, Jossey-Bass, San Francisco, 2006.

Dale, B.G. and Plunkett, J.J. *Quality Costing*, Chapman and Hall, London, 1991.

Dixon, J.R., Nanni, A. and Vollmann, T.E. *The New Performance Challenge – Measuring Operations for World Class Competition*, Business One Irwin, Homewood, 1990.

Hall, R.W., Johnson, J.Y. and Turney, P.B.B. *Measuring Up – Charting Pathways to Manufacturing Excellence*, Business One Irwin, Homewood, 1991.

Karim, K., Davis, S., Marosszeky, M. and Naik, N. Designing an effective framework for process performance measurement in safety and quality, *Construction Information Quarterly*, 2003.

Kaplan, R.S. (ed.). *Measures for Manufacturing Excellence*, Harvard Business School Press, Boston, MA, 1990.

Kaplan, R.S. and Norton, P. *The Balanced Scorecard*, Harvard Business School Press, Boston, MA, 1996.

Marosszeky, M. Performance measurement in construction process management, keynote address to *Rethinking and Revitalizing Construction Safety, Health, Environment and Quality*, CIBW99 4th Triennial International Conference, South Africa, 2005.

Marosszeky, M. and Ward, M. *Public Sector Experience with the Use of KRA and KPI Frameworks on Alliances*, Evans & Peck, Sydney, 2010.

Marosszeky, M., Karim, K., Davis, S. and Naik, N. Lessons learnt in developing effective performance measures for construction safety management, *Proceedings 12th Conference of IGLC* Helsingor, Denmark, 2004.

Neely, A. *Measuring Business Performance* (2nd edn), Economist Books, London, 2002.

Neely, A., Adams, C. and Kennerly, M. *Performance Prism: The Scorecard for Measuring and Managing Business Services*, FT Prentice Hall, Englewood Cliffs, NJ, 2002.

Parasuraman, A., Zeithaml, V.A. and Berry, L.L. SERVQUAL: a multiple item scale for measuring consumer perceptions of service quality, *Journal of Retailing*, 64(1), 1988.

Porter, L.J. and Rayner, P. Quality costing for TQM, *International Journal of Production Economics*, 27: 69–81, 1992.

Trethewy, R., Cross, J., Marosszeky, M. and Gavin, I. Safety measurement: a positive approach towards best practice, *Journal of Occupational Health and Safety*, 16(3): 237–246, 2000.

Wood, D.C. (ed.). *Principles of Quality Costs: Financial Measures for Strategic Implementation of Quality Management* (4th edn), ASQ Press, Milwaukee, 2008.

Zairi, M. *TQM-Based Performance Measurement*, TQM Practitioner Series, Technical Communication (Publishing), Letchworth, 1992.

Zairi, M. *Measuring Performance for Business Results*, Chapman and Hall, London, 1994.

CHAPTER HIGHLIGHTS

Performance measurement and the improvement cycle

- Traditional performance measures based on cost-accounting information provide little to support lean quality because they do not map process performance and improvements seen by the customer.
- Measurement is important in tracking progress, identifying opportunities and comparing performance internally and externally. Measures, typically non-financial, are used in process control and performance improvement.
- Some financial indicators, such as ROI, are often inaccurate, irrelevant and too late to be used as measures for performance improvement.
- The Deming cycle of Plan Do Check Act is a useful design aid for measurement systems; but firstly four basic questions about measurement should be asked: why, what, where and how.
- In answering the question 'how to measure?' progress is important in five main areas: effectiveness, efficiency, productivity, quality and impact.
- Activity-based costing (ABC) is based on the activities performed being identified and costs traced to them. ABC uses cost drivers which reflect the demands placed on activities.

Costs of quality

- A competitive product or service based on a balance between quality and cost factors is the principal goal of responsible management.
- The analysis of quality-related costs may provide a method of assessing the effectiveness of the management of quality and of determining problem areas, opportunities, savings and action priorities.
- Total quality costs may be categorized into prevention, appraisal, internal failure and external failure costs, the P-A-F model.
- Prevention costs are associated with doing it right the first time, appraisal costs with checking it is right and failure costs with getting it wrong.

- When quality awareness in an organization is low, the total quality-related costs are high and the failure costs predominate. After an initial rise in costs, mainly through investment in training and appraisal, increasing investment in prevention causes failure, appraisal and total costs to fall.

The process model for quality costing

- The P-A-F model or quality costing has a number of drawbacks, mainly due to estimating the prevention costs and its association with an 'optimized' or minimum total cost.
- An alternative – the process costs model – rationalizes cost of quality (COQ) into the costs of conformance (COC) and the cost of non-conformance (CONC). COQ = COC + CONC at each process stage.
- Process cost modelling calls for the choice of a process and its definition; construction of a process diagram; identification of outputs and customers, inputs and suppliers, controls and resources; flowcharting the process and identifying owners; allocating activities as COC or CONC; and calculating the costs. A process cost report with summaries and results is produced.
- The failure costs of CONC should be prioritized for improvements.

A performance measurement framework

- A suitable performance measurement framework (PMF) has four elements related to strategy development: goal deployment, process management, individual performance management and review.
- The key to successful performance measurement at the strategic level is the identification of a set of critical success factors (CSFs) and associated key performance indicators (KPIs).
- The key to success at the process level is the identification and translation of customer requirements and strategic objectives into a process framework with process performance measures.
- The key to success at the individual level is performance appraisal and planned formal reviews through integrated performance management
- The key to success in the review stage is the use of appropriate innovative techniques to identify improvement opportunities. Of course, these need to be followed by good implementation.
- The following number of factors are critical to the success of performance measurement systems which include top management support for non-financial performance measures: the identification of the vital few measures; the involvement of all individuals in the development of performance measurement; the clear communication of strategic objectives; the inclusion of customers and suppliers in the measurement process; and the identification of the key drivers of performance.

The implementation of performance measurement systems

- The value of any measure must be compared with the cost of producing it. All critical parts of the process must be measured, but often it is better to start with the simple measures and then improve them.
- Process-owners should take part in defining the performance measures, and they must reflect customer requirements.
- Prior to introducing lean quality measurement, an audit of existing systems should be carried out to establish their effectiveness, compatibility, relationship and closeness to the customer.

- Following the audit, there are twelve basic steps for implementation – six of which are planning steps. The measurement system then must be designed, planned and implemented to reflect customer requirements. Measures should give visibility to the processes and progress made, communicate the total quality effort and drive continuous improvement. They must also be periodically reviewed.

Self-assessment, audits and reviews

10

ASSESSMENTS OF LEAN QUALITY IN CONSTRUCTION

In this chapter we look at the ideas of self-assessment and assessment by independent parties who may be expert and experienced in the task. To be effective any assessment must be undertaken against a well-structured and well-documented framework of criteria, otherwise it cannot be compared to assessments conducted by others, nor can it be repeated at different times with any consistency.

Organizations everywhere are under constant pressure to improve their business performance, and they measure themselves against world-class standards and focus their efforts on the customer. To help in this process, many are turning to total quality models such as the European Foundation for Quality Management's (EFQM) Excellence Model (see Chapter 3).

Businesses in the construction sector are no different, although relatively few organizations have evaluated their performance against this type of standard. Some of the case studies in this book provide examples of construction sector enterprises that have an outstanding record of achievement when measured against such criteria. They have succeeded, often against enterprises from other sectors, to win some of the most prestigious awards.

FRAMEWORKS FOR SELF-ASSESSMENT

'Total quality' or, as we have preferred in this book, 'lean quality' may be the goal of many organizations but it has been difficult until relatively recently to find a universally accepted definition of what these terms actually mean. For some people 'total quality' means a combination of a quality management system, statistical process control and the involvement of the workforce through teamwork. In some organizations, it has been replaced by the terms Business Excellence or Lean Six Sigma. In fact, a number of people from the construction sector have asked the authors what the difference is between these many conceptualizations of process improvement.

Clearly, there are many different views on what constitutes 'excellence' in an organization and even with an understanding of a framework there exists the difficulty of calibrating the performance or progress of any organization towards it.

The so-called excellence models now available recognize that customer satisfaction, business objectives, safety and environmental considerations are mutually dependent and are applicable in any organization. Clearly the application of these ideas involves investment primarily in people and time: time to implement new concepts, time to train and time for people to recognize the benefits and move forward into new or different organizational cultures. But how will organizations know when they are getting close to 'excellence' or whether they are even on the right road? How will they measure their progress and performance?

There have been many recent developments and there will continue to be many more in the search for a standard or framework against which organizations may be assessed or measure themselves and carry out the so-called 'gap analysis'. To many the ability to judge progress against an accepted set of criteria would be most valuable and informative.

Most lean and quality-based approaches strongly emphasize measurement while some insist on the use of costs of quality or waste. The value of a structured discipline using a points system has been well established in quality and safety assurance systems (e.g. ISO 9000 and vendor auditing). The extension of this approach to a total quality auditing process has been long established in the Japanese 'Deming Prize', perhaps the most demanding and intrusive auditing process. There are of course other excellence models and standards used throughout the world. In the US the Baldrige 'Criteria for Performance Excellence' is the best known. In Europe and other parts of the world the EFQM Excellence Model is better known both in terms of identifying 'enablers' and 'results' of focused management attention, and through its recognition of 'society' as a key stakeholder with an interest in outcomes. Many companies have realized the necessity to assess themselves against the Baldrige, EFQM and Deming criteria, if not to enter for the awards or prizes then certainly as an excellent basis for self-audit and review to highlight areas for priority attention and provide internal and external benchmarking.

The European Excellence Model for self-assessment

It has been recognized that the technique of self-assessment is very useful for any organization wishing to monitor and improve its performance. In 1992 the European Foundation or Quality Management (EFQM) launched a European Quality Award that is now widely used for systematic review and measurement of operations. The EFQM Excellence Model recognizes that processes are the means by which an enterprise harnesses and releases the talents of its people to produce results performance.

One feature of the EFQM Excellence Model that differentiates it from others and makes it particularly suitable for organizations in the construction sector is its inclusion of *Society Results*. Given the great and critical impact that the construction sector has on the community and the environment, the EFQM Excellence Model has a most suitable structure against which construction sector-based organizations should assess themselves. It is no surprise then that several of the case study companies have placed a very major emphasis on increasing the safety of their workforce, and also on their environmental and whole-of-life performance.

Figure 10.1 displays graphically the principle of the full Excellence Model. As described in Chapter 3, customer results, people results and favourable society results are achieved through leadership driving strategy, people, partnerships, resources, processes, products and services so that ultimately excellence in key performance results are achieved. The enablers deliver the results which in turn drive learning, creativity and innovation. Assessments using the EFQM Excellence Model usually use a weighting for each criteria which may be used in scoring self-assessments and making awards. The weightings should not be rigid and may be modified to suit specific organizational needs, allowing an organization to bias its scoring towards the priority improvement goals that it has set.

The EFQM have thus built a model of criteria and a review framework against which an organization may face and measure itself: and against which it can examine any 'gaps'. Such a process is known as self-assessment, and organizations such as the EFQM, and in

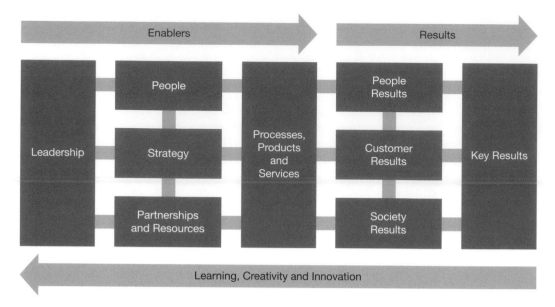

FIGURE 10.1 The EFQM Excellence Model

the UK the BQF (British Quality Foundation), publish guidelines for self-assessment including specific ones directed at public sector organizations.

Many managers feel the need for a rational basis on which to measure progress in their organization and would like answers to questions such as: 'Where are we now? Where do we need/want to be?' and 'What have we got to do to get there?' These questions need to be answered from internal employees' views, the customers' views, the views of suppliers and the views of society.

Self-assessment promotes lean quality by involving a regular and systematic review of processes and results. It highlights strengths and improvement opportunities and drives continuous improvement.

Enablers

In the Excellence Model, the enabler criteria of leadership, strategy, people, resources, partnerships, processes, products and services focus on what is needed to be done to achieve results. The structure of the enabler criteria is shown in Figure 10.2. Enablers (see Figure 10.3, Chart 1, The Enablers) are assessed on the basis of the combination of two factors:

1 the degree of excellence of the approach; and
2 the degree of deployment of the approach.

The elements of the detailed criteria are as follows.

Leadership

This is how leaders develop and facilitate the achievement of the vision and mission; develop values required for long-term success and implement these via appropriate actions and behaviours; and are personally involved in ensuring that the organization's management system is developed and implemented.

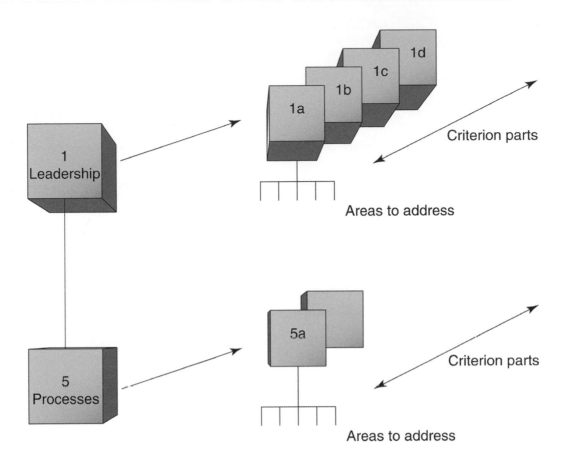

FIGURE 10.2 Structure of the criteria

Self-assessment should demonstrate how leaders:

- develop the vision, mission and values and are role models of a culture of excellence;
- are personally involved in ensuring the organization's management systems are developed, implemented and continuously improved;
- are involved with customers, partners and representatives of society; and
- motivate, support and recognize the organization's people.

Strategy

This is how the organization implements its vision and mission via a clear stakeholder-focused strategy, supported by relevant policies, plans, objectives, targets and processes. Self-assessment should demonstrate how strategy is:

- based on the present and future needs and expectations of stakeholders;
- based on information from performance measurement, research, learning and creativity-related activities;
- developed, reviewed and updated;
- deployed through a framework of key processes; and
- communicated and implemented.

Approach	Score	Deployment, assessment and review
Anecdotal or no evidence.	0%	Little effective usage.
Some evidence of soundly based approaches and prevention-based processes/systems.	25%	Implemented in about one-quarter of the relevant areas and activities.
Some evidence of integration into normal operations.		Some evidence of assessment and review.
Evidence of soundly based systematic approaches and prevention-based processes/systems.	50%	Implemented in about half the relevant areas and activities.
Evidence of integration into normal operations and planning well established.		Evidence of assessment and review.
Clear evidence of soundly based systematic approaches and prevention-based processes/systems.	75%	Applied to about three-quarters of the relevant areas and activities.
Clear evidence of integration of approach into normal operations and planning.		Clear evidence of refinement and improved business effectiveness through review cycles.
Comprehensive evidence of soundly based systematic approaches and prevention-based processes/systems.	100%	Implemented in all relevant areas and activities.
Approach has become totally integrated into normal working patterns. Could be used as a role model for other organizations.		Comprehensive evidence of refinement and improved business effectiveness through review cycles.

For *Approach, Deployment, Assessment* and *Review* the assessor may choose one of the five levels 0%, 25%, 50%, 75% or 100% as presented in the chart, or interpolate between these values.

FIGURE 10.3 Scoring within the self-assessment process: Chart 1, the enablers

People

This is how the organization manages, develops and releases the knowledge and full potential of its people at an individual, team and organization-wide level and plans these activities in order to support its strategy and the effective operation of its processes.

Self-assessment should demonstrate how people:

- resources are planned, managed and improved;
- knowledge and competencies are identified, developed and sustained;
- are involved and empowered;
- and the organization have a dialogue; and
- are rewarded, recognized and cared for.

Partnerships and resources

This is how the organization plans and manages its external partnerships and internal resources in order to support its strategy and the effective operation of its processes.

Self-assessment should demonstrate how it manages:

- external partnerships;
- finances;
- buildings, equipment and materials;
- technology; and
- information and knowledge.

Processes, products and services

This is how the organization designs, manages and improves its processes, products and services in order to support its strategy and fully satisfy and generate increasing value for its customers and stakeholders.

Self-assessment should demonstrate how:

- processes, products and services are systematically designed and managed;
- processes are improved as needed, using innovation in order to fully satisfy and generate increasing value for customers and other stakeholders;
- products and services are designed and developed based on customer needs and expectations;
- products and services are produced, delivered and serviced; and
- customer relationships are managed and enhanced.

Assessing the enablers criteria

The criteria are concerned with how an organization or business unit achieves its results.

Self-assessment asks the following questions in relation to each criterion:

- What is currently done in this area?
- How is it done? Is the approach systematic and prevention-based?
- How is the approach reviewed and what improvements are undertaken following review?
- How widely used are these practices?

Results

The EFQM Excellence Model's result criteria for customers, people and society and the key performance results focus on what the organization has achieved and is achieving in relation to its:

- external customer;
- people;
- local, national and international society, as appropriate; and
- planned performance.

Although these can be expressed as discrete results, ideally they should be viewed as trends over a period of years. (The structure of the results criteria is shown in Figure 10.4.)

'Performance excellence' is assessed relative to the organization's business environment and circumstances, based on information which sets out the:

- organization's actual performance; and
- organization's own targets.

and wherever possible, the:

- performance of competitors or similar organizations;
- performance of 'best in class' organizations.

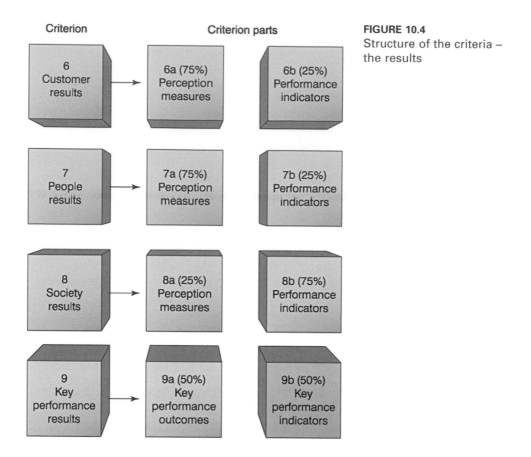

Criterion	Criterion parts	
6 Customer results	6a (75%) Perception measures	6b (25%) Performance indicators
7 People results	7a (75%) Perception measures	7b (25%) Performance indicators
8 Society results	8a (25%) Perception measures	8b (75%) Performance indicators
9 Key performance results	9a (50%) Key performance outcomes	9b (50%) Key performance indicators

FIGURE 10.4
Structure of the criteria –
the results

Results (see Figure 10.5, Chart 2) are assessed on the basis of the combination of two factors:

1 the degree of excellence of the results; and
2 the scope of the results.

Customer results

What is the organization achieving in relation to its external customers?

Self-assessment should demonstrate the organization's success in satisfying the needs and expectations of its external customers.

Areas to consider are the:

- results achieved for the measurement of customer perception of the organization's products, services and customer relationships; and
- internal performance indicators assessing the organization's customers.

People results

What the organization is achieving in relation to its people.

Self-assessment should demonstrate the organization's success in satisfying the needs and expectations of its people.

Areas to consider are the:

- results of people's perception of the organization; and
- internal performance indicators relating to people.

Results	Score	Scope
No results or anecdotal information.	0%	Results address few relevant areas and activities.
Some results show positive trends and/or satisfactory performance. Some favourable comparisons with own targets/external organizations. Some results are caused by approach.	25%	Results address some relevant areas and activities.
Many results show strongly positive trends and/or sustained good performance over the last 3 years. Favourable comparisons with own targets in many areas. Some favourable comparison with external organizations. Many results are caused by approach.	50%	Results address many relevant areas and activities.
Most results show strong positive trends and/or sustained excellent performance over at least 3 years. Favourable comparisons with own targets in most areas. Favourable comparisons with external organizations in many areas. Most results are caused by approach.	75%	Results address most relevant areas and activities.
Strongly positive trends and/or sustained excellent performance in all areas over at least 5 years. Excellent comparisons with own targets and external organizations in most areas. All results are clearly caused by approach. Positive indication that leading position will be maintained.	100%	Results address all relevant areas and facets of the organization.

For both *Results* and *Scope* the assessor may choose one of the five levels 0%, 25%, 50%, 75% or 100% as presented in the chart, or interpolate between these values.

FIGURE 10.5 Scoring within the self-assessment process: Chart 2, the results

Society results

What the organization is achieving in relation to local, national and international society as appropriate.

Self-assessment should demonstrate the organization's success in satisfying the needs and expectations of the community at large.

Areas to consider are the:

- society's perception of the organization;
- internal performance indicators relating to the organization and society;
- impact of the organization's operations on the environment; and
- whole-of-life efficiency of products.

Key performance results

What the organization is achieving in relation to its planned performance.
Areas to consider are the:

- key performance outcomes, including financial and non-financial; and
- key indicators of the organization's performance which might predict likely key performance outcomes.

Assessing the results criteria

These criteria are concerned with what an organization has achieved and is achieving. Self-assessment addresses the following issues:

- measures used to indicate performance;
- extent to which the measures cover the range of the organization's activities;
- relative importance of the measures presented;
- organization's actual performance;
- organization's performance against targets;

and wherever possible:

- comparisons of performance with similar organizations;
- comparisons of performance with 'best in class' organizations.

Self-assessment against the Excellence Model may be performed generally using the so-called RADAR system:

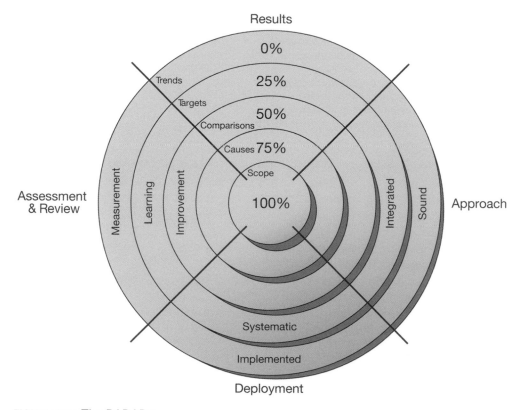

FIGURE 10.6 The RADAR screen

- Results;
- Approach;
- Deployment;
- Assessment;
- Review.

The RADAR 'screen' with the net level of detail is shown in Figure 10.6.

METHODOLOGIES FOR LEAN QUALITY ASSESSMENT

A flow diagram of the general steps involved in undertaking self-assessment is shown in Figure 10.7.

There are a number of approaches to carrying out self-assessment including:

- discussion group/workshop methods;
- surveys, questionnaires and interviews (peer involvement);
- pro formas;
- organizational self-analysis matrices such as the lean quality assessment or maturity model in Figure 10.8;
- an award simulation;
- activity or process audits; and
- hybrid approaches.

Whichever method is used, the emphasis should be on understanding the organization's strengths and areas for improvement rather than the score. The scoring charts provide a consistent basis for establishing a quantitative measure of performance against the model; gaining consensus through the use of a consistent framework promotes discussion and development of the issues facing the organization. It should also gain the involvement, interest and commitment of the senior management, but the scores should not become an end in themselves.

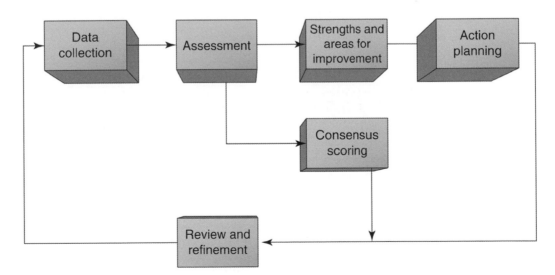

FIGURE 10.7 The key steps in self-assessment

Using assessment

There is great overlap between the criteria used by the various awards, and it may be necessary for an organization to rationalize them. The main components, however, must be the organization's leadership developing and driving its strategy through processes, management systems, people management and results to deliver customer

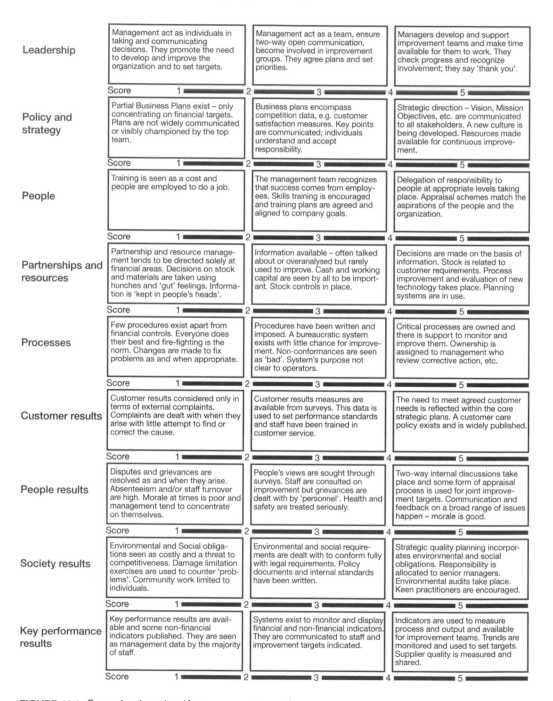

FIGURE 10.8 Organizational self-assessment matrix

results and key performance results. Self-assessment can provide an organization in the construction sector with vital information in monitoring its progress towards its goals and business 'excellence'. The external assessments used in the processes of making awards should be based on these self-assessments that are performed as prerequisites for improvement.

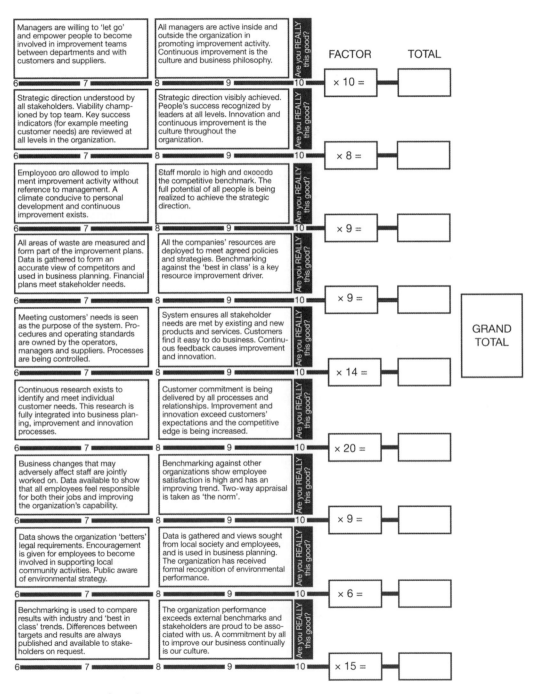

FIGURE 10.8 *continued*

Whatever the main 'motors' for driving an organization towards its vision or mission are, they must be linked to the five stakeholder groups recognized by the values of every organization, namely:

- Customers;
- Employees;
- Suppliers;
- Shareholders;
- Community.

In any normal business or organization, measurements are continuously being made, often in retrospect, by the leaders of the organization to reflect the value put on the organization by its five stakeholders. Too often, these continuous readings are made by internal, biased agents with short-term priorities; this is not always in the best long-term interests of the organization or its customers: narrow, fire-fighting scenarios that can blind the organization's strategic eye. Third-party agents, however, can carry out or facilitate periodic assessments from the perspective of one or more of the key stakeholders, with particular emphasis on forward priorities and needs. These reviews allow the realignment of the principle driving motors to focus on the critical success factors and the continuous improvement necessary to maintain the balanced and powerful general thrust that moves the whole organization towards its mission.

The relative importance of the five stakeholders may vary in time but all are important. The first three, customers, employees and suppliers, comprise the core value chain and are the determinant elements. The application of lean quality principles in these areas will provide satisfaction as a result to the shareholders and the community. The resulting added value will benefit the community and the environment. The ideal is a long way off in most organizations however, and active attention to the needs of the shareholders and/or community remain a priority, for one major reason – they are the 'customers' of most organizational activities and are vital stakeholders.

CAPABILITY MATURITY MODEL INTEGRATION (CMMI) ASSESSMENTS

One of the ways organizations may pursue operational excellence is to consider how mature their operations are. A set of requirements for increasing levels of maturity are defined and assessments are made to determine how mature the assessed entity is. Frameworks exist, such as CMMI for software development (see www.cmmiinstitute.com). Maturity is defined in a framework that sets out the 'criteria' that make up excellence in the area of focus. For each criterion, a set of requirements are arranged in a hierarchy, usually with five levels: basic at level 1 and world-class at level 5, with possible intermediate steps in between. Gaining consensus within the organization on criteria/requirements can take time as it is about understanding what is critical to success and then clearly defining what 'good' looks like.

'Good' definitions of criteria/requirements are essential and they should:

- require quantitative support where possible;
- seek evidence of actual deployment in the entity;
- be mutually exclusive and collectively exhaustive;
- be written in simple language to allow global coverage; and
- be accompanied by detailed assessor guidelines.

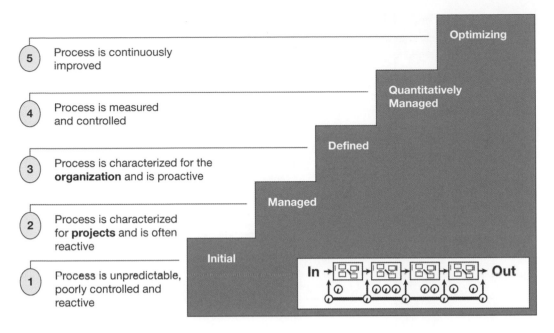

FIGURE 10.9 CMMI – maturity levels

Capability Maturity Model Integration (CMMI) is an approach which can be used to guide improvements across a project, a division or an entire organization. It helps to 'integrate separate organizational functions, appraise the effectiveness of current processes and priorities improvement activity'. The tools and techniques can be used at the:

- process capability level (continuous representation) to satisfy specified product quality, service quality and individual process performance objectives; and
- organizational maturity level (staged representation) to support the application and use of a defined set of capable processes in line with the organization's overall business objectives.

CMMI uses a scale of four levels of capability or five levels of maturity to provide a series of well-defined evolutionary plateaus for improvement (Figure 10.9). Attaining specific levels will form a firm foundation for further development and embed continuous process improvement (see Figure 10.10 for a process example).

Usually, assessments are by either local self-assessment or 'audits' from a 'centre':

- local self-assessment has the advantage of local ownership but with a large training challenge; and
- audits from a centre have the advantage of global consistency but can meet local resistance.

A synthesis of these two approaches often works using a small central and/or independent core team (objectivity and consistency) working with nominated local assessors (ownership). Every effort should be made to make the assessment a positive learning experience for the entity assessed to develop the maturity of the team and to ensure the group is motivated to improve.

5 – Optimizing	Continuous process improvement	Organizational Performance Management (OPM) Causal Analysis and Resolution (CAR)
4 – Quantitatively Managed	Quantitative management	Organizational Process Performance (OPP) Quantitative Project Management (QPM)
3 – Defined	Process standardization	Requirements Development (RD) Technical Solution (TS) Product Integration (PI) Verification (VER) Validation (VAL) Organizational Process Focus (OPF) Organizational Process Definition (OPD) Organizational Training (OT) Integrated Project Management (IPM) Risk Management (RSKM) Decision Analysis and Resolution (DAR)
2 – Managed	Basic process/ project management	Requirements Management (REQM) Project Planning (PP) Project Monitoring and Control (PMC) Supplier Agreement Management (SAM) Measurement and Analysis (MA) Process and Product Quality Assurance (PPQA) Configuration Management (CM)
1 – Initial		

FIGURE 10.10 CMMI for development – process areas

Maturity assessments using CMMI type frameworks should prompt appropriate action by showing both the current status and what needs to be done next to advance up the maturity curve. The assessments often lead to standardized reports which fulfil three main purposes:

1   show the level of maturity of each entity against the set criteria / requirements;
2   indicate where the priorities are for the entity to progress further up the maturity levels; and
3   show the results from multiple entities so that best practices can be identified as a basis for organizational learning.

Actions should be generated from these reports and mechanisms established to ensure implementation. Periodic re-assessments may need to be scheduled. Clear reporting of all entities' maturity status and progress should be visible to all and reviewed by senior management regularly. The overall goal of any maturity assessments should be to engage people and channel their efforts into activities that will most efficiently improve the operations in their organization.

Highways England (HE) have developed a detailed and thorough Lean Maturity Assessment Tool kit (HELMA), the aim of which is to encourage their supply chain 'to adopt lean principles to help foster a culture of continuous improvement for mutual advantage'. HELMA covers the following areas:

- integration of lean in business strategy;
- lean leadership and engagement;
- deployment management / lean infrastructure;
- understanding customer value;
- understanding of processes and value streams;

- use of methodologies and tools;
- organizational coverage, activity and capability;
- performance improvement/benefit realization and delivery;
- lean collaboration, climate, and culture; and
- supplier maturity – using assessment to shape supply chain capability (see also section on supply chain development).

For each of these, there is a series of 'Lean Adoption Questions' corresponding with four capability level descriptions and indicative evidence. The HE case study (CS14) illustrates the structure of the HE approach to improving the capacity of its suppliers. This approach is open to any general contractor wishing to improve the capability of its supply chains.

SECURING LEAN QUALITY BY AUDIT AND REVIEW OF THE MANAGEMENT SYSTEMS

Error or defect prevention is the process of removing or controlling error/defect causes in the lean quality management systems. There are two major elements of this:

1 checking the systems; and
2 error/defect investigation and follow-up.

These have the same objectives: to find, record and report *possible* causes of error or waste and to recommend future preventive or corrective action.

Checking the systems

There are six methods in general use:

1 *Lean quality audits and reviews*. These subject each area of an organization's activity to a systematic critical examination. Every component of the total system is included: strategy, attitudes, training, processes, decision features, operating procedures and documentation. Audits and reviews, as in the field of accountancy, aim to disclose the strengths and the main areas of vulnerability or risk – the areas for improvement.
2 *Lean quality survey*. This is a detailed, in-depth examination of a narrower field of activity: major key areas revealed by system audits, individual sites/plants, procedures or specific problems common to an organization as a whole.
3 *Lean quality inspection*. A routine scheduled inspection of a unit or department. The inspection should check standards, employee involvement and working practices, and that work is carried out in accordance with the agreed processes and procedures.
4 *Lean quality tour*. An unscheduled examination of a work area to ensure, for example, that the standards of operation are acceptable, obvious causes of defects or errors are removed and in general lean quality standards are maintained.
5 *Lean quality sampling*. This measures by random sampling, similar to activity sampling, the error/defect/waste potential. Trained observers perform short tours of specific locations by prescribed routes and record the number of potential errors, defects or waste seen. The results may be used to portray trends in the general lean quality situation.
6 *Lean quality scrutiny*. This is the application of a formal, critical examination of the process and technological intentions for new or existing facilities, and for assessing the potential for maloperation or malfunction of equipment and the consequential effects for lean quality. There are similarities between a lean quality scrutiny and FMECA studies (see Chapter 8).

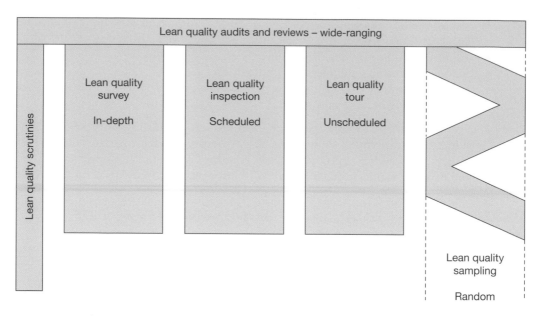

FIGURE 10.11 A prevention programme combining various elements of 'checking' the system

The design of a prevention programme, combining all these elements, is represented in Figure 10.11.

Error, defect, waste investigations and follow-up

The investigation of errors, defects and waste can provide valuable error and waste prevention information. The general method is based on:

- *collecting* data and information relating to the error, defect or waste;
- *checking* the validity of the evidence; and
- *selecting* the evidence without making assumptions or jumping to conclusions.

The result of the analysis is then used to:

- *decide* the most likely cause(s) of the error, defect or waste;
- *notify* immediately the person(s) able to take corrective action;
- *record* the findings and outcomes; and
- *record* them to everyone concerned, to prevent recurrence.

The investigation should not become an inquisition to apportion blame, but a focus on the positive preventive aspects.

Hopefully errors, defects and waste are not investigated so frequently that the required skills are rote learned. Since these skills are not easily learned in a classroom, one way to overcome this gap in experience is the development of a sequence of questions to form the basis of an error/defect/waste investigation questionnaire. The proposed structure for such an investigation could be:

- *People*: duties, information, supervision, instruction, training, attitudes, etc.
- *Systems*: procedures, instructions, monitoring, control methods, etc.
- *Plant/equipment*: description, condition, controls, maintenance, suitability, etc.
- *Environment*: climatic, space, humidity, noise, etc.

A formal improvement process to find a better method, perhaps using Kaizen events, is recommended in some instances (see Chapter 15 on Continuous Improvement).

INTERNAL AND EXTERNAL MANAGEMENT SYSTEM AUDITS AND REVIEWS

A good quality management system (QMS see Chapter 14) will not function without adequate audits and reviews. The system reviews, which need to be carried out periodically and systematically, are conducted to ensure that the system achieves the required effect, whilst audits are carried out to make sure that actual methods are adhering to the QMS. The reviews should use the findings of the audits, for failure to operate according to the plan often signifies difficulties in doing so. A re-examination of the processes actually being used may lead to system improvements unobtainable by other means.

A schedule for carrying out the *audits* should be drawn up: different activities perhaps requiring different frequencies. All processes should be audited at least once during a specified cycle, but not necessarily all at the same audit. For example, every three months a selected random sample of the processes could be audited, with the selection designed so that each process is audited at least once per year. However, there must be a facility to adjust this on the basis of the audit results.

A QMS *review* should be instituted, perhaps every 12 months, with the aims of:

- ensuring that the system is achieving the desired results;
- revealing defects or irregularities in the system;
- indicating any necessary improvements and/or corrective actions to eliminate waste or loss;
- checking on all levels of management;
- uncovering potential danger areas; and
- verifying that improvements or corrective action procedures are effective.

Clearly, the methods for carrying out the audits and reviews, and the results, should be documented and subjected to review. Useful guidance on quality management system audits is given in the international standard, ISO 19011, 'Guidelines for auditing management systems'.

The assessment of a QMS against a particular standard or set of requirements by internal audit and review is known as a *first-party* assessment or approval scheme. If an *external* customer makes the assessment of a supplier against either its own or a national or international standard, a *second-party* scheme is in operation. The assessment by an independent organization (not connected with any contract between customer and supplier, but acceptable to them both) is known as an *independent third-party* assessment scheme. This type of assessment often results in some form of certification or registration from the assessment body.

One advantage of the third-party schemes is that they obviate the need for customers to make their own detailed checks, potentially saving both suppliers and customers time and money as well as avoiding issues of commercial confidentiality. Just one knowledgeable organization has to be satisfied rather than a multitude with varying levels of competence. This method can be used to certify suppliers for contracts without further checking, but good customer/supplier relations often include second-party extensions to the third-party requirements and audits.

Each certification body has its own recognized mark; once a company has passed an assessment, they may use the registered mark of the certification body on their literature, letterheads and marketing. There are also publications containing lists of organizations whose quality management systems and/or products and services have been assessed. To be of value, the certification body must itself be recognized and, usually, assessed and registered with a national or international accreditation scheme.

Many organizations have found that the effort of designing and implementing a QMS good enough to stand up to external independent third-party assessment has been extremely rewarding in:

- better staff involvement and improved morale;
- better process control and improvement;
- reduced wastage and costs; and
- reduced customer service costs.

This is also true for organizations that have obtained third-party registrations and for supply companies which still insist on their own second-party assessment. The reason for this is that most of the standards on QMS whether national, international or company-specific are now very similar indeed. For example, a system that meets the requirements of the ISO 9001:2015 standard (see Chapter 14) should meet the requirements of most other standards, with only the slight modifications and small emphases here and there required for specific customers. It is the authors' experience that an assessment carried out by one of the good, independent certified assessment bodies is a rigorous and delving process.

Internal system audits and reviews should be positive and conducted as part of the preventive strategy and not as a matter of expediency resulting from problems. They should not be carried out only prior to external audits, nor should they be left to the external auditor – whether second- or third-party. An external auditor discovering discrepancies between actual and documented systems will be inclined to ask why the internal review methods did not discover and correct them.

Any management team needs to be fully committed to operating an effective QMS for all the people within the organization not just the staff in the 'quality department'. The system must be planned for it to be effective and achieve its objectives in an uncomplicated way. Having established and documented the processes, it is necessary to ensure that they are working and that everyone is operating in accordance with them. The system once established is not static; it should be flexible to enable the constant seeking of improvements or streamlining.

Quality auditing standard

The growing use of standards internationally emphasizes the importance of auditing as a management tool for this purpose. There are several guides available to management systems auditing (e.g. ISO 19011); and the guidance provided in these can be applied equally to any one of the three specific and yet different auditing activities:

1 **First-party or internal audits** – are carried out by an organization on its own systems either by staff who are independent of the systems being audited, or by an outside agency.
2 **Second-party audits** – are carried out by one organization (a purchaser or its outside agent) on another with which it either has contracts to purchase products or services, or intends to do so.
3 **Third-party audits** – are carried out by independent agencies to provide assurance to existing and prospective customers for the product or service.

Audit objectives and responsibilities, including the roles of auditors and their independence and those of the 'client' or auditee, should be understood. The generic steps involved are as follows:

- **initiation,** including the audit scope and its frequency;
- **preparation,** including review of documentation, programme and working documents;
- **execution,** including the opening meeting, examination and evaluation, and collecting evidence, observations and closing the meeting with the auditee;
- **report,** including its preparation, content and distribution; and
- **completion,** including report submission and retention.

Attention should be given at the end of the audit to corrective action and follow-up, and the improvement process should be continued by the auditee after the publication of the audit report. This may include a call by the client for a verification audit of the implementation of any corrective actions specified.

Any instrument which is developed for assessment, audit or review may be used at several stages in an organization's history:

- before starting an improvement programme to identify 'strengths' and 'areas for improvement', and focus attention (at this stage a parallel cost of quality exercise may be a powerful way to overcome scepticism and get 'buy-in');
- as part of a programme launch, especially using a 'survey' instrument; and
- every one or two years after the launch to steer and benchmark.

The systematic measurement and review of operations is one of the most important management activities of any organization. Self-assessment, audit and review should lead to clearly discerned strengths and areas for improvement by focusing on the relationship between the planning, people, processes and performance. Within any successful organization these will be regular activities.

BIBLIOGRAPHY

Ahern, D.M., Clouse, A. and Turner, R. *CMMI Distilled: A Practical Introduction to Integrated Process Improvement* (2nd edn), Addison-Wesley, Boston, 2004.

Arter, D.R., Cianfrani, C.A. and West, J.E. *How to Audit the Process Based BMS* (2nd edn), ASQ, Milwaukee, 2012.

Blazey, M.L. *Insights to Performance Excellence 2013–2014: Understanding the Integrated Management System and the Baldrige Criteria*, NIST, Gaithersburg, 2013.

BQF. *The X-factor, Winning Performance Through Business Excellence*, European Centre for Business Excellence/British Quality Foundation, London, 1999.

BQF. *The Model in Practice and The Model in Practice 2*, British Quality Foundation, London, 2000 and 2002 (Prepared by the European Centre for Business Excellence).

British Standard 6143. *Guide to the Determination and Use of Quality Related Costs*, BSI, London, 1981.

British Standards Institute (and International Standards Organisation), *BS EN ISO 19011 Guidelines for Auditing Management Systems*, BSI, London, 2015

EFQM. *Assessing for Excellence – A Practical Guide for Self-assessment*, EFQM, Brussels, 2013.

Feigenbaum, A.V. *Total Quality Control* (4th edn), McGraw-Hill, New York, 2004.

International Standards Organization. *ISO 19011: Guidelines on Quality and/or Environmental Management Systems Auditing*, ISO, 2003.

JUSE (Union of Japanese Scientists and Engineers). *Deming Prize Criteria*, JUSE, Tokyo, 2015.

Keeney, K.A. *The ISO 9001:2000 Auditor's Companion*, Quality Press, Milwaukee, 2002.

NIST (National Institute of Standards and Technology), *The Malcolm Baldrige National Quality Award Criteria*, NIST, Gaithersburg, 2015.

Porter, L.J. and Tanner, S.J. *Assessing Business Excellence* (2nd edn), Butterworth-Heinemann, Oxford, 2003.

Porter, L.J., Oakland, J.S. and Gadd, K.W. *Evaluating the European Quality Award Model for Self-Assessment*, CIMA, London, 1998.

Pronovost, D. *Internal Quality Auditing*, Quality Press, Milwaukee, 2000.

Tricker, R. *ISO 9001 Audit Procedures*, Butterworth-Heinemann, Oxford, 2008.

CHAPTER HIGHLIGHTS

Frameworks for self-assessment

- Many organizations are turning to lean quality models to measure and improve performance. These frameworks include the Japanese Deming Prize, the US Baldrige Award and in Europe the EFQM Excellence Model.

- The nine components of the Excellence Model are: leadership, strategy, people, partnerships and resources, and processes, products and services (ENABLERS), people results, customer results, society results and key performance results (RESULTS).

- The various award criteria provide rational bases against which to measure progress towards lean quality in organizations. For example, self-assessment against the EFQM Excellence Model should be a regular activity as it identifies opportunities for improvement in performance through processes and people.

Methodologies for lean quality assessment

- Self-assessment against the Excellence Model may be performed using RADAR: Results, Approach, Deployment, Assessment and Review.

- There are a number of approaches for self-assessment, including groups/workshops, surveys, pro formas, matrices, award simulations, activity/process audits or hybrid approaches.

Capability Maturity Model Integration (CMMI) assessments

- In CMMI a set of requirements for increasing levels of maturity are defined, and assessments are made to determine how mature the assessed entity is.

- Maturity is defined in a framework that sets out the 'criteria' that make up excellence in the area of focus; for each criterion, a set of requirements are arranged in a hierarchy, usually with five levels: basic at level 1 and world-class at level 5, with possible intermediate steps in between.

- CMMI assessments should prompt appropriate action by showing both the current status and what needs to be done next to advance up the maturity curve; the assessments often lead to standardized reports.

Securing lean quality by audit and review of the management systems

- There are two major elements of error/defect/waste prevention: checking the system, and error/defect/waste investigations and follow-up. Six methods of checking the quality systems are in general use: audits and reviews, surveys, inspections, tours, sampling and scrutinize.

- Investigations proceed by collecting, checking and selecting data, and analysing it by deciding causes, notifying people, recording and reporting findings and outcomes.

Internal and external management system audits and reviews

- A good management system will not function without adequate audits and reviews. Audits make sure the actual methods are adhering to documented procedures. Reviews ensure the system achieves the desired effect.
- System assessment by internal audit and review is known as first-party, by external customer as second-party and by an independent organization as third-party certification. For the last to be of real value the certification body must itself be recognized.

Benchmarking and change management

11

THE WHY AND WHAT OF BENCHMARKING

Product, service and process improvements can take place only in relation to established standards, with the improvements then being incorporated into new standards. *Benchmarking*, one of the most transferable aspects of Rank Xerox's approach to total quality management, and thought to have originated in Japan, measures an organization's operations, products and services against those of its competitors in a ruthless fashion. It is a means by which targets, priorities and operations that will lead to competitive advantage can be established.

As with performance measurement, the construction sector has had some difficulty in implementing benchmarking to full effect. Projects vary widely in terms of scope, process and technique, and because each project has its own unique set of circumstances in terms of the negotiated time frame, price and contract, comparisons between projects can be difficult to make. Furthermore, the sector has historically focused on projects and when seen through this lens, construction is different to other sectors of the economy: each project is seen as separate and unique, and this is an obstacle to the transfer of management innovation from other sectors. Only relatively recently have leading academics and managers developed a process view of construction. This perspective sees projects within a continuum of process change, with key performance attributes being compared between projects.

Sutter Health, whose VNGC Campus case study (CS2) is featured in the book, as a client, has introduced a range of measures with which it has benchmarked its projects for more than ten years. These give an insight into how a lean quality-based organization can create a focus for improvement on its projects. Sutter has reported the following areas of measurement for its projects at conferences over this period:

- comparison to schedule and budget;
- total on-site time labour hours;
- labour productivity overall;
- labour productivity in specific trades compared to trade baselines;
- rework 15–80 per cent compared to trade baselines;
- accuracy of the as-built building by laser scanning compared to the BIM model;
- numbers of RFIs;
- numbers of change orders;
- percentage of failed inspections.

In all industries it can be a challenge to find similar companies to benchmark against, unless it is done through a third-party mediated benchmarking exercise. The authors have come across two examples within the construction sector that are worth reflection. A truss company in Australia sought out a truss company using similar technology in the US to benchmark against; this was possible because they were not competitors in the same market. They openly compared all the critical aspects of performance. The other case is of a construction business wanting to benchmark its administrative efficiency. It worked out that it could compare the cost of paying staff salaries against industry leaders in banking and insurance. Of course this was not assessing the efficiency of the entire administrative function, but gave an accurate comparison of the efficiency of a single function against market leaders in other industries, and hence a glimpse into the efficiency of its own administration. It is often useful to find simple measures which can act as surrogate indicators for a broader area of performance. These examples show ways in which benchmarking can be creatively used to gauge performance.

There are many drivers for benchmarking, including the external ones:

- customers continually demand better quality, lower prices, shorter lead-times, etc.;
- competitors are constantly trying to get ahead and steal markets;
- legislation – changes in our laws place ever greater demands for improvement.

Internal drivers include:

- targets which require improvements on our 'best ever' performance;
- technology – to benefit fully from introducing new technologies (e.g. BIM) a fundamental change in processes is often required; and
- self-assessment results provide opportunities to learn from adapting best practices.

The word 'benchmark' is a reference or measurement standard used for comparison, and benchmarking is the continuous process of identifying, understanding and adapting best practice and processes that will lead to superior performance.

Benchmarking is *not*:

- a panacea to cure the organization's problems, but simply a practical tool to drive up process performance;
- a cost reduction exercise, though many benchmarking studies will result in improved financial performance;
- industrial tourism because though study tours have their place, proper benchmarking goes beyond 'tourism' and promotes a real understanding of the enablers to outstanding results;
- spying because the benchmarking code of conduct ensures the work is done with the agreement and openness of all parties;
- catching up with the best because the aim is to reach out and extend the current best practice (by the time we have caught up, the benchmark will have moved anyway).

There may be many reasons for carrying out benchmarking. Some of them are set against various objectives in Table 11.1. The links between benchmarking and lean quality are clear: establishing objectives based on industry best practice should directly contribute to the better meeting of the internal and external customer requirements.

The benefits of benchmarking can be numerous, but include:

- creating a better understanding of the current position;
- heightening sensitivity to changing customer needs;
- encouraging innovation;
- developing realistic stretch goals; and
- establishing realistic action plans.

TABLE 11.1 Reasons for benchmarking

Objectives	Without benchmarking	With benchmarking
Becoming competitive	• Internally focused • Evolutionary change	• Understanding of competitiveness • Ideas from proven practices
Industry best practices	• Few solutions • Frantic catch up activity	• Many options • Superior performance
Defining customer requirements	• Based on history or gut feeling • Perception	• Market reality • Objective evaluation
Establishing effective goals and objectives	• Lacking external focus • Reactive	• Credible, unarguable • Proactive
Developing true measures of productivity	• Pursuing pet projects • Strengths and weaknesses not understood • Route of least resistance	• Solving real problems • Understanding outputs • Based on industry best practices

The American Productivity and Quality Center (APQC) provides an 'Open Standards Benchmarking' service that was launched in 2004 (see www.apqc.org/benchmarking). This is powered by a database underwritten by organizations that support the creation of common, open frameworks to measure processes and is based on APQC's widely adopted *Process Classification Framework* (see Chapter 12). The database contains more than 1,200 standardized measures, spanning people, process and technology. The same source has identified that the average return from benchmarking is typically **five times** the cost of the study, in terms of reduced costs, increased sales, greater customer retention and enhanced market share.

There are four basic categories of benchmarking:

- **Internal** – the search for best practice of internal operations by comparison, a multi-site comparison of process management performance in terms of planning reliability, safety and quality management outcomes.
- **Functional** – seeking functional best practice outside an industry, for example a mining company benchmarking preventative maintenance of pneumatic/hydraulic equipment with Disney; a construction company benchmarking its salary payment processes with an insurance company; or a mechanical contractor benchmarking service call outs with a white goods maintenance company.
- **Generic** – comparison of outstanding processes irrespective of industry or function, for example a restaurant chain benchmarking kitchen design with US nuclear submarine fleet to improve restaurant to kitchen space ratios.
- **Competitive** – specific competitor to competitor comparisons for a product, service or function of interest retail outlets comparing price performance and efficiency of internet ordering systems; design firms benchmarking financial ratios through an independent benchmarking club; or an Australian truss manufacturer benchmarking its processes with a similar business in the USA.

THE PURPOSE AND PRACTICE OF BENCHMARKING

The evolution of benchmarking in an organization is likely to progress through four focuses. Initially attention may be concentrated on competitive products or services, including, for example, design, development and operational features. This should develop into a focus on industry best practices and may include, for example, aspects of distribution or service. The real breakthroughs are when organizations focus on all aspects of the in total business performance, across all functions and aspects, and address current *and projected* performance gaps. This should lead to the focus on processes and true continuous improvement.

At its simplest, competitive benchmarking, the most common form, requires every department or function to examine itself against the counterpart in the best competing companies. This includes a scrutiny of all aspects of their activities. Benchmarks which may be important for *customer satisfaction*, for example, might include:

- product or service quality and consistency;
- correct and on-time delivery;
- speed of response on new product development; and
- correct billing.

For internal *impact* the benchmarks may include:

- waste, rejects or errors;
- inventory levels/work in progress;
- costs of operation; and
- staff turnover.

The task is to work out what has to be done to improve on the competition's performance in each of the chosen areas.

Benchmarking is very important in the 'administration' areas, since it continuously measures services and practices against the equivalent operation in the toughest direct competitors, or organizations renowned as leaders in the areas, even if they are in the same organization. An example of quantitative benchmarks in absenteeism is given in Table 11.2.

All of the case study companies use benchmarking in numerous ways to drive process improvement. The JB Henderson case study (CS8) illustrates how value-adding labour time was being measured to drive a focus on productive time. The High-Technology Manufacturing case study (CS4) indicates that the company set cost and time to market improvement targets for its building projects and measured itself against those. To reduce the amount of documentation used to describe its requirement, the company set 'doc diet' reduction targets and measured its performance against those goals as well.

Technologies and conditions vary between different industries and markets, but the basic concepts of measurement and benchmarking are of general validity. The objective should be to produce products and services that conform to the requirements of the customer in a never-ending improvement environment. The way to accomplish this is to use a continuous improvement cycle in all the operating departments – nobody should be exempt. Benchmarking is not a separate science or unique theory of management but rather another strategic approach to getting the best out of people and processes – to deliver improved performance.

The purpose of benchmarking then is predominantly to:

- **change** the perspectives of executives and managers;
- **compare** business practices with those of world-class organizations;
- **challenge** current practices and processes;
- **create** improved goals and practices for the organization.

TABLE 11.2 Quantitative benchmarking in absenteeism

Organization's absence level (%)	Productivity opportunity
Under 3	This level matches an aggressive benchmark that has been achieved in 'excellent' organizations.
3–4	This level may be viewed within the organization as a good performance – representing a moderate productivity opportunity improvement.
5–8	This level is tolerated by many organizations but represents a major improvement opportunity
9–10	This level indicates that a serious absenteeism problem exists.
Over 10	This level of absenteeism is extremely high and requires immediate senior management attention

As a managed process for change, benchmarking uses a disciplined structured approach to identify **what** needs to change; **how** it can be changed; and the **benefits** of the change. It also creates the desire for change in the first place. Any process or practice that can be defined can be benchmarked but the focus should be on those which impact on customer satisfaction and/or business results – financial or non-financial.

For organizations which have not carried out benchmarking before, it may be useful initially to carry out a simple self-assessment of their readiness in terms of:

- how well processes are understood;
- how much customers are listened to; and
- how committed the senior team is.

Table 11.3 provides a simple pro forma for this purpose. The score derived gives a crude guide to the readiness of the organization for benchmarking:

0–15   Ready for benchmarking.
0–16   Some further preparation required before the benefits of benchmarking can be fully derived.
0–17   Some help is required to establish the foundations and a suitable platform for benchmarking.

The benchmarking process has five main stages which are all focused on trying to measure comparisons and identify areas for action and change (Figure 11.1). The detail is as follows:

PLAN the study

- Select the process(es) for benchmarking.
- Bring together the appropriate team to be involved and establish roles and responsibilities.
- Identify and define benchmarks and measures for data collection.
- Identify best competitors or operators of the process(es), perhaps using customer feedback or industry observers.
- Document the current process(es).

- Plan
  - Initial research
  - Record current performance

- Collect
  - Identify benchmarking partners
  - Site visits

- Analyse
  - Identify best practices
  - Identify enablers

- Adapt
  - Share best practice learning
  - Adapt and implement best practice

- Review
  - Post completion review

FIGURE 11.1 The benchmarking methodology

TABLE 11.3 Is the organization ready for benchmarking?

After studying the statements below tick one box for each to reflect the level to which the statement is true for the organization.

	Most	*Some*	*Few*	*None*
Processes have been documented with measures to understand performance.	☐	☐	☐	☐
Employees understand the processes that are related to their own work.	☐	☐	☐	☐
Direct customer interactions, feedback or studies about customers influence decisions about products and services.	☐	☐	☐	☐
Problems are solved by teams.	☐	☐	☐	☐
Employees demonstrate by words and deeds that they understand the organization's mission, vision and values.	☐	☐	☐	☐
Senior executives sponsor and actively support quality improvement projects.	☐	☐	☐	☐
The organization demonstrates by words and by deeds that continuous improvement is part of the culture.	☐	☐	☐	☐
Commitment to change is articulated in the organization's strategic plan.	☐	☐	☐	☐
Add the columns:	☐	☐	☐	☐
	x 6 =	x 4 =	x 2 =	Zero
Multiply by the factor	☐	☐	☐	☐
Obtain the grand total?	══	══	══	══

COLLECT data and information

- Decide information and data collection methodology, including desk research.
- Record current performance levels.
- Identify benchmarking partners.
- Conduct a preliminary investigation.
- Prepare for any site visits and interact with target organizations.
- Use chosen data collection methodology.
- Carry out site visits.

ANALYSE the data and information

- Normalize the performance data, as appropriate.
- Construct a matrix to compare current performance with benchmarking competitors'/partners' performance.
- Identify outstanding practices.
- Isolate and understand the process enablers as well as the performance measures.

ADAPT the approaches

- Catalogue the information and create a 'competency profile' of the organization.
- Develop new performance level objectives/targets/standards.
- Vision alternative process(es) incorporating best practice enablers.
- Identify and minimize barriers to change.
- Develop action plans to adapt and implement best practices, make process changes and achieve goals.
- Implement specific actions and integrate them into the organization.

REVIEW performance and the study

- Monitor the results/improvements.
- Assess outcomes and learnings from the study.
- Review benchmarks.
- Share experiences and best practice learnings from implementation.
- Review relationships with target/partner organizations.
- Identify further opportunities for improving and sustaining performance.

In a typical benchmarking study involving several organizations, the study will commence with the *Plan* phase. Participants will be invited to a 'kick-off' meeting where they will share their aspirations and objectives for the study and establish roles and responsibilities. Participants will analyse their own organization to understand the strengths and areas for improvement. They will then agree to appropriate measures for the study.

It is important to include participants from all levels in the organization to get the appropriate level of 'buy-in'. So, for example, if a construction company were to introduce benchmarking between projects, it is important to involve the key stakeholders at the head office, project engineering and management as well as the project supervisory level, otherwise implementation may become stalled down the track.

In the *Collect* phase, participants will collect data on their current performance based on the agreed measures. The benchmarking partners will be identified using a suitable screening process, and the key learning points will be shared. The site visits will then be planned and conducted with appropriate training. Five to seven site visits might take place in each study.

Data collected from the site visits will be *Analysed* in the next phase to identify best practices and the enablers which deliver outstanding performance. The reports from this phase will capture the learning and key outcomes from the site visits and present them as the main process enablers linked to major performance outcomes.

In the *Adapt* phase, the participants will attend a feedback session where the conclusions from the study will be shared, and they will be assisted in adapting them to their own organization. Reports to partners should be issued after this session. A 'subject expert' is often useful in benchmarking studies to ensure good learning and adaptation at this stage.

The final phase of the study will be a post-completion *Review*. This will give all the participants and partners valuable feedback and establish, above all else, what actions are required to sustain improved performance. Best practice databases may be created to enable further sharing and improvement amongst participants and other members of the organization.

THE ROLE OF BENCHMARKING IN CHANGE

One aspect of benchmarking is to enable organizations to gauge how well they are performing against others who undertake similar tasks and activities. But a more important aspect of best practice benchmarking is gaining an understanding of *how* other organizations achieve superior performance. For example, a good benchmarking study in customer satisfaction and retention will provide its participants with data and ideas on how excellent organizations undertake their activities and demonstrate best practices that may be adopted or adapted and used.

This new knowledge will result in the benchmarking team being able to judge the gap between leading and lesser performances as well as plan considered actions to bring about changes to bridge that gap. These changes may be things that can be undertaken quickly, with little adaptation and at a minimum of cost and disruption. Such changes, often brought about by the effected operational team, are called 'quick wins'. This type of change is incremental and carries low levels of risk but usually also lower levels of benefit.

Quick wins will often give temporary or partial relief from the problems associated with poor performance and tend to address symptoms not the underlying 'diseases'. They can have a disproportionate favourable physiological impact upon the organizations. Used well, quick wins should provide a platform from which longer lasting changes may be made, having created a feeling of movement and success. All too often, however, once quick wins are implemented there is a tendency to move on to other areas, without either fully measuring the impact of the change or getting to the root cause of a performance issue.

Quick wins are clearly an important weapon in effecting change but must be followed up properly to deliver sustainable business improvement through the adoption of best or, at least, good practice. The changes needed to do this will usually be of a more fundamental nature and require investment in effort and money to implement. Such changes will need to be carefully planned and systematically implemented as a discrete change project or programme of projects. They carry substantial risk, if not systematically managed and controlled, but they have the potential for significant improvement in performance. These types of change projects are sometimes referred to as 'step change' or 'breakthrough' projects/programmes (see Figure 11.2).

The Highways England Lean Maturity Assessment approach (HELMA) introduced in Chapter 10 provides a basis for benchmarking the whole supply chain and combines lean and quality in its approaches.

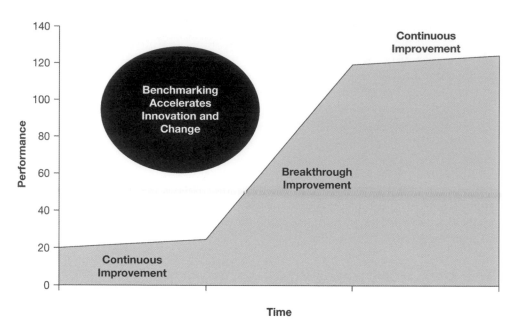

FIGURE 11.2 Benchmarking; breakthrough and continuous improvement

Whatever type of change is involved, a key ingredient of success is taking the people along. A first-class communication strategy is required throughout and beyond any change activity as well as the linked activity of stakeholder management. The benchmarking efforts need to fit into the change model deployed (see Figure 11.3). Many change models exist in diagrammatic form and are often quite similar in both intent and structure. Such a model may be considered as a 'footprint' that will lead to the chosen destination: in this case, the desired performance improvements through adoption of best practice. The footprint in Figure 11.3 demonstrates where benchmarking activities link into the general flow of change activity leading to better results.

The success and benefits derived from any benchmarking and change-related activity are directly related to the excellence of the preparation. It is necessary to consider both the 'hard' and 'soft' aspects represented in Figure 11.3 and to systematically plan to meet and overcome any difficulties and challenges identified.

COMMUNICATING, MANAGING STAKEHOLDERS AND LOWERING BARRIERS

The importance of first-class communication during benchmarking and resulting change can never be over-emphasized. A vital element of excellent communication is targeting the right audience with the right message in the right way at the right time. A scattergun approach to communication rarely has the intended impact.

In any benchmarking study it is a wise and well-founded investment in time and effort to define and understand the key stakeholders. The increased use of the term 'stakeholder' in business language used to describe any group or individual that has some, however small, vested interest or influence in the proposed change is to be welcomed. Stakeholders are frequently referred to in generic groupings and may be either internal or external to an organization or business. The importance of forming, managing and maintaining good working relations with these groups is widely acknowledged and accepted.

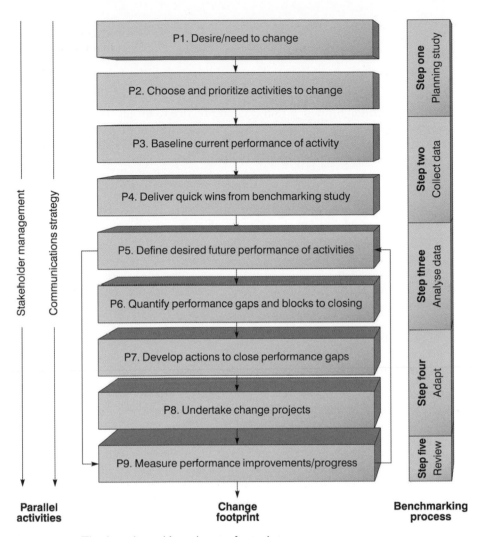

FIGURE 11.3 The benchmarking change footprint

The reality is that this activity is frequently not performed well in benchmarking. A disgruntled or ignored stakeholder with high direct organizational power or influence can easily derail the intent and hard work of others. Stakeholders with less direct power or influence can, at best, provide an unwelcome and costly distraction from the main objectives of a benchmarking study. The art of stakeholder management is to proactively head off any major confrontations. This means really understanding the stakeholders' needs and their potential to do both good and ill.

The burden of effective stakeholder management rests with the benchmarking team charged with stimulating change. They may need the ongoing patronage and support of people outside their direct control. In any good benchmarking study, early thought will be given to who the stakeholders are, and this will be valuable input to developing a robust stakeholder management strategy.

The elements of successful stakeholder management should include:

1   defining and mapping the stakeholder groupings;
2   analysing and prioritizing these groupings;

3   researching the key players in the most important groupings;
4   developing a management strategy;
5   deploying the strategy by tactical actions;
6   reviewing effectiveness of the strategy and improving the future approach.

Objective measurement is also key to targeting change activity wisely. Benchmarking project budgets are often limited, and it is good practice to target such discretionary spend at changes and improvements that will deliver the best return for their investment. Systematic measurement will provide a reliable baseline for making such decisions. By relating current performance against desired performance it should be possible to define both the gaps and appreciate the scale of improvements required to achieve the desired change.

Benchmarking studies add an extra dimension by understanding the levels of performance that best practices and leading organizations achieve. This allows realistic and sometimes uncomfortable comparisons with what an organization is currently able to achieve and what is possible. This is especially useful when setting stretch but realistic targets for future performance.

Base lining performance will allow teams to monitor and understand how successful they have been in delivering beneficial change. Used with care, as part of an overall communications strategy, successes on the road to achieving superior performance through change is a powerful motivator and useful influencing tool. Many organizations have clearly defined sets of performance measures, some self-imposed and some statute-based – these should be used, if in existence. If the interest is in customer satisfaction and retention, for example, a generic but good starting point might be:

- internal measures (the lead/predictor measures) – production cycle times, unit costs, defect rate found (quality) and complaints resolved; and
- external measures (the lag/reality measures) – customer satisfaction (perception), customer retention and complaints received, time to market with a new product/service.

The benchmarking activity may provide teams with ideas on how they might change the way goods or services are produced and delivered. They will need to prioritize this opportunity, however, to deliver best value for time and money invested and to ensure the organization does not become paralysed by initiative overload – whilst making improvements the day job has to continue!

The benchmarking data collected will give a clear steer to the areas that require the most urgent attention but decisions will still have to be made. Measurement and benchmarking are tools not substitutes for management and leadership – the data on its own cannot make the decisions.

CHOOSING BENCHMARKING-DRIVEN CHANGE ACTIVITIES WISELY

As we have seen, benchmarking studies should fuel the desire to undertake change activities, but the excitement generated can allow the desire for change to take on a life of its own and irrational and impractical decisions can follow. These can result in full or partial failure to deliver the desired changes and waste of the valuable financial and people resources spent on the benchmarking itself.

Organizations should resist the temptation to start yet another series of improvement initiatives without any consideration of their impact upon existing initiatives and the 'business as usual' activities as well as the overall strategy, of course. It is important to target the change wisely, and a number of key questions need to be answered, including:

TABLE 11.4 Simple decision tool for choosing change activities

No.	Filter Test	Yes	No
1	Does the benchmarking-driven proposed change support the achievement of one or more of the defined business goals?	Allow the opportunity to move forward for consideration	Decline the opportunity or defer taking forward and schedule a review
2.	Does the change require financial and people resources above those agreed for the current budget round?	Prepare a business case within a project definition for consideration by senior management	Pass the opportunity to local operational management to undertake the changes as 'quick win' initiative
3	Will current improvement activity be adversely impacted by the envisaged new changes?	Consider the relative merits and benefits of new and existing change initiates and amalgamate or amend or cancel existing initiates	Allow change project to proceed and add to the controlled list of overall change projects
4.	Is the required additional financial and people resource needed to undertake new change projects available?	Senior management agree and sign off project definition and project begins	Senior management prioritize change activity agreeing necessary slippage or deferment or cancellation of some change projects

- Do we fully understand the scale of the change?
- Do we have the financial resources to support the change?
- Do we have the people resources to undertake the change?
- Do we have the right skills available to undertake the change?
- Do we fully understand the operational impact during the change?
- Can beneficial changes be made without major disruption to the business?
- Will the delivered change support achievement of our business goals?
- What will the new changes do to existing change initiatives?
- Is the organization culturally ready for change?

Table 11.4 shows a simple decision-making tool to help consider the opportunities that are presented. The process may be viewed as a series of filters. It is assumed that the organization has defined business goals.

Work to improve quality management systems (QMS) benefits from the use of benchmarking. This perhaps should include making the QMS more process-based than previously and making better use of web-based technologies.

Benchmarking studies in various sectors have provided insight on, for example, the potential for new technology to radically change existing product development processes. Benchmarking should be an integral part of each process re-engineering project that is undertaken. The external perspective provided by the benchmarking studies help employees see how things could be different ('thinking outside the box'), and provide valuable input to the steps required to implement new processes (see also Chapters 12 and 13).

The drivers of change are everywhere but properly conducted systematic benchmarking studies can assist in defining clear objectives and help their effective deployment through well-executed change management. Best practice benchmarking and change management clearly are bedfellows. If well understood and integrated they can deliver lasting improvements in performance, satisfying all stakeholder needs. Benchmarking is an efficient way to promote effective change by learning from the successful experiences of others and putting that learning to good effect.

A FRAMEWORK FOR ORGANIZATIONAL CHANGE

Based on research carried out by The Oakland Institute (www.theoaklandinstitute.com) and other sources, an organizational change framework has been developed by the authors and their colleagues. This is a powerful aid for organizations wishing to undertake any change programme or are in the processes of delivering change and want to increase their chances of success.

It identifies two main constructs of change management, which can be better understood within the overall framework for change as shown in 'the figure of 8' diagram depicted in Figure 11.4. Based on the results of the research, the change framework has two interacting cycles:

- readiness for change; and
- implementing change.

The experiences of many organizations that have launched change programmes such as Lean, Six Sigma, or a combination of the two, or wish to implement change following benchmarking is that the first part – readiness – is often not at all well understood or developed. This can be caused by a desire to rush into implementation, with huge emphasis on training programmes and projects without particular attachment to strategy.

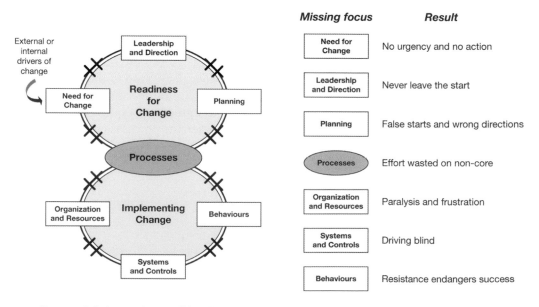

Successful change is possible when all 7 key elements are addressed and readiness is achieved before starting changing processes, organization, systems, other changes

FIGURE 11.4 The organizational change framework

To break into the top circle we need to start with the *Drivers of Change*. It is important to understand what the key drivers for change inside or outside the organization are. In order to focus stakeholders' desire for change the *Need for Change* must be understood and articulated. This is where leaders give meaning to the required change without which, as many organizations later discover, initial enthusiasm and energy quickly dissolves. For example, what were the drivers for the introduction of digital technology into BBC World Service: reduced costs, better programme reception and more effective programme making?

Good output from the *need for change* stage will provide focus for the stakeholders' *desire for change*. Clear expression of what drives the *need for change* is key, as this is what will drive clear and consistent *Leadership and Direction*. This in turn converts needs into expectations – vision, goals, measured objectives and targets, which will set the expectations for change. Robust *Planning* allows the priorities to emerge and focuses people's minds on the strategic objectives.

The 'implementation of change' stage is where the risk of failure is greatest; this is a minefield for the unsuspecting. Worse than that, most managers tend to think they have entered the minefield at the wrong point. Trying to change behaviours, such as attitudes and empowerment for example, is a frequent and highly risky starting point for many 'change programmes'. Behavioural change must be bedded in the reality of the business, and the performance improvements that are being sought.

Following clarity on need, clear and unambiguous leadership and direction and good detailed planning the first port of call must be the organization's *Processes* in which people live and work. Whether we like it or not, and whether we have worked them out or not, the processes drive the way the *Organization and Resources* work – the structure, roles, competencies and resources deployed. Performance measures and technology then support the organization's *Systems and Controls*. This is where *Behaviour* comes in – all of the above drive behaviour – the way the organization is structured: who my boss is; how I am measured; the processes and systems, good or bad, that I live and work in. When managers talk about attitudes of the people and their assumptions, it might be interesting for them to understand where these come from. Attitudes stem from beliefs and values, both of which are management's responsibility to influence. Most people start work for an organization with positive attitudes and behaviours, and it is frequently the systems, the environment and the assumptions and unwritten 'rules' that cause problems and deterioration (see Chapter 13 and the section on 'Assumption Busting').

With the wrong attitudes and behaviours, the 'figure of 8' improvement process shown in Figure 11.4 stalls. It is our behaviour which makes the processes work, or not, resulting in achievements in quality and on-time delivery, or not, and reinforcing the change, or not. Taking another circuit round the Readiness for Change portion of the 'figure of 8' model will strengthen the foundations underpinning the change strategy, validate the change protocols and strengthen the motivation for change.

It is often difficult for managers to stand back and view their work on change in a holistic fashion. Personal agendas or simply perceptions can lead to a push on Human Resource issues or Information Technology issues, distorting the holistic view of the change initiative. An output of the 'leadership' box may be scenarios for reacting to the need for change. There's usually more than one way to address the need, and managers often benefit from a structured approach that drives questions that lead to alternative routes to change.

Figure 11.4 also shows on the right-hand side the likely results of missing focus in any of the seven key elements. This can act as a simple diagnostic, starting with the result or effect and working back to the likely area(s) lacking time and attention. The figure of 8 framework can be adapted to any aspect of change management.

BIBLIOGRAPHY

Bendell, T. *Benchmarking for Competitive Advantage*, Longman, London, 1993.

Camp, R.C. *Business Process Benchmarking: Finding and Implementing Best Practice*, ASQ Quality Press, Milwaukee, 1995.

Camp, R.C. *Global Cases in Benchmarking: Best Practices from Organizations around the World*, Quality Press, Milwaukee, 1998.

Spendolini, M.J. *The Benchmarking Book*, ASQ, Milwaukee, 1992.

Zairi, M. *Benchmarking for Best Practice*, Butterworth-Heinemann, Oxford, 1996.

Zairi, M. *Effective Management of Benchmarking Projects*, Butterworth-Heinemann, Oxford, 1998.

CHAPTER HIGHLIGHTS

The why and what of benchmarking

- Benchmarking measures an organization's products, services and processes to establish targets, priorities and improvements, leading, in turn, to competitive advantage and/or cost reductions.
- Benefits of benchmarking can be numerous and include creating a better understanding of the current position, heightening sensitivity to changing customer needs, encouraging innovation, developing stretch goals and establishing realistic action plans.
- Data from APQC suggests an average benchmarking study takes six months to complete, occupies a quarter of the team members' time, and the average return was five times the costs.
- Review the areas of benchmarking that Sutter Health use to compare performance on their projects; it gives a guide to how projects can be compared at the process level to drive continuous improvement.
- Consider the two examples of benchmarking with other businesses and think of how you can set up benchmarking in critical areas with comparable businesses or business functions to gauge your competitiveness against best practice.
- The four basic types of benchmarking are: internal, functional, generic and competitive. The evolution of benchmarking in an organization is likely to be driven by a focus on continuous improvement.

The purpose and practice of benchmarking

- Benchmarking is in the first instance a mechanism for creating realistic perceptions about the performance of the business. Improvement initiatives are often stimulated by the understanding of how poor performance actually is.
- The evolution of benchmarking is likely to progress through four phases: an initial focus on competitive products/services; then on industry best practices; followed by an analysis of the business to define performance gaps; and finally a focus on processes and true continuous improvement.
- The purpose of benchmarking is predominantly to change perceptions, compare business practices, challenge current practices and processes, and to create improved goals and practices with the focus on customer satisfaction and business results.
- A simple scoring pro forma may help an organization to assess whether it is ready for benchmarking, if it has not engaged in it before. Help may be required to establish the right platforms if low scores are obtained.
- The benchmarking process has five main stages: plan, collect, analyse, adapt and review. These are focused on trying to measure comparisons and identify areas for action and change.

The role of benchmarking in change

- An important aspect of benchmarking is gaining an understanding of how other organizations achieve superior performance. Some of this knowledge will result in 'quick wins' with low risk but relatively low levels of benefit.
- Step changes are of a more fundamental nature, usually require further investment in time and money, will need to be carefully planned and systematically implemented, and typically carry a higher risk.
- A change model or 'footprint' should lead to the chosen destination – improved performance through the adoption of best practice – and show the role of benchmarking.

Communicating, managing stakeholders and lowering barriers

- Communication is vital during change, and a vital element is targeting the right audience with the right message in the right way at the right time.
- Defining and understanding the key stakeholders is a wise investment of time. This should be followed by building and managing good relationships. This falls on the benchmarking team.
- Elements of successful stakeholder group management include: defining and mapping; analysing and prioritizing; researching key players/groups; developing and deploying a strategy; and reviewing effectiveness.
- Objective measurement is key to targeting change wisely and provides a reliable baseline for decisions. Baselining performance allows teams to monitor and understand success in delivering beneficial change.

Choosing benchmarking-driven change activities wisely

- Organizations should start benchmarking-driven improvement activities only after consideration of their impact on existing improvement initiatives. Questions to be asked include those related to the scale of the change, the financial and people resources (including skills) required, the impact and disruption aspects, the degree of support to the business goals and the cultural implications.
- Benchmarking may be used to drive revisions in business management systems, facilitate the application of new technologies and generally to help people to see how processes might be different.
- Properly conducted systematic benchmarking studies can aid the definition of clearer objectives and help their deployment through well-executed change management.

A framework for organizational change

- Based on research, a 'figure of 8' organizational change framework has been developed to provide an effective basis for organizations wishing to undertake any change programme, or to increase the success of an existing change initiative that has lost steam.
- The framework identifies two main constructs of change management in the form of two interacting cycles: readiness for change (strategic) and implementing change (operational).
- The figure of 8 framework may be used as a simple diagnostic by analysing problems in the change process and using the *missing focus* list of potential weaknesses listed in Figure 11.4 to identify areas for improvement.

Part III Discussion questions

1   a   Using the expression: 'if you don't measure you can't improve', explain why measurement is important in service delivery improvement, giving examples.

    b   Using your knowledge of process management, show where measurement should take place in a global building facade manufacturing and erection company and how it should be conducted.

2   Discuss the important features of a performance measurement system based on a lean quality approach. Suggest an implementation strategy for a performance measurement system in a progressive design and construct residential development company which is applying lean quality principles to its business processes.

3   It is often said that 'you can't control what you can't measure and you can't manage what you can't control'. Measurement is, therefore, considered to be at the heart of managing business processes, activities and tasks. What do you understand by improvement-based performance measurement? Why is it important? Suggest a strategy of introducing lean quality-based performance measurement for a public sector client organization of your own choice.

4   List the main categories of the US Baldrige Performance Excellence Model. How may such criteria be used as the basis for a self-assessment process? Compare this method to the lean quality maturity model and note the differences.

5   Self-assessment using the EFQM Excellence Model criteria enables an organization to systematically review its business processes and results. Briefly describe the criteria and discuss the main aspects of self-assessment. Compare this method to the lean quality maturity model and note the differences.

6   You are a large multinational high-tech manufacturer; describe how you would use the lean quality business maturity framework to drive improvement in the practices of your construction services supply chain.

7   Self-appraisal or assessment against a hybrid 'Excellence Model' can be used by organizations to monitor their progress. Design the criteria for such a hybrid framework and explain the steps that an organization would have to follow to carry out a self-assessment. How could self-assessment against the model be used in a large multi-site organization to drive continuous improvement? What additional requirements would be introduced for an organization that was asked by customers to assess against the CMMI framework?

8   Benchmarking is an important component of many companies' improvement strategy. What do you understand by benchmarking? How does benchmarking link with performance measurement? Suggest a strategy for integrating benchmarking into a lean quality approach.

9   a   Some people would argue that benchmarking is not different from competitor analysis and is a practice that organizations have always carried out. Do you agree with this? How would you differentiate benchmarking and what are its key elements?

b   Suggest and describe in full an approach to change management that would be suitable for implementing the findings from benchmarking studies for a progressive company that has no previous knowledge or experience of doing this.

10 a   What are the major limitations of the 'Prevention-Appraisal-Failure (PAF)' costing model? Why would the process cost model be a better alternative?

b   Discuss the link between benchmarking and quality costing.

11 A construction company is concerned about its record of completing projects on time. Considerable penalty costs are incurred if the company fails to meet the agreed contractual completion date. How would you investigate this problem and what methodology would you adopt?

Part IV

Processes

Process management

12

THE PROCESS MANAGEMENT VISION

Organizations create value by delivering their products and/or services to customers. Everything they do in that whole chain of events is a process. So to perform well in the eyes of the customers and the stakeholders all organizations need very good process management – underperformance is primarily caused by poor processes and/or their interaction with people and technology. This process-centred view of operations is as important at the project level as at strategic levels within the organization. Work on site is also planned and executed through myriad processes with identifiable inputs and outputs.

In 1999 Fujio Cho, President of the Toyota Motor Company sent this chilling message to the marketplace, 'We get brilliant results from average people managing brilliant processes – while our competitors get average or worse results from brilliant people managing broken processes'. This comparison between organizations is as true today as it was then, yet we still see many executives and managers in organizations all over the world, large and small, who are struggling to find how to implement the relatively simple but ubiquitously elusive concepts behind this message.

In recent times, we have seen organizations adopting a host of different approaches to improving performance. In addition to Total Quality Management (TQM), there has been Statistical Process Control (SPC), Business Process Reengineering (BPR), Lean, Six Sigma (and Lean Six Sigma!), Hoshin Kanri, Taguchi Methods, Business Process Improvement (BPI), etc. Many of these approaches are associated with their own 'technical' jargon; some fall out of favour and others seem to gain ground in particular organizations. What all these methodologies have in common, however, is a focus on processes. At a fundamental level, performance improvement is about changing the way that organizations create and deliver value to their customers through processes, regardless of what labels may be used to describe the approach.

This is recognized, of course, in the EFQM Excellence Model, in which the Processes criterion is the central 'anchor' box linking the other enablers and the results together. Performance can be improved often by improving or changing processes, but the devil is in the detail and successful exponents of process management understand all the dimensions related to:

- process strategy – particularly deployment;
- operationalizing processes – including definition and design systems;
- process performance – measurement and improvement;
- people and leadership roles – values, beliefs, responsibilities, accountabilities, authorities and rewards;
- information and knowledge – capturing and leveraging throughout the supply chains.

Where process management is established and working, executives no longer see their organizations as sets of discrete vertical functions with silo-type boundaries. Instead, they visualize things from the customer perspective as a series of inter-connected work and information flows that cut horizontally across the business. As the supply chain for a product becomes more fragmented through outsourcing, the challenge for management is to design, visualize and implement efficient processes across organizational boundaries.

Executives in such organizations picture the customer as 'taking a walk' through some or all of these 'end-to-end' processes and interfacing with the company or service organization wherever it lies within the supply chain, experiencing how it generates demand for products and services, how it fulfils orders, services products, etc. (Figure 12.1). All these processes need managing (planning, measuring and improving), sometimes discontinuously.

In monitoring process performance, measurement will inevitably identify necessary improvement actions. In many process managed companies, they have shifted the focus of the measurement systems from functional to process goals and even based remuneration and career advancement on process performance.

In the construction sector – especially general contractors engaged in the delivery of design and construction services – not only is some 85 per cent of the work outsourced, but it is outsourced to a different group of companies on every project. A major challenge

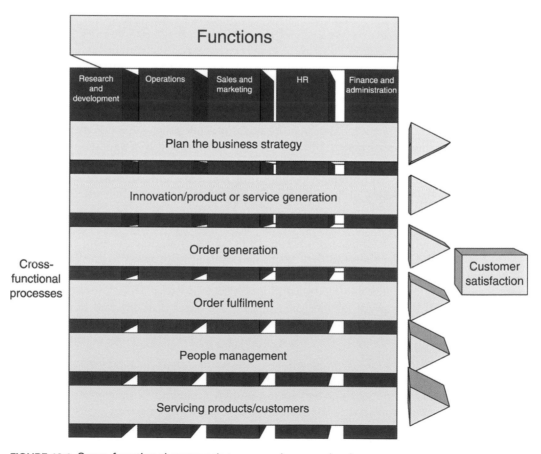

FIGURE 12.1 Cross-functional approach to managing core business processes

is posed by the short-term relationships that contractors the world over have with their supply chains. Very many of the risks in design and construction operations at the project level are in the hands of subcontractor managers and employees. The management of those risks is further challenged by the fact that general contractors only have a relationship with their subcontractors when working together on a project. This inevitably means that they generally have relatively shallow, transaction-based relationships with their suppliers.

Operationalizing process management

Top management in many organizations now base their approach to business on the effective management of 'key or core business processes'. These are well-defined and developed sequences of steps with clear rationale, which add value by producing required outputs from a variety of inputs. Moreover these management teams have aligned the core processes with their strategy, combining related activities and eliminating those that do not add value. This has led in some cases to a fundamental change in the way enterprises are managed, and the changes required have caused these organizations to emerge as true 'process enterprises'. There are many such organizations, including those featured in the case studies section of this book.

The Costain case study very clearly illustrates how a major construction company has defined its core processes on the basis of the EFQM framework. It then developed an information system on a smartphone-based application to support its staff in their compliance with the company's requirements and even developed a rating system to assess and benchmark its projects on the basis of compliance with its core business processes.

Companies comprising a number of different business units, such as outsourcing companies, face an early and important strategic decision when introducing process management: should all the business units follow the same process framework and standardization, or should they tailor processes to their own particular and diverse needs? Each organization must consider this question carefully, and there can be no one correct approach.

Deployment of a common high-level process framework throughout the organization gives many benefits, including presenting 'one company' to the customers and suppliers, lower costs and increased flexibility particularly in terms of resource allocation. An example of a high-level process framework for a large complex outsourcing company is shown in Figure 12.2.

In research on award-winning companies, the authors and their colleagues identified process management best practices as:

- Identifying the key business processes
  - by prioritizing on the basis of the value chain, customer needs and strategic significance, and using process models and definitions.
- Managing processes systematically
  - by giving process ownership to the most appropriate individual or group and resolving process interface issues through meetings or ownership models.
- Reviewing processes and setting improvement targets
  - by empowering process owners to set targets and collect data from internal and external customers.
- Using innovation and creativity to improve processes
  - by adopting self-managed teams, business process improvement and idea schemes.
- Changing processes and evaluating the benefits
  - through process improvement or re-engineering teams, project management and involving customers, and suppliers.

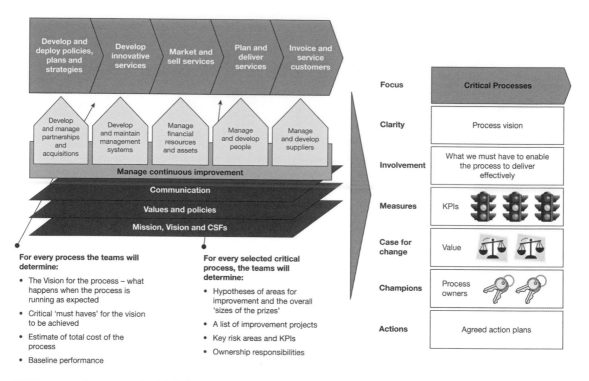

FIGURE 12.2 Example of a high-level process framework

Too many businesses are still not process oriented but focus instead on tasks, jobs, the people who do them and structures.

The Process Classification Framework and process modelling

In establishing a high level or core process framework, many organizations have found inspiration in the Process Classification Framework developed by the American Productivity and Quality Center (APQC) with the assistance of several major international corporations. The APQC have created and developed a high-level, generic enterprise model, a taxonomy of cross functional business processes that should encourage businesses and other organizations to see their activities from a cross-industry, process viewpoint rather than from a narrow functional viewpoint. The intention is to allow the objective comparison of performance within and among organizations.

The Process Classification Framework supplies a generic view of business processes often found in multiple industries and sectors: manufacturing and service companies, health care, government, education, and others. It seeks to represent major processes and sub-processes through its structure (Figure 12.3) and vocabulary (available in detail as a download from APQC). The framework does not list all processes within any specific organization; likewise, not every process listed in the framework is present in every organization.

The sub-processes listed under the high level processes shown in Figure 12.3 are as follows:

1 Develop Vision and Strategy
    1.1 Define the business concept and long term vision

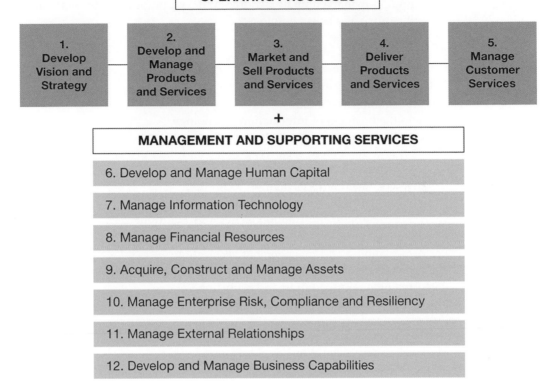

FIGURE 12.3 APQC Process Classification Framework overview

    1.2 Develop business strategy
    1.3 Manage strategic initiatives
  2 Develop and Manage Products and Services
    2.1 Manage product and service portfolio
    2.2 Develop products and services
  3 Market and Sell Products and Services
    3.1 Understand markets, customers and capabilities
    3.2 Develop marketing strategy
    3.3 Develop sales strategy
    3.4 Develop and manage marketing plans
    3.5 Develop and manage sales plans
  4 Develop Products and Services
    4.1 Plan for and align supply chain resources
    4.2 Procure materials and services
    4.3 Produce/manufacture/deliver product
    4.4 Deliver service to customer
    4.5 Manage logistics and warehousing
  5 Manage Customer Services
    5.1 Develop customer care/customer service strategy
    5.2 Plan and manage customer service operations
    5.3 Measure and evaluate customer service operations

6 Develop and Manage Human Capital
  6.1 Develop and manage human resources (HR) planning, policies and strategies
  6.2 Recruit, source and select employees
  6.3 Develop and counsel employees
  6.4 Reward and retain employees
  6.5 Redeploy and retire employees
  6.6 Manage employees information

7 Manage Information Technology
  7.1 Manage the business of information technology
  7.2 Develop and manage IT customer relationships
  7.3 Develop and implement security, privacy and data protection controls
  7.4 Manage enterprise information
  7.5 Develop and maintain information technology
  7.6 Deploy information technology solutions
  7.7 Deliver and support information technology solutions

8 Manage Financial Resources
  8.1 Perform planning and management accounting
  8.2 Perform revenue accounting
  8.3 Performa general accounting and reporting
  8.4 Manage fixed-asset project accounting
  8.5 Process payroll
  8.6 Process accounts payable and expense reimbursements
  8.7 Manage treasury operations
  8.8 Mange internal; controls
  8.9 Manage taxes
  8.10 Manage international funds/consolidation

9 Acquire, Construct and Manage Assets
  9.1 Design and construct/acquire non-productive assets
  9.2 Plan maintenance work
  9.3 Obtain and install assets, equipment and tools
  9.4 Dispose of productive and non-productive assets

10 Manage Enterprise Risk, Compliance and Resiliency
  10.1 Manage enterprise risk
  10.2 Manage business resiliency
  10.3 Manage environmental health and safety

11 Manage External Relationships
  11.1 Build investor relationships
  11.2 Manage government and industry relationships
  11.3 Manage relations with board of directors
  11.4 Manage legal and ethical issues
  11.5 Manage public relations programme

12 Develop and Manage Business Capabilities
  12.1 Manage business processes
  12.2 Manage portfolio, programme and project
  12.3 Manage quality
  12.4 Manage change
  12.5 Develop and manage enterprise-wide knowledge management (KM) capability
  12.6 Measure and benchmark

The Process Classification Framework can be a useful tool in understanding and mapping business processes. Those elements of this framework that relate to issues such as communications, commercial transactions and compliance in areas such as safety and quality are developed at the head office level. However, in construction organizations, those that relate to operational issues such as work planning and work coordination are left to the on-site project management team.

In particular, a number of organizations have used the framework to classify both internal and external information for the purpose of cross-functional and cross-divisional communication. It is a continually evolving document and the APQC will continue to enhance and improve it on a regular basis. To that end, the Center welcomes your comments, suggestions for improvement and any insights you gain from applying it within your organization. The APQC would like to see the Process Classification Framework receive wide distribution, discussion and use. Therefore, it grants permission for copying the framework as long as acknowledgement is made to the American Productivity and Quality Center (www.apqc.org; email: pcf_feedback@apqc.org).

PROCESS MODELLING

As we saw in Chapter 2, a process is simply something that converts a set of inputs into outputs. We have already seen examples of process modelling described in relation to design management (Chapter 8) and cost modelling (Chapter 9). The inputs can include materials or information and are supplied into the process externally or internally. The outputs of a process go to a customer – again external or internal. This simple explanation may be represented as the flow of SIPOC: (**Supplier–Inputs–PROCESS–Outputs–Customer**). Any substantial process may be broken down into its main steps or subprocesses, all of which have the same SIP. Many years ago, the United States Air Force adopted 'Integration DEFinition Function Modelling' (IDEFØ) as part of its Integrated Computer-Aided Manufacturing (ICAM) architecture. IDEFØ is a method designed to model the decisions, actions and activities of an organization or system. IDEFØ was derived from a well-established graphical language, the Structured Analysis and Design Technique (SADT). The US Air Force 'commissioned the developers of SADT to develop a function modelling method for analysing and communicating the functional perspective of a system'.

In December 1993, the Computer Systems Laboratory of the National Institute of Standards and Technology (NIST) released IDEFØ as a standard for Function Modelling in Federal Information Processing Standards (FIPS) Publication 183. This provides a useful structured graphical framework for describing and improving business processes. The associated 'Integration Definition for Information Modelling' (IDEFIX) language allows the development of a logical model of data associated with processes such as measurement.

These techniques are widely used in business process re-engineering (BPR) and business process improvement (BPI) projects, and to integrate process information. A range of specialist software (including Windows/PC-based) is also available to support the applications. IDEFØ may be used to model a wide variety of new and existing processes, define the requirements, and design an implementation to meet the requirements.

An IDEFØ model consists of a hierarchical series of diagrams, text and glossary cross-referenced to each other through boxes (process components) and arrows (data and objects). The method is expressive and comprehensive and is capable of representing a wide variety of business, service and manufacturing processes. The relatively simple language allows coherent, rigorous and precise process expression, and promotes consistency. Figure 12.4 shows the basis of the approach.

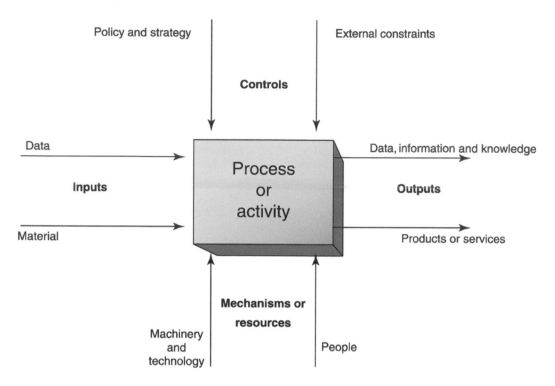

FIGURE 12.4 IDEFØ model language

The ADePT method for design management, presented in Chapter 8, gives an illustration of the IDEFØ modelling technique being used to analyse the design process, this then leads to the Dependency Structure Matrix – the core of the AdePT process for the management of complex problems.

For a full description of the IDEFØ methodology, it is necessary to consult the FIPS PUBS standard (NIST, 1993). It should be possible, however, from the simple description given here, to begin process modelling (or mapping) using the technique.

Processes can be any combination of things, including people, information, software, equipment, systems, products or materials. The IDEFØ model describes what a process does, what controls it, what things it works on, what means it uses to perform its functions and what it produces. The combined graphics and text comprise:

Boxes – which provide a description of what happens in the form of an active verb or verb phrase;

Arrows – which convey data or objects related to the processes to be performed (they do not represent flow or sequence as in the traditional process flow model).

Each side of the process box has a standard meaning in terms of box/arrow relationships. Arrows on the left side of the box are **inputs** that are transformed or consumed by the process to produce **output** arrows on the right side. Arrows entering the top of the box are **controls** that specify the conditions required for the process to generate the correct outputs. Arrows connected to the bottom of the box represent **'mechanisms'** or **'resources'**. The abbreviation ICOR (inputs, controls, outputs, resources) is sometimes used.

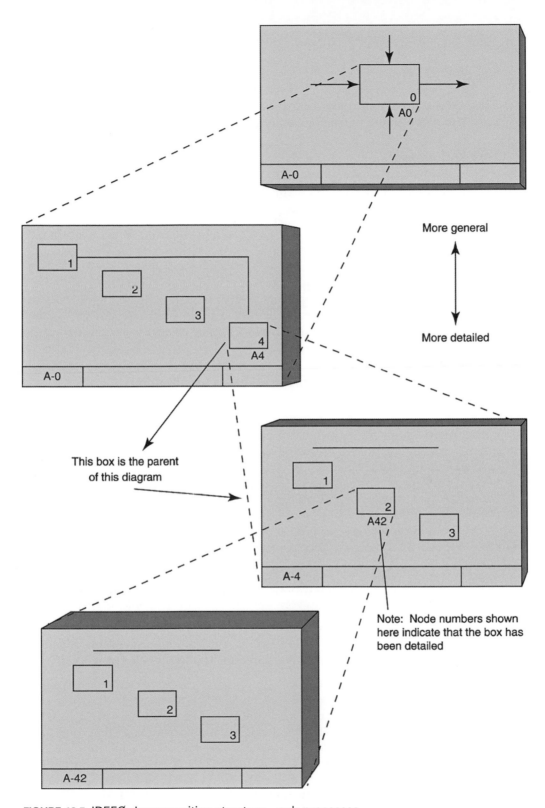

FIGURE 12.5 IDEFØ decomposition structure – sub-processes

Using these relationships, process diagrams are broken down or decomposed into more detailed diagrams, the top-level diagram providing a description of the highest level process. This is followed by a series of 'child' diagrams providing details of the sub-processes (see Figure 12.5).

Each process model has a top-level diagram on which the process is represented by a single box with its surrounding arrows. Each sub-process is modelled individually by a box, with parent boxes detailed by child diagrams at the next lower level.

Text and glossary

An IDEFØ diagram may have associated structured text to give an overview of the process model. This may also be used to highlight features, flows and inter-box connections and to clarify significant patterns. A glossary may be used to define acronyms, key words and phrases used in the diagrams.

Arrows

Arrows on high-level IDEFØ diagrams represent data or objects as constraints. Only at low levels of detail can arrows represent flow or sequence. These high-level arrows may usefully be thought of as pipelines or conduits with general labels. An arrow may branch, fork or join indicating that the same kind of data or object may be needed or produced by more than one process or sub-process.

IDEFØ process modelling, improvement and teamwork

The IDEFØ methodology includes procedures for developing and critiquing process models by a group or team of people. The creation of an IDEFØ process model provides a disciplined teamwork procedure for process understanding and improvement. As the group works on the process following the discipline, the diagrams are changed to reflect corrections and improvements. More detail can be added by creating more diagrams, which in turn can be reviewed and altered. The final model represents an agreement on the process for a given purpose and from a given viewpoint and can be the basis of new process or system improvement projects.

Such an approach would be ideal, for example, to investigate well-known problem areas such as the causes of defective work in construction projects, variations or poor documentation within an organization. The IDEFØ/ICOR methodology could be used to model current processes in terms of inputs, controls, resources and outputs, and then to develop improved systems.

In using such techniques in process management there can be a propensity for maps to assume disproportionate importance. This can result in participants becoming distracted in pursuit of accuracy, or even the overall purpose of process improvement being supplanted by the modelling process itself – this is to be avoided at all cost!

IDEFIX

This is used to produce structural graphical information models for processes that may support the management of data, the integration of information systems and the building of computer databases. It is described in detail in the FIPS PUB 184 (December 1993, NIST). Its use is facilitated by the introduction of IDEFØ modelling for process understanding and improvement.

A number of commercial software packages that support IDEFØ and IDEFIX implementation are available.

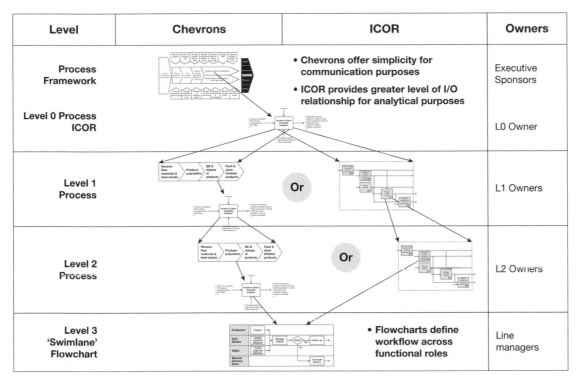

Level	Chevrons	ICOR	Owners
Process Framework **Level 0 Process ICOR**		• **Chevrons offer simplicity for communication purposes** • **ICOR provides greater level of I/O relationship for analytical purposes**	Executive Sponsors L0 Owner
Level 1 Process		*Or*	L1 Owners
Level 2 Process		*Or*	L2 Owners
Level 3 'Swimlane' Flowchart		• **Flowcharts define workflow across functional roles**	Line managers

FIGURE 12.6 Summary of process mapping approaches

Levels of process detail

The management and ownership of processes take place at various levels in an organization. For example, the high-level process framework for the organization and strategic processes such as 'defining company direction' are owned at the executive level, while tactical processes such as 'dealing with an enquiry' are at lower levels. The detail required at each level can be provided by process chevron diagrams that simply show and list activities and process analysis techniques such as IDEFØ/ICOR maps, or flowcharting, and swim lane diagrams (see Figure 12.6). As one chief executive commented, 'I need to understand the process, but I don't want to see pages of wiring diagrams!' A key challenge in making best use of process management tools is to pitch the definition at the most appropriate level, bearing in mind the nature of the improvement likely.

PROCESS FLOWCHARTING

Another powerful method of describing a process is flowcharting. This owes much to computer programming, where the technique is used to arrange the sequence of steps required for the operation of the programme. It has a much wider application, however, than computing.

Certain standard symbols are used on flowcharts which are shown in Figure 12.7. The starting point of the process is indicated by a circle. Each processing step, indicated by a rectangle, contains a description of the relevant operation. Where the process ends is indicated by an oval. A point where the process branches because of a decision is shown by a diamond. A parallelogram relates to process information but is not a processing step.

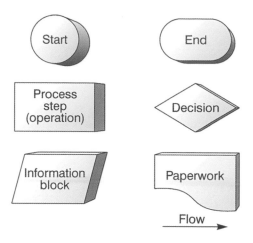

FIGURE 12.7 Flowcharting symbols

The arrowed lines are used to connect symbols and to indicate direction of flow. For a complete description of the process, all operation steps (rectangles) and decisions (diamonds) should be connected by pathways to the start (circle) and end (oval). If the flowchart cannot be drawn in this way, the process is not fully understood.

It is a salutary experience for most people to sit down and try to draw the flowchart for a process in which they take part every working day. It is often found that:

- the process flow is not fully understood; and
- a single person is unable to complete the flowchart without help from others.

The very act of flowcharting will improve knowledge of the process and will begin to develop the teamwork necessary to find improvements. In many cases, the convoluted flow and octopus-like appearance of the chart will highlight unnecessary movements of people, materials and information and lead to common-sense suggestions for waste elimination.

Figures 12.8 and 12.9 provide a before and after example of flowcharting in use to improve a travel booking procedure in a company. The total time taken for the starting or 'current state' procedure, excluding the correction of any errors and the preparation of overview reports was 23 minutes per travel request, the flowchart for the process shown in Figure 12.8. An improvement team was set up to analyse the process and make recommendations for improvement using brainstorming and questioning techniques. They made proposals to change the procedure and the flowchart for the improved or 'to-be' process is shown in Figure 12.9. The proposal reduced the total administrative effort per travel request (or per travel arrangement, because the travel request was eliminated) from 23 minutes to 5 minutes.

The details that appear on a flowchart for an existing process should be obtained from direct observation of the process, not by imagining what is done or what should be done. The latter may be useful, however, in the planning phase, or for outlining the stages in the introduction of a new concept. Such an application is illustrated in Figure 12.10 for the installation of statistical process control (SPC) charting systems (see Chapter 15). Similar charts may be used in planning the implementation of quality management systems.

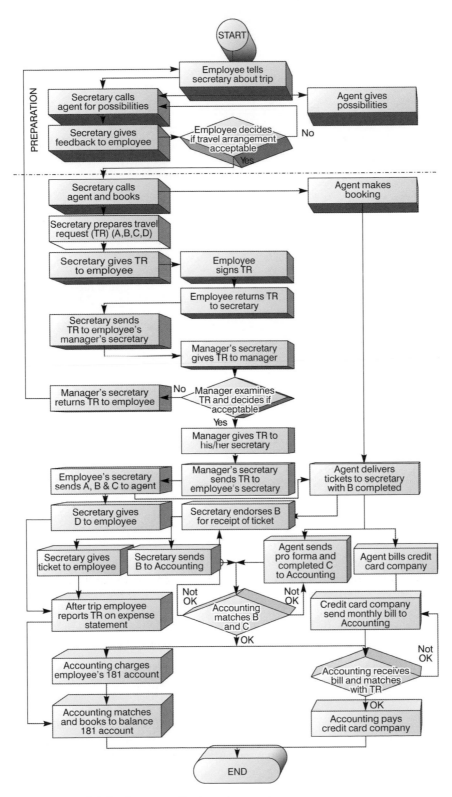

FIGURE 12.8 Original process for travel procedure

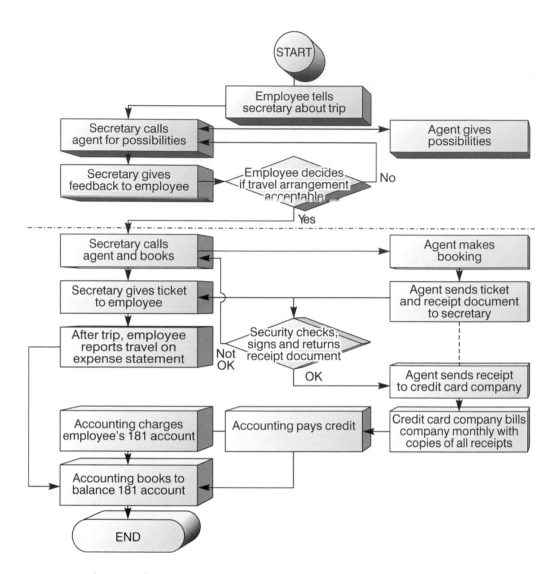

FIGURE 12.9 Improved travel procedure

It can be surprisingly difficult to draw flowcharts for even the simplest processes, particularly managerial ones, and following the first attempt, it is useful to ask whether:

- the facts have been correctly recorded;
- any over-simplifying assumptions have been made; and
- all the factors concerning the process have been recorded.

The authors have seen too many process flowcharts that are so incomplete as to be grossly inaccurate. Flowcharts should provide excellent documentation and be useful trouble-shooting tools to determine how each step is related to the others. By reviewing the flowchart, it should be possible to discover inconsistencies and determine potential sources of variation and problems. For this reason, flowcharts are very useful to process improvement teams when they are examining an existing process to highlight the problem

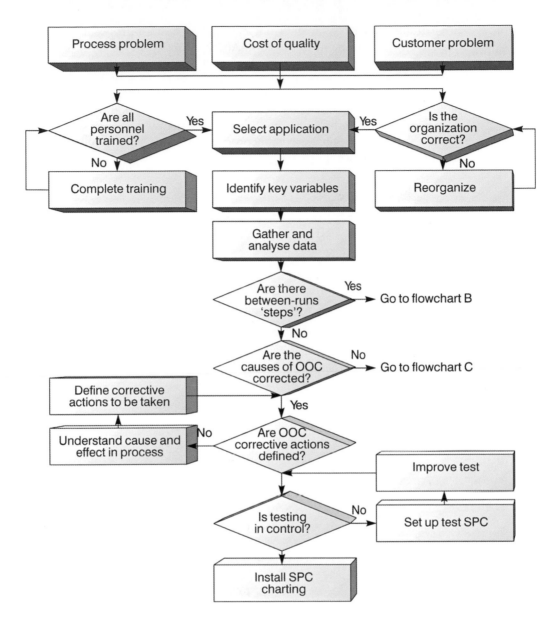

FIGURE 12.10 Flowchart for SPC implementation

areas. A group of people, with the knowledge about the process, should take the following simple steps:

1 Draw a flowchart of the existing process.
2 Draw a second chart of the flow the process could or should follow.
3 Compare the two to highlight the changes necessary.

A number of commercial software packages which support process flowcharting are available.

One of the authors was asked to work with a service provider in the oil and gas sector to assist in the simplification of a job-start process. The owner was complaining that the three months that it took to get from an initial request for proposal (RFP) for a small brownfield improvement to the commencement of work on site was far too long. In between the RFP and the commencement of work on site there was a scoping stage, estimation and several review and approval steps both within the service provider and the owner organization. As with many routine processes, this one had become clunky over a number of years. It is common, as a reaction to the occurrence of problems from time to time, for people to insert another approval gate, or a review step into an existing process hoping that in future a similar problem will not escape notice.

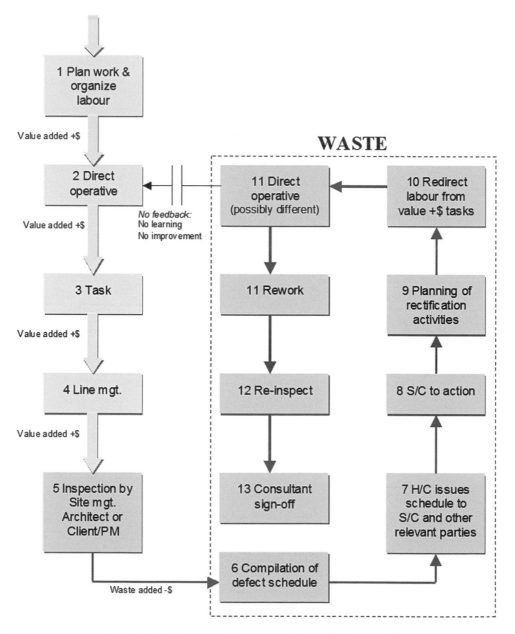

FIGURE 12.11 Flowchart showing current process for dealing with defects

The work site was remote, so producing the current state flowchart was done off-line. A senior person involved in the service provider organization was nominated to lead this task. After consultation with colleagues, the existing process map was produced.

Some help was required to establish the drafting method and format, but otherwise step 1 was prepared, circulated, reviewed and refined off-line. Step 2 was achieved through a half-day workshop. A group of key stakeholders from the owner and service provider organizations participated. The goal was to brainstorm how the overall process could be simplified. This exercise was undertaken using sticky notes on the wall. The various key sub-process owners went through their parts of the process and the overall team discussed

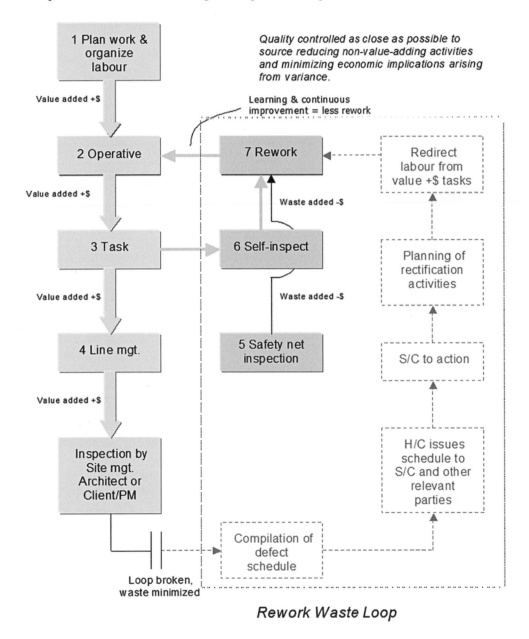

FIGURE 12.12 Flowchart showing ideal process for dealing with defects

alternatives and possible changes. By the end of the afternoon, many redundant steps had been removed and the new process was mapped on the wall. The period of the entire process had been halved to six weeks.

The graphic flowcharts were developed for a research project at the Australian Centre for Construction Innovation to depict the process typically found for defect identification and rectification on Sydney high-rise residential construction sites. Figure 12.11 depicts the current state indicating all the areas of waste that could be removed by avoiding quality defects or as a minimum solving quality issues at source.

Figure 12.12 depicts the proposed ideal state. The difference between the two process maps is stark; and this is a very effective way of communicating the waste in the existing process and demonstrating the logic of the change. Of course, to make the change requires many detailed changes in management processes, organizational relationships and skills.

A major Australian building contractor who was a key sponsor of this research subsequently established standard detailing in critical risk areas and set up a training centre to train its subcontractors. It took control of how it wanted work to be detailed in the field to avoid the waste incurred through the repetition of quality errors on its sites. It recently reported in a conversation with one of the authors that that particular source of waste has been stopped.

LEADERSHIP, PEOPLE AND IMPLEMENTATION ASPECTS OF PROCESS MANAGEMENT

There are many top executives who have famously used process management to great effect, issuing statements such as:

> Business processes are designed to be customer driven, cross functional, and value based. They create knowledge, eliminate waste, and abandon unproductive work, yielding world-class productivity and higher perceived service levels for customers.

and:

> Publishing the process on the wall helps people understand their place in the big picture ... the continuously improving profits earned by TNT Express are a consequence of superlative performance that is derived from well thought out processes, ongoing measurement of a few carefully selected key indicators, good communications, and the full involvement everywhere of all people working in the company.

These approaches manifest themselves in the complete understanding, by everyone, of the end-to-end process, which is described, for example, by TNT Express as the 'perfect transaction' (Figure 12.13).

Perhaps the most visible difference between a process management enterprise and a more functionally based one is the existence of process owners – management layers with end-to-end responsibilities for individual processes (see Table 12.1). They have real responsibility for and authority over the process design, operation and measurement of performance. This may require change from being functionally driven and usually means a major cultural challenge for the organization. Process owners who simply adopt a command and control style will fail, so their ability to work with their team is at least as important in the conversion to process management as the structure of the management system.

This is a change in management structure with which organizations in the construction sector struggle. In part, this is driven by the project focus of the industry. Output is generally conceptualized as a series of projects through which value is created. Senior

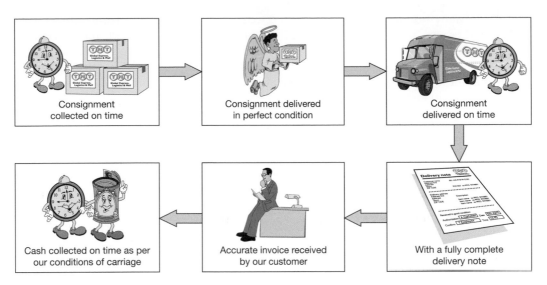

FIGURE 12.13 TNT Express Delivery Services – the perfect transaction process

TABLE 12.1 Summary of key process roles

Roles	Descriptions
Executive Sponsor	• Member of the Executive team • Advocate and ambassador for the overall process • Exerts influence inside and outside of the organization
Process Owner	• Member of the senior management team • End-to-end responsibility and accountability for the performance of the process • Ownership of the overall process
Functional (Line) Manager	• Member of the operational staff • Specific functional (line) responsibilities and accountabilities • Impact within the overall process

managers in general contracting organizations where nearly all the actual production is outsourced, see the reduction of site-based *prelims* or *project overheads* as one of their key challenges. Head office functions are primarily oriented towards marketing and administration, and the ability of head office to add value to project delivery is questioned. Senior head office managers find it difficult to integrate process-based thinking into their organizational structure. The lessons learned in other sectors are particularly relevant in construction as the major opportunities for improved efficiency and effectiveness are process-based.

Managing the people who work in the processes demands attention to:

- designing, developing and delivering training programmes;
- setting performance targets;
- regular communication, preferably face-to-face;
- keeping them informed of changing customer needs;
- listening to concerns and ideas; and
- negotiation and collaboration.

Many organizations have realized that to be cost-effective, competitive and, indeed, world-class, they must ensure that all processes are understood, measured and under control. Almost everyone needs to be trained in process management and improvement and shown how they are part of a supplier–process–customer chain. The training reinforces that these chains are interdependent and all processes support the delivery of products or services to customers.

Operators of every process need to be properly trained, have necessary work instructions available and have the appropriate tools, facilities and resources to perform the process to its optimum capability. This applies to all processes throughout the organization, including those in Finance and Human Resources and generally shared services areas.

In many process managed organizations, this type of approach has changed the way they assign and train employees, emphasizing the whole process rather than narrowly focused tasks. It has made fundamental changes to cultures: stressing process-based teamwork and customers rather than functionally driven command and control. Creativity and innovation in process improvement are recognized as core competencies, and the annual performance reviews and personal development plans are linked to these.

The first thing that top management must recognize is that moving to process management requires much more than re-drawing the organizational chart or structure. The changes needed are fundamental, leading to new ways of working and managing, and they will challenge any company or public service organization.

Many organizations today are facing a large number of changes and initiatives, often driven by public–private transitions, customer or government demands, technology, and so on. Therefore, before implementing process management, a senior management team needs to examine closely all its current change initiatives using simple frameworks such as the Figure of 8 Change Model shown in Chapters 11 and 19 or the Total Organizational Excellence Model (see Chapter 19). Those that are not relevant to a process managed business need to be pruned while those that are may need combining or rationalizing.

The introduction of process management is often driven or directly connected to a strategic initiative such as reducing cycle times, increasing customer satisfaction, reducing working capital (perhaps tied up in work-in-progress), an enterprise resource planning (ERP) implementation, changes in technology or introduction of e-business. The application of new enabling technologies is an ideal time to review the design and configuration of key processes. Failure to do so could lead to missed opportunities to extract maximum benefit from the technology.

Implementing process management, like many change initiatives, cannot be a quick fix, and it will not happen overnight. Top management need the resolution and commitment for major changes in the way things are structured, carried out and measured. As with all such implementation, it needs careful planning and an understanding of what needs to be done first. Things high on the list will be the establishment of a core process framework, aligned to the needs of the business, and the appointment of key process owners. A process-based performance measurement framework then needs to be set up to track progress. As with all change initiatives, delivering some tangible measurable benefits early on will help overcome the inevitable resistance. In one pharmaceutical company, for example, the success of the work on the product development and product promotion processes helped significantly the cause of process management, and the company extended its approach into the supply chain management and other processes.

As companies and public service organizations move inexorably towards the wider introduction of e-commerce to do business, this will place a premium on rapid and fault-free execution of business processes. Putting a website in front of an inefficient, ineffectual or even broken process will soon bring it to its knees, together with everyone working in it and around it. This will also bring 'back-office' mistakes to the attention of the marketplace. Some of these processes will, of course, need to be redesigned: from customer order fulfilment to procurement. They will need to change 'shape' as demands, technologies and markets change. Without good process management in place this is going to be very difficult for functionally driven organizations.

A relevant cautionary tale comes from a conversation witnessed by one of the authors between the project manager of an EPC contractor and his client on a very major energy project. The client raised an issue regarding invoicing during the wide-ranging conversation. The essence of this was that if the EPC's back-office could not get its invoicing correct, there were people in the client organization who wanted to find a new service provider. Note that no complaint was made about service provision, they were pleased with the work; however the failure of back-office processes was putting the entire relationship at risk.

BIBLIOGRAPHY

Besterfield, D. *Total Quality Management*, Pearson Education, New York, 2011.

BQF. *The X-factor, Winning Performance through Business Excellence*, European Centre for Business Excellence/British Quality Foundation, London, 1999.

BQF. *'The Model in Practice' and 'The Model in Practice 2'*, British Quality Foundation, London, 2000 and 2002 (Prepared by European Centre for Business Excellence).

Dimaxcescu, D. *The Seamless Enterprise – Making Cross-functional Management Work,* Harper Business, New York, 1992.

Francis, D. *Unblocking the Organizational Communication*, Gower, Aldershot, 1990.

Harrington, H.J., *Total Improvement Management*, McGraw-Hill, New York, 1995.

Jeston, J. and Nelis, J. *Business Process Management – Practical Guidelines to Successful Implementations*, Elsevier, Oxford, 2008.

National Institute of Standards and Technology (NIST). *Federal Information Processing Standard Publication 184,* NIST, Gaithersburg, 1993.

Oakland, J.S. *Total Organizational Excellence*, Butterworth-Heinemann, Oxford, 1999 (paperback, 2001).

Rummler, G.A. and Brache, A.P. *Improving Performance: How to Manage the White Space on the Organization Chart* (2nd edn), Jossey-Bass Publishing, San Francisco, CA, 1998.

Senge, P.M. *The Fifth Discipline; The Art and Practice of the Learning Organization,* Random House, London, 2006.

Senge, P.M., Roberts, C., Ross, R.B., Smith, B.J. and Kleiner, A. *The Fifth Discipline Fieldbook – Strategies and Tools for Building a Learning Organization,* Nicholas Brearley, London, 1994.

Weske, M. *Business Process Management*, Springer, New York, 2012.

CHAPTER HIGHLIGHTS

Process management vision

- Everything organizations do to create value for customers of their products or services is a process. Process management is key to improving performance.
- Process managed organizations see things from a customer perspective as a series of inter-connected work and information flows that cut horizontally across the business functions.
- The key or core business processes are well-defined and developed sequences of steps with clear rationale that add value by producing required outputs from a variety of inputs.

- Deployment of a common high-level process framework throughout the organization gives many benefits, including reduced costs and increased flexibility.
- Process management best practices include: identifying the key business processes; managing processes systematically; reviewing processes and setting improvement targets; using innovation and creativity to improve processes; changing processes; and evaluating the benefits.

Process classification framework and process modelling

- The APQC's Process Classification Framework creates a high-level generic, cross-functional process view of an enterprise – a taxonomy of business processes.
- The IDEF (Integrated Definition Function Modelling) language provides a useful structured graphical framework for describing and improving business processes. It consists of a hierarchical series of diagrams and a text, cross-referenced to each other through boxes. The processes are described in terms of Inputs, Controls, Outputs and Resources (ICOR).

Process flowcharting

- Flowcharting is a method of describing a process in pictures, using symbols: *rectangles* for operation steps; *diamonds* for decisions; *parallelograms* for information; and *circles/ovals* for the start/end points. *Arrow lines* connect the symbols to show the 'flow'.
- Flowcharting improves knowledge of the process and helps to develop the team of people involved.
- Flowcharts document processes and are useful as trouble-shooting tools and in process improvement. An improvement team would flowchart the existing process and the improved or desired process, comparing the two to highlight the changes necessary.

Leadership, people and implementation

- Top management who have used process management to great effect recognize its contribution in creating knowledge and eliminating waste, they understand the importance of involving people, measurement and good communications.
- Process owners are key to effective process management. They have responsibility for and authority over process design, operation and measurement of performance.
- Managing the people who work in the processes requires attention to training programmes, performance targets, communicating changing customer needs, negotiation and collaboration.
- Moving to process management requires some challenging fundamental changes, leading to new ways of working and managing. Current initiatives should be carefully examined to ensure good planning and an understanding of what needs to be done first.
- As with all change initiatives, delivering some tangible measurable benefits early on will help overcome the inevitable resistance.
- With the wider introduction of e-commerce systems, there will be greater pressure to run rapid, fault-free business processes. Some of the processes will need to change 'shape' as demands, technologies and markets change.

Process re-design/ engineering

13

RE-ENGINEERING THE ORGANIZATION?

When it has been recognized that a major business process requires radical re-assessment, business process re-engineering or re-design (BPR) methods are appropriate. In their book *Re-Engineering the Corporation* (1993), Hammer and Champy talked about re-inventing the nature of work, 'starting again – re-inventing our corporations from top to bottom'. BPR was launched on a wave of organizations needing to completely re-think how and why they do what they do in order to cope with the ever-changing world, particularly the development of technology based solutions.

The reality is, of course, that many processes in many organizations are very good and do not need re-engineering, re-designing or re-inventing, not for a while anyway. These processes should be subjected to a regime of continuous improvement (Chapter 15) at least until we have dealt with the very poorly performing processes that clearly do need radical review.

Some businesses and industries more than others have been through some pretty hefty changes – technological, political, financial and/or cultural. Customers of these organizations may be changing and demanding certain new relationships. Companies are finding leaner competitors encroaching into their market place, increased competition from other countries where costs are lower, and start-up competitors which do not share the same high bureaucracy and formal structures.

Enabling an organization, whether in the public or private sector, to be capable of meeting these changes is not a case of working harder but working differently. There have been many publicized BPR success stories and, equally, there have been some abject failures. In some cases, radical changes to major business processes have brought corresponding radical improvements in productivity. However, knowing how to reap such benefit, or indeed knowing if and how to apply BPR, has proven difficult for some organizations.

Many companies adopted lean quality initiatives in the 1980s hoping to win back business lost to Japanese competition. When Ford benchmarked Mazda's accounts payable department, they discovered a business process being run by five people compared to Ford's 500. Even with the difference in scale of the two companies, this still demonstrated the relative inefficiency of Ford's accounts payable process. At Xerox, taking a customer's perspective, the company identified the need to develop systems rather than stand-alone products, thus highlighting Xerox's own inefficient office systems.

Both Ford and Xerox realized that incremental improvement alone was not enough. They had developed high infrastructure costs and bureaucracies that made them relatively unresponsive to customer service. Focussing on internal customer-supplier interfaces improved quality but preserved the current process structure, and they could not hope to achieve in a few years what had taken the Japanese 30 years. To achieve the necessary improvements required a radical rethink and redesign of these processes.

What was being applied by organizations such as Ford and Xerox was **discontinuous improvement.** In order to respond to the competitive threats of Canon and Honda, Xerox and Ford needed lean quality to catch up; but to get ahead, they felt they required radical breakthroughs in performance. Central to these breakthrough improvements was information technology (IT).

Information technology as a driver for BPR

BPR is often based on new possibilities for breakthrough performance provided by the emergence of new enabling technologies. The most important of these, the one that is the nominal ingredient in many BPR recipes, is IT. Explosive advances in IT have enabled the dissemination, analysis and use of information from and to customers and suppliers and within enterprises in new ways and in time frames that impact processes, organization designs and strategic competencies. Computer networks, open systems, client-server architecture, groupware and electronic data interchange have opened up the possibilities for the integrated automation of business processes. Neural networks, enterprise analysis approaches, computer-assisted software engineering and object-oriented programming now facilitate systems design around many processes in most organizations.

The pace of change has, of course, been enormous, and IT systems unavailable just ten to fifteen years ago have enabled sweeping changes in business process improvement, particularly in office systems and in design development and documentation systems. Just as statistical process control (SPC) enabled manufacturing processes to be improved by controlling variation and improving efficiency so IT is enabling the myriad non-manufacturing processes in areas such as logistics, design documentation and communication to be fundamentally restructured.

IT in itself, however, does not offer all the answers. Many companies putting in major new computer or software systems have achieved only the automation of existing processes. Frequently, different functions within the same organization have systems that are incompatible with each other. Locked into traditional functional structures, managers have spent large amounts on IT systems that have not been used cross-functionally. Yet it is in this cross-functional area that the big improvement gains through IT are to be made. Once a process view is taken to designing and installing an IT system, it becomes possible to automate cross-functional, cross-divisional and even cross-company processes.

In a research report for government, co-authored by one of the authors (Yetton and Marosszeky, 1998), the main findings were that while specialist companies had automated many of their core processes and had gained competitive benefits as individual firms, the challenge and opportunity was to re-engineer processes and drive genuine inter-organizational collaboration across the supply chain. While significant progress has taken place since that time, through the use of project intranets to share documents and the use of BIM modelling to jointly develop federated 3D models of what is to be built, there are still many opportunities to further improve collaboration. A major impediment is the transaction-based approach to business in the sector; because there are limits to how far parties will trust each other, companies are reluctant to share too much information in case they lose competitive advantage.

The supply chain opportunity

The potential of IT when used across industry sectors is best described relative to its use within the sector. It was found that most applications of IT have been confined to a single sector. Consequently, benefits to date have been restricted. The greatest potential for transforming the industry beyond another round of cost reduction lies in the re-engineering of the supply chain to deliver increased value for the client. This potential can be perceived at three levels and requires cross-sectoral, inter-organizational collaboration.

Level 1: IT can be and, typically, is used to improve the efficiency, speed and quality of communication across sectors, thereby reducing cycle times and making a small gain in quality for the whole supply chain.

Level 2: IT can be used to facilitate the creation of a transformed supply chain. By taking a different approach to cross-sectoral relationships (e.g. by encouraging greater concurrency between tasks conducted by firms in different sectors through greater sharing of information) it may be possible to achieve substantial savings in time and money for the client.

Level 3: In a supply chain characterized by the sharing of information and knowledge, the potential exists to increase the total value to the end-client (the developer/operator of the building or plant) by improving performance on multiple dimensions, including operational manageability and return on the asset. For example, if architects, engineers, contractors and clients would start to share information when a design is first conceived through appropriately rich communication channels, it may become possible to design and build more efficiently. There would be fewer difficulties for the designers and builders and far greater benefits to the customer because new kinds of solution would be developed collaboratively. These solutions would be safe, more aesthetic, easier to build and would perform better for the client.

When Frank Gehry designed the Guggenheim Museum in Bilbao, he was able to create a totally innovative landmark because his design process was tightly linked through IT to his suppliers. This meant that he was able to ensure the feasibility of his design as he developed it. The detailed design collaboration described in the VNGC case study illustrates how constructability can be optimized through a joint approach by the designer and constructor (steel-fixer) working together using 3D modelling.

Such potential is enabled by IT but requires more than mere adoption of the technology. Successful implementation requires 'buy-in' by those who will use the technology. The achievement of competitive benefits typically requires organizational change. For example, the productivity gains from CAD are maximized for architects only when drafting is integrated with the design task. Leading companies have done this. Successful transformation of businesses is achieved incrementally over several years through a cycle of learning and organizational change.

For organizations within the supply chain to capture benefits available as a result of inter-organizational integration, they need to go through a shared cycle of learning and change. For example, through the shared use of BIM models, the collaborating organizations need to develop protocols for co-operative work practices as well as go through a period of shared learning. Only then can information be passed seamlessly throughout the supply chain and the potential of efficient computer integrated manufacture be realized in the sector.

In construction today, the greatest opportunity for re-engineering processes comes from the adoption of shared BIM modelling. This technology enables closer and more effective collaboration between designers and with stakeholders. It enables front-end designers to join with fabricators and constructors to refine constructability; it enables the automatic take-off of quantities and generation of estimates, fundamentally redefining several professional roles; and it allows construction workers to build from the model as seen in the DPR case study creating a virtually paperless construction site. While all the changes described in this paragraph are taking place to some extent, the eventual integration of all these processes and the associated redefinition of profession roles around the use of the BIM model will lead to the re-engineering of the construction process at an industry level. This change process is underway and although it will not be quick, the authors believe it is inevitable.

WHAT IS BPR AND WHAT DOES IT DO?

There are almost as many definitions of BPR as there are of TQM and lean systems! However most of them boil down to the same substance: the fundamental rethink and radical redesign of a business process, its structure and associated management systems designed to deliver major or step improvements in performance (which may be in process, customer or business performance terms).

Of course, BPR and lean quality programmes are complementary under the umbrella of process management. The continuous and step change improvements must live side by side – when does continuous change become a step change anyway? There has been over the years much debate, including some involving the authors, about this issue. Whether it gets resolved is not usually the concern of the organization facing today's uncertainties, realizing that, 'business as usual' will not do, and some major changes in the ways things are done are required.

Put into a strategic context, BPR is a means of aligning work processes with customer requirements in a dynamic, flexible way in order to achieve long-term corporate objectives. This requires the involvement of customers and suppliers, and thinking about future requirements. Indeed, the secret to redesigning a process successfully lies in thinking about how to reshape it for the future.

BPR then challenges managers to rethink their traditional methods of doing work and commit to customer-focused processes. Many outstanding organizations have achieved and/or maintained their leadership through process re-engineering, especially where they found processes which were not customer focused. Companies using these techniques have reported significant bottom-line results, including better customer relations, reductions in cycle time to market, increased productivity, fewer defect/errors and increased profitability. BPR uses recognized methods for improving business results and questions the effectiveness of the traditional organizational structure. Defining, measuring, analysing and re-engineering work processes to improve customer satisfaction can pay off in many different ways.

For example, Motorola had set stretch goals of ten-fold improvement in defects and two-fold improvement in cycle time within five years. The time period was subsequently revised to three years and the now famous Six Sigma goal of 3.4 defects per million became a slogan for the company and probably one of the real drivers (see also Chapter 15). In case study 4, we see a similarly challenging stretch target set by the High-Tech-Manufacturer for the reduction of its construction costs and increase of construction speed.

Such stretch goals represent a focus on discontinuous improvement, and there are many examples of other companies that have made dramatic improvements following major organizational and process redesign as part of lean quality initiatives, including approaches such as the 'clean sheet' design of a 'green field' plant around work cells and self-managed teams. Earlier on in the computer revolution, a construction consortium known as IDC realized that it could create an opportunity for itself by preparing for the inevitable tall demands of the high pressure computer industry, a sector where time to market with new products can meet the difference between growth and stagnation. IDC created a set of super-effective teams, which included all members of the subcontract supply chain and cut across organizational boundaries. They developed independent multifunction workgroups; they adopted the latest IT-based technologies to integrate their planning and work processes and developed their own management tools so that when approached and asked to do the impossible, they would be in a position to deliver it. On their first project they were able to reduce time and cost by more than 25 per cent compared with their previous similar project.

Most organizations have vertical functions: experts of similar backgrounds grouped together in a pool of knowledge and skills capable of completing any task in that discipline. This focus, however, fosters a vertical view and limits the organization's ability to operate effectively. Barriers to customer satisfaction evolve, resulting in unnecessary work, restricted sharing of resources, limited synergy between functions, delayed development time, and no clear understanding of how one department's activities affect the total process of attaining customer satisfaction. Managers remain tied to managing singular functions with rewards and incentives for their narrow missions inhibiting a shared external customer perspective.

BPR breaks down these internal barriers and encourages the organization to work in cross-functional teams with a shared horizontal view of the business. As we have seen in earlier chapters, this requires shifting the work focus from managing functions to managing processes. Process owners, accountable for the success of major cross-functional processes, are charged with ensuring that employees understand how their individual work processes affect customer satisfaction. The interdependence between one group's work and the next becomes quickly apparent when everyone understands who the customer is and the value they add to the entire process of satisfying that customer.

The ConXtech case study (CS3) illustrates a new product system that has created a step change in the processes of building steel structural frames. This company has developed a clip in, high precision, high strength connection for structural frames. It makes work at heights in assembling structural steel much faster and safer; it reduces crane time as large sections simply clip in and produces a high precision frame. On oilfield pipe racks a more than five-fold labour productivity improvement has been achieved. The system is being used in residential, commercial health and industrial buildings. In 2008, a five-storey, 15,600 sqm building frame was erected in 14 working days.

The ConXtech system shows how the radical reengineering of building subsystems has the potential to accelerate construction of entire buildings. However, it raises the obvious challenge for other subsystem designers and fabricators to accelerate their processes to match ConXtech's speed, as the greatest benefit is realized when all the systems move at the same rate.

Ultimately the challenge for the ConXtech technology is whether it can spark a revolution in steel frame fabrication and erection that is so efficient that it becomes industry-wide and, with matching service subsystem development, can accelerate building construction five-fold, to match its own pace.

Processes for redesign

IT provided the means to achieve the breakthroughs in process performance in some organizations. The inspiration, however, came from understanding both the current and potential processes. This required a more holistic view than that taken in traditional lean quality programmes involving wholesale redesigns of the processes concerned.

Ford estimated a 20 per cent reduction in head count if it automated the existing processes in accounts payable. Taking an overall process perspective, Ford achieved a 75 per cent reduction in one department. Xerox took an organizational view and concentrated on the cross-functional processes to be re-engineered, radically changing the relationship between supplier and external customer.

Clearly, the larger the scope of the process, the greater and farther reaching are the consequences of the redesign. At a macro level, turning raw materials into a product used by a delighted customer is a process made up of subsets of smaller processes. The aim of the overall process is to add value to the raw materials. Taking a holistic view of the process makes it possible to identify non-value-adding elements and remove them. It enables people to question why things are done and to determine what should be done.

The case study of the re-engineering of timber floor construction (CS13) for dwellings on sloping sites in the Australian residential sector describes process re-engineering across the supply chain. Such changes are more difficult to achieve than process changes which are entirely within a single organization. However, some areas of construction innovation require such change. It may be best achieved through partnerships among the organizations within the supply chain or by the major players who have the intellectual and financial capacity to make the change and to capture the benefits from them.

Some of the re-engineering literature advised starting with a bank sheet of paper and redesigning the process anew. The problems inherent in this approach are:

*   the danger of designing another inefficient system; and
*   not appreciating the scope of the problem.

Therefore, the authors recommend a thorough understanding of current processes before embarking on a re-engineering project.

Current processes can be understood and documented by process mapping and flow-charting. As processes are documented, their interrelationships become clear and a map of the organization emerges. Figure 13.1 shows a much simplified process map. As the aim of BPR is to make discontinuous, major improvements, this invariably means organizational change, the extent of which depends on the scope of the process re-engineered.

Taking the organization depicted in Figure 13.1 as an example, if the decision is made to redesign the processes in finance, the effect may be that in Figure 13.2a eight individual processes have become three. There has been no organizational effect on the processes in the other functions, but finance has been completely restructured. In Figure 13.2b, a chain

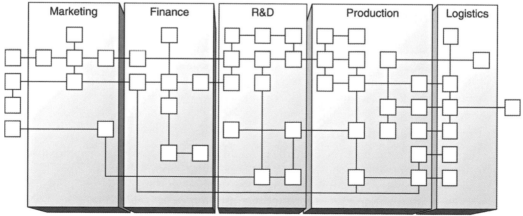

Each box is a process. Lines identify linked processes

FIGURE 13.1 Simplified process map

of processes crossing all the functions has been re-engineered. The effect has been the loss of redundant processes and possibly many heads, but much of the organization has been unaffected. Figure 13.2c shows the organization after a thorough re-engineering of all its processes. Some elements may remain the same, but the effect is organization-wide.

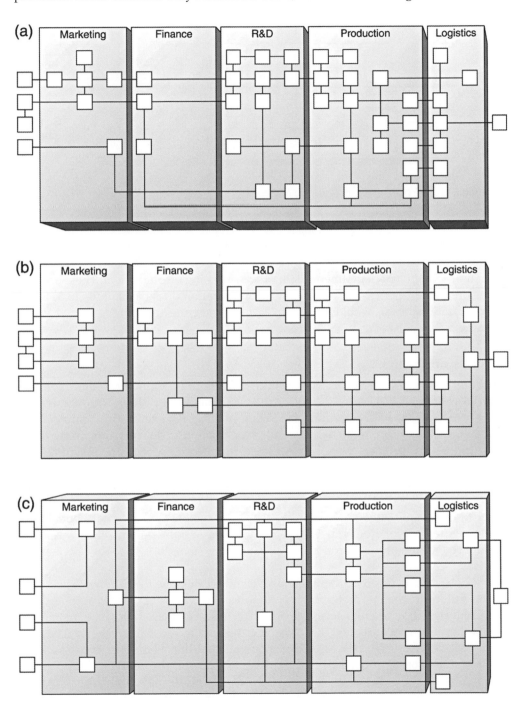

FIGURE 13.2 (a) Process redesign in finance, (b) cross-functional process design, (c) organizational process redesign

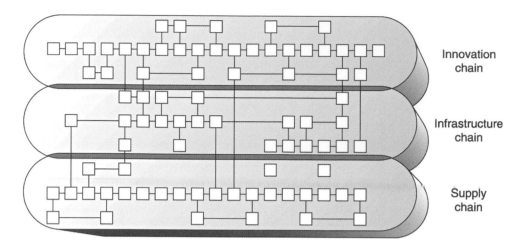

FIGURE 13.3 Process organization

Whatever the scope of the redesign, head count is not the only change. When work processes are altered, the way people work alters. Figures 13.1 and 13.2 show an organization's functional departments with process running through them. These are the handful of core processes that make up what an organization does (see Figure 13.3) and in many organizations these would benefit from re-engineering to improve added value output and efficiency.

Focus on results

BPR is not intended to preserve the status quo but to fundamentally and radically change what is done; it is *dynamic*. Therefore, it is essential for a BPR effort to focus on required customers. This will determine the scope of the BPR exercise. A simple requirement may be a 30 per cent reduction in costs or a reduction in delivery time of two days. These would imply projects with relatively narrow scope, which are essentially inwardly focused and probably involve only one department, for example, the finance department in Figure 13.2a.

When Wal-Mart focused on satisfying customer needs as an outcome, it started a redesign that not only totally changed the way it replenished inventory but also made this the centrepiece of its competitive strategy. The system put in place was radical, and required tremendous vision. That or similar systems are used by all major retailers today.

Focusing on results rather than just activities can make the difference between success and failure in change projects. The measures used, however, are crucial. At every level of redesign and re-engineering, a focus on results gives direction and measurability, whether it be cost reduction, head-count reduction, increase in efficiency, customer focus, identification of core processes and non-value-adding components, or strategic alignment of business processes. Benchmarking is a powerful tool for BPR and is the trigger for many BPR projects, as it was in Ford's accounts payable process. As shown in Chapter 11, the value of benchmarking does not lie in what can be copied but in its ability to identify goals. If used well, benchmarking can shape strategy and identify potential competitive advantage.

THE REDESIGN PROCESS

Central to BPR is an objective overview of the processes to be redesigned. Whereas information needs to be obtained from the people directly involved in those processes it is never initiated by them. Even at its lowest level, BPR has a top-down approach, and most BPR efforts therefore take the form of a project. There are numerous methodologies proposed, but all share common elements. Typically, the project takes the form of seven phases, shown in Figure 13.4.

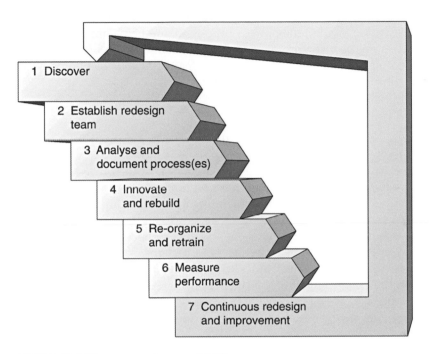

FIGURE 13.4 The seven phases of BPR

1 Discover and define

This involves firstly identifying a problem or unacceptable outcome, followed by determining the desired outcome. This usually requires an assessment of the business need and will certainly include determining the processes involved, including the scope, identifying process customers and their requirements, and establishing effectiveness measurements.

2 Establish redesign team

Any organization, even a small company, is a complex system: there are customers, suppliers, employees, functions, processes, resources, partnerships, finances, etc.; and many large organizations are incomprehensible: no one person can easily get a clear picture of all the separate components. Critical to the success of the redesign is the make-up of a redesign team.

The team should comprise as a minimum the following:

- senior manager as a sponsor;
- steering committee of senior managers to oversee the overall re-engineering strategy;
- process owner;

- team leader; and
- redesign team members.

It is generally recommended that the redesign team:

- has between five and ten people;
- represents the scope of the process (that is, if the process to be re-engineered is cross-functional, so must the team be); and
- only works on one redesign at a time.

Typically a redesign team is made up of both insiders and outsiders – insiders are people currently working within the process concerned and help gain credibility with co-workers. Outsiders are people from outside the organization who bring objectivity and can ask the searching questions necessary for the creative aspects of the redesign. Many companies use consultants for this purpose.

3 Analyse and document process(es)

Making visible the invisible, documenting the process(es) through mapping and/or flowcharting is the first crucial step that helps an organization see the way work really is done and not the way one thinks or believes it is done. Seeing the process as it is provides a baseline from which to measure, analyse, test and improve.

Collecting supporting process data, including benchmarking information and IT possibilities, allows people to weigh the value each task adds to the total process, rank and select areas for the greatest improvement and spot unnecessary work and points of unclear responsibility. Clarifying the root causes of problems, particularly those that cross department lines, safeguards against quick-fix remedies and assures proper corrective action, including the establishment of the right control systems.

4 Innovate and rebuild

In this phase, the team rethinks and redesigns the new process using the same process mapping technique as they used to map the current state in an iterative approach, involving all the stakeholders including senior management. A powerful method for challenging existing practices and generating breakthrough ideas is 'assumption busting' (see later section in this chapter).

5 Re-organize and retrain

This phase includes piloting the changes and validating their effectiveness. The new process structure and operation/system will probably lead to some re-organization. This may be necessary in order to reinforcement of the process strategy and to achieve the new levels of performance. Training and/or retraining for the new technology and roles play a vital part in successful implementation. People need to be equipped to assess, re-engineer and support (with the appropriate technology) the key processes that contribute to customer satisfaction and corporate objectives. Therefore, BPR efforts can involve substantial investment in training but they also require considerable top management support and commitment.

6 Measure performance

It is necessary to develop appropriate metrics for measuring the performance of the new process(es), sub-processes, activities and tasks. These must be meaningful in terms of the inputs and outputs of the processes, and in terms of the customers of and suppliers to the process(es) (see Chapter 9).

7 Continuous redesign and improvement

The project approach to BPR suggests a one-off approach. When the project is over, the team is disbanded and business returns to normal, albeit a radically different normal. It is generally recommended that an organization does not attempt to re-engineer more than one major process at a time, because of the disruption and stress caused. Therefore, in major re-engineering efforts of more than one process, as one team is disbanded another is formed to redesign yet another process. Considering that Ford took five years to redesign its accounts payable process, BPR on a large scale is clearly a long-term commitment.

In a rapidly changing, ever more competitive business environment, it is becoming more likely that companies will re-engineer one process after another. Once a process has been redesigned, continuous improvement of the new process by the team of people working in the process should become the norm.

ASSUMPTION BUSTING

Within BPR is a powerful method for challenging existing practices and generating breakthrough ideas for improvement. 'Assumption busting' (Hammer and Champy, 1993) aims to identify the **rules** that govern the way we do business and then uncover the real underlying **assumptions** behind the adoption of these rules. Business processes are governed by a number of rules that determine the way the process is designed, how it interfaces with other activities within the organization and how it is operated. These rules can exist in the form of explicit policies and guidelines or, what is more often the case, in the mind of the people who operate the process. These unwritten rules are the product of assumptions about the process environment that have been developed over a number of years and often emerge from uncertainties surrounding trading relationships, capabilities, resources, authorities, etc. Once these underlying assumptions are uncovered, they can be challenged for relevance and, in many cases, can be found to be false. This opens up new opportunities for process redesign and, as a consequence, the creation of new value and improved performance.

For example, Resources Ltd, a supplier of TV and radio studio and outside broadcast resource services, faced the requirement to improve business performance. The business was losing money and still faced competition from independent providers. They needed to improve the efficiency of their processes whilst retaining their core capability that created competitive advantage. In order to stimulate breakthrough thinking, a team was commissioned to review the core, value-adding processes and set challenging targets for improvements in performance. The team decided to take a more radical approach by using assumption busting and prepared a six-week programme of work. Within that time frame they used an established eight-step method (shown in Figure 13.5) to redesign the core, end-to-end service delivery processes for two major business units – Studios and Outside Broadcasts. The work involved identifying the key areas of cost consumption, challenging the rules and assumptions that governed the existing process and generating a set of improvement opportunities. When they had evaluated their findings, the team presented ideas to deliver an improvement in excess of 15 per cent in process efficiency.

One of the process rules concerned the use of a highly technically qualified member of staff for the planning and delivery of all of the programmes supported. The core underlying assumption was that all of the programmes were complex in nature. When this

FIGURE 13.5 The assumption busting cycle

assumption was challenged, the team in Resources Ltd realized that, since not all programmes were so complex in nature, less technically qualified members of staff could be utilized at lower cost to the business.

Application of the technique

In practice the authors and colleagues have found this technique to be of greatest value when applied by a cross-functional group of process operators and supervisors who are given a specific problem to fix. In using the technique care must be exercised in the use of terms such as 'rule' and 'assumption'. They often cause initial confusion, and there can be real difficulty in uncovering the core underlying assumptions. Rules should be clearly stated and tested for validity before proceeding down what eventually could become a blind alley. Furthermore, a rigorous approach to the identification of the core assumption is vital to uncovering the real opportunities for improvement. An assumption by definition 'is a statement/belief that is accepted or supposed to be true without proof or demonstration'. In some cases, rules are created from specific knowledge about the business and its environs, and not based on assumptions. Assumptions spring from our beliefs about the environment and not our specific knowledge.

Other applications of assumption busting

Assumption busting is of particular benefit when applied by partners within a supply chain. The trading relationships and practices that exist in a modern supply chain such as a supermarket and its multiple tiers of suppliers are the product of a number of assumptions made by the supply chain partners about what is possible. Once teams from each of the partnering businesses work collaboratively to uncover the rules and assumptions that govern their trading relationships, the door is unlocked to new methods and economies.

The method can also be immensely powerful when companies are introducing new technology. Breakthrough technologies can lead to breakthrough performance as they make possible what is considered impossible today. Hence, a number of current rules and assumptions are there to be challenged as processes are redesigned to take advantage of the new technology. We are often just as constrained by our lack of imagination regarding the possibilities of tomorrow as we are by our knowledge of what is possible today. One example of this was in BBC World Service, where the introduction of digital technology to replace analogue was accompanied by assumption busting to let process redesign take advantage of the new technological capability.

While assumption busting has been primarily applied to the generation of new process designs, it exists in its own right as a method for developing more 'lateral' solutions to problems. In the early 1970s, Dr Edward de Bono introduced the concept of lateral thinking as an alternative method of generating ideas to that of the more traditional logical or 'vertical' thinking. Dr de Bono argued that our thinking is constrained by patterns that form in our minds over time and channel our future thoughts. Assumption busting helps people break out of this 'channelled thinking' to develop creative ideas. Managers could benefit from applying *assumption busting* to a number of problems or opportunities in their businesses – assumptions constrain us everywhere, not just within our business processes.

Whether it is in response to specific customer requirements, new technology or in the quest for competitive advantage, assumption busting provides a simple but effective method for breaking into new areas of adding value. World-class performance will not be achieved by effort alone: creativity and innovation are cornerstones of future success. Innovative ways of delivering new value will be rewarded. Assumption busting provides a powerful method for generating new ideas from looking at today and tomorrow in a different way.

BPR – THE PEOPLE AND THE LEADERS

For an organization to focus on its core processes almost certainly requires an understanding of its core competencies. Moreover, core process redesign can channel an organization's competencies into an outcome that gives it strategic competitive advantage. The key element is visioning that outcome. Visioning the outcome may not be enough, however, since many companies' 'vision' desires without simultaneously 'visioning' the systems that are required to generate them. Without a clear vision of the systems, processes, methods and approaches that will allow achievement of the desired results, dramatic improvement is frequently not obtained as the organization fails to align around a common tactical strategy. Such an 'operational' vision is lacking in many organizations.

The fallout from BPR has profound impacts on the employees in any enterprise at every level – from executives to operators. Therefore, for BPR to be successful, significant changes in organization design and enterprise culture are also often required. Unless the leaders of the enterprise are committed to undertake these changes, the BPR initiative will flounder. The point is, of course, that organization design and culture changes are much more difficult than modifying processes to take advantage of new IT.

While the enabling IT is often necessary and is clearly going to play a role in many BPR exercises, it is by no means sufficient, nor is it the most difficult hurdle on the path to success. Thanks to IT we can radically change the processes an organization operates and, hopefully, achieve dramatic improvements in performance. However, in any BPR project there will be considerable risk attached to building the information system that will support the new, redesigned processes. Information systems should be, but rarely are, described so that they are easy for people to understand.

While BPR may be a distinct, short-term activity for a specific business function, the record indicates that BPR activities are most successful when they occur within the framework of a long-term thrust for excellence. Hence within a lean quality culture, a BPR effort is more likely to find the process focus, supportive workforce, organization design and mindset changes needed for its success.

Process improvement is sometimes positioned as a bottom-up activity. In some contrast, lean quality involves setting longer-term goals at the top and modifying the business as necessary to achieve the goals. Often, the modifications to the business required to achieve the goals are extensive and ground-breaking. The history of successful lean quality thrusts in award-winning companies throughout the world is replete with new organization designs, flattened structures and empowered employees in the service of end customers. In many successful organizations, BPR has been an integral part of the culture – a process-driven change dedicated to the ideals and concepts of lean quality. That change must create something that did not exist before: namely, a 'learning organization' capable of adapting to a changing competitive environment. When processes, or even the whole business, need to be re-engineered, the radical change may not and probably will not be readily accepted.

ACKNOWLEDGEMENT

The authors are grateful to the contribution made by colleagues Ken Gadd and Mike Turner to the preparation of this chapter.

BIBLIOGRAPHY

Arbulu, R.J. and Tommelein, I.D. Alternative supply-chain configurations for engineered or catalogued made-to-order components: case study on pipe supports used in power plants, *Proceedings IGLC-10*, August, Gramado, Brazil, 2002, www.cpgec.ufrgs.br/norie/iglc10/

Elfving, J., Tommelein, I.D. and Ballard, G. An international comparison of the delivery process of power distribution equipment, *Proceedings IGLC-11*, August, Blacksburg, USA, 2003, http://strobos.cee.vt.edu/IGLC11/

Hammer, M. and Champy, J. *Re-engineering the Corporation*, Nicholas Brearley, London, 1993.

Hammer, M. and Stanton, S.A. *The Re-engineering Revolution – The Handbook*, BCA, Glasgow, 1995.

Koskela, L. and Howell, G. The theory of project management: explanation to novel methods, *Proceedings IGLC-10*, August, Gramado, Brazil, 2002, www.cpgec.ufrgs.br/norie/iglc10/

Marosszeky, M., Karim, K., Davis, S. and Naik, N. Lessons learnt in developing effective performance measures for construction safety management, a case study, *Proceedings IGLC-12*, August, Ellsinore, Denmark, 2004.

Marosszeky, M., Sauer, C., Johnson, K., Karim, K. and Yetton, P. Information technology in the building and construction industry: the Australian experience, *INCITE 2000 – Implementing IT to Obtain a Competitive Advantage in the 21st Century,* The Hong Kong Polytechnique University, Hong Kong, 2000.

Marosszeky, M., Thomas, R., Karim, K., Davis, S. and McGeorge, D. Quality management tools for lean production: moving from enforcement to empowerment, *Proceedings IGLC-10,* August, Gramado, Brazil, 2002, www.cpgec.ufrgs.br/norie/iglc10/frame_proceedings.htm

Saurin, T., Formoso, C., Guimarães, L.M. and Soares, A. Safety and production: an integrated planning and control model, *Proceedings IGLC-10,* August, Gramado, Brazil, 2002, www.cpgec.ufrgs.br/norie/iglc10/frame_proceedings.htm

Tommelein, I.D., Akel, N.G. and Boyers, J.C. Application of lean supply chain concepts to a vertically-integrated company: a case study, *Proceedings IGLC-12,* August, Ellsinore, Denmark, 2004, www.iglc2004.dk/13729

Yetton, P. and Marosszeky, M. *Information Technology in the Building and Construction Industry: Current Status and Future Directions,* Fujitsu Centre AGSM and Building Research Centre, UNSW, 1998.

CHAPTER HIGHLIGHTS

Re-engineering the organization

- Lean focuses on the client, the elimination of waste and the maximization of value.
- The primary forms of waste are overproduction, waiting, excess conveyance, extra processing, excessive inventory, unnecessary motion and defects requiring rework or scrapping. To these, Koskela added for the construction sector – reduction (the proportion of non-value-adding activities): reduction in lead time, reduction in variability, simplification of processes, increasing flexibility and increasing transparency.
- When a major business process requires radical re-assessment, perhaps through the introduction of new technology, discontinuous methods of business process re-engineering or redesign (BPR) are appropriate.
- The opportunity for radical change in construction processes may involve collaboration across the supply chain and this might be best achieved through partnerships among organizations.
- Drives for process change include information technology (IT), political, financial, competitive aspects of culture. These often require a change of thinking about the ways processes are and could be operated.
- IT often creates opportunities for breakthrough performance, but BPR is needed to deliver it. Successful practitioners of BPR have made striking improvements in customer satisfaction and productivity in short periods of time.
- Inter-organizational integration of IT is one of the greatest opportunities and challenges facing the sector – it has the potential to unlock significant value.

What is BPR and what does it do?

- There are many definitions of BPR but the basic elements involve a fundamental re-think and radical redesign of a business process, its structure and associated management systems to deliver step improvements in performance.
- BPR and lean quality are complementary under the umbrella of process management – the continuous and discontinuous improvements living side by side. Both require the involvement of customers and suppliers, and their future requirements.
- BPR challenges managers to rethink their traditional methods of doing work and to commit to customer-focused processes. This breaks down organizational barriers and encourages cross-functional teams.

Processes for redesign/focus on results

- Much larger savings and head-count reductions are possible through properly applied BPR than simply automating existing processes. The larger the scope of the process the greater and farther reaching the consequences of the redesign.
- A thorough understanding of the current process is needed before embarking on a re-engineering project. Documentation of processes through mapping and flowcharting allows interrelationships to be clarified.
- Focusing on results rather than activities can make the difference between success and failure in BPR and other change projects, but the measures used are critical. Bench-marking is a powerful tool for BPR and often the trigger for many projects.

The redesign process/assumption busting

- BPR has a top-down approach and needs an objective overview of the process to be redesigned to drive the project.
- Typically a BPR project will have seven phases: discover – identifying the problem or unacceptable outcome; establish redesign team; analyses and document processes; innovate and rebuild; re-organize and retrain; measure performance; continuous redesign and improvement.
- Assumption busting is a useful eight-step BPR method which aims to identify and challenge the 'rules' and assumptions that govern and underlie the way business is done. A team is formed to: identify the core value to be delivered to customers and stakeholders; map the process at high level; select problems to resolve and collect performance data; brainstorm and test the rules; rigorously review each rule to uncover underlying assumptions; identify modified assumptions and process rules; identify impact and construct new sets of process principles; develop revised processes and test validity.

BPR – the people and the leaders

- For an organization to focus on its core processes requires an understanding of its core competencies and the channelling of these into outcomes that deliver strategic competitive advantage.
- BPR has profound impacts on employees from the top to the bottom of an organization. In order to be successful, significant changes in organization design and enterprise culture are also often required. This requires commitment from the leaders to undertake these changes.
- Lean quality ideals and concepts provide a perfect platform for BPR projects and the creation of a 'learning organization' capable of adapting to a radically changing environment.

Quality management systems

14

WHY A QUALITY MANAGEMENT SYSTEM?

In the construction sector, worldwide, the value of formal quality management systems is still widely questioned and it is small wonder. In some countries, when the international standard was introduced, government policy simply mandated certification against the ISO 9000 family of standards for all government suppliers. Consequently, the managers within some companies saw the need to obtain ISO 9000 certification only as a necessary hurdle to getting on to government tender lists and, hence, primarily as a marketing problem. In these cases, where senior management gave the issue scant attention, marketing and/or other managers hired a consultant to develop and deliver a compliant system.

In the early 1990s, quality consultants generally understood little or nothing about construction. However, they did know how to put together a top-down generic ISO 9000 compliant management system that satisfied the needs of senior construction management faced with an urgent need to obtain certification. Hence, there was little or no workplace involvement in early management system development in the construction sector. Frequently, system 'quality manuals' were very thick and did not reflect the business goals or management needs of the enterprise. Because of this inappropriate start, to this day, there are construction organizations worldwide that have still not grasped the strategic significance of the total quality philosophy for their businesses. The case studies in this book bring together the stories of companies that have understood and have succeeded in meeting this challenge.

In earlier chapters we have seen how the keystone of quality management is the concept of customer and supplier working together for their mutual advantage. For any particular organization this becomes 'total' quality management if the supplier/customer interfaces extend beyond the immediate customers, back inside the organization, and beyond into the contractors and supply chain. In order to achieve this, a company must organize itself in such a way that the human, administrative and technical factors affecting quality will be under control. This leads to the requirement for the development and implementation of a quality management system that enables the objectives set out in the quality policy to be accomplished. Clearly, for maximum effectiveness and to meet individual customer requirements, the management system in use must be appropriate to the type of activity and product or service being offered.

It may be useful to reflect on why such a device is necessary to achieve control of processes. John Oakland still remembers being at a table in a restaurant with eight senior managers who all ordered the 'Chef's Special Individual Soufflé'. All eight soufflés arrived together at the table, magnificent in their appearance and consistency, each one exhibiting an almost identical size and shape – a truly remarkable demonstration of culinary skill.

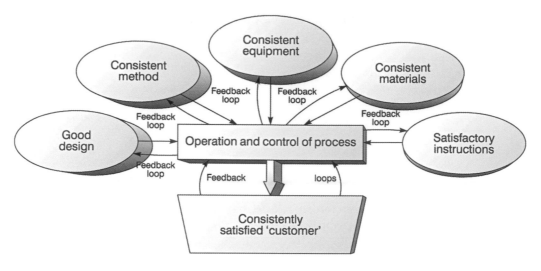

FIGURE 14.1 The systematic approach to process management

How had this been achieved? The chef had *managed* such consistency by making sure that, for each soufflé, he used the same ingredients (materials), the same equipment (plant), the same method (procedure) in exactly the same way every time. The process was under control. This is the aim of a good quality management system, to provide the 'operator' of the process with consistency and satisfaction in terms of methods, materials, equipment, etc. (Figure 14.1). Two feedback loops are also required: the 'voice' of the customer (marketing activities) and the 'voice' of the process (measurement activities).

The chef's soufflés were not British Standard, NIST Standard, Australian Standard or ISO Standard soufflés – they were the 'chef's special soufflés'. It is not conceivable that the chef sat down with a blank piece of paper to invent a soufflé recipe. Why re-invent wheels? He probably used a standard formula and changed it slightly to make it his own. This is exactly the way in which successful organizations use the international standards on quality management systems that are available. The 'wheel' has been invented but it must be built in a way that meets the specific organizational and product or service requirements. The international family of standards, ISO 9000 'Quality Management Systems', specifies systems which can be implemented in an organization to ensure that all the product/service performance requirements and needs of the customer are fully met.

Let us return to the chef in the restaurant and propose that his success leads to a desire to open eight restaurants in which his special soufflés are served. Clearly he cannot rush from each one of these establishments to another every evening making soufflés. The only course open to him, to ensure consistency of output in all eight restaurants, is for him to define the system he uses, and then make sure that it is used on all sites, every time a soufflé is produced. Moreover, he must periodically visit the different sites to ensure that:

1 The people involved are operating according to the designed system (a system audit).
2 The soufflé system still meets the requirements (a system review).

If in his system audits and reviews, he discovers that an even better product or less waste can be achieved by changing the method or one of the materials, then he may wish to effect a change. To maintain consistency, he must ensure that the appropriate changes are made to the management system *and* that everyone concerned understands the revision and begins to operate accordingly.

A good quality management system will ensure that two important requirements are continually met:

- *The customer's requirements* – for confidence in the ability of the organization to deliver the desired product or service consistently.
- *The organization's requirements* – both internally and externally including the statutory and regulatory context (within which the organization operates) and at an optimum cost, with efficient utilization of the resources deployed – material, human, technological, environmental (including working environment) and information.

The requirements can be truly met only if objective evidence is provided, in the form of information and data, which supports the system activities from the ultimate supplier through to the ultimate customer.

A *Quality Management System (QMS)* may be defined, then, as an assembly of components, such as the leadership/management responsibilities, processes and resources for implementing total quality. These components interact and are affected by being in the system, so the isolation and study of each one in detail will not necessarily lead to an understanding of the system as a whole. Often the interactions between the components – such as materials and processes, people and responsibilities – are just as important as the components themselves, and problems can arise from these interactions as much as from the components. Clearly, if one of the components is removed from the system, the whole thing will change.

The adoption of a QMS is, of course, a strategic decision and its design should be influenced by the organization's environment, operating context, objectives, structure and size, the products or services offered, its processes and subcontract and supply chain decisions. The QMS should help the organization improve its overall performance and provide a sound basis for sustainable development initiatives.

ISO 9000 quality management systems

'BS EN ISO 9001:2015 *Quality Management Systems – Requirements*' is the current International Standard which superseded ISO 9001:2008. Detailed information on the ISO 9000:2015 family of standards, may be found on the websites: www.iso.ch and www.bsigroup.com. ISO 9001:2015 now employs a process approach, which incorporates the Plan-Do-Check-Act (PDCA) cycle and risk-based thinking. The potential benefits to an organization of adopting this standard are:

- the ability to consistently provide products and services that meet customer and applicable statutory and regulatory requirements;
- facilitating opportunities to enhance customer satisfaction;
- addressing risks and opportunities associated with the business' context and objectives;
- the ability to demonstrate conformity to specified quality management system requirements.

ISO 9001:2015 is one of three core **9000** series standards developed by the International Standards Organization:

- ISO 9000 *Quality management systems – Fundamentals and vocabulary* provides an essential background for the proper understanding and implementation of ISO 9001. The quality management principles are described in detail in ISO 9000 and have been taken into consideration during the development of 9001. These principles are not requirements in themselves, but they form the foundation of the requirements specified. ISO 9000 also defines the terms, definitions and concepts used in ISO 9001.

- ISO 9001 specifies requirements aimed primarily at giving confidence in the products and services provided by an organization and thereby enhancing customer satisfaction. Its proper implementation can also be expected to bring other organizational benefits, such as improved internal communication, better understanding and control of the organization's processes.
- ISO 9004 *Managing for the sustained success of an organization – A quality management approach* provides guidance for organizations that choose to progress beyond the requirements of ISO 9001, to address a broader range of topics that can lead to improvement of the organization's overall performance. ISO 9004 includes guidance on a self-assessment methodology for an organization to be able to evaluate the level of maturity of its quality management system

The additional International Standards **10000** series outlined below can provide assistance to organizations when they are establishing or seeking to improve their quality management systems, their processes or their activities (more detail is given in Annexe B of ISO 9001):

- ISO 10001 Quality management – Customer satisfaction – Guidelines for codes of conduct for organizations.
- ISO 10002 Quality management – Customer satisfaction – Guidelines for complaints handling in organizations.
- ISO 10003 Quality management – Customer satisfaction – Guidelines for dispute resolution external.
- ISO 10004 Quality management – Customer satisfaction – Guidelines for monitoring and measuring.
- ISO 10005 Quality management systems – Guidelines for quality plans.
- ISO 10006 Quality management systems – Guidelines for quality management in projects – of particular relevance to the construction supply chain as it is applicable to projects from the small to large, from simple to complex, from an individual project to being part of a portfolio of projects. ISO 10006 is intended to be used by personnel managing projects who need to ensure that their organization is applying the practices contained in the ISO quality management system standards.
- ISO 10007 Quality management systems – Guidelines for configuration management.
- ISO 10008 Quality management – Customer satisfaction – Guidelines for business-to-consumer electronic commerce transactions.
- ISO 10012 Measurement management systems – Requirements for measurement processes and measuring equipment.
- ISO 10013 Guidelines for quality management system documentation.
- ISO 10014 Quality management – Guidelines for realizing financial and economic benefits.
- ISO 10015 Quality management – Guidelines for training.
- ISO 10017 Guidance on statistical techniques for ISO 9001.
- ISO 10018 Quality management – Guidelines on people involvement and competence.
- ISO 10019 Guidelines for the selection of quality management system consultants and use of their services.

ISO 19011 *Guidelines for auditing management systems* provides guidance on the management of an audit programme, on the planning and conducting of an audit of a management system, as well as on the competence and evaluation of an auditor and an audit team. ISO 19011 is intended to apply to auditors, organizations implementing management systems and organizations needing to conduct audits of management systems.

Quality management principles and ISO 9000

The ISO 9001:2015 standard is based on the adoption of seven quality management principles. These are:

1 Customer Focus;
2 Leadership;
3 Engagement of people;
4 Process approach;
5 Improvement;
6 Evidence-based decision-making;
7 Relationship management.

As stated in the standard, understanding and managing interrelated processes as a system contributes to an organization's effectiveness and efficiency in achieving its intended results. This 'process' approach enables the organization to control the inter-relationships and interdependencies among the processes of the system, so that the overall performance of the organization can be enhanced.

The process approach involves the systematic definition and management of processes and their interactions, so as to achieve the intended results in accordance with the quality policy and strategic direction of the organization. Management of the processes and the system as a whole can be achieved using the PDCA cycle with an overall focus on risk-based thinking aimed at taking advantage of opportunities and preventing undesirable results. These principles are, of course, generic in the fields of Lean–Quality.

The diagram in Figure 14.2, taken directly from the ISO 9001:2015 standard, shows the adoption of the Supplier-Input-Process-Output-Customer (SIPOC) methodology, a widely used and respected means of understanding the design and interdependencies of the processes of an organization.

The schematic shows the interaction of the elements of a single process. The check points necessary to monitor and control the process are specific to each case and are dependent on the risks inherent in the process.

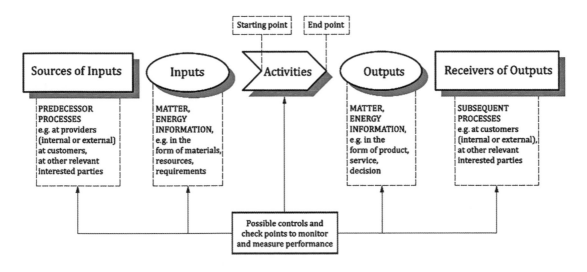

FIGURE 14.2 Schematic representation of the elements of a single process

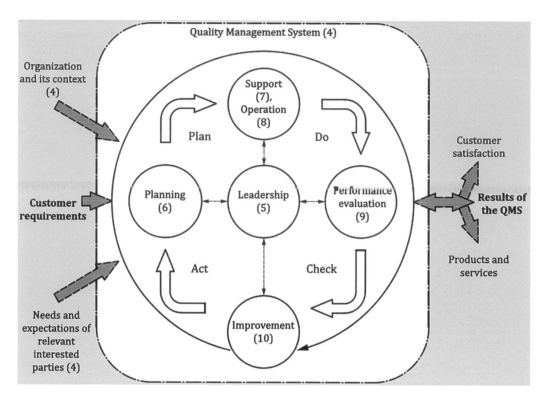

FIGURE 14.3 Representation of the structure of ISO 9001 is the PDCA cycle

The integration of the PDCA process into the ISO 9001:2015 standard dovetails perfectly with the process-based approach in the standard. PDCA can be applied to all processes and to the QMS as a whole. The standard includes a schematic illustrating how its clauses can be considered to form a high-level PDCA process for the organization. The seven activity areas are numbered 4–10 in Figure 14.3, reflecting the clause numbers within the standard.

This approach now formally brings together Deming's Cycle of continuous improvement – PLAN DO CHECK ACT – and quality management systems. The steps involved in Deming's PDCA cycle, in the context of a QMS, guide the organization to:

- **Plan**: Establish the objectives of the system, its processes and the resources needed in order to deliver results in accordance with customer requirements and the organization's policies and identify and address any risks or opportunities.
- **Do**: Implement what was planned.
- **Check**: Monitor and measure processes and the resulting products and services for compliance with policies, objectives, requirements and planned activities and report on these results.
- **Act**: Take actions to improve performance, as necessary.

Risk-based thinking is essential for establishing and then maintaining the effectiveness of a QMS. The concept of risk-based thinking had been implicit in previous versions of the ISO 9001 standard, for example, in the requirement for implementation of preventive actions to eliminate potential nonconformities and analysing any nonconformities that do occur to understand and eliminate the causes to prevent their recurrence.

The 2015 version of the standard goes much further than this and requires that organizations plan and implement actions to address risks and opportunities. This provides a basis for improving the effectiveness of the QMS by exploiting opportunities as they appear, improving performance and results and preventing negative effects.

Opportunities might relate to circumstances that allow an organization to, for example, attract new customers, develop new products and services, reduce waste or improve productivity. Actions to address opportunities should also include the consideration of any associated risk. Risks are essentially the potential effects of uncertainty and can have positive or negative implications.

QUALITY MANAGEMENT SYSTEM DESIGN AND ISO 9000

A good and effective QMS should provide the structured approach through which all of the organization's processes are designed, managed and continuously improved to optimize their effectiveness in satisfying customer requirements and their efficiency. It begins with the identification of the customer requirements and ends with their satisfaction, at every transaction interface, and it should be driven by the purpose of the business or organization. The activities may be classified in several ways – generally as processing, communicating and controlling, but more usefully and specifically as shown in the quality management process model described in ISO 9001:2015 (Figure 14.3). This reflects graphically the integration of the seven major elements represented by the clauses of the ISO 9001:2015 standard.

The seven clauses that now form ISO 9001:2015 are as follows:

- Context of the organization;
- Leadership;
- Planning;
- Support;
- Operation;
- Performance evaluation;
- Improvement.

The management system requirements under these headings are, of course, specified and detailed in the standard.

A series of fundamental changes to the structure and focus of ISO 9001 have continued through each new standard release and the latest 2015 version is no exception. Compared to the original version of the standard, ISO 9001:2015 is now much more effective and the improvements are both relevant and important for the construction sector. The 1994 version of the standard was management system focused so that compliance could be obtained by a company on the basis of its management system, regardless of the quality of the actual product and services they provided. The new version of the standard links product and service quality to system quality through a focus on customer satisfaction and the use of a process-focused approach. It builds into the system the need for performance evaluation and improvement. The new standard is a far more solid basis for a quality system and compliance can no longer be achieved by the rather cynical purchase of a *'quality manual'*. Hence, its potential value to organizations within the construction sector has increased immeasurably.

As well as the fundamental changes to the principles by which the latest version of the ISO 9001 standard requires that an organization's QMS is managed, there have been further changes that expand the scope of the system's application. ISO 9001:2015 incorporates a requirement to evaluate the context in which the organization operates, which means applying effort to understand the environment in which it operates, considering such things as legislative changes, evolving customer or market requirements, competitor activity and the development of technologies that might improve the organization's effectiveness, efficiency or product and service offerings.

Consider an example similar to the one mentioned earlier of the restaurant serving the high-quality, repeatable soufflé dish. If the restaurant was perhaps a 'gastro pub', more dependent on keeping up with the changing tastes of the clientele, then the advantages of the expansion to the requirement of the ISO 9001 standard can be easily visualized.

The restaurant may have been in a position of strength, benefiting from the improved quality and efficiency provided by their accreditation to a QMS standard. If, however, they remain unaware of changes in customer taste or demands (maybe their perfectly prepared 'burger' has lost its attraction when all of the competitor restaurants now offer cheese-burgers or bacon and cheese-burger options), or perhaps they fail to adopt the use of microwave oven technology or pre-packed portioning from their suppliers and become increasingly uncompetitive or maybe they remain ignorant of changes to legislation requiring that their menu display calorific values for their dishes and suffer prosecution.

Any of the above are likely to result in diminishing customers and economic problems, even if the QMS had ensured that all dishes retain their consistent, optimum quality. The QMS might have sustained the 'quality' delivered by the organization but their product and service offerings, and the QMS that assured them, might have become irrelevant or obsolete.

The new standard's requirement, to investigate, collect data, analyse and understand the risks and opportunities surrounding the ever-changing context and environment in which the organization operates provides security in that it will ensure that the organization is aware of and responds to any changes to its context of operation in an effective way. It should prevent the organization being left behind by their competitors and might provide the opportunity to exploit the 'first mover' advantages associated with the opportunities identified.

ISO 9001:2015 also removes the requirement in previous standards for a 'quality manual' and for 'documented procedures'. Indeed, previously familiar words such as *document* and *record* are absent, replaced by terms such as *documented information*, which is intended to mean information describing the management system and its processes. This now applies to internally and externally produced information, in all possible formats, ranging from the top-level policies through to basic instructions and check sheets. The standard states that documented information must be provided to ensure effectiveness of the management system and its processes, so it would be wrong to interpret that the need for written procedures has been removed or reduced.

In many organizations established methods of working already exist around identified processes, and all that is required is the *documenting of what is currently done*. In some instances, companies may not have procedures to satisfy the requirements of a good standard, and they may have to begin to devise them. Alternatively, it may be found that two people, supposedly performing the same task, are working in different ways and there is a need to standardize procedures. In the context of QMS, some organizations use the effective slogan '*If it isn't written down, it doesn't exist*'. This can be a useful discipline, provided it doesn't lead to an overly bureaucratic approach.

One person alone cannot document a QMS; the task is the job of all personnel who have responsibility for any part of it. This means that the system, by definition, has to be built from the operational level up and cannot be imposed by one 'controlling mind'. The QMS must be a *practical working one* – that way it ensures that consistency of operation is maintained and it may be used as a training aid.

In the operation of any process, a useful guide is:

- No process without data collection (**Measurement**).
- No data collection without **Analysis**.
- No analysis without **Decisions**.
- No decisions without **Actions** (Improvement) – which can include doing nothing.

This excellent discipline is built into any good quality management system.

Probably the most significant change to the requirements brought about with the process-based format of the ISO 9001:2015 standard is the requirement for performance evaluation, along with a factual approach to decision-making. This embeds the good process guidance mentioned above as a clear requirement.

This is particularly important in how it effects and delivers process improvements. In previous versions of the standard much of the requirement for improvement was aligned to the findings of the internal and external audit processes. Whilst these might be able to identify useful opportunities to reduce the risk of non-conformance they do not provide, in themselves, an adequate source of performance improvement. A couple of immediate weaknesses of the audit approach to process improvement are:

- In many organizations the internal audit system provided the only guaranteed means of review for any particular organizational process. However, a specific process might only be subject to audit once every six or twelve months. This is unlikely to provide a rich supply of feedback on the effectiveness and efficiency of any process.
- The audit, as a technique, carries the inherent weakness that the audit itself is always likely to affect the activity that it is evaluating. The process as observed during the audit might differ significantly from the way in which the activities are discharged on every other day of the year on which the area is not being audited.

The requirement for performance evaluation for every process of the organization should ensure that data is available regarding the ongoing performance of the processes. This should enable the processes to be *performance managed*, meaning that ongoing, real-time feedback on effectiveness and efficiency is available for analysis and performance improvement action on a regular basis. Not only is the data sourced through ongoing operation of the process much more likely to be representative, but it should also be available in a volume that allows analysis to be completed statistically, opening up new opportunities for the enhancement of process performance.

QUALITY MANAGEMENT SYSTEM REQUIREMENTS

The QMS that needs to be documented and implemented will be determined by the nature of the processes carried out to ensure that the product or service conforms to customer requirements. Certain fundamental principles are applicable, however, throughout industry, commerce and the services. These fall into generally well-defined categories which are detailed in ISO 9001:2015.

ISO 9001:2015 specifies requirements for a quality management system when an organization:

1 needs to demonstrate its ability to consistently provide products and services that meet customer and applicable statutory and regulatory requirements; and
2 aims to enhance customer satisfaction through the effective application of the system, including processes for improvement of the system and the assurance of conformity to customer and applicable statutory and regulatory requirements.

All the requirements of ISO 9001:2015 are generic and are intended to be applicable to any organization, regardless of its type or size, or the products and services it provides.

Context of the organization

Understanding the organization and its context
All organizations in the construction supply chain will need to identify, monitor and review the external and internal issues (both positive and negative) that are relevant to its purpose and strategic direction and can affect its ability to achieve the intent of its QMS. Understanding the external context should include the consideration of issues arising from the legal, technological, competitive, market, cultural, social and economic environments in which the business operates, whether international, national, regional or local. The internal context might include issues related to the values, culture, knowledge and performance of the organization.

Understanding the needs and expectations of interested parties
For many organizations in the sector, the ability to consistently provide products and services that meet all of their customer and regulatory requirements can be influenced by third parties. There is a need to identify, monitor and review information about all interested parties that are relevant to its operations. It is essential that any specific requirements of these interested parties, which are relevant to the QMS, are understood and addressed.

Interested parties obviously include employees, local community groups, neighbours, industry groups, clubs or trade associations. This is not an exhaustive list and within the construction supply chain there may be a need to consider the expectations, timescales and schedules of planning committees, surveyors and site inspectors and ensure that these do not negatively impact on organizational commitments.

The Graniterock case study (CS5) shows how the company surveys its customers annually at every interface, at both internal team interfaces and with external customers, to identify feedback that can help it refocus its quality improvement efforts. It also scans for opportunities which emerge as a result of developments in technology.

The Heathrow case study also shows how passenger requirements are driving the business's quality strategy.

Determining the scope of the quality management system
Organizations will need to determine the boundaries and applicability of the QMS to establish its scope and needs to consider:

- the external and internal issues relevant to the organization's context;
- the requirements of relevant interested parties;
- the products and services of the organization.

All the requirements of ISO 9001:2015 need to be considered if they are applicable within the determined scope of the QMS, which must be available and be maintained as documented information. The scope needs to state the types of products and services covered, and provide justification for any requirement of the standard that it is determined is not applicable to the scope of the QMS.

The Crossrail case study (CS7) shows how a major client has brought quality improvement by its supply chain into its overall strategy at a similar level to safety. Crossrail has taken a very 'hands-on' approach, the strategy is led by the CEO and the board, and it has defined CSFs and metrics which its supply chain must report to.

Quality management system and its processes

The ISO 9000 standard requires that organizations establish, maintain and continually improve a QMS which must incorporate all of the processes necessary to ensure that the product or service conforms to customer requirements.

Hence, there is a need to identify and define the processes needed for the QMS and the requirements of their effective application throughout the organization, i.e.:

- determine the input and output requirements of the processes;
- determine the sequence order for the processes and the characteristics of their interactions;
- determine and apply criteria and methods (monitoring, measurement and performance indicators) appropriate to ensuring the effective operation and control of the processes;
- determine the resources needed for the processes to work effectively and ensure their deployment;
- assign appropriate ownership, responsibilities and authorities for the processes;
- address any risks and opportunities identified;
- evaluate the effectiveness of the processes and implement any changes needed to ensure that they achieve their intended results;
- improve the processes and the quality management system.

The Costain case study (CS12) describes how ISO 9000 was used as the basis for defining the Costain Way framework and its 33 performance areas. For each of these performance areas, online support defines what must be done, how it should be done with associated tools. It also defines where approval and review gates occur throughout the system.

One effective methodology for collecting and documenting information regarding the processes which will form the QMS of the organization, along with their sequence and interactions, is a flowchart model of planning/improvement process; an example is given in the Graniterock case study.

Leadership

Leadership and commitment (see also Chapter 4)

The top management need to demonstrate leadership and commitment with respect to the QMS. This should be evident through their:

- taking accountability for the effectiveness of the QMS;
- ensuring that the quality policy and quality objectives are established for the QMS and are compatible with the context and strategic direction of the organization;
- ensuring the integration of the QMS requirements into the organization's business processes;
- promoting the use of the process approach and risk-based thinking;
- ensuring that the resources needed for the QMS are available;
- communicating the importance of effective quality management and of conforming to the QMS requirements;
- ensuring that the QMS achieves its intended results;
- engaging, directing and supporting persons to contribute to the effectiveness of the QMS;
- promoting improvement; and
- supporting other relevant management roles to demonstrate their leadership as it applies to their areas of responsibility.

The aim of the necessity to focus on customer needs and specify them as defined requirements for the organization is clearly to achieve customer confidence in the products and/or services provided. It is also necessary to ensure that the defined requirements are understood and fully met.

Top management need to demonstrate leadership and commitment with respect to customer focus by ensuring that:

- customer and applicable statutory and regulatory requirements are determined, understood and consistently met;
- the risks and opportunities that can affect conformity of products and services and the ability to enhance customer satisfaction are determined and addressed;
- the focus on enhancing customer satisfaction is maintained.

Policy (see also Chapter 5)

The top management of the organization should also define, implement and maintain a *quality policy*, which forms one element of the corporate policy. Full commitment is required from the most senior management to ensure that the policy is communicated, understood, implemented and maintained at all levels in the organization.

For every project, the quality plans must fully reflect the company quality policy and the project leadership must be responsible for implementing company quality policy and goals within the supply chain at the project level.

The company quality policy should be authorized by top management and signed by the chief executive, or equivalent, who must also ensure that it:

- is suitable for the purpose and context of the organization and aligned to its strategic direction;
- provides a framework for establishing and reviewing quality objectives;
- includes commitment to meeting all applicable requirements;
- includes a commitment to continual improvement for all levels of the organization and its QMS.

The organization's quality policy should be available and maintained as documented information and should be communicated effectively, understood and applied as appropriate throughout the organization. It should also be made available to any relevant interested parties, as appropriate.

Organizational roles, responsibilities and authorities

Top management must ensure that the responsibilities and authorities for all relevant roles relating to the QMS are defined, assigned, communicated and understood throughout the organization. The scope of these assigned responsibilities and authorities could include:

- ensuring that the QMS conforms to the requirements of the International ISO 9001 Standard;
- ensuring that the processes are delivering their required outputs;
- reporting on the performance of the QMS and opportunities for improvement;
- ensuring the promotion of customer focus throughout the organization;
- ensuring the continued integrity of the QMS during the planning for and implementation of changes to its scope or content.

At the project level, responsibility for conformance with the quality policy and its implementation lies fully with line management and should not be separated out as a special responsibility. Doing so would only lead to a conflict within the project organization.

The Heathrow case study (CS10) clearly articulates the quality responsibilities for quality control as belonging to the framework contractors while its own project managers have a QA role.

Planning

Actions to address risk and opportunity

Top management must consider the issues identified relating to the context of the organization when planning for and designing the QMS, determining the risks and opportunities that should be addressed in order to:

- provide assurance that the QMS is capable of delivering the requirements identified;
- enhance any desirable effects identified;
- prevent or reduce the risk of occurrence or implications of any undesirable effects identified;
- achieve improvement.

There must be a planning and delivery of actions to address these risks and opportunities and to incorporate them into the QMS and its processes. Further consideration and planning should be undertaken relating to how the effectiveness of the actions will be evaluated.

The actions taken to address risks might include avoiding risk, taking risk in order to pursue an opportunity, eliminating the source of the risk, changing the likelihood or consequences of the risk, sharing the risk or making an informed judgement that the risk can be accepted. Any action planned should be appropriately proportionate to the potential impact on the conformity and integrity of the products and services.

Quality objectives and planning to achieve them

Organizations should establish quality objectives at relevant functions, levels and processes in the QMS, maintaining documented information regarding these quality objectives which should be:

- consistent with the quality policy;
- measurable;
- appropriate to applicable requirements;
- appropriate to the conformity of products and services and the achievement and enhancement of customer satisfaction;
- monitored;
- communicated;
- updated as appropriate.

When planning for the delivery of quality objectives there is a need to define how the action will be achieved. This will include the detail of the action, who will complete it, what resources will be required, when it will be completed and how the effectiveness of the delivered action will be evaluated.

Planning for change

When a need for change to the QMS has been identified, the changes must be delivered in a planned and organized manner and consider:

- the purpose of the changes and their potential consequences;
- the integrity of the QMS;
- availability of resources;
- potential reallocation of responsibilities and authorities where appropriate.

Support

Resources

The top management of the organization must determine and then provide all of the resources necessary for the establishment, implementation, effective operation, maintenance and continual improvement of the QMS.

This will require consideration of the capabilities of and constraints on any existing resources deployed to the QMS and the specific requirements demanded of external provider organizations.

One methodology that has become popular for its ability to assist in the facilitation of the consideration of resource requirements is the resource management diagram or Turtle diagram. This can be an effective tool for capturing the requirements for the effective operation of a process across a range of different resource types. A basic example, showing the content and concept of a Turtle diagram, is shown in Figure 14.4.

The Turtle diagram can be very effective at prompting a thorough consideration of the resource needs of each process. It clarifies what the process is expected to deliver in terms of transforming specified inputs into defined outputs and challenges the organization to think about the capability and capacity requirements of each type of resource as well as the supporting information, guidance and environment needs of delivering the process. A physical or mental walk through of the process considering the who, when, where and how's of the process is a good start point.

A number of specific categories of resources are discussed within the requirements of the ISO 9001 standard, as follows, with a short summary of the scope discussed against each:

- **People** – identify and provide the people necessary for the effective implementation and operation of the QMS and for the operation and control of processes.

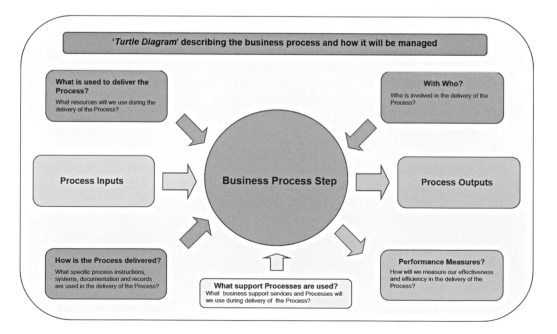

FIGURE 14.4 Example of a Turtle diagram for a business process step

- **Infrastructure** – identify, provide and maintain the infrastructure necessary for the operation of the processes and the achievement of conformity of products and services; considerations should include buildings, utilities, plant and equipment, transportation resources and information technology.
- **Environment for the operation of processes** – identify, provide and maintain the environment necessary for the operation of the processes and the achievement of conformity of products and services; considerations should include both human and physical factors such as the social, psychological and physical working environment.
- **Monitoring and measuring resources** – determine and provide the resources needed to ensure valid and reliable results when verifying the conformity of products and services through measurements; resources provided must be suitable for the specific measurement and monitoring activities being undertaken and appropriately maintained to ensure their continued fitness for purpose.
- **Organizational knowledge** – determine the knowledge necessary for the operation of the processes and the achievement of conformity of products and services. Organizational knowledge can be considered to be knowledge specific to the organization, gained internally or externally and is used and shared to achieve the organization's objectives.

The DPR case study (CS9) shows how a process of defining client requirements in terms of 'Key Distinguishing Features' which define in detail what each trade group must do to satisfy the client's requirement. Compliance with the system provides a leading indicator of the quality effort.

Competence

Clearly, the organization needs to select and assign people who are competent, on the basis of applicable education, training, skills and experience, to those activities which impact the conformity of products and/or services. On construction sites, where most of the work is actually undertaken by subcontractors, this includes workers and supervisors across the supply chain.

To achieve this, the organization needs to:

- determine the competency levels necessary for persons doing work under its control that could affect the performance or effectiveness of the QMS;
- ensure that the people deployed are suitably competent on the basis of appropriate education, training or experience;
- when appropriate, take action to provide the necessary competence and evaluate the effectiveness of the actions taken;
- retain appropriate evidence in the form of documented information as evidence of competence.

In the construction sector it may be beneficial to conduct joint training of supervisors and managers on a project to ensure that they are working within a consistent framework of values and expectations. Other actions to secure the necessary levels of competency might include training, mentoring, staff redeployment, hiring or contracting.

Awareness

It is necessary to ensure that all persons undertaking work under the organization's control are aware of:

- the quality policy;
- all relevant quality objectives;

- the requirements of their contribution to the effectiveness of the QMS, including the benefits of improved performance;
- the implications of not conforming to the QMS requirements.

Communication

There is a need to determine the internal and external communication requirements appropriate to the QMS. These should define what content should be communicated, when and to whom. The mechanism of communication should be defined as well as who within the organization will deliver the communication.

Documented information

The organization's QMS needs to include documented information as required by the ISO 9001 standard and determined to be necessary to the effectiveness of the QMS. The extent of documented information necessary will vary significantly dependent on the size of the organization, its types of activities, processes, products and services, the complexity of their processes and their interactions and the competency of the persons employed.

When creating documented information, appropriate identification and description, format and review and approval mechanism are important to ensure their suitability and adequacy.

The documented information required of the QMS and the international standard must be controlled effectively to ensure that it is available and suitable for use, where and when needed and that it is adequately protected. To achieve an appropriate level of control of the documented information the organization should address the following considerations, as necessary:

- distribution, access, retrieval and use;
- storage and preservation;
- control of change;
- retention and disposition.

Documented information of external origin necessary for the planning and operation of the QMS should be identified and controlled appropriately. Documented information retained as evidence of product or service conformity must be protected from unintended alteration.

Operation

Operational planning and control

As we have seen in Figure 14.2, any construction organization needs to determine the processes required to convert customer requirements into customer satisfaction, in order to consistently deliver the required product and/or service. In determining such processes, the organization needs to consider the outputs from the quality planning process.

The sequence and interaction of these processes need to be determined, planned and controlled to ensure they operate effectively, and there is a need to assign responsibilities for the operation and monitoring of the product/service generating processes. These processes clearly need to be operated under controlled conditions and produce outputs which are consistent with the organization's quality policy and objective and it is necessary to:

- determine the requirements of the products and services;
- establish criteria for the processes;

- establish criteria for the acceptance of products and services;
- determine the resources needed to achieve conformity to the product and service requirements;
- implement control of the processes in accordance with the criteria;
- determine, maintain and retain documented information appropriate to providing confidence that the processes have been carried out as planned and to demonstrate the conformity of the products and services to the requirements.

The outputs from the planning activities must be appropriate and suitable for the organization's operations; for external operations the organization needs to make equivalent provisions to ensure that all outsourced processes are controlled.

Requirements for products and services

It is important to determine appropriate mechanisms for communication with customers, including:

- providing information relating to products and services;
- handling enquiries, contracts or orders, including changes;
- obtaining customer feedback relating to products and services, including customer complaints;
- handling or controlling customer property;
- establishing specific requirements for contingency actions, when relevant.

When determining the requirements for the products and services to be offered to customers, the organization should ensure that:

- the requirements for the products and services are defined, including any applicable statutory and regulatory requirements and those considered necessary by the organization;
- the organization can meet the claims for the products and services it offers.

The organization must ensure that it has the ability to meet the requirements for products and services to be offered to customers. The organization therefore needs to conduct a review before committing to supply products and services to a customer, to include:

- requirements specified by the customer, including the requirements for delivery and post-delivery activities;
- requirements not stated by the customer, but necessary for the specified or intended use, when known;
- requirements specified by the organization;
- statutory and regulatory requirements applicable to the products and services;
- contract or order requirements differing from those previously expressed.

The organization also needs to ensure that contract or order requirements differing from those previously defined are resolved so that the customer's requirements are confirmed by the organization before acceptance, when the customer does not provide a documented statement of their requirements.

In some situations, such as internet sales, a formal review is impractical for each order. Instead, the review can cover relevant product information, such as catalogues.

The organization needs to retain documented information, as applicable:

- on the results of the review;
- on any new requirements for the products and services.

There is also a need to ensure that relevant documented information is amended, and that relevant persons are made aware of the changed requirements, when the requirements for products and services are changed.

Where an organization is supervising or using customer property, care needs to be exercised to ensure verification, storage and maintenance. Any customer product or property that is lost, damaged or otherwise found to be unsuitable for use should, of course, be recorded and reported to the customer. Customer property may include intellectual property, e.g. information provided in confidence.

Design and development of products and services

The organization needs to establish and maintain a process to control design and development of products and services in order to ensure that they are compliant with relevant requirements.

In order to ensure the appropriate definition of design and development stages and controls, consideration should be given to the:

- nature, duration and complexity of the design and development activities;
- required process stages, including applicable design and development reviews;
- required design and development verification and validation activities;
- responsibilities and authorities involved in the design and development process;
- internal and external resource needs for the design and development of products and services; the need to control interfaces between persons involved in the design and development process;
- need for involvement of customers and users in the design and development process;
- requirements for subsequent provision of products and services;
- level of control expected for the design and development process by customers and other relevant interested parties;
- documented information needed to demonstrate that design and development requirements have been met.

In the design and development input requirements essential for the specific types of products and services to be designed and developed, consideration should be given to:

- functional and performance requirements;
- information derived from previous similar design and development activities;
- statutory and regulatory requirements;
- standards or codes of practice that the organization has committed to implement;
- potential consequences of failure due to the nature of the products and services.

Inputs need to be adequate for design and development purposes, complete and unambiguous, and conflicting design and development inputs resolved. Documented information on design and development inputs need to be retained.

Controls need to be applied to the design and development process to ensure that:

- the results to be achieved are defined;
- reviews are conducted to evaluate the ability of the results of design and development to meet requirements;
- verification activities are conducted to ensure that the design and development outputs meet the input requirements;

- validation activities are conducted to ensure that the resulting products and services meet the requirements for the specified application or intended use;
- any necessary actions are taken on problems determined during the reviews, or verification and validation activities;
- documented information of these activities is retained.

The organization should ensure that design and development outputs:

- meet the input requirements;
- are adequate for the subsequent processes for the provision of products and services;
- include or reference monitoring and measuring requirements, as appropriate, and acceptance criteria;
- specify the characteristics of the products and services that are essential for their intended purpose and their safe and proper provision.

Changes made during, or subsequent to, the design and development of products and services need to be identified, reviewed and controlled. This means retain documented information on:

- design and development changes;
- the results of reviews;
- the authorization of the changes;
- the actions taken to prevent adverse impacts.

Control of externally provided processes, products and services

Externally provided processes, products and services need to conform to requirements, which means controls should be applied to externally provided processes, products and services when:

- products and services from external providers are intended for incorporation into the organization's own products and services;
- products and services are provided directly to the customer(s) by external providers on behalf of the organization;
- a process, or part of a process, is provided by an external provider as a result of a decision by the organization.

Criteria for the evaluation, selection, monitoring of performance and re-evaluation of external providers, need to be determined and applied based on their ability to provide processes or products and services in accordance with requirements. Documented information on these activities and records of any necessary actions arising from the evaluations need to be kept.

There is a need to ensure that externally provided processes, products and services do not adversely affect the organization's ability to consistently deliver conforming products and services to its customers. In determining the type and extent of control necessary:

- ensure that externally provided processes remain within the control of its QMS;
- define both the controls that it intends to apply to an external provider and those it intends to apply to the resulting output;
- take into consideration the potential impact of the externally provided processes, products and services on the organization's ability to consistently meet customer and applicable statutory and regulatory requirements and on the effectiveness of the controls applied by the external provider themselves;

- determine the verification, or other activities, necessary to ensure that the externally provided processes, products and services meet requirements.

Production and service provision

Organizations need to implement production and service provision under controlled conditions, including, as applicable:

- availability of documented information that defines the characteristics of the products to be produced, the services to be provided or the activities to be performed as well as the results to be achieved;
- availability and use of suitable monitoring and measuring resources;
- implementation of monitoring and measurement activities at appropriate stages to verify that criteria for control of processes or outputs, and acceptance criteria for products and services, have been met;
- use of suitable infrastructure and environment for the operation of processes;
- appointment of competent persons, including any required qualification;
- validation, and periodic revalidation, of the ability to achieve planned results of the processes for production and service provision, where the resulting output cannot be verified by subsequent monitoring or measurement;
- implementation of actions to prevent human error;
- implementation of release, delivery and post-delivery activities.

Suitable means are required to identify outputs when it is necessary to ensure the conformity of products and services, including the status of outputs with respect to monitoring and measurement requirements throughout production and service provision.

Documentation is needed to be retained for control of unique identification of outputs, when traceability is a requirement.

There is a need to exercise care with property belonging to customers or external providers while it is under the organization's control or being used with a requirement to identify, verify, protect and safeguard customers' or external providers' property provided for use or incorporation into the products and services. When the property of a customer or external provider is lost, damaged or otherwise found to be unsuitable for use, the organization is required to report this to the customer or external provider and retain documented information on what has occurred.

The organization needs to meet requirements for post-delivery activities associated with the products and services, considering:

- statutory and regulatory requirements;
- the potential undesired consequences associated with its products and services;
- the nature, use and intended lifetime of its products and services;
- customer requirements;
- customer feedback.

Post-delivery activities can include actions under warranty provisions, contractual obligations, such as maintenance services, and supplementary services such as recycling or final disposal.

Some leading residential builders include in their contractual arrangements inspections at 6 months, 1 and 2 years, and request permission to inspect at 5 and 10 years. These companies seek to develop a lifetime relationship involving regular five-yearly inspections and maintenance. This ongoing relationship helps to separate construction defects from maintenance neglect and provides direct feedback into product improvement. A further by-product of such ongoing contact is that it enables a company to build long-term relationships with buyers; it attracts customer recommendations and improves market perceptions of the company's products and service.

Release of products and services

Planned arrangements, implemented at appropriate stages, are needed to verify that the product and service requirements have been met. The release of products and services to the customer should not proceed until these arrangements have been satisfactorily completed, unless otherwise approved by a relevant authority or by the customer. Documented information on the release of products and services may include:

- evidence of conformity with the acceptance criteria;
- traceability to the person(s) authorizing the release.

Control of nonconforming outputs

Outputs that do not conform to their requirements need to be identified and controlled to prevent their unintended use or delivery. Appropriate action is needed, based on the nature of the nonconformity and its effect on the conformity of products and services. This also applies to nonconforming products and services detected after delivery of products, during or after the provision of services. The organization needs to deal with nonconforming outputs in one or more of the following ways:

- correction;
- segregation, containment, return or suspension of provision of products and services;
- informing the customer;
- obtaining authorization for acceptance under concession.

Conformity to the requirements must be verified when nonconforming outputs are corrected, with documented information that:

- describes the nonconformity;
- describes the actions taken;
- describes any concessions obtained;
- identifies the authority deciding the action in respect of the nonconformity.

Performance evaluation

Monitoring, measurement, analysis and evaluation

The organization should determine:

- what needs to be monitored and measured;
- the methods for monitoring, measurement, analysis and evaluation needed to ensure valid results;
- when the monitoring and measuring needs to be performed;
- when the results from monitoring and measurement will be analysed and evaluated.

The organization must monitor customer satisfaction in terms of the customers' perceptions of the degree to which their needs and expectations have been fulfilled. The methods for obtaining, monitoring and reviewing this information need to be determined, of course. Examples of monitoring customer perceptions include customer surveys, customer feedback on delivered products and services, meetings with customers, market-share analysis, compliments, warranty claims and dealer reports.

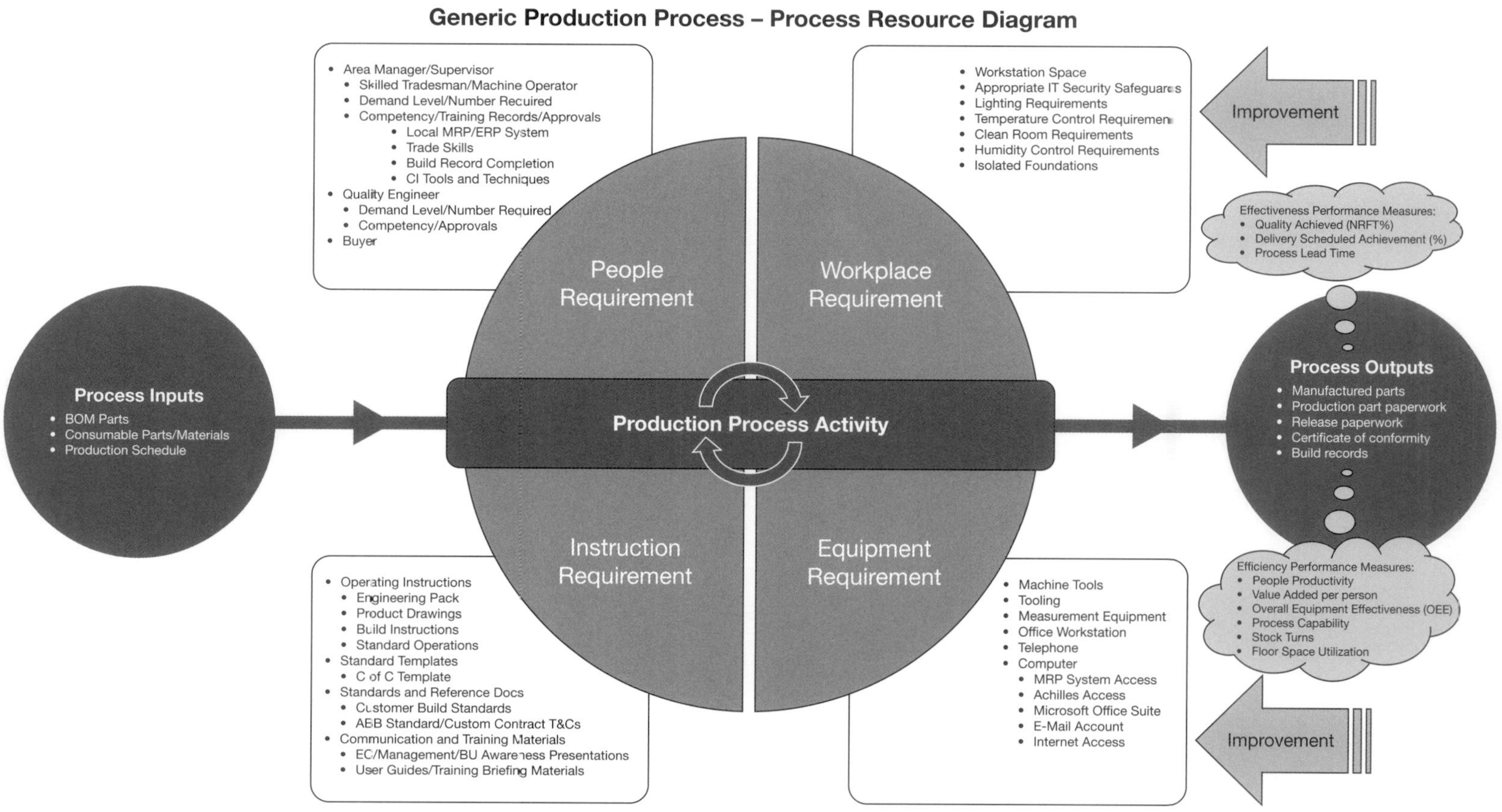

Generic Production Process – Process Resource Diagram

Process Inputs
- BOM Parts
- Consumable Parts/Materials
- Production Schedule

People Requirement
- Area Manager/Supervisor
- Skilled Tradesman/Machine Operator
- Demand Level/Number Required
- Competency/Training Records/Approvals
  - Local MRP/ERP System
  - Trade Skills
  - Build Record Completion
  - CI Tools and Techniques
- Quality Engineer
  - Demand Level/Number Required
  - Competency/Approvals
- Buyer

Instruction Requirement
- Operating Instructions
  - Engineering Pack
  - Product Drawings
  - Build Instructions
  - Standard Operations
- Standard Templates
  - C of C Template
- Standards and Reference Docs
  - Customer Build Standards
  - AEB Standard/Custom Contract T&Cs
- Communication and Training Materials
  - EC/Management/BU Awareness Presentations
  - User Guides/Training Briefing Materials

Production Process Activity

Workplace Requirement
- Workstation Space
- Appropriate IT Security Safeguards
- Lighting Requirements
- Temperature Control Requirements
- Clean Room Requirements
- Humidity Control Requirements
- Isolated Foundations

Equipment Requirement
- Machine Tools
- Tooling
- Measurement Equipment
- Office Workstation
- Telephone
- Computer
  - MRP System Access
  - Achilles Access
  - Microsoft Office Suite
  - E-Mail Account
  - Internet Access

Process Outputs
- Manufactured parts
- Production part paperwork
- Release paperwork
- Certificate of conformity
- Build records

Effectiveness Performance Measures:
- Quality Achieved (NRFT%)
- Delivery Scheduled Achievement (%)
- Process Lead Time

Efficiency Performance Measures:
- People Productivity
- Value Added per person
- Overall Equipment Effectiveness (OEE)
- Process Capability
- Stock Turns
- Floor Space Utilization

Improvement

FIGURE 14.5 Generic process resource diagram

Analysis and evaluation of appropriate data and information arising from monitoring and measurement must be performed and used to determine:

- conformity of products and services;
- degree of customer satisfaction;
- performance and effectiveness of the QMS;
- if planning has been implemented effectively;
- effectiveness of actions taken to address risks and opportunities;
- performance of external providers;
- need for improvements to the QMS.

Returning again to the Turtle, or process resource, diagram this can be useful for assisting an organization to think about what performance measures are appropriate to a particular process and how to direct analysis and improvement. An example of a process resource diagram, for a manufacturing activity is shown in Figure 14.5.

This is a modified version Oakland Consulting have used successfully to understand processes and achieve effective control for a range of clients.

Another useful methodology for managing the performance and analysis necessary to evaluate and appropriately respond to the performance of the processes is of course the P-D-C-A cycle. An example of an advanced version clearly articulating its fit to the management of an organization's processes is shown in Figure 14.6.

Again, the use of such a schematic can provide clarity and facilitate better understanding throughout the organization of where performance measurement and the evaluation of the objective data produced fits with the day job activity of every employee's role within the company.

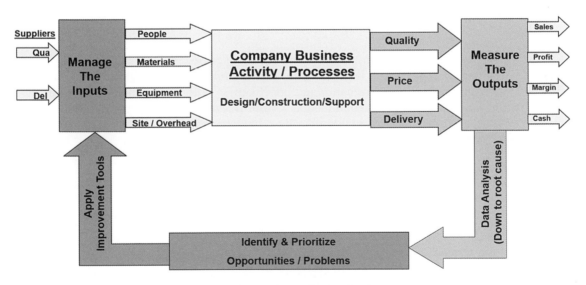

FIGURE 14.6 Advanced version of the PDCA cycle applied to business processes

Internal audit

The organization needs to conduct internal audits at planned intervals to provide information on whether the QMS:

- conforms to the organization's own requirements for its QMS and the requirements of the ISO 9000 series standard;
- is effectively implemented and maintained.

The needs include:

- plan, establish, implement and maintain an audit programme(s) including the frequency, methods, responsibilities, planning requirements and reporting, which shall take into consideration the importance of the processes concerned, changes and the results of previous audits;
- define the audit criteria and scope for each audit;
- select auditors and conduct audits to ensure objectivity and the impartiality of the audit process;
- ensure that the results of the audits are reported to relevant levels of management;
- take appropriate corrective actions without undue delay;
- retain documented information as evidence of the implementation of the audit programme and the audit results.

See ISO 19011 – Guidelines for auditing management systems for further guidance.

Management review

Top management must review the organization's QMS at planned intervals to ensure its continuing suitability, adequacy, effectiveness and alignment with the strategic direction of the organization.

The management review needs to be planned and carried out taking into consideration:

- the status of actions from previous management reviews;
- changes in external and internal issues that are relevant to the QMS;
- information on the performance and effectiveness of the QMS, including trends in:
    - customer satisfaction and feedback from relevant interested parties;
    - the extent to which quality objectives have been met;
    - process performance and conformity of products and services;
    - nonconformities and corrective actions;
    - monitoring and measurement results;
    - audit results;
    - the performance of external providers;
- the adequacy of resources;
- the effectiveness of actions taken to address risks and opportunities;
- opportunities for improvement.

The outputs of the management review should include decisions and actions related to:

- opportunities for improvement;
- any need for changes to the QMS;
- resource needs.

The organization should retain documented information as evidence of the results of management reviews.

Improvement

The organization needs to determine and select opportunities for improvement and implement any necessary actions to meet customer requirements and enhance customer satisfaction, including:

- improving products and services to meet requirements, as well as to address future needs and expectations;
- correcting, preventing or reducing undesired effects;
- improving the performance and effectiveness of the QMS.

Examples of improvement can include correction, corrective action, continual improvement, breakthrough change, innovation and re-organization.

Nonconformity and corrective action

When a nonconformity occurs, including any arising from complaints, the organization must:

- react to the nonconformity by taking action to control and correct it and dealing with the consequences, as applicable;
- evaluate the need for action to eliminate the cause(s) of the nonconformity, in order that it does not recur or occur elsewhere, by:
  - reviewing and analysing the nonconformity;
  - determining the causes of the nonconformity;
  - determining if similar nonconformities exist, or could potentially occur;
- implement any action needed;
- review the effectiveness of any corrective action taken;
- update risks and opportunities determined during planning, if necessary;
- make changes to the QMS, if necessary.

Corrective actions should be appropriate to the effects of the nonconformities encountered and documented information be retained as evidence of the:

- nature of the nonconformities;
- any subsequent actions taken;
- results of any corrective action.

Continual improvement

ISO 9001:2015 now includes a requirement to continually improve the suitability, adequacy and effectiveness of the QMS considering the results of analysis and evaluation, and the outputs from management reviews, to determine if there are needs or opportunities that are to be addressed as part of continual improvement.

OTHER MANAGEMENT SYSTEMS

Organizations of all kinds are increasingly concerned to achieve and demonstrate sound environmental performance. Many have undertaken environmental audits and reviews to assess this. To be effective these need to be conducted within a structured management system, which in turn is integrated with the overall management activities dealing with all aspects of desired environmental performance.

Such a system should establish processes for setting environmental policy and objectives, and achieving compliance to them. It should be designed to place emphasis on the prevention of adverse environmental effects, rather than on detection after occurrence. It should also identify and assess the environmental effects arising from the organization's existing or proposed activities, products, or services and from incidents, accidents and potential emergency situations. The system should also identify the relevant regulatory requirements, the priorities and pertinent environmental objective and targets. In addition it needs to facilitate planning, control, monitoring, auditing and review activities to ensure that the policy is complied with, that it remains relevant and that it is capable of evolution to suit changing circumstances.

The international standard ISO 14001 contains a specification for environmental management systems for ensuring and demonstrating compliance with stated policies and objectives. The standard is designed to enable any organization to establish an effective management system, as a foundation for both sound environmental performance and participation and environmental auditing schemes. Like ISO 14001, other ISO standards that look at different types of management systems, including ISO 9001 for quality management (see above for detail) and ISO 45001 for occupational health and safety, all use a high-level structure. This means that ISO 14001 can be integrated easily into any existing ISO-based management system. ISO 14001 is suitable for organizations of all types and sizes, be they private, not-for-profit or governmental. It requires that an organization considers all environmental issues relevant to its operations, such as air pollution, water and sewage issues, waste management, soil contamination, climate change mitigation and adaptation, and resource use and efficiency.

Like all ISO management system standards now, ISO 14001 includes the need for **continual improvement** of an organization's systems and approach to environmental concerns. The standard was revised in 2015, with key improvements such as the increased prominence of environmental management within the organization's strategic planning processes, greater input from leadership and a stronger commitment to proactive initiatives that boost environmental performance.

ISO 14001:2015 now requires:

- environmental management to be more prominent within the organization's strategic direction;
- a greater commitment from leadership;
- the implementation of proactive initiatives to protect the environment from harm and degradation, such as sustainable resource use and climate change mitigation;
- a focus on lifecycle thinking to ensure consideration of environmental aspects from development to end-of-life;
- the addition of a stakeholder-focused communication strategy.

It also allows for easier integration into other management systems thanks to the same structure, terms and definitions. Accredited certification to ISO 14001 is not a requirement, of course, and organizations can reap many of the benefits from using the standard without going through the accredited certification process. However, third-party certification – where an independent certification body audits practices against the requirements of the standard – is a way of signalling to buyers, customers, suppliers and other stakeholders that the standard has been implemented properly. Moreover, for some organizations, it helps to show how they meet regulatory or contractual requirements.

The ISO 14000 family comprises a number of standards that complement ISO 14001, some of which are listed below. More information can be found in the brochure *Environmental management – The ISO 14000 family of International Standards*, a basic introduction to the ISO 14000 series of standards:

- **ISO 14004** provides guidance on the establishment, implementation, maintenance and improvement of an environmental management system and its coordination with other management systems.

- **ISO 14006** is intended to be used by those organizations that have implemented an environmental management system in accordance with ISO 14001, but can help integrate eco-design into other management systems.
- **ISO 14064–1** specifies principles and requirements at the organizational level for the quantification and reporting of greenhouse gas (GHG) emissions and removal.

Management systems are needed in all areas of activity, whether large or small businesses, manufacturing, service or public sector. The advantages of systems in manufacturing are obvious, but they are just as applicable in areas such as marketing, sales, personnel, finance, research and development, as well as in the service industries and public sectors. No matter where it is implemented a good management system will improve process control, reduce wastage, lower costs, increase market share (or funding), facilitate training, involve staff and raise morale.

ACKNOWLEDGEMENT

The authors are grateful to the contribution made by Karl Smith, Associate Consultant with Oakland Consulting LLP, to the preparation of this chapter.

Permission to reproduce extracts from British Standards is granted by BSI. British Standards can be obtained in PDF or hard copy formats from the BSI online shop: www.bsigroup.com/Shop or by contacting BSI Customer Services for hardcopies only: Tel: +44 (0)20 8996 9001, Email: cscrvices@bsigroup.com.

BIBLIOGRAPHY

Arter, D.R., Cianfrani, C.A. and West, J.E. *How to Audit the Process Based QMS* (2nd edn), ASQ, Milwaukee, 2012.

British Standards Institute (BSI and International Standards Organization), BS EN ISO 9001:2015, *Quality Management Systems Requirements*, BSI, London, 2015 (see also references in text on related ISO 9000 and ISO 10000 series standards).

British Standards Institute (and International Standards Organisation), BS EN ISO 14001:2015, *Environmental Management Systems – Specification with Guidance for Use*, BSI, London, 2015.

Dissanayaka, S.M., Kumaraswamy, M.M., Karim, K. and Marosszeky, M. Evaluating outcomes from IS0 9000 certified quality systems of Hong Kong constructors, *Total Quality Management Journal*, 12(1):29–40, 2000.

Gillett, J., Simpson, P. and Clarke, S. *Implementing ISO 9001:2015*, Infinite Ideas, Oxford, 2015.

Hoyle, D. *ISO 9000 Quality Systems Handbook*, Butterworth-Heinemann, Oxford, 2010.

Jarvis, A. and Palmes, P. *ISO 9001:2015 – Understand, Implement, Succeed*, Prentice Hall, New York, 2016.

Tricker, K. *ISO 9001:2015 in brief* (4th edn), Taylor & Francis, London, 2016.

Whitelaw, R. *ISO 14001 Environmental Systems Handbook* (2nd edn), Elsevier Butterworth-Heinemann, Oxford, 2004.

CHAPTER HIGHLIGHTS

Why a quality management system?
- While formal quality management systems may have a poor name in construction due to the cynical way in which they were first implemented, construction industry leaders the world over have demonstrated the benefits they have derived from them.
- An appropriate quality management system (QMS) will enable the objectives set out in the policies and strategies to be accomplished.

- The International Organization for Standardization (ISO) 9000:2015 series sets out methods by which a QMS can be implemented to ensure that the specified customer requirements are met.
- A QMS may be defined simply as an assembly of components which include leadership/management responsibilities, process and resources.

ISO 9000 quality management systems

- ISO 9000:2015 'BS EN ISO 9001:2015 *Quality Management Systems – Requirements*' is the current International Standard which superseded ISO 9001:2008. Detailed information on the ISO 9000:2015 family of standards may be found on the websites: www.iso.ch and www.bsigroup.com.
- ISO 9001:2015 now employs a process approach, which incorporates the Plan-Do-Check-Act (PDCA) cycle and risk-based thinking; it links QMS to product/service and process quality through a focus on demonstrating customer satisfaction and continuous improvement.
- ISO 19011: *'Guidelines for auditing management systems'* provides guidance on the management of an audit programme, on the planning and conducting of an audit of a management system and on the competence and evaluation of an auditor and an audit.

Quality management principles and ISO 9000

- The ISO 9001:2015 standard is based on the adoption of seven quality management principles: customer focus, leadership, engagement of people, process approach, improvement, evidence-based decision-making and relationship management.
- Understanding and managing interrelated processes as a system contributes to an organization's effectiveness and efficiency in achieving its intended results and this is a major feature of the revised standard.
- The Supplier-Input-Process-Output-Customer (SIPOC) methodology is recommended as the means of understanding the design and interdependencies of the processes of an organization.
- The integration of the PDCA process into the ISO 9001:2015 standard dovetails perfectly with the process-based approach in the standard; PDCA can be applied to all processes and to the QMS as a whole.
- Risk-based thinking is essential for establishing and then maintaining the effectiveness of a QMS; the 2015 version of the standard requires that organizations plan and implement actions to address risks and opportunities.

Quality management system design and ISO 9000

- The seven clauses that form ISO 9001:2015 are: context of the organization, leadership, planning, support, operation, performance evaluation and improvement; the management system requirements under these headings are specified and detailed in the standard.
- A significant change with the process-based format of the 2015 standard is the requirement for performance evaluation, along with a factual approach to decision-making; this embeds the good process guidance mentioned above as a clear requirement.

Other management systems

- The international standard ISO 14001 contains a specification for environmental management systems for ensuring and demonstrating compliance with stated policies and objectives; the standard helps to establish a foundation for both sound environmental performance and participation, and environmental auditing schemes.

- ISO 14001:2015 can be integrated easily into any existing ISO-based management system and is suitable for organizations of all types and sizes, be they private, not-for-profit or governmental.
- Like all ISO management system standards, ISO 14001 now includes the need for continual improvement of an organization's systems and approach to environmental concerns, together with increased prominence within the organization's strategic planning processes, greater input from leadership and a stronger commitment to proactive initiatives that boost environmental performance.

Continuous improvement

15

APPROACHES, METHODOLOGIES AND TOOLS

Continuous improvement is a term in common use throughout most industries including the construction sector. It can become a meaningless term unless it is linked to organizational strategy, driven by performance measurement, has a defined structure, a chosen approach, a methodology and an associated toolkit. Figure 15.1 provides a generic eight-stage structure that may be applied to most organizations. It begins unsurprisingly with leadership and the top-down cascade of objectives, particularly those related to achieving on-quality, on-time, on-cost delivery of products and services chosen to boost customer confidence and drive efficiency improvements.

Unfortunately, at this point the senior management in many organizations rush straight to the 'instruction manual' (often from other industries) to implement approaches such as 'Lean' and 'Six Sigma', without establishing the much needed managerial infrastructure and resources required in stage 2. Good programme and systems management will be essential, as will communications with people inside and outside on what the chosen approach and even 'brand' means. The third stage requires the careful choice of a pragmatic, fact-based improvement methodology which, hopefully, will yield rapid results using proven methodologies, supported by fact and data-based, hands-on tools and techniques to be recommended.

If this choice is based on the need to improve on-quality, on-time and on-cost delivery (Figure 15.2), then approaches that reduce time wastage and reduce variation to improve quality will be appropriate. Moreover, as these approaches also reduce costs, overt cost reduction programmes can be avoided since these often merely knock out capability and render the organization incapable of functioning properly – the equivalent of going to the doctor for help in reducing body weight only for him to chop off your left arm!

As we have seen throughout this book, lean quality approaches are designed to eliminate non-value-added time, activities and processes while six sigma approaches are designed to reduce variation, so a combination of the two can provide an excellent ready-made solution for the CI approach – the so-called Lean Six Sigma or even Lean–Sigma. To go with the overall approach, a closed loop fact-based improvement methodology is needed. This can be provided by the traditional six sigma method of DMAIC stages: Define, Measure, Analyse, Improve and Control (see later section in this chapter).

The fourth stage of the structure outlined in Figure 15.1 is skilled improvers – people trained, qualified and 'living in the organization' as opposed to being part of a separate improvement task force of some kind. We are not all naturally born with the ability to make improvements in a structured and effective way and just as we need to be trained and developed to be civil engineers, charge-hands, accountants or lawyers, we also need

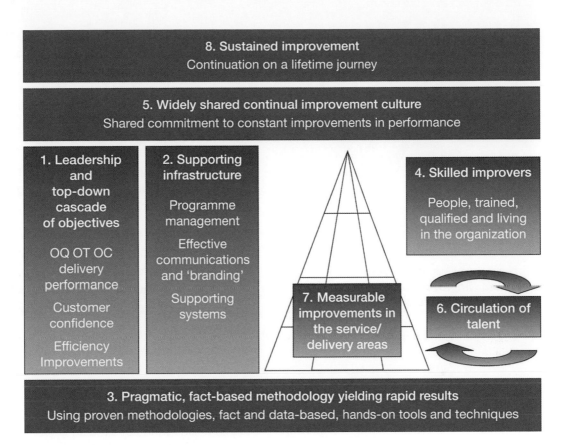

FIGURE 15.1 An overall approach structure for continuous improvement

to be trained and developed to be improvers, whether we are civil engineers, charge-hands, accountants or lawyers. The best people to work on improving an engineering design process are the engineers who work in the process, who have been trained and who have developed how to make those improvements effectively rather than an 'outside team' who are not fully conversant with how the process operates. That is not to say there is no place for an 'external task force' to tackle certain situations, but the norm should be that we expect the 'experts' who function in the process every day to make the changes required.

The development of a widely shared continual improvement culture, with shared commitment to constant improvements in performance, is the fifth stage in Figure 15.1. Of course, the leadership plays a key role here in establishing the 'norms' by asking for each and every improvement project to be properly defined and tackled using the chosen methodology: led by properly trained improvers and targeting specific expected outcomes – people soon change their behaviour when they get clear messages about what the top people want and expect.

The sixth stage of circulation of talent refers to situations where you do have specifically trained improvement task forces such as the 'Black Belts' trained and developed in many companies. It is necessary to ensure they do not see their role as a dead-end job. Jack Welsh famously made the GE Six Sigma programme successful by insisting that the HR policies not only supported these trained and experienced improvers returning to line management but they also stipulated that managers would not get promoted unless they had served time as a Black Belt in the company.

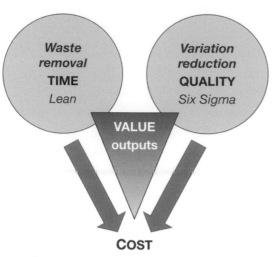

Reduce process time
- Apply approaches such as 'Lean' to remove the non-value-adding waste from a process, pathway or service

Reduce variation
- Use techniques such as '**Six Sigma**' to reduce undesirable variation in patient-facing and back office processes

Bust assumptions
- **Assumption busting** leads to paradigm shifts – that can drastically reduce costs and create step changes in quality

FIGURE 15.2 CI approach to delivering OQ OT OC

Only now are we ready to move to the seventh stage of making measurable improvements in the design, building, operations or service areas. Rushing into this without the establishment of the previous six stages usually results in disappointment and the claims that 'SPC is too statistical to work here; Six Sigma is Greek for us'; or 'lean is too *mean* for our organization'.

Continuous improvement does mean sustained improvements on a lifetime journey. In the short-term world we live in, this can be a challenge but do look around and accept that truly great organizations have developed a solid culture of doing the right things, right first time – every time. In a rapidly changing world, the only way this can be achieved is by having a culture of continually improving what is done for all customers and other stakeholders.

DRIVER: A CONTEXT-DEPENDENT PROCESS VIEW OF LEAN

John Oakland and his colleagues have developed a fully closed loop improvement methodology which brings together the best of Lean, Six Sigma and Cost of Quality approaches – DRIVER with the stages of Define, Review, Investigate, Verify, Execute and Reinforce. This well-established approach, which has been used in literally hundreds of organizations, prevents people jumping from the problem to the solution without considering the improvement options.

Many organizations make mistakes in implementing lean quality because they focus on specific tools and techniques that are not well adapted to their organizational setting. When this happens, not only does the lean initiative fail to deliver what was expected, but this becomes another example of 'how lean does not work in our environment'. However, those organizations that have *succeeded* in implementing lean approaches have done so by adapting or developing approaches based on the underlying principles of lean by focusing on what the customer sees as being of value and then identifying activities that do not add value and taking steps to eliminate them.

Organizations that have repeatedly delivered major sustainable benefits from lean implementations have tended to follow a structured approach to improvement that has focused on what will work for them in their context rather than trying to follow a set 'recipe book' based on specific tools and techniques. By adopting a pragmatic approach, it becomes possible to tailor specific tools and interventions from the vast armoury that is available to deliver what is required within that specific organizational setting.

In today's world, where even services are packaged as products, some organizations have created 'recipes' blending certain lean quality tools into a package which they market as an off the shelf 'solution' to solve every organization's problems. The importance of a pragmatic approach (rather than slavishly following a specific model or recipe) needs to be emphasized. Experience has shown that a bottom-up approach, which first of all identifies the specific improvement to be made and then selects the most suitable lean quality tools and strategies, is a more reliable and successful approach than simply buying a 'silver bullet' lean recipe off someone's shelf. That approach harkens back to the early quality systems that were brought off the shelf simply to get onto a government tender list. Just as the 'recipe' approach failed at that time, it is guaranteed to lead to disappointment and disillusionment with lean quality thinking.

The continuous improvement cycle of DRIVER, setting out a six-phase approach to lean, is shown in Figure 15.3. Organizations that take a pragmatic path to lean implementation tend to focus on these six key phases.

This view of DRIVER is further elaborated in Figure 15.4 which describes the issues that should be considered in each phase of implementation.

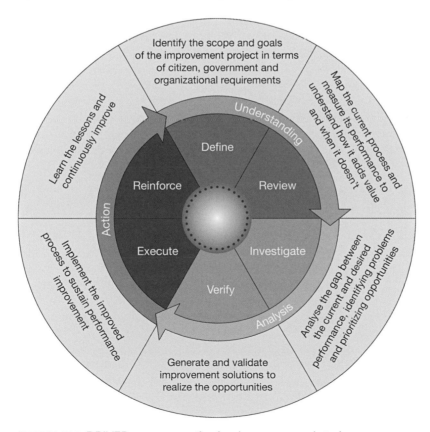

FIGURE 15.3 DRIVER – a pragmatic six-phase approach to lean

Phase 1: Define	Phase 2: Review	Phase 3: Investigate	Phase 4: Verify	Phase 5: Execute	Phase 6: Reinforce
• Project scope and definition • Strategic alignment • Stakeholder requirements • Voice of the business • Process families • Voice of the customer • Project governance • Project planning • Team selection • Communication strategy	• Current state (as-is) • Value stream mapping (inc. measures) • Customer value proposition • Business value proposition • Customer touch-points • Supporting processes	• Challenge status quo • Identify non-value-adding activity • Identify value-destroying activity at customer touch-points • Identify systemic effects/impacts • Build potential future states	• Assess impact and implications of potential future states • Test different scenarios for stakeholder impacts • Assess alignment of potential future states with strategy and values • Identify needs for pilot studies • Develop prioritized future state • Identify potential gains	• Establish senior management and stakeholder buy-in and commitment • Develop and implement implementation plan • Develop and implement communications plan • Identify training and development needs • Develop and implement training and development programme	• Continuous improvement philosophy • Develop and sustain ongoing capability • Ongoing communication of successes • Reward and recognition systems • Link of future Lean initiatives with strategy and business planning and performance measurement

FIGURE 15.4 DRIVER – a six-phase approach to lean

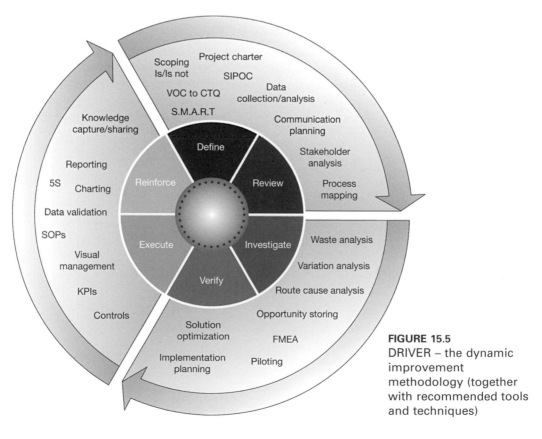

FIGURE 15.5
DRIVER – the dynamic improvement methodology (together with recommended tools and techniques)

The basic model of the DRIVER process (Figure 15.5) has been successfully used in many private and public sector organizations as an improvement methodology and is tried and tested. It is clear that this approach embraces the concepts of lean whilst allowing sufficient flexibility for the detailed tools and techniques that could and should be used. It is therefore presented as a pragmatic improvement approach for lean interventions.

Figures 15.6–15.11 elaborate the detail of how the DRIVER process is applied throughout each phase of an improvement project.

Define

At the start of any improvement intervention, it is important to go through a number of key steps. The goals of the project should be articulated in terms of desired outcomes, and these should ideally be expressed in terms of the requirements of relevant stakeholders. For any one project the individual stakeholders may vary. For public sector organizations, for example, one would expect them to include at the least citizen, government and organizational stakeholders.

Stakeholder requirements also include two specific inputs to the intervention:

- the voice of the customer in terms of what value means to the customer; and
- the voice of the business in terms of driving out non-value-adding activities from the value stream so as to reduce time and waste and thereby cost.

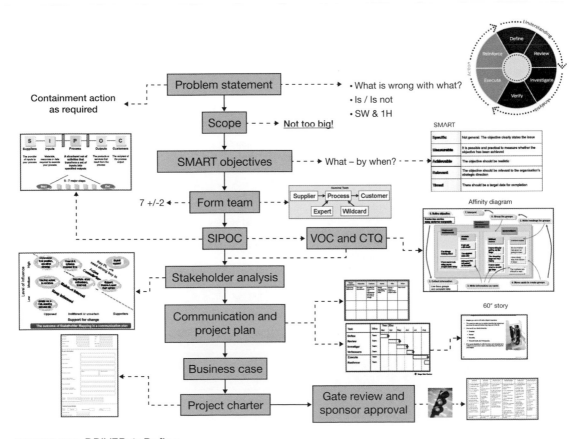

FIGURE 15.6 DRIVER 1: Define

The complex multiple-stakeholder perspective in many organizations means that gaining a clear understanding of the value proposition for all stakeholders will be challenging, but requirements tend to be easier to articulate when expressed in terms of value.

Other approaches have highlighted the need to identify first the process that the lean intervention will be conducted on and then to understand the process family within which it sits; in other words, what other processes exist within the organization that have the same underlying series of steps and activities. The process family perspective allows multiple gains to be made by leveraging the benefits of any solution across a number of different processes.

The scope of the process is critical to the outcomes, and it must be very clear from the outset what is in scope and what is out of scope. As a general guideline, one should always scope the process at its largest and then work down to what is feasible and achievable. It is important to look at end-to-end processes (field to fork) if possible to avoid the dangers of scoping at a too low a level and thereby missing the most critical areas for removal of non-value-adding activity.

Other key considerations at this stage include project management, team membership, governance at the intervention level, establishing performance metrics related to stakeholder requirements, communications strategies and a direct, visible and clearly articulated link between this lean intervention and the vision and values of the organization.

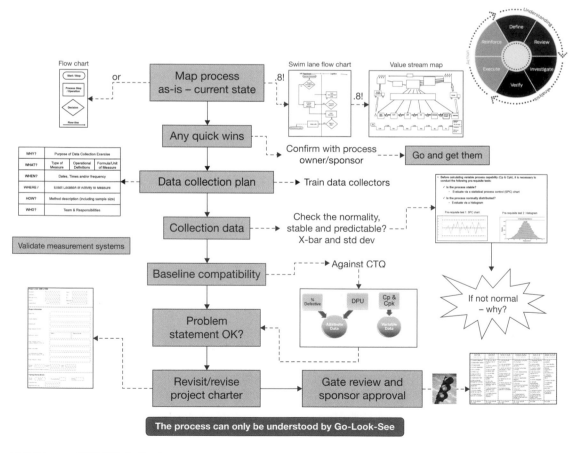

FIGURE 15.7 DRIVER 2: Review

Review

The Review phase sees a description and analysis of the current state ('as-is'). At this stage, it is appropriate to start to build the Value Stream map based on the team's knowledge of the process. Clearly team involvement is crucial to get a realistic and meaningful view of 'reality' and, as the view is built, it may need to be checked and refined. Measures need to collected and mapped onto the process steps to identify time and costs.

It is important that Voice of the Business and Voice of the Customer data continue to be developed during this phase as the view is built up, as this will challenge the current state analysis. The authors and their colleagues have found, in service environments, where the customer has direct contact with the value stream, it is important to identify these customer 'touch-points' and, likewise, to understand the value propositions that exist at each 'touch-point'. It is also highly likely that supporting processes will be identified as impacting on customer value at these 'touch-points'. These will need to be explored to provide the cause and effect analysis that will take place in the next phase.

Investigate

Once the current state picture has been built and metrics have been added to the Value Stream Map to provide a clear view of the time taken for each process step and the costs incurred to deliver both value-adding and non-value-adding activities, we move into the

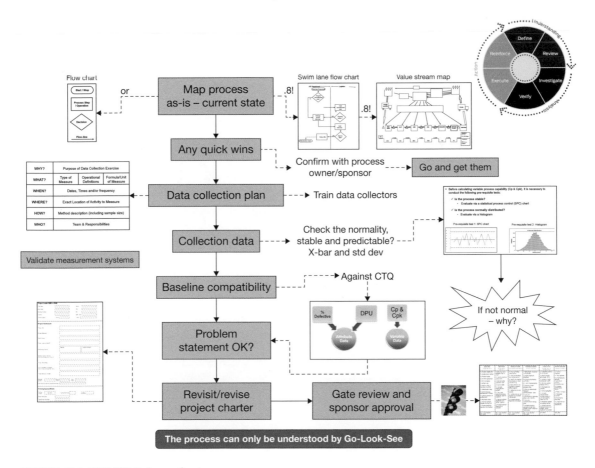

FIGURE 15.8 DRIVER 3: Investigate

'investigate' phase. The aim in this phase is to find ways by which we can reduce the time taken and reduce non-value-adding costs whilst still delivering the value add to customers and even increasing it, if required. This phase requires an in-depth challenging of the status quo and any assumptions that might be present in order to generate possible solutions to eliminate non-value-adding activity and 'dead time' when nothing is actually being done.

If the customer interacts with the process, touch-points need to be explored in detail to understand what adds value, what is not adding value and what is actually destroying value in the customer's eyes.

Any other processes interacting with this process, such as supply from other parties, will also need to be investigated for impacts and mapped as appropriate. Similarly, any other processes that are impacted should also be identified, as any changes made could have effects elsewhere in the 'system'.

This phase is really about understanding where and how changes might be made and what the effects of those changes will be. A range of possible solutions may be developed and all possibilities should be considered in order to generate potential 'future states'.

Verify

The investigation comes to an end when all possibilities have been explored and further ideas are exhausted. We now need to decide on the solutions that will be implemented by verifying the impacts of the various ideas and determining which are the most appropriate for the specific context we are in.

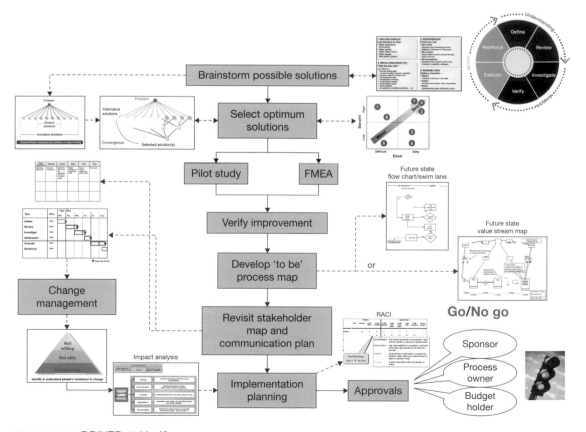

FIGURE 15.9 DRIVER 4: Verify

In multiple stakeholder environments, we need to be sure of the impact of any changes on different stakeholder groups, and some of the ideas generated previously may need to be modified, or even discarded, if they present problems to key stakeholder groups. It is also important to ensure that the proposed solutions fit with the organization's strategy and values.

The aim of this phase is to generate the definitive 'future state' map that will then be implemented and, in doing so, there may be substantial testing and retesting of assumptions. In some cases, there may be a need for pilot studies to test the feasibility of some of the proposed changes. The proposed 'future state' process design then should generate confidence in that it can be successfully implemented and will deliver the potential gains identified.

Execute

Once the new process has been determined, the changes will need to be implemented. This will require a robust change management plan. Implementing new ways of working is always a challenge, and there are a number of documented methods specifically relating to this (see figure of 8 framework in Chapter 11). Implementation needs to be tailored to the organization's culture and specific requirements and is usually best carried out as a 'phased' approach.

Senior management commitment to the proposed changes is critical as is a broader buy-in from key stakeholders. Staff affected by the changes also need to be on board with

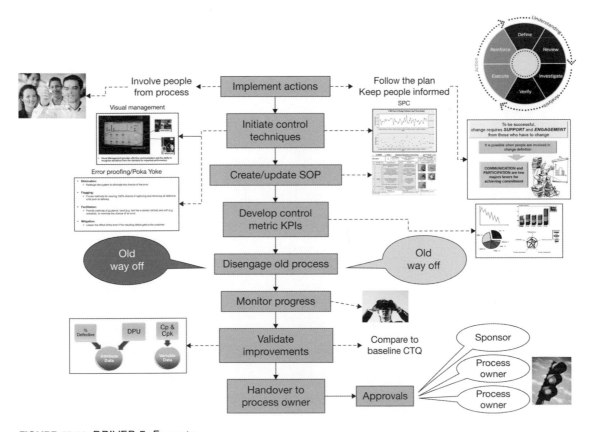

FIGURE 15.10 DRIVER 5: Execute

the proposed changes and, therefore, it is essential that confidence that the required results can and will be achieved by the changes is established at all levels and with all constituencies involved. Clearly, excellent two-way communications will be an essential component of the implementation.

The implementation will require appropriate documentation to be developed. Training requirements will need to be determined and appropriate interventions developed and delivered. These will undoubtedly include training staff in lean tools and techniques, and if consultants have been involved a full skills transfer programme should be in place, including mentoring and coaching, where appropriate. If a pilot has not been conducted in the Verify stage, it is advisable to establish a 'safe test' facility before full rollout.

Some lean tools described originally by Engineer Ohno from Toyota are now part of the lean 'toolkit' and play a key role. For example, mistake-proofing systems can be built-in and tools identified in the Investigate and Verify phases now need to be implemented along with the new process.

Reinforce

Having made changes that will create a better value proposition for both the organization and its customers (and other stakeholders), it is important that the changes are held and that the process does not 'slip back' to its previous state. It is here that the concepts

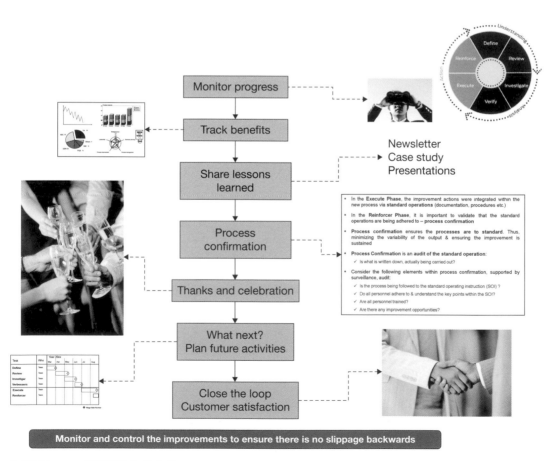

FIGURE 15.11 DRIVER 6: Reinforce

of Kaizen as a continuous improvement philosophy come in. Techniques such as 5S can considerably impact the culture of the organization, and people should be encouraged to 'think lean' in their everyday operations.

Whereas initial capability should have been built in the Execute phase, there will be a need for ongoing capability development to ensure that knowledge is not lost. Many successful lean initiatives have included processes for communicating learning points and establishing communities of lean 'champions' and/or 'practitioners' who are skilled and knowledgeable in lean and take a lead in continuing lean interventions. The development of 'Lean Academies' is also popular in some industries.

Ongoing communication of successes is a good way of ensuring that the initial momentum and key messages are not lost. Building-in an appropriate reward and recognition system to ensure that positive lean behaviours are encouraged and are also effective ways to maintain momentum and stay on message.

Companies and organizations that succeeded in establishing an ongoing commitment to Lean have embedded the thinking in the day-to-day culture and have ensured that there is a direct link between lean initiatives and the organizational strategy and performance measurement framework.

This chapter provides further detail on the specific tools and techniques, while the final section returns to a more detailed discussion of the implementation of the DRIVER technique.

THE NEED FOR DATA AND SOME BASIC TOOLS AND TECHNIQUES

The most obvious feature of the management practices described in the case studies is the use of performance measurement to drive continuous improvement. In the never-ending quest for improvement in business processes and outcomes, numbers and information should always form the basis for understanding, decisions and actions, and a thorough data gathering, recording and presentation system is essential:

- *Record* data: all processes can and should be measured; all measurements should be recorded;
- *Use* data: if data are recorded and not used they will be abused;
- *Analyse* data: data analysis should be carried out by means of some basic systematic tools; and
- *Act* on the results: recording and analysis of data without action leads to frustration.

Within the construction sector, there is a general perception that the sector deals with one-of-a-kind production and that numerical data analysis is either of very limited or no value. They have been developed in manufacturing and, therefore, have very limited application in construction. This may seem to be logical while the industry is viewed as delivering an unrelated sequence of projects. However, as soon as the industry is regarded as the producer of generically similar though superficially different products, through a set of largely repetitive processes, an entirely different rationale must be applied. Suddenly the tools for presenting and analysing data are just as critical to any process improvement exercise in construction as in any other sector of the economy. The companies in the case studies all use data analysis extensively to help achieve excellent outcomes. The challenge for every senior manager in a construction industry organization is to discover how he/she can use the management tools and numerical techniques presented in this chapter to improve his/her processes.

In addition to the basic elements of a quality management system that provide a framework for recording, there exists a set of methods the Japanese quality guru Ishikawa called the 'seven basic tools'. These should be used to interpret and derive the maximum use from data. The simple methods listed below, of which there are clearly more than seven, will offer any organization means of collecting, presenting and analysing most of its data:

- process mapping or flowcharting – what is done;
- check sheets/tally charts – how often is it done;
- histograms – what do overall variations look like;
- scatter diagrams – what are the relationships between factors;
- stratification – how is the data made up;
- Pareto analysis – which are the big problems;
- root cause and effect analysis and brainstorming (including CEDAC, NGT, RCA and the Five Whys) – what causes the problems;
- force field analysis – what will obstruct or help the change or solution;
- control charts – which variations to control and how;
- affinity diagram – to sort ideas;
- interrelationship diagraph – logical linking of ideas;
- matrix diagrams – to outline interrelationships and indicate relative importance.

Sometimes more sophisticated techniques such as analysis of variance, regression analysis and design of experiments need to be employed.

The effective use of the tools requires their application by the people who actually work on the processes. Their commitment to this will be possible only if they are assured that management cares about improving quality. Managers must show they are serious by establishing a systematic approach and providing the training and implementation support required.

Improvements cannot be achieved without specific opportunities (commonly called problems) being identified or recognized. A focus on improvement opportunities leads to the creation of teams whose membership is determined by their work on and detailed knowledge of the process, and their ability to take improvement action. The teams must then be provided with good leadership and the right tools to tackle the job.

The systematic approach of Figure 15.1 should lead to the use of factual information collected and presented by means of proven techniques, to open a channel of communications not available to the many organizations that do not follow this or a similar structured approach to problem solving and improvement. Continuous improvements in the quality of products, services and processes can often be obtained without major capital investment, if an organization marshals its resources, through an understanding and breakdown of its processes in this way.

By using reliable methods, creating a favourable environment for team-based problem solving and continuing to improve using systematic techniques the never-ending improvement helix (see Chapter 4) will be engaged. This approach demands the real-time management of data and actions focused on processes and inputs rather than outputs. It will require a change in the language of many organizations from percentage defects, percentage 'prime' product and number of errors to *process capability*. The climate must change from the traditional approach of: 'If it meets the specification, there are no problems and no further improvements are necessary.' The driving force for this will be the need for better internal and external customer satisfaction levels that will lead to the continuous improvement question: 'Could we do the job better?'

Basic tools and techniques for data analysis

Understanding processes so that they can be improved by means of the systematic approach requires knowledge of a simple kit of tools or techniques. What follows is a brief description of each technique; a full description and further examples of some of them may be found in Oakland's SPC book (2008).

Process mapping/flowcharting

The use of these techniques, which are described in Chapter 12, ensures a full understanding of the inputs, outputs and flow of the process. Without that understanding, it is not possible to draw the correct map or flowchart of the process. In flowcharting, it is important to remember that in all but the smallest tasks no single person is able to complete a chart without help from others. This makes flowcharting a powerful improvement team forming exercise.

Check sheets or tally charts

A check sheet is a tool for data gathering and a logical point to start in most process control or problem-solving efforts. It is particularly useful for recording direct observations and helping to gather facts rather than opinions about the process. In the recording process, it is essential to understand the difference between data and numbers.

Data are pieces of information, including numerical, that are useful in solving problems or provide knowledge about the state of a process. Numbers alone often represent meaningless measurements or counts that tend to confuse rather than to enlighten. Numerical data on quality will arise either from counting (attributes) or measurement (variables).

The use of simple check sheets or tally charts aids the collection of the right type of data, in the right form, at the right time. The objectives of the data collection will determine the design of the record sheet used.

Histograms

Histograms show, in a very clear pictorial way, the frequency with which a certain value, or group of values, occurs. They can be used to display both attribute and variable data and are an effective means of letting the people who operate the process know the results of their efforts. Data gathered on truck turn-round times are drawn as a histogram in Figure 15.12.

Scatter diagrams

Depending on the technology, it is frequently useful to establish the association, if any, between two parameters or factors. A technique to begin such an analysis is a simple X-Y plot of the two sets of data. The resulting grouping of points on scatter diagrams (e.g. Figure 15.13) will reveal whether or not a strong or weak, positive or negative correlation exists between the parameters. The diagrams are simple to construct and easy to interpret, and the absence of correlation can be as revealing as finding that a relationship exists.

Stratification

Stratification is simply the dividing of a set of data into meaningful groups. It can be used to great effect in combination with other techniques, including histograms and scatter diagrams. If, for example, three shift teams are responsible for a certain product output,

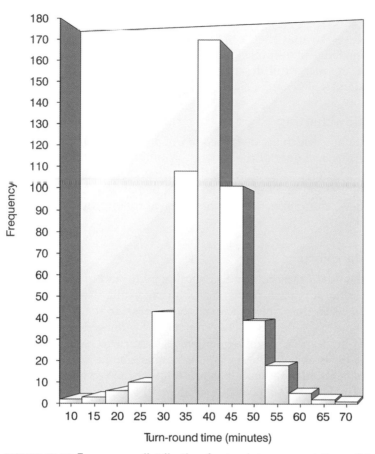

FIGURE 15.12 Frequency distribution for truck turn-round times (histogram)

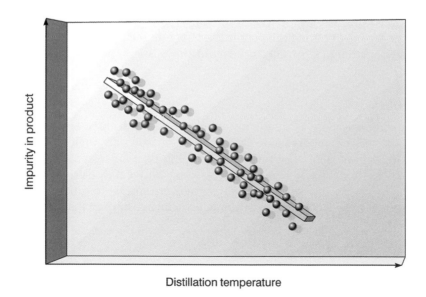

Distillation temperature

FIGURE 15.13 Scatter diagram showing a negative correlation between two variables

'stratifying' the data into the shift groups might produce histograms that will indicate that 'process adjustments' were taking place at shift changeovers.

Pareto analysis

If the symptoms or causes of defective output, or some other 'effect', are identified and recorded, it will be possible to determine what percentage can be attributed to any cause; and the probable results will be that the bulk (typically 80 per cent) of the errors, waste or 'effects', will be derived from a few of the causes (typically 20 per cent). For example, Figure 15.14 shows a *ranked frequency distribution* of incidents in the distribution of a certain product. To improve the performance of the distribution process, therefore, the major incidents (broken bags/drums, truck scheduling and temperature problems) should be tackled first. An analysis of data to identify the major problems is known as *Pareto analysis,* after the Italian economist who realized that approximately 90 per cent of the wealth in his country was owned by approximately 10 per cent of the people. Without an analysis of this sort, it is far too easy to devote resources to addressing one symptom only because its cause seems immediately apparent.

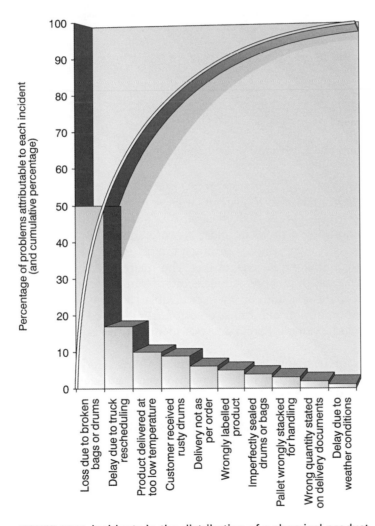

FIGURE 15.14 Incidents in the distribution of a chemical product

Root cause and effect analysis and brainstorming

A useful way of mapping the inputs that affect quality is the *cause and effect diagram*, also known as the Ishikawa diagram (after its originator) or the fishbone diagram (after its appearance, Figure 15.15). The effect or incident being investigated is shown at the end of a horizontal arrow. Potential causes are then shown as labelled arrows entering the main cause arrow. Each arrow may have other arrows entering it as the principal factors or causes are reduced to their sub-causes and sub-sub-causes by *brainstorming*.

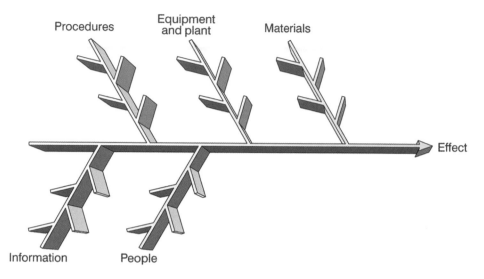

FIGURE 15.15 The cause and effect, Ishikawa or fishbone diagram

Brainstorming is a technique used to generate a large number of ideas quickly and may be used in a variety of situations. Each member of a group in turn may be invited to put forward ideas concerning a problem under consideration. Wild ideas are safe to offer as criticism or ridicule is not permitted during a brainstorming session. The people taking part do so with equal status to ensure this. The main objective is to create an atmosphere of enthusiasm and originality. All ideas offered are recorded for subsequent analysis. The process is continued until all the conceivable causes have been included. For example, the proportion of nonconforming output attributable to each cause is then measured or estimated and a simple Pareto analysis identifies the causes that are most worth investigating.

A useful variant on the technique is negative brainstorming. Here the group brainstorms all the things that would need to be done to ensure a negative outcome. For example, in the implementation of lean quality, it might be useful for the senior management team to brainstorm what would be needed to make sure lean quality *was not* implemented well. Having identified in this way the potential roadblocks, it is easier to dismantle them.

CEDAC

A variation on the cause and effect approach, which was developed at Sumitomo Electric and now is used by many major corporations across the world, is the cause and effect diagram with addition of cards (CEDAC). The effect side of a CEDAC chart is a quantified

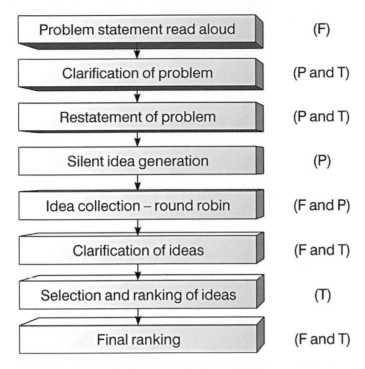

Problem statement read aloud	(F)
Clarification of problem	(P and T)
Restatement of problem	(P and T)
Silent idea generation	(P)
Idea collection – round robin	(F and P)
Clarification of ideas	(F and T)
Selection and ranking of ideas	(T)
Final ranking	(F and T)

FIGURE 15.16 Nominal group technique (NGT)

description of the problem, with an agreed and visual quantified target and continually updated results on the progress of achieving it. The cause side of the CEDAC chart uses two different coloured cards for writing facts and ideas. This ensures that the facts are collected and organized before solutions are devised. The basic diagram for CEDAC has the classic fishbone appearance.

Nominal group technique (NGT)

The nominal group technique (NGT) is a particular form of team brainstorming used to prevent domination by particular individuals. It has specific application for multi-level, multidisciplined teams where communication boundaries are potentially problematic.

In NGT a carefully prepared written statement of the problem to be tackled is read out by the facilitator (F). Clarification is obtained by questions and answers and then the individual participants (P) are asked to restate the problem in their own words. The group then discusses the problem until its formulation can be satisfactorily expressed by the team (T). The method is set out in Figure 15.16. NGT results in a set of ranked ideas that are close to a team consensus view obtained without domination by one or two individuals.

The 5 Whys

A method used to get to the root cause of a problem is known as the 5 Whys, so called because you simply ask the question why something happened at least five times; you only stop when you can go no further. A brief example relating to work that was not done as planned follows; the authors have often heard the first reason accepted in a meeting without delving further:

1   Why didn't we get the work done yesterday?
    The truck delivering the materials only arrived at 4pm.
2   Why was the delivery so late?
    It was only ordered the day before.
3   Why was it ordered so late?
    The designers only sent through the detail on Monday.
4   Why were the designers so late?
    The client only made the final decision on the detail last Friday.
5   Why was the client decision so late?
    We only realized that the original detail would not work a week ago.
6   Why did we only realize that the original detail would not work a week ago?
    We are running late with our entire design review process.

Stopping at the first answer will never help the team to improve its performance; however, digging deep by asking at least 5 Whys will create the opportunity for improvement next time. The team had to get on top of its design reviews and blaming the transport company would never drive the required changes.

Force field analyses

Force field analysis is a technique used to identify the forces that either obstruct or help a change that needs to be made. It is similar to negative brainstorming-cause/effect analysis and helps to plan how to overcome the barriers to change or improvement. It may also provide a measure of the difficulty in achieving the change.

The process begins with a team describing the desired change or improvement and defining any objectives or solution. Having prepared the basic force field diagram through brainstorming, the favourable/positive/driving forces and the unfavourable/negative/restraining forces are identified. These forces are placed in opposition on the diagram and, if possible, rated for their potential influence on the ease of implementation. The results are evaluated and an action plan to overcome some of the restraining forces and increase the driving forces is prepared. Figure 15.17 shows a force field analysis produced by a senior management team considering the implementation of lean quality in its organization.

Control charts

Charts should be made easy to understand and interpret and, with experience, they can become sensitive diagnostic tools to be used by operating staff. Once first-line supervisers become skilled in their use, these charts will help to prevent errors or defective output being produced. Time and effort spent to explain the working of the charts to all concerned are never wasted.

The most frequently used control charts are simple run-charts, where the data is plotted on a graph against time or sample number. The CUSUM chart is a time-weighted control chart that displays the cumulative sums (CUSUMs) of the deviations of each sample value from the target value. Because it is cumulative, even a minor drift in a process mean (quality, reliability) will lead to steadily increasing or decreasing cumulative deviation values. This is a graph that takes a little longer to draw than the conventional control chart but gives a lot more information. It is particularly useful for plotting the evolution of processes because it presents data in a way that enables the eye to separate true changes from a background of random variation.

Figure 15.18 shows a comparison of an ordinary run-chart and a CUSUM chart that have been plotted from the same data-errors in samples of 100 invoices. The change, which

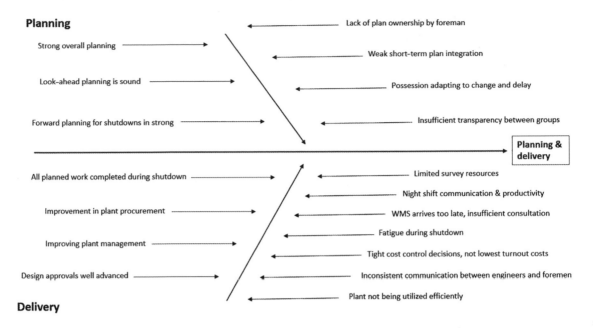

FIGURE 15.17 Force field analysis showing factors working for and against the reliable planning and delivery of work on a project

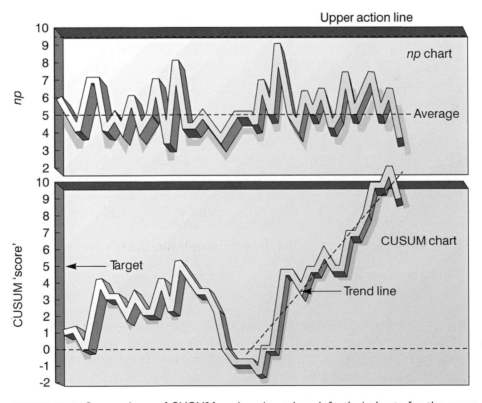

FIGURE 15.18 Comparison of CUSUM and *np* (*number defective*) charts for the same data

is immediately obvious on the CUSUM chart, is difficult to detect on the conventional control chart.

The range of type and use of control charts is now very wide, and within the present text it is not possible to indicate more than the basic principles underlying such charts. All of them can be generated electronically using the various software tools available. A full treatment of all control chart methods is given in *Statistical Process Control* (Oakland, 2008).

Affinity diagram

This is used to gather large amounts of language data (ideas, issues, opinions) and organize them into groupings based on the natural relationship between the items. In other words, it is a form of brainstorming.

The steps for generating an affinity diagram are as follows:

1 Assemble a group of people familiar with the problem of interest. Six to eight members in the group works best.
2 Phrase the issue to be considered. It should be vaguely stated so as not to prejudice the responses in a predetermined direction.
3 Give each member of the group a stack of cards and allow 5–10 minutes for everyone individually in the group to record ideas on the cards, writing down as many ideas as possible.
4 At the end of the 5–10 minutes, each member of the group in turn reads out one of his/her ideas and places it on the table for everyone to see, without criticism or justification.
5 When all ideas are presented, members of the group place all the cards with related ideas together, repeating the process until the ideas are clustered in a few groups.
6 Look for one card in each group that captures the meaning of that group.

The output of this exercise is a compilation of a maximum number of ideas under a limited number of major headings. This data can then be used with other tools to define areas for attack. One of these tools is the interrelationship diagraph.

Interrelationship diagraph

This tool is designed to take a central idea, issue or problem and map out the logical or sequential links among related factors. While this still requires a very creative process, the interrelationship diagraph begins to draw the logical connections that surface in the affinity diagram. In designing, planning and problem solving it is obviously not enough to just create an explosion of ideas. The affinity diagram allows some organized creative patterns to emerge but the interrelationship diagraph lets *logical* patterns become apparent (Figure 15.19).

Matrix diagrams

The purpose of the matrix diagram is to outline the interrelationships and correlations between tasks, functions or characteristics and to show their relative importance. There are many versions of the matrix diagram but the most widely used is a simple L-shaped matrix known as the *quality table,* in which customer demands (the Whats) are analysed with respect to substitute quality characteristics (the Hows; see Figure 15.20). Correlations between the two are categorized as strong, moderate and possible. The customer demands shown on the left of the matrix are determined in co-operation with the customer in a joint meeting, if possible.

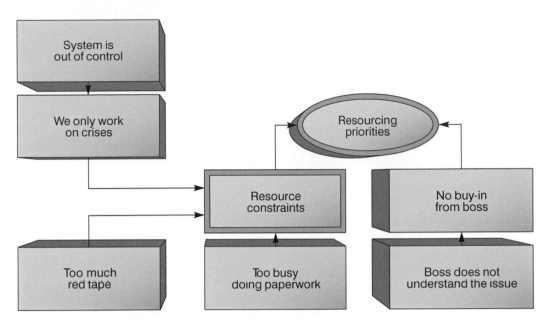

FIGURE 15.19 Example of the interrelationship diagraph

	Substitute quality characteristics								
	MFR	Ash	Importance	Current	Best competitor	Plan	IR	SP	RQW
No film breaks	◯ 17	▲ 6	4	4	4	4	1	◯	5.6
High rates	◉ 23		3	3	4	4	1.3		4.6
Low gauge variability	◉ 37	▲ 7	4	3	4	4	1.3	◯	7.3

Customer demands

◉ Strong correlation
◯ Some correlation
▲ Possible correlation
IR Improvement ratio
SP Sales point
RQW Relative quality weight

FIGURE 15.20 Example of the matrix diagram

The right side of the chart is often used to compare current performance to competitors' performance, company plan and potential sales points with reference to the customer demands. Weights are given to these items to obtain a 'relative quality weight', which can be used to identify key customer demands. The relative quality weight is then used with the correlations identified on the matrix to determine key quality characteristics.

A T-shaped matrix is nothing more than the combination of two L-shaped matrix diagrams. Figure 15.21 shows one application – the relationship between a set of courses in a curriculum and two important sets of considerations: who should do the training for each course and which would be the most appropriate groups of people and roles within the business to attend each of the courses? There are other matrices that deal with ideas such as product or service function, cost, failure modes, capabilities, etc. There are at least 40 different types of matrix diagrams available.

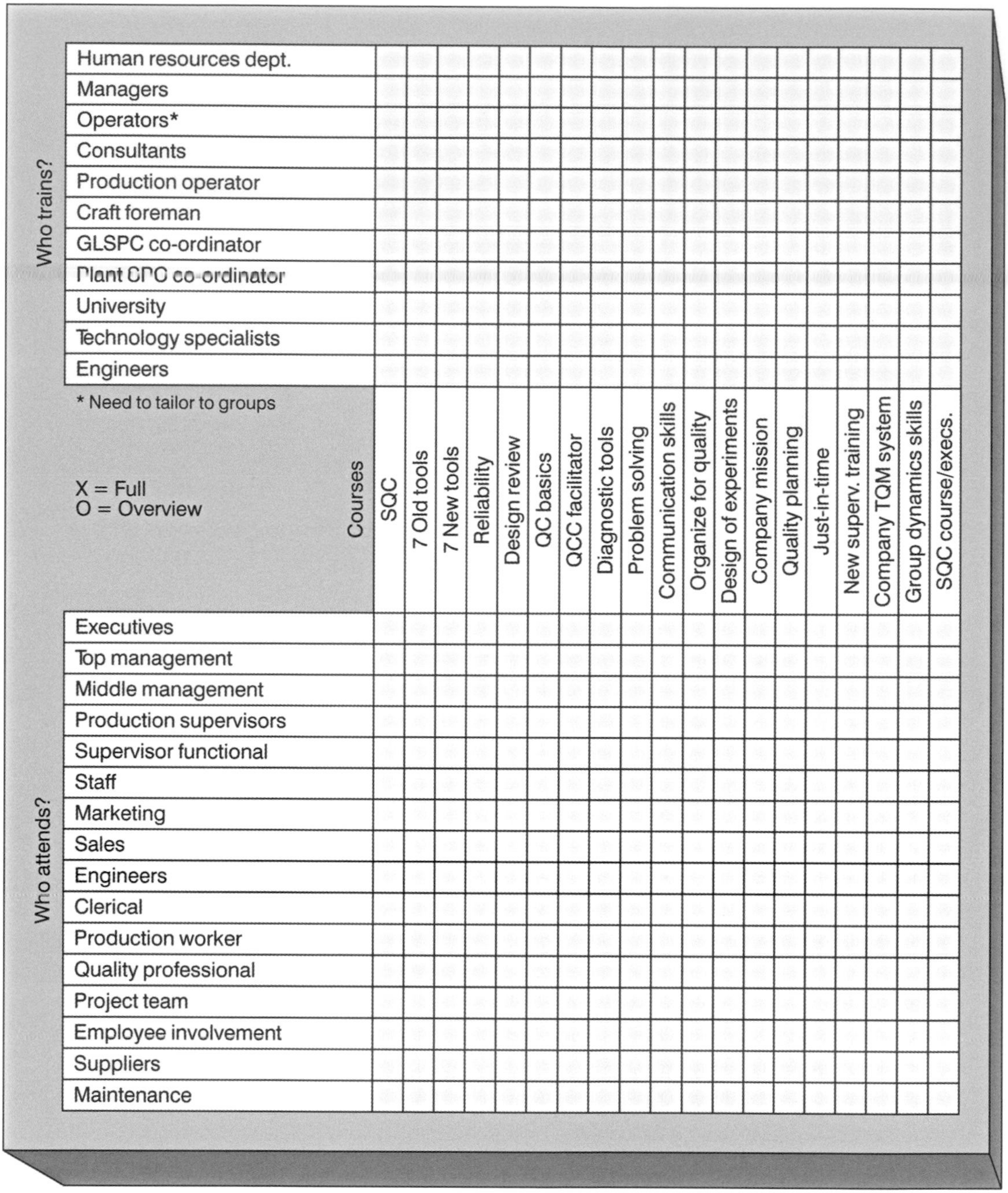

Who trains?	Courses	SQC	7 Old tools	7 New tools	Reliability	Design review	QC basics	QCC facilitator	Diagnostic tools	Problem solving	Communication skills	Organize for quality	Design of experiments	Company mission	Quality planning	Just-in-time	New superv. training	Company TQM system	Group dynamics skills	SQC course/execs.
Human resources dept.																				
Managers																				
Operators*																				
Consultants																				
Production operator																				
Craft foreman																				
GLSPC co-ordinator																				
Plant SPC co-ordinator																				
University																				
Technology specialists																				
Engineers																				

* Need to tailor to groups

X = Full
O = Overview

Who attends?		SQC	7 Old tools	7 New tools	Reliability	Design review	QC basics	QCC facilitator	Diagnostic tools	Problem solving	Communication skills	Organize for quality	Design of experiments	Company mission	Quality planning	Just-in-time	New superv. training	Company TQM system	Group dynamics skills	SQC course/execs.
Executives																				
Top management																				
Middle management																				
Production supervisors																				
Supervisor functional																				
Staff																				
Marketing																				
Sales																				
Engineers																				
Clerical																				
Production worker																				
Quality professional																				
Project team																				
Employee involvement																				
Suppliers																				
Maintenance																				

FIGURE 15.21 T-matrix on company-wide training

Statistical process control (SPC)

The responsibility for quality in any process must lie with the operators of that process. To fulfil this responsibility, however, people must be provided with the tools necessary to:

- know whether the process is capable of meeting the requirements;
- know whether the process is meeting the requirements at any point in time; and
- make correct adjustment to the process or its inputs when it is not meeting the requirements.

Lean quality management requires that the processes should be improved continually by reducing variability. This is brought about by studying all aspects of processes using the basic question: 'Could we do this job more consistently and on target?' The answer drives the search for improvements.

Statistical process control procedures exist because there is variation in the characteristics of all material, articles, services and people. For many construction companies, SPC will bring a new approach, a new 'philosophy', and the importance of the numbers and information should not be disguised. Simple presentation of data using diagrams, graphs and charts should become the means of communication concerning the state of control of processes. It is on this understanding that improvements are based.

Traditionally on construction sites, SPC has primarily been applied to assessments of concrete batch quality and limited analysis of safety incidents. However, there is a far greater opportunity for using these techniques to analyse process reliability, error levels and productivity rates. For example, with the increasing use of prefabrication, the control of variance in the size of fabricated products is an ideal area for SPC techniques. Statistically based process control procedures are designed to divert attention from individual pieces of data and focus on the characteristics of the process as a whole. SPC techniques may be used to measure and control the degree of variation of any purchased materials, services, processes and products, and to compare these, if required, to previously agreed specifications. In essence, we can use SPC techniques to select and analyse a representative, simple, random sample from the 'population'; the data can be an input to or an output from a process. From the analysis of the sample data, it is possible to make decisions regarding the current performance of the process.

Construction-related organizations that embrace the lean quality concepts should recognize the value of SPC techniques in areas such as workmanship quality on site, particularly in areas such as coating and membrane thickness and reinforcement cover depth as well as in sales, purchasing, invoicing, finance, distribution, training and services areas generally. These are outside the traditional areas for SPC use, but it needs to be seen as an organization-wide approach to reducing variation within key techniques throughout the business.

A Pareto analysis, a histogram, a flowchart or a control chart are all vehicles for communication. Data are data and, whether the numbers represent defects or invoice errors, weights or delivery times, or whether the information relates to machine settings, process variables, prices, quantities, discounts, sales or supply points, is irrelevant – the techniques can always be used.

In the authors' experience, some of the most exciting applications of SPC have emerged from organizations and departments which, when first introduced to the methods, could see little relevance in them to their own activities.

Following appropriate training, however, they have learned how to, for example:

- *Pareto analyse* errors on invoices to customers and industry injury data;
- *brainstorm and cause and effect analyse* reasons for late payment and poor purchase invoice matching;
- *histogram* defects in building components and arrival of trucks at certain times during the day; and
- *control chart* the weekly absenteeism or consumption of a certain material.

Construction management and staff have used control charts to monitor the proportion of late deliveries, and Pareto analysis and force field analysis to look at causes of defects and complaints about the construction process.

Those organizations that have made most progress in implementing continuous improvement have recognized at an early stage that SPC is for the whole organization. Restricting it to traditional manufacturing or operational activities means that a window of opportunity for improvement has been closed. Applying the methods and techniques outside the usual operations areas will make it easier, not harder, to gain maximum benefit from SPC. A full treatment of statistical methods is given in *Statistical Process Control* (Oakland, 2008).

ADDITIONAL TOOLS FOR PROCESS IMPROVEMENT

Value Stream Mapping

The concept of the Value Stream is fundamental to lean. In many manufacturing industries, this can be seen as an internal core process of value-adding activity that ultimately delivers value (in the form of a product) to the customer. Customers very rarely get involved in the delivery of that product, be it a car or a computer (Figure 15.22).

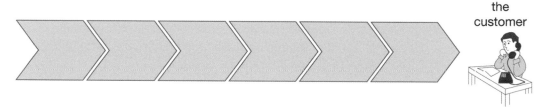

FIGURE 15.22 Value Stream 1

However, in services, as in many support processes in design, construction or manufacturing, the customer does interact at various 'touch-points' along the value delivery chain (Figure 15.23).

In *Lean Solutions* (2005), Womack and Jones expanded their approach again to give a more thorough view of the customer, stressing the need to understand 'consumption', as in the customer's demand requirements, and 'provision', in terms of the organization's capability to deliver these requirements.

Value Stream Mapping (VSM) studies the set of specific actions required to bring a product family from raw material to finished goods as per customer demand, concentrating on information management and physical transformation tasks.

The outputs of a VSM-based study are a current state map, future state map and implementation plan for getting from the current to the future state. Using VSM it should be

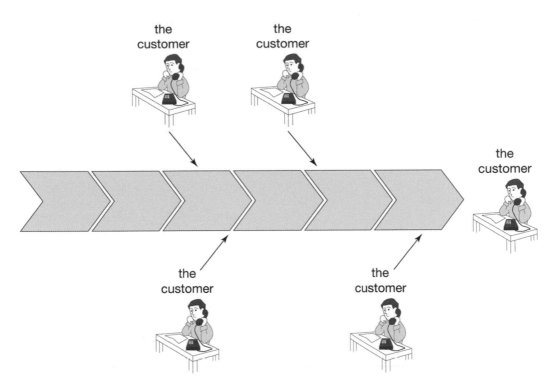

FIGURE 15.23 Value Stream 2

possible to bring the lead time closer and closer to the actual value-added processing time by attacking the identified bottlenecks and constraints. Bottlenecks addressed could include long set-up times, unreliable equipment, unacceptable first pass yield, or high work or process inventories.

The VSM technique has been central to the approach advocated by Womack and Jones from their original work with Toyota and is still the mainstay of lean interventions. The method is to describe the current process, looking at both material and information flows, on a flowchart and to indicate measures that relate to value and waste such as time taken, time between steps or cost to each process step. By also identifying which activities add value and which do not, it is possible to analyse the process from a value creating perspective and determine the potential gains from eliminating non-value-adding activity.

Figure 15.24 shows a Value Stream Map from a manufacturing context where the original time taken has been identified and then altered to show what has been made possible through understanding what can be changed.

In this example, the time taken for information to get from the customer to the manufacturer prior to the start of manufacture was 66 days and amounted to 1360 minutes of actual process time. Manufacturing time took 21 days (1075 minutes' actual process time).

By identifying these timings and then challenging how much of that time was actually spent adding value and what was not, it was possible to reduce the total lead time from 87 (66 + 21) days to 20 (15 + 5) – a saving of 67 days throughput time. Although this is from a manufacturing environment, the overall process (in terms of process steps and flow) is very similar to many service processes, in that much of the non-value-added time is spent 'in transit' waiting for things to be done – often on administrative tasks.

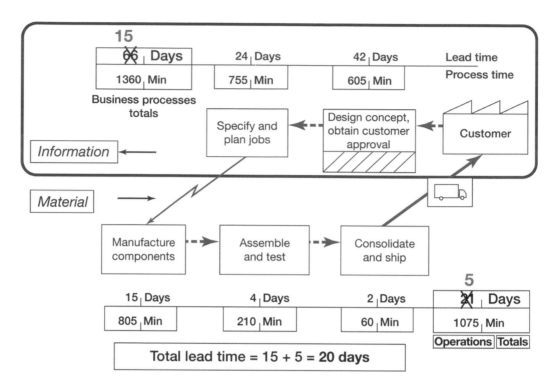

FIGURE 15.24 Value Stream Map (example 1)

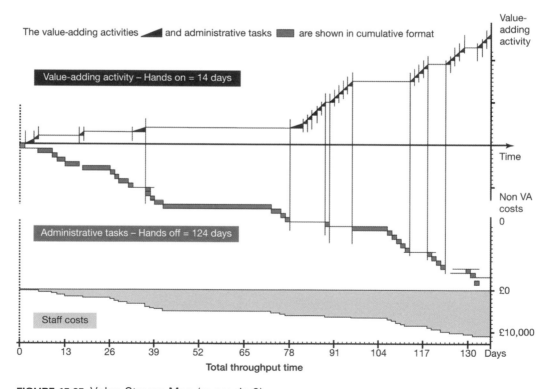

FIGURE 15.25 Value Stream Map (example 2)

VSM has been used successfully in the health sector in reducing patient wait times, where process delays are often due to a non-alignment of the process steps – in the language of Womack and Jones' model they are not arranged in 'continuous flow' and therefore are not adding value and, in fact, are increasing waste and cost.

Although there are some standardized approaches to VSM, there are also different approaches in operation. Figure 15.25 shows a Value Stream map for outpatient heart failure services at a hospital. The map takes the form of a horizontal timeline showing the sequential process steps, but also plotted on a vertical axis showing the cost of value-adding activity (above the timeline) and the cost of non-value-adding activity (below the timeline). The addition of staff costs provides a powerful diagnostic tool by which to address how non-value-adding activities might be eliminated.

Although VSM would appear to be a simple tool/technique, it is important to understand how the various activities make up the process. In complex service environments, there may be some apparent non-value-adding steps that are essential to another process that is in some way linked to the process being investigated. It is important that these dependencies are understood, and it is therefore essential that Value Stream Maps are not created by individuals but by teams of people working in the process who know what is going on and can challenge each other's perspectives. Lean quality initiatives should always be undertaken by cross-functional teams. A development of VSM by the author and his colleagues is 'Carbon Stream Mapping' (CSM), a process mapping tool in which processes are studied to identify carbon emission at various stages; Figure 15.26 is an example of a CSM.

5S

5S has been described as a system that creates a disciplined, clean and well-ordered work environment. Its original role in the manufacturing context was that of ensuring that work areas were free from clutter, clean, and laid out in such a way that tools were not

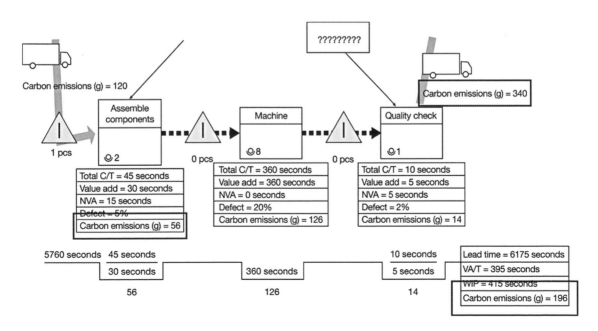

FIGURE 15.26 Carbon Stream Mapping (CSM) example

lost and time was not wasted on non-value-adding activities to do with general house-keeping. The secret to this was contained in five simple steps:

1 Sort
The first step is to free up the workplace by getting rid of everything that is not required for the work – this reduces problems and decreases lead times.

2 Set in Order
What is left is then organized so that it is easily accessible and time is not lost looking for things that are misplaced.

3 Shine
This step relates to cleanliness and keeping things in good working order. Although this is easy to visualize in a shop floor manufacturing environment, the basic concepts can still apply in services.

4 Standardize
In this step, the above three steps are standardized so that they become a routine part of day-to-day work. Management should create the time to allow resources for people to perform the first three steps.

5 Sustain
The fifth and last step is about embedding the practice in business as usual by involving people at all levels, through peer and leadership behaviours, 5S audits, 5S goals and providing feedback on performance.

The case study companies working on site all have introduced 5S practices both in their fabrication shops and on their construction sites. The JB Henderson case study (CS8) and the Southland Industries (CS6) have several photos of 5S results.

Interestingly, the 5S process is as relevant in the office and on the computer hard drive as it is in the workshop or on a construction site. It is an approach to creating an ordered environment wherever you are.

Kaizen

The well-known team-based improvement process known as Kaizen is presented in Chapter 16.

SIX SIGMA

Since the early 1980s, most of the world has been in what the authors call a 'quality revolution'. Based on the simple premise that organizations of all kinds exist mainly to serve the needs of the customers of their products or services, good quality management has assumed great importance. Competitive pressures on companies and government demands on the public sector have driven the need to find more effective and efficient approaches to managing businesses and non-profit-making organizations.

In the early days of the realization that improved quality was vital to the survival of many companies, especially in manufacturing, senior managers were made aware through national campaigns and award programmes that the basic elements had to be right. They learned through adoption of quality management systems, the involvement of improvement teams and the use of quality tools that improved business performance could be achieved only through better planning, capable processes and the involvement of people. These are the basic elements of a lean quality approach, and this has not changed no matter how many sophisticated approaches and techniques come along.

The development of lean quality has seen the introduction and adoption of many dialects and components including quality circles, international systems and standards, statistical process control, business process re-engineering, lean manufacturing, continuous improvement, benchmarking and business excellence.

An approach finding favour in some companies is Six Sigma, most famously used in Motorola, General Electric and Allied Signal. This operationalized lean quality into a project-based system based on delivering tangible business benefits often directly to the bottom line. Strange combinations of the various approaches have led to Lean Six Sigma and other company specific acronyms such as 'Statistically Based Continuous Improvement (SBCI)'!

The six sigma improvement model

There are five fundamental phases or stages in applying the six sigma approach to improving performance in a process: Define, Measure, Analyse, Improve and Control (DMAIC in Figure 15.27). These form an improvement cycle grounded in Deming's original Plan, Do, Check, Act (PDCA). In the six sigma approach, DMAIC (like *DRIVER*) provides a breakthrough strategy and disciplined methods of using rigorous data gathering and statistically based analysis to identify sources of errors and ways of eliminating them. It has become increasingly common in so-called 'six sigma organizations' for people to refer to 'DMAIC Projects'. These revolve around the three major strategies for processes to bring about rapid bottom-line achievements – design/redesign, management and improvement.

Table 15.1 shows the outline of the DMAIC steps.

There is a whole literature on the subject of Six Sigma, for more detail refer to John Oakland's *Total Quality Management and Operational Excellence* (4th edn): Chapter 14 (2014). The major contribution of Six Sigma has not been in the creation of new technology or method but in bringing to the attention of senior management the need for a disciplined structured approach and their commitment, if real performance and bottom-line improvements are to be achieved.

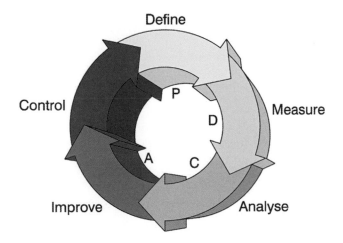

FIGURE 15.27 The Six Sigma improvement model – DMAIC

TABLE 15.1 The DMAIC steps

D	Define the scope and goals of the improvement project in terms of customer requirements and the process that delivers these requirements – inputs, outputs, controls and resources
M	Measure the current process performance – input, output and process – and calculate the short and longer-term process capability – the sigma value
A	Analyse the gap between the current and desired performance, prioritize problems and identify root causes of problems. Benchmarking the process outputs, products or services, against recognized benchmark standards of performance may also be carried out
I	Generate the improvement solutions to fix the problems and prevent them from reoccurring so that the required financial and other performance goals are met
C	This phase involves implementing the improved process in a way that 'holds the gains'. Standards of operation will be documented in systems such as ISO 9000 and standards of performance will be established using techniques such as statistical process control (SPC)

Technical note

Sigma is a statistical unit of measurement that describes the distribution about the mean of any process or procedure. A process or procedure that can achieve plus or minus *six sigma* capability can be expected to have a defect rate of no more than a few parts per million, even allowing for some shift in the mean. In statistical terms, this approaches *zero defects*.

BIBLIOGRAPHY

Fukuda, R. *CEDAC – A Tool for Continuous Systematic Improvement*, Productivity Press, Cambridge, MA., 1990.

Harry, M. and Schroeder, R. *Six-Sigma – The Breakthrough Management Strategy Revolutionising the World's Top Corporations*, Doubleday, New York, 2000.

Oakland, J.S. *Statistical Process Control: A Practical Guide* (6th edn), Butterworth-Heinemann, Oxford, 2008.

Oakland, J.S. *Total Quality Management and Operational Excellence; Text and Cases* (4th edn), Routledge, Oxford, 2014.

Ohno, T. and Rosen, C.B. 1988, *Toyota Production System: Beyond Large-scale Production*, Productivity Press, Stanford, CT, 1988.

Pande, P.S., Neumann, R.P. and Cavanagh, R.R. *The Six-Sigma Way – How GE, Motorola and Other Top Companies are Honing their Performance*, McGraw-Hill, New York, 2000.

Pyzdek, T. and Keller, P.A. *The Six Sigma Handbook* (3rd edn), ASQ Press, Milwaukee, 2009.

Scherkenbach, W.W. *Deming's Road to Continual Improvement*, SPC Press, Knoxville, 1991.

Shingo, S. *Zero Quality Control: Source Inspection and the Poka-yoke System*, Productivity Press, Stamford, CT., 1986.

Wheeler, D.J. *Understanding Statistical Process Control* (2nd edn), SPC Press, Knoxville, 1992.

Wheeler, D.J. *Understanding Variation*, SPC Press, Knoxville, 1993.

Womack, J.P. and Jones, D.T. From lean production to the lean enterprise, *Harvard Business Review*, 72(2): 93–103, 1994.

Womack, J.P. and Jones, D.T. Beyond Toyota: how to root out waste and pursue perfection, *Harvard Business Review*, 74(5): 140–158, 1996.

Womack, J.P. and Jones, D.T. *Lean Thinking: Banish Waste and Create Wealth in Your Corporation*, Simon & Schuster, New York, 1996.

Womack, J.P. and Jones, D.T. *Lean Solutions*, Simon & Schuster UK Ltd, London, 2005.

Womack, J.P., Jones, D.T. and Roos, D. *The Machine that Changed the World*, Rawson MacMillan, New York, 1990.

CHAPTER HIGHLIGHTS

Approaches, methods and tools

- Continuous improvement must be linked to organizational strategy, have a defined structure, a chosen approach, a method and an associated toolkit.
- A generic eight stage structure for continuous improvement may be applied to most organizations; it involves: Leadership and Top-Down Cascade of Objectives, Supporting Infrastructure, Pragmatic, Fact-based Method Yielding Rapid Results (Lean Six Sigma), Skilled Improvers, Widely Shared Continual Improvement Culture, Circulation of Talent, Measurable Improvements in the Service/Delivery Areas and Sustained Improvement.

The 'DRIVER' framework for continuous improvement

- A fully closed loop improvement method that brings together the best of Lean, Six Sigma and Cost of Quality approaches is 'DRIVER' with the stages of Define, Review, Investigate, Verify, Execute and Reinforce.

The need for data and some basic tools and techniques

- Numbers and information will form the basis for understanding, decisions, and actions in never-ending improvement – record data, use/analyse data, act on results.
- A set of simple tools is needed to interpret fully and derive maximum use from data. More sophisticated techniques may need to be employed occasionally; the effective use of the tools requires the commitment of the people who work on the processes, which in turn needs management support and the provision of training.
- The basic tools and the questions answered are:
  - process mapping or flowcharting – what is done;
  - check/tally charts – how often is it done;
  - histograms – what do overall variations look like;
  - scatter diagrams – what are the relationships between factors;
  - stratification – how is the data made up;
  - Pareto analysis – which are the big problems;
  - root cause and effect analysis and brainstorming (including CEDAC, NGT, RCA and the Five Whys) – what causes the problems;
  - force field analysis – what will obstruct or help the change or solution;
  - emphasis curve – which are the most important factors;
  - control charts (including CUSUM) – which variations to control and how?;
  - affinity diagram – to sort ideas;
  - interrelationship diagraph – logical linking of ideas;
  - matrix diagrams – to outline interrelationships and indicate relative importance.
- People operating a process must know whether it is capable of meeting the requirements; know whether it is actually doing so at any time; and make correct adjustments when it is not – SPC techniques will help here.

- Before using SPC, it is necessary to identify what the process is; what the inputs/outputs are; and how the suppliers and customers and their requirements are defined. The most difficult areas for this can be in non-manufacturing.
- All processes can be monitored and brought 'under control' by gathering and using data. SPC methods with management commitment provide objective means of controlling quality in any transformation process.
- SPC is not only a toolkit; it is a strategy for reducing the variability to drive never-ending improvement. This is achieved by answering the following questions:
  - Are we capable of doing the job correctly?
  - Do we continue to do the job correctly?
  - Have we done the job correctly?
  - Could we do the job more consistently and on target?
- SPC provides knowledge and control of process capability.
- SPC techniques have value in the service sector and in the non-manufacturing areas such as marketing and sales, purchasing, invoicing, finance, distribution, training and personnel.

Six Sigma

- Six Sigma is not a new technique; its origins may be found in TQM, Lean and SPC. It is a framework through which powerful TQM, Lean and SPC tools flourish and reach their full potential. It delivers breakthrough benefits in the short term through the intensity and speed of change. The Excellence Model is a useful framework for mapping the key six sigma breakthrough strategies.
- A process that can achieve six sigma capability (where sigma is the statistical measure of variation) can be expected to have a defect rate of a few parts per million, even allowing for some drift in the process setting.
- Six Sigma is a disciplined approach for improving performance by focusing on enhancing value for the customer and eliminating costs which add no value.
- There are five fundamental phases/stages in applying the six sigma approach: Define, Measure, Analyse, Improve, and Control (DMAIC). These form an improvement cycle similar to Deming's Plan, Do, Check, Act (PDCA). Six Sigma delivers the strategies of process design/redesign, management and improvement leading to bottom-line achievements.
- Six sigma approaches question organizational cultures and the measures used. Six sigma organizations, in addition to focusing on understanding customer requirements, identify core processes, involve all employees in continuous improvement, are responsive to change, base management on fact and metrics and obtain outstanding results.
- Properly implemented six sigma strategies comprise: leadership involvement and sponsorship, organization-wide training, project selection tools and analysis, improvement methods and tools for implementation, measurement of financial benefits, communication, control and sustained improvement.
- Six sigma process improvement experts are known as Black Belts and Green Belts. They perform the training, lead teams and carry out the improvements. Mature six sigma programmes have about 1 per cent of the workforce as Black Belts.

Part IV Discussion questions

1 Explain what is meant by taking a business process management (BPM) approach to running an organization outlining the main advantages of adopting BPM successfully. What would be the key components of an implementation plan for BPM?

2 Develop a high-level process framework for a construction organization of your choice identifying the 'value-adding' processes and the main support processes. Give a breakdown to the first level sub-processes of one value-adding process and one support process.

3 Using an appropriate process modelling technique show the core processes for a company manufacturing and selling fast moving materials such as stock sizes of timber and plasterboard sheeting. Identify the key inputs and outputs for the processes and explain how you would engage the senior management of the company in the development of the process framework for the business.

4 Explain the basic philosophy behind quality management systems such as those specified in the ISO 9000 series. How can an effective quality management system contribute to continuous improvement in an international construction operation?

5 Explain what is meant by independent third-party certification to a standard such as ISO 9000 and discuss the merits of such a scheme for an organization.

6 Compare and contrast the role of quality management systems in the following organizations:

 a building a cross-London underground railway system;
 b a medium-sized road maintenance/engineering company;
 c a regional branch of a major national house building operation.

7 A major multinational contractor is concerned about its poor quality and project completion performance on its new projects. Considerable costs are incurred in rework near and after practical completion. As the company's new Chief Quality and Business Improvement Officer you have been asked to lead an improvement programme to reduce this significant business risk. Describe the approaches and methodologies you would adopt and list some of the tools and techniques that might be used in a systematic approach.

8 The Marketing Department of a large housing developer is reviewing its sales forecasting activities. Over the last three years the sales forecasts have been grossly inaccurate. As a result, a Process Improvement Team has been formed to look at this problem. Give an account of how you would advise that team in this situation and outline a programme of work for them to consider.

9 It has been suggested by Deming and Ishikawa that statistical techniques can be used by staff at all levels within an organization. Comment on this view and explain how such techniques could help the following people in a construction business:

 a Senior managers to assess performance.
 b Design managers to demonstrate process capability to customers.
 c Process teams to achieve quality improvement on site.

10 'Lean thinking and systems' have been used widely in many sectors to bring about performance improvement. Prepare a presentation on 'Lean' for the senior management team of an organization of a construction organization of your choice, so that they may understand the concept and its building blocks. Recommend an appropriate systematic approach and toolkit. Make proposals on how they should go about implementing Lean in their business.

Part V

People

Human resource management

16

STRATEGIC ALIGNMENT OF HRM POLICIES

Technological advances and variation in demand for products and services worldwide since the turn of the century has created relative instability, cyclic hiring and downsizing in many organizations. However, during these times the way in which people are managed and developed at work has become recognized as one of the primary keys to improved and sustained organizational performance. This is reflected by popular idioms such as 'people are our most important asset' or 'people make the difference'. Indeed, such axioms now appear in the media and on corporate public relations documents with such regularity that the accuracy and integrity of such assertions has begun to be questioned. This chapter draws on some of the research undertaken by the European Centre for Business Excellence (now The Oakland Institute for Business Research and Education, www.theoakland institute.com). The research focused on world-class, successful and, in many cases, award-winning organizations. It describes the main people management activities that are currently being used in these leading edge organizations.

The cyclic variation of demand for construction sector products and services is perhaps among the more volatile among all industry sectors; investment in the property sector varies widely. It is at the whim of investors; share market buoyancy, taxation policy and government interest rate policy are just three of its main drivers. This has led to extreme highs and lows in demand, the variation between peaks and troughs at times exceeding 100 per cent. This naturally places a consequent strain on resources at the peaks, forcing many in the sector to look for work elsewhere during the slumps.

There is an overwhelming amount of evidence that successful organizations pay much more than lip service to the claim that people are their most important resource. This is consistent with the recognition in the past decade that intellectual capital reflects a significant part of any company's value and that knowledge management (Chapter 18) is a key strategic activity, especially if the tacit knowledge within an organization is to be properly leveraged for the benefit of the company.

On a general level, successful organizations share a fundamental philosophy to value and invest in their employees. More specifically, world-class organizations value and invest in their people through the following activities:

- strategic alignment of human resource management (HRM) policies;
- effective communication;
- employee empowerment and involvement;
- motivation through recognition of excellence;
- training and development;
- teams and teamwork;

- review and continuous improvement; and
- fostering of social cohesion.

All the case study companies stand out for their commitment to their employees. Many have been recognized as the most preferred employers locally while DPR and Graniterock have been recognized as being among the ten most preferred employers nationally in the USA by Fortune 500. Examine the case studies carefully and gain some insight into the way in which people are central to the development strategy of excellent companies. It is clear that leading edge organizations adopt a common approach or plan (illustrated in Figure 16.1) to align their HR policies to the overall business strategy.

Key elements of the HR strategy (e.g. skills, recruitment and selection, health and safety, appraisal, employee benefits, remuneration, training, etc.) are first identified usually by the HR Director, who then reports regularly to the Board. The HR plan, typically spanning three years, is aligned with the overall business objectives and is an integral part of company strategy. For example, if a business objective is to expand at a particular site, then the HR plan provides the necessary additional manpower with the appropriate skills profile and training support. The HR plan is revised as part of the overall strategic planning process. Divisional boards then liaise with the HR Director to ensure that the HR plan supports and is aligned with overall policy.

In addition, the HR Director holds regular meetings with key personnel from employee relations, health and safety, training and recruitment, etc. to review and monitor the HR plan, drawing upon published data and benchmarking activities in all relevant areas of policy and practice. Divisional managing directors and the HR Director report progress on how the HR plan is supporting the business to the quality committee or Board. An overview of this Human Resource process is illustrated in Figure 16.2.

Although it is beyond the scope of this chapter to make a detailed examination of HR policy, it is prudent to outline briefly some of the common practices that emerged from the identified best practices relating to selection and recruitment, the development of skills and competencies, employee appraisal and reward, recognition and benefits.

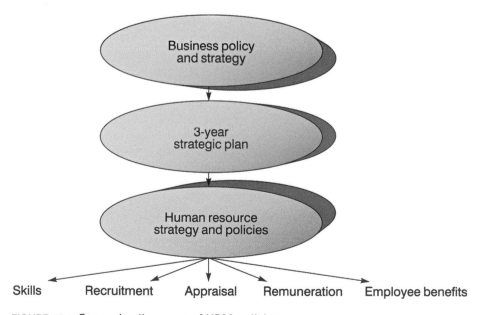

FIGURE 16.1 Strategic alignment of HRM policies

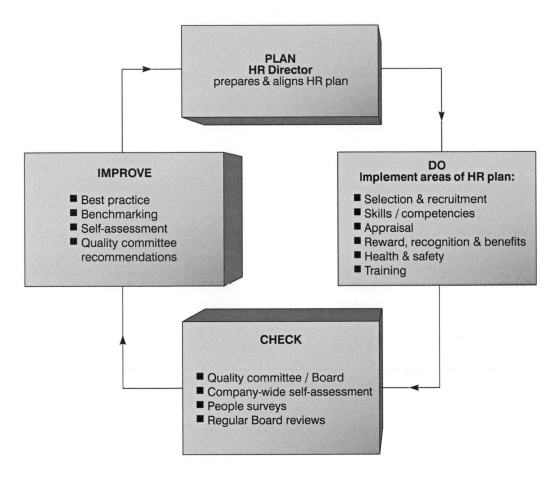

FIGURE 16.2 Human resources process

Selection and recruitment

The following practices are common amongst the organizations studied regarding selection and recruitment:

1 Ensure fairness by using standard tools and practices for job descriptions and job evaluations.
2 Enhance 'transparency' and communication through jargon-free booklets that provide detailed information to new recruits about performance, appraisal, job conditions and so on.
3 Ensure that job descriptions are responsibility- rather than task-oriented.
4 Train all managers and supervisors in interviewing and other selection techniques.
5 Align job descriptions and competencies so that people with the appropriate skills and attributes for the job are identified.
6 Compare the organization's employment terms and conditions (on a regular basis) with published data on best practice and documents to ensure the highest standards are being met.
7 Review HR policies regularly to ensure that they fully reflect legislative and regulatory changes together with known best practice.

8   In the recruitment of new graduates, start early; some leading companies in the construction sector recruit cadets at the end of second-year university, thus giving the company and the cadet time to assess and get to know each other, while providing support to the cadets. The company, by getting in early, is able to pick the best talent available.

Skills/competencies

Since the publication of *The Competent Manager* (Boyatsis, 1982), the terms competence and competency have been widely used and underpin the work of the specific bodies in any country associated with vocational qualifications and occupational standards. In line with this, good organizations have skills/competence based human resource management policies underpinning selection and recruitment, training and development, promotion and appraisal.

Although numerous lists of generic management competencies have been published, in essence, they are all very similar and are closely allied to the core management competencies underpinning HR policies: leadership, motivation, people management skills, teamworking skills, comprehensive job knowledge, planning and organizational skills, customer focus, commercial and business awareness, effective communication skills – oral and written – and change management skills coupled with a drive for continuous improvement.

Appraisal process

As with other HR policies, the main thrust of the appraisal process is alignment – of personal, team and corporate goals coupled with appraisals to help individuals achieve their full potential (see Figure 16.3).

Without exception, the appraisal systems described in world-class organizations were based on objectives. Agreed objectives were also time-based so that completion dates provided the opportunity for automatic review processes. Typically, employees are appraised annually and the managers conducting appraisals attend training in appraisal skills. Before each appraisal, the appraisee and appraiser each complete preparation forms thus making the interview a two-way discussion on performance against objectives during,

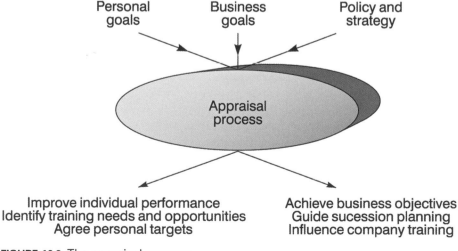

FIGURE 16.3 The appraisal process

say, the previous twelve months. Training and development work to achieve the objectives are agreed upon and, if necessary, additional help is available in the form of advice and counselling.

Employee reward, recognition and benefits

Reward and recognition

Although an in-depth study of the policies and practices relating to financial reward and recognition was beyond the scope of the research, it is possible to highlight the following activities that were common amongst the organizations:

- rewards are based on consistent, quality-based performance;
- awards are given to employees but also to customers, suppliers, universities, colleges, students, etc.;
- financial incentives are offered for company-wide suggestions and new idea schemes;
- internal promotion, for example, from non-supervisory roles to divisional managing directors encourage a highly motivated workforce and enhance job security;
- commendations include ad hoc recognition for length of service, outstanding contributions, etc.;
- recognition is given through performance feedback mechanisms, development opportunities, pay progressions and bonuses; and
- recognition systems operate at all levels of the organization but with particular emphasis on informal recognition, ranging from a personal 'thank you' to recognition at team meetings and events.

With regard to employee benefits, it is well documented that benefits are seen as a tangible expression of the psychological bond between employers and employees. However, to maximize effectiveness, benefits packages should be able to be selected on the basis of what is good for the employee as well as the employer. Moreover, when employees can design their own benefits package, both they and the company benefit.

Leading edge organizations favour a 'cafeteria' approach to employee benefits. In recent years, there has been increasing interest in this idea of cafeteria benefits to maximize flexibility and choice, particularly in the area of fringe benefits, and these can make up a high proportion of the total remuneration package. Under this scheme, the company provides a core package of benefits to all employees including a base salary and a 'menu' of other costed benefits (e.g. personal medical care, dental care, company car, health insurance, etc.) from which the employee can select their personal package.

Some of the ideas underpinning cafeteria benefits sit well with the literature on motivation – emphasizing that different people have different needs and expectations from work. Moreover, by setting out the benefits package and providing employees with benefit flexibility, the positive impact is further increased: not only are employees more likely to get the benefits they want, but they are also made aware of all the different benefits they can chose from, thus increasing morale.

EFFECTIVE COMMUNICATION

Effective communication emerges from the research as an essential facet of people management – be it communication of the organization's goals, vision, strategy and policies or the communication of facts, information and data. For business success, regular, two-way communication, particularly face-to-face with employees, is an important factor in establishing trust and a feeling of being valued. Two-way communication is regarded as both a core management competency and as a key management responsibility. For example, a typical list of management responsibilities for effective communication is to:

- regularly meet all their people;
- ensure people are briefed on key issues in language free of technical jargon;
- communicate honestly and as fully as possible on all issues which affect their people;
- encourage team members to discuss company issues and give upward feedback; and
- ensure issues from team members are fed back to senior managers and timely replies given.

Regular two-way communication also involves customers, shareholders, financial communities and the general public.

Communications process

Successful organizations follow a systematic process for ensuring effective communications as shown in Figure 16.4.

Plan

Typically, the HR Director is responsible for the communication process. He/she assesses the communication needs of the organization and liaises with divisional directors, managers

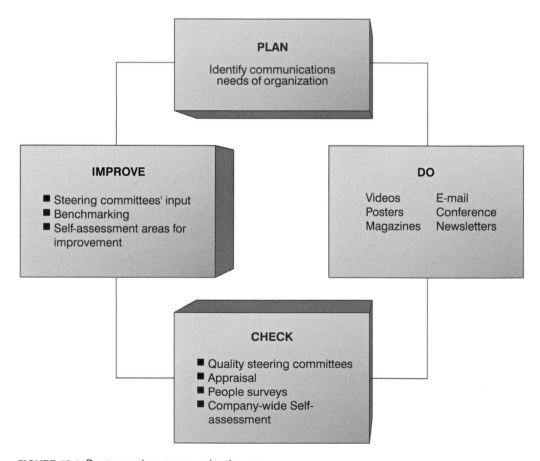

FIGURE 16.4 Best practice communications process

or local management teams to ensure that the communication plans are in alignment with overall policy and strategy. A communication programme accompanies any major changes in organization policy or objectives.

Do

A comprehensive mix of diverse media is used to support effective communication throughout any organization. These include:

Videos	Posters	Open-door policies
Surveys	Campaigns	Email
Magazines	Briefings	Notice boards
Newsletters	Conferences	Internet/Intranet
Appraisals	Meetings	Focus groups

It is evident that the introduction of electronic systems has brought about radical changes in communications. Typically, employees are able to access databases, spreadsheets, word processing, email and diary facilities. Information on business performance, market intelligence and quality issues can also be easily and quickly cascaded. Further, video conferencing is used to facilitate internal face-to-face communications with major customers across the world, resulting in substantial savings on travel and associated costs. Provision is made for depots, units, regions, divisions, departments, etc. to hold 'virtual' meetings and conferences. Feedback questionnaires then check that events are valuable and help the planning of future events.

Check

Quality steering or review committees, people surveys, appraisal and company-wide self-assessment are used to review the effectiveness of the communications process. Appraisal and staff survey data are analysed to ensure that the communications process is continuing to deliver effective upward, downward and lateral communications. Reports are then made on a quarterly, six-monthly and/or annually basis to the chief executive and/or the most senior team on the effectiveness and relevance of the communications process. The people survey data are also used to ascertain employee perceptions and to keep in touch with current opinion.

Improve

The results of the various review processes highlight areas for improvement, and results are verified by benchmarking against, for instance, a national survey. Quality steering committees then put forward recommendations for future planning and continuous improvements.

Communications structure

Successful organizations place great emphasis on communication channels that enable people at all levels in the organization to feel able to talk to each other. Consequently, managers are not only trained but 'are committed to being open-minded, honest, more visible and approachable'. Many formal and informal communication mechanisms exist, all designed to foster an environment of open dialogue, shared knowledge, information and trust in an effective upward, downward, lateral and cross-functional structure such as the one illustrated in Figure 16.5.

(See also Chapter 18 for more detail on the communication process.)

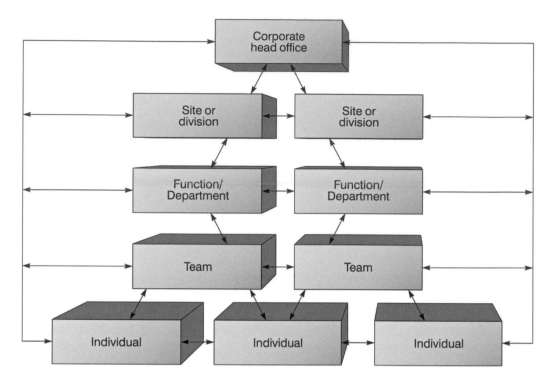

FIGURE 16.5 Multi-directional communications structure

EMPLOYEE EMPOWERMENT AND INVOLVEMENT

To encourage employee commitment and involvement, successful organizations place great importance on empowering their employees. The positive effects of employee empowerment are well documented but the notion has been challenged, with some writers claiming that it is not possible to empower people; rather, it is possible only to create a climate and a structure in which people will take responsibility. Nonetheless, it is clear that the organizations studied in the research considered empowerment to be a key issue and made efforts to create a working environment that was conducive to the employees taking responsibility.

Employee encouragement and empowerment to identify waste, to identify and solve problems and to develop innovations are key themes running through many of the case studies. For example, in the JB Henderson case study, management attributes most of the business improvements since the company adopted lean quality to its workers. Management says, 'We created the environment; they have come up with the ideas'. Similarly, many of the case study contractors encourage their workers to produce two-minute videos to describe their innovations. These are posted on the company intranet to share the lessons learned across the business.

Many companies with impressive customer satisfaction scores subscribe to the importance of employee empowerment by encouraging employees to:

1   set their own goals;
2   judge their own performance;

3 take ownership of their actions; and
4 identify with the company by becoming stock/shareholders.

Graniterock's HR policy is to work with employees to identify and plan their long-term career aspirations through an *individual person's development plan* (IPDP) and to assist them to identify the training needs that will help them achieve their individual goals which are consistent with the needs of the company.

TNT report that 'all employees are empowered to respond to normal and extraordinary situations without further recourse' and that they have 'worked hard to create a no blame culture, where our people are empowered to take decisions to achieve their objectives'. Along the same lines, Hewlett-Packard advocate teamwork and high levels of empowerment combined with a strong setting of objectives and freedom for employees to achieve them. To address the above issues, management map out processes to provide employees with the necessary authority and skills. In addressing the issue of not caring, employee surveys reveal that appraisal systems can be a major roadblock and the appraisal process may need to be revised.

Common initiatives

There are three widely accepted initiatives which successful organizations employ:

- *Corporate employee suggestion schemes* – these provide a formalized mechanism for promoting participative management, empowerment and employee involvement.
- *Company-wide culture change programmes* – in the form of workshops, ceremonies and events used to raise awareness and to empower individuals and teams to practise continuous improvement.
- *Measurement of Key Performance Indicators* – whereby the effectiveness of staff involvement and empowerment is measured by improvements in Human Resource Key Performance Indicators (KPIs) such as labour turnover, accident rate, absenteeism and lost time through accidents. Typically, KPI measurements, coupled with appraisal feedback and survey results, are regularly reviewed by the HR Director, and he/she uses the information as the basis for reports and suggestions for improvements to the Board.

On a more general level, successful organizations increase commitment by empowering and involving more and more of their employees in formulating plans that shape the business vision. As more people understand the business and where it is planned to go, the more they will become involved in and committed to developing the organization's goals and objectives.

The Graniterock case study demonstrates this kind of employee involvement in process improvement. Many quality improvement teams are at work each year improving processes and solving problems.

Motivation through recognition of excellence

Publicly recognizing excellent contributions by individuals and teams is an essential part of a HR programme – it is key to motivating people and engendering an ongoing commitment to continuous improvement. This, however, requires that processes be put in place to identify and recognize excellence and that the forums and media for recognizing outstanding achievements be planned rather than ad hoc.

Line managers have the responsibility for mentoring those immediately under them: setting training and development goals and reviewing performance. This also places them in the ideal position to identify outstanding achievements. However, guidance has to be provided by senior HR management, and the process of identification has to be regular and ongoing.

For example, case study company Graniterock and client Crossrail both have an annual recognition day to recognize outstanding performance. Senior executives attend these presentations and participate in award ceremonies.

Typically on this day, people's contribution to the company and their life achievements are publicly recognized. Naturally, recognition is also afforded through the company's regular newsletter; this type of recognition can also be given through a company intranet. The most important aspects of a recognition day are that senior management is involved and that the process is planned and held regularly.

TRAINING AND DEVELOPMENT

The training and development of people at work have increasingly come to be recognized as an important part of successful management. Major changes in an organization, the introduction of new technology and the widening of range of tasks all require training provision. The approaches and systems set down in ISO 9000, benchmarking and self-assessment against frameworks such as the EFQM Excellence Model have further high-lighted the need for properly trained employees.

It is widely acknowledged that many writers and practising managers sing the praises of training, saying it is a symbol of the employers' commitment to staff or that it shows an organization's strategy is based on adding value rather than lowering costs. However, others claim that a lack of effective training can still be found in many organizations today and that serious doubts remain as to whether or not management actually *does* invest in the training of their human resources.

It is perhaps not surprising then that research on successful and award-winning organizations revealed an ongoing commitment to investing in the provision of planned, relevant and appropriate training. In such organizations training was found to be care-fully planned through training needs analysis processes that linked the training needs with those of the organization, groups, departments, divisions and individuals. To maintain training relevancy and currency, databases of training courses are widely available and, to encourage diversification, employees are able to realize their full potential by training in quality, job skills, general education, health and safety, and so on through exams, qualifications, assessor training, etc. Typically, training strategies in these organizations require managers to:

- play an active role in training delivery (cascade training) and support (including quality tools and techniques);
- receive training and development based on personal development plans;
- fund training and improvement activities to allow autonomy at 'local' levels for short payback investments; and
- coordinate discussions and peer assessments to develop tailored training plans for individuals.

As a result of their investments, these companies boast business benefits such as:

- increases in sales volumes;
- not losing customers to competitors; and
- low employee turnover.

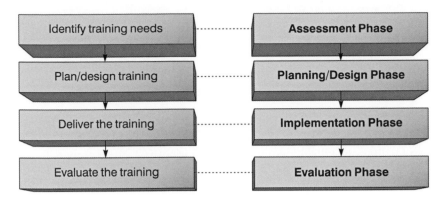

FIGURE 16.6 A systematic model of training

What is particularly noteworthy about the training activities identified is that they are almost identical to those processes and activities commonly found in the management literature on the theory of training. Many writers have developed models of the training process which can be summarized into the four phases shown in Figure 16.6.

The *Assessment Phase* identifies what is needed (the content of the training) at the organizational, group, and individual levels. This may involve some overlap, for example, an individual's poor sales performance may be a symptom of production problems at the group level as well as a need for more product innovation at the organizational level. The Assessment Phase thus involves identifying training needs by assessing the gaps between future requirements of a job and the current skills, knowledge or attitudes of the person in the job. So the organization looks at what is presently happening and what should or could be happening. Any differences between the two will give some indications of training needs.

The *Planning/Design Phase* identifies where and when the training will take place and involves such questions as:

- who needs to be trained;
- what competencies are required;
- how long will training take;
- what are the expected benefits of training;
- who/how many will undertake the training; and
- what resources are needed – money, equipment, accommodation, etc.?

Typically, the organizations involved in the studies review their training programmes annually and make plans according to the needs of the business, circulating lists of available courses well in advance of the training dates. The information ensures that all managers are aware of what is provided so that they are able to schedule attendance by staff. The strategic training plan is supported by an annual budgeting and planning system with quarterly meetings to monitor and review performance. The budgeting and planning process with its integral HR element is cascaded throughout the organization to teams at all locations.

The *Implementation Phase* involves the actual delivery of the training. This might be on site or away from the premises and will include training techniques: simulators, business games, case studies, coaching and mentoring, planned experience and computer assisted instruction. Demonstration or 'sitting next to Nellie' is another commonly used training technique.

Induction and devolved training form an integral part of the training implementation phase. New employees attend induction courses and are issued with personal development documents giving details of what training and assessment will take place in the first few months of employment as well as copies of the vision and mission statements. At regular intervals throughout the induction period (say, every three or four weeks) new employees are then reviewed to identify training and development needs for the remainder of the year.

In addition, much of the training is devolved to line managers through facilitation and facilitator packs. This requires the training and development of all levels of managers and supervisors in facilitation skills. Line managers then identify team members to be trained as facilitators. Adopting this approach is said to create an environment in which everyone is aware of training and development issues for themselves and their colleagues.

The *Evaluation Phase* is widely acknowledged as one of the most critical steps in the training process and can take many forms: observation, questionnaires, interviews, etc. For example, in this phase the overall effectiveness of training is evaluated and this provides feedback for the trainers, future improvements to the programme, senior managers and the trainees themselves. Providing trainees with a set of training objectives will help them know what they need to learn and give them feedback on their progress. This will then influence their attitudes towards future training and even the company itself.

In sum, it seems that successful organizations approach training and development in a planned and systematic way, involving training needs analysis, assessment of training content, carefully planned implementation and continuous evaluation and review – a convincing argument for the value of theory when it is put into practice.

A challenge in the construction environment is posed by the day-to-day pressure of work on construction sites. So, while the head office HR department is responsible for developing the training opportunities, work pressures on the site can restrict the ability of staff to attend. One of the authors had a personal experience of this when he was involved in providing training to the staff of a leading construction contractor: it was common for only half of the registrants to arrive at training on any day.

TEAMS AND TEAMWORK

It is clear that leading edge organizations place great emphasis on the value of people working together in teams. This is hardly surprising as a great deal of theory and research indicates that people are motivated and work better when they are part of a team. Teams can also achieve more through integrated efforts and problem solving.

Teams are a management tool and are most effective when team activity is clearly linked to organizational strategy. For this, the strategy must be communicated to influence team direction, which then links to the production of team mission statements and the use of team agendas and scorecards. Importantly though, many people emphasize the value of cross-functional teams, which have proven to be a common feature in many of the organizations studied. Here, teams which have originally evolved out of the old functional departments or units within an organization gain experience and benefit from team building and become cross-functional. Each team is required to identify its customers, the customer requirements and what measures need to be used to ensure that those requirements are being satisfied.

The fostering of social cohesion

We live at a time of high employee mobility and low levels of corporate loyalty. It is said that modern employees are increasingly self-interested and that they develop their competencies and CVs so that they can attract the highest remuneration in the market. While there is evidence to support that many employees do have such a world-view, there is a counter cyclical trend among employees in world-class companies. These people identify with their work organizations and enjoy being with their colleagues. They do not seek to move at the first opportunity because they enjoy where they are; they receive recognition and they are given challenges.

All the factors in this section combine to form a highly satisfying work environment; however, social cohesion is of equal importance to those already discussed. World-class companies take the same planned and structured approach to social activity as they do to everything else they do (see also Chapter 17 for more detail on teams and teamwork).

Many of the case study companies have active engagement programmes with the communities within which they operate. They get involved in community projects and raise money for local charities. One of the case study companies, Graniterock, assesses the success of its social programme by the level of employee participation, seeing its organization as having a social role that extends beyond the company; hence the company directly supports its employees' efforts in the community.

ORGANIZING PEOPLE FOR LEAN QUALITY

In some organizations, management systems are still viewed in terms of the internal dynamics between marketing, design, sales, production/operations, distribution, accounting, etc. A change is required from this to a larger process-based system that encompasses and integrates the business interests of customers and suppliers. Management needs to develop an in-depth understanding of these relationships and how they may be used to cement the partnership concept. A lean quality function can be the organization's focal point in this respect, and should be equipped to gauge internal and external customers' expectations and degree of satisfaction. It should also identify deficiencies in all business functions, and promote improvements.

The role of the lean quality function is to make lean quality an inseparable aspect of every employee's performance and responsibility. The transition in many companies from quality departments with line functions will require careful planning, direction and monitoring. Quality professionals have developed numerous techniques and skills focused on product or service quality. In many cases there is a need to adapt these to broader, process improvement applications. The first objectives for many traditional 'quality managers' will be to gradually disengage themselves from line activities, which will then need to be dispersed throughout the appropriate operating departments. This should allow lean quality to be understood as a 'process' at a senior level and to be concerned with the following throughout the organization:

- encouraging and facilitating improvement;
- monitoring and evaluating the progress of improvement;
- promoting the 'partnership' in relationships with customers and suppliers;
- designing, planning, managing, auditing and reviewing lean quality management systems;
- planning and providing training and counselling, or consultancy;
- giving advice to management on:
  - establishment of process management and control;

- relevant statutory/legislation requirements with respect to quality;
- quality and process improvement programmes; and
- inclusion of lean quality elements in all processes, job instructions and procedures.

Lean quality directors and managers may however have an initial task to help those who control the means to implement this concept – the leaders of industry and commerce – to really believe that lean quality must become an integral part of all of the organization's operations.

The authors have a vision of lean quality as a strategic business management function that will help organizations to change their cultures. To make this vision a reality, existing quality professionals must expand the application of lean quality concepts and techniques to all business processes and functions and develop new forms of providing assurance of quality at every supplier–customer interface. They will need to know the entire cycle of products or services from concept to the *ultimate* end user. An example of this was observed in the case of a company manufacturing pharmaceutical seals, whose customer expressed concern about excess aluminium projecting below and round a particular type of seal. This was considered a cosmetic defect by the immediate customer, the Health Service, but a safety hazard by a blind patient – the *customer's customer*. The prevention of this 'curling' of excess metal meant changing practices at the mill that rolled the aluminium – the *supplier's supplier*. Clearly, the lean quality professional dealing with this problem needed to understand the supplier's processes and the ultimate customer's needs in order to judge whether the product was indeed capable of meeting the requirements.

The shifts in 'philosophy' will require considerable staff education in many organizations. Not only must people in other functions acquire quality management skills, but quality personnel must change old attitudes and acquire new skills: replacing the inspection, calibration, specification-writing mentality with knowledge of defect prevention, waste elimination, process improvement as well as wide-ranging lean quality management systems design and audit. Clearly the challenge for many quality professionals is not so much making changes in their organization as recognizing the changes required in themselves. It is more than an overnight job to change the attitudes of an inspection police force into those of a consultative, team-oriented improvement resource. This emphasis on prevention and improvement-based systems elevates the role of lean quality professionals from a technical one to that of general management. A narrow departmental view of quality is totally out of place in an organization aspiring to lean quality, and many existing quality directors and managers will need to widen their perspective and increase their knowledge to encompass all facets of the organization.

To introduce the concepts of process management required for lean quality will require not only a determination to implement change but sensitivity and skills in interpersonal relations. This will depend very much of course on the climate within the organization. Those whose management is truly concerned with co-operation and concern for the people will engage strong employee support for the lean quality manager or director in his/her catalytic role in the improvement process. Those with aggressive, confrontational management will create for the quality professional impossible difficulties in obtaining support from the rank and file.

Quality appointments

Perhaps the most interesting initial findings from a recent research project by the authors and their colleagues on 'Quality in 21st Century' are to do with the perceived requirement for **'Lean Quality Leaders'**. The study found that:

- they will need first and foremost to be **business people** with multifunctional experience who understand and speak the language of business and can relate to those running the business; and
- the lack of that type of person in senior lean quality appointments is the root cause of complaints about top management from most quality people – it is because they are **not the right people**!

Many organizations have realized the importance of the contribution a senior, qualified director of lean quality, senior vice-president of quality and operational excellence, chief quality (and business improvement) officer, or other similar titles, can make to the prevention strategy. Smaller organizations may well feel that the cost of employing a full-time lean quality manager is not justified other than in certain very high-risk areas. In these cases, a member of the management team may be appointed to operate on a part-time basis, performing the lean quality management function in addition to his/her other duties. To obtain the best results from a lean quality director/manager, he/she should be given sufficient authority to take necessary action to secure the implementation of the organization's lean quality policy, and must have the personality to be able to communicate the message to employees at all levels: staff, management and directors. Occasionally the lean quality director/manager may require some guidance and help on specific technical quality matters. However, the major attribute they will require is the attitude to realize that they need to acquire the necessary information and assistance.

In large organizations, it may be necessary to make several specific appointments, or to assign detailed responsibilities to certain managers. The following actions may be deemed to be necessary.

Assign a lean quality director, manager or coordinator

This person will be responsible for the planning and implementation of lean quality. He or she will be chosen first for process, project and people management abilities rather than detailed knowledge of quality assurance matters. Depending on the size and complexity of the organization and its previous activities in quality management, the position may be either full- or part-time but the appointed person must report directly to the Chief Executive.

Appoint a lean quality manager adviser

A professional expert on lean quality management may be required to advise on the 'technical' aspects of planning and implementing lean quality. This is a consultancy role and may be provided from within or without the organization, full- or part-time. This person needs to be a persuader, philosopher, teacher, adviser, facilitator, reporter and motivator. He or she must clearly understand the organization, its processes and interfaces, be conversant with the key functional languages used in the business and be comfortable operating at many organizational levels. On a more general level this person must fully understand and be an effective advocate and teacher of lean quality, be flexible and become an efficient agent of change.

Steering committees and teams

Devising and implementing lean quality management in an organization takes considerable time and ability. It must be given the status of a senior executive project. The creation of cost-effective performance improvement is difficult because of the need for full integration with the organization's strategy, operating philosophy and management

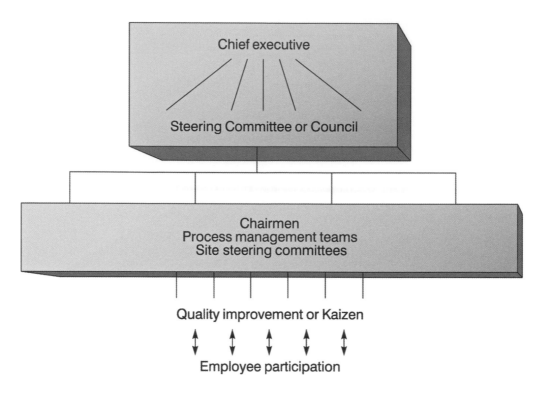

FIGURE 16.7 Employee participation through the team structure

systems. It may require an extensive review and substantial revision of existing systems of management and ways of operating. Fundamental questions may have to be asked, such as 'do the managers have the necessary authority, capability and time to carry this through?'

Any review of existing management and operating systems will inevitably 'open many cans of worms' and uncover problems that have been successfully buried and smoothed over, perhaps for years. Therefore, authority must be given to those charged with lean quality to follow through with actions that they consider necessary to achieve the goals. The commitment will continually be questioned and will be weakened, perhaps destroyed, by failure to delegate authoritatively or insufficient visible support from senior management.

The following steps are suggested in general terms. Clearly, different types of organization will need to make adjustments to the detail, but the component parts are the basic requirements.

A disciplined and systematic approach to continuous improvement may be established in a lean quality or Business Excellence 'Steering Committee or Council' (Figure 16.7). The Committee/Council should meet at least monthly to review strategy, implementation progress and improvement. It should be chaired by the Chief Executive, who must attend every meeting – only death or serious illness should prevent him/her being there. Clearly, postponement may be necessary occasionally, but the council should not continue to meet without the Chief Executive being present. The council members should include the top management team and the chairmen of any 'site' steering committees or process management teams, depending on the size of the organization.

The objectives of the council are to:

- provide strategic direction on lean quality for the organization;
- establish plans for lean quality on each 'site';
- set up and review the process teams that will own the key or critical business processes; and
- review and revise lean quality plans for implementation.

The process management teams and any site steering committees should also meet monthly, shortly before the senior steering committee/council meetings. Every senior manager should be a member of at least one such team. This system provides the 'top-down' support for employee participation in process management and development. It also ensures that the commitment to quality at the top is communicated effectively through the organization.

The three-tier approach of steering committee, process management teams and lean quality improvement teams allows the steering committee to concentrate on lean quality strategy, rather than become a senior problem-solving group. Progress is assured if the team chairmen are required to present a status report at each meeting.

The process management teams or steering committees control all the lean quality improvement teams and have responsibility for:

- the selection of projects for the teams;
- providing an outline and scope for each project to give to the teams;
- the appointment of team members and leaders; and
- monitoring and reviewing the progress and results from each team project.

As the focus of this work will be the selection of projects, some attention will need to be given to the sources of nominations. Projects may be suggested by:

- steering committee/council members representing their own departments, process management teams, their suppliers or their customers, internal and external;
- lean quality improvement teams;
- Kaizen teams or lean quality circles (if in existence);
- suppliers; and
- customers.

The process team members must be given the responsibility and authority to represent their part of the organization in the process. The members must also feel that they represent the team to the rest of the organization. In this way the team will gain knowledge and respect and be seen to have the authority to act in the best interests of the organization, with respect to their process.

QUALITY CIRCLES OR KAIZEN TEAMS

No book on lean quality would be complete without a mention of Kaizen and quality circles.

Kaizen is a philosophy of continuous improvement of all the processes in an organization so that the employees can perform their tasks a little better each day. It is a never-ending journey centred on the concept of starting anew each day with the principle that methods can always be improved.

Kaizen Teian is a Japanese system for generating and implementing employee ideas. Japanese suggestion schemes have helped companies to improve quality and productivity, and reduced prices to increase market share. They concentrate on participation and the rates of implementation rather than on the 'quality' or value of the suggestion. The emphasis is on encouraging everyone to make improvements quickly, there and then.

Kaizen Teian improvements are usually small-scale, quick 'hot' solutions in the worker's own area and are easy and cheap to implement. Key points are that the objectives are clear and implementation is rapid, resulting in many small improvements that can accumulate to massive total savings and improvements.

One of the most publicized aspects of the Japanese approach to lean quality has been these quality circles or Kaizen teams. The quality circle may be defined then as a group of workers who do similar work and meet:

- voluntarily;
- regularly;
- in normal working time; and
- under the leadership of their 'supervisor' to identify, analyse and solve work-related problems and recommend solutions to management.

Where possible, quality circle members should implement the solutions themselves.

The quality circle concept first originated in Japan in the early 1960s, following a post-war reconstruction period during which the Japanese placed a great deal of emphasis on improving and perfecting their quality control techniques. As a direct result of work carried out to train foremen during that period, the first quality circles were conceived, and the first three circles registered with the Japanese Union of Scientists and Engineers (JUSE) in 1962. Since that time, the growth rate has been phenomenal. The concept has spread to Taiwan, the USA and Europe. Although circles in many countries have been successful, there have been some failures.

In the early days, it was very easy to regard quality circles as the magic ointment to be rubbed on the affected spot. Unfortunately, many managers in the West at first saw them as a panacea for all ills; there are no panaceas. To place this concept into perspective, Juran, who was an important influence in Japan's improvement in quality, stated that quality circles represented only 5–10 per cent of the canvas of the Japanese success. The rest is concerned with understanding lean quality, its related costs together with the organization, and systems and techniques necessary for achieving customer satisfaction.

Given the right sort of commitment by top management, the correct introduction and environment in which to operate, quality circles can produce the 'shop floor' motivation to achieve quality performance at that level. Circles should develop out of *an understanding and knowledge of quality on the part of senior management*. They must not be introduced as a desperate attempt to do something about poor quality. There are a number of different names for quality circles, but regardless of the names, the basic concepts and operational aspects are to be found in many organizations.

The structure of a quality circle organization

The unique feature about quality circles or Kaizen teams is that people are asked to join and not told to do so. Consequently, it is difficult to be specific about the structure of such a concept. It is, however, possible to identify four elements in a circle organization:

- members;
- leaders;
- facilitators or coordinators; and
- management.

Members form the prime element of the concept. They will have been taught the basic problem-solving and process control technique and, hence, will possess the ability to identify and solve work-related problems.

Leaders are usually the immediate supervisors or foremen of the members. They will have been trained to lead a circle and bear the responsibility of its success. A good leader, one who develops the abilities of the circle members, will benefit directly by receiving valuable assistance in tackling nagging problems.

Facilitators are the managers of the quality circle programmes. They, more than anyone else, will be responsible for the success of the concept, particularly within an organization. The facilitators must coordinate the meetings, training and energies of the leaders and members and form the link between the circles and the rest of the organization. Ideally the facilitator will be an innovative industrial teacher capable of communicating with all levels and with all departments within the organization.

Management support and commitment are necessary to quality circles or, like any other concept, they will not succeed. Management must retain its prerogatives, particularly regarding acceptance or non-acceptance of recommendations from circles; however, the quickest way to kill a programme is to ignore a proposal arising from it. One of the most difficult facts for management to accept, and yet one forming the cornerstone of the quality circle philosophy, is that the real 'experts' on performing a task are those who do it day after day.

Training Kaizen teams or quality circles

The training of quality circle/Kaizen leaders and members is the foundation of all successful programmes. The whole basis of the training operation is that the ideas must be easy to adopt and express in a way that facilitates understanding. Simplicity must be the key word, with emphasis being given to the basic techniques. Essentially there are eight segments of training:

- introduction to quality circles;
- brainstorming;
- data gathering and histograms;
- cause and effect analysis;
- Pareto analysis;
- sampling and stratification;
- control charts; and
- presentation techniques.

Managers should also be exposed to some training in the part they are required to play in the quality circle philosophy. A quality circle programme can only be effective if management believes in it and supports it. Since changes in management style may be necessary, managers' training is essential.

Operation of quality circles/Kaizen teams

There are no formal rules governing the size of a quality circle/Kaizen team. Membership usually varies from three to fifteen people, with an average of seven to eight. It is worth remembering that as the circle becomes larger than this it becomes increasingly difficult for all members of the circle to participate.

Meetings can be held in the work area or away from it so that members are free from interruptions and are mentally and physically at ease. If away from the work space, the room should be arranged in a manner conducive to open discussion and any situation that physically emphasizes the leader's position should be avoided. To a large extent, the nature of the problems selected will determine the nature of the meetings, the interval between them and the venue.

Great care is needed to ensure that every meeting is productive, no matter how long it lasts or how frequently it is held. Any of the following activities may take place during a circle meeting:

- training – initial or refresher;
- problem identification;
- problem analysis;
- preparation and recommendation for problem solution;
- management presentations; and
- quality circle/Kaizen team administration.

It is sometimes necessary for quality circles to contact experts in a particular field: engineers, quality experts, safety officers, maintenance personnel. This communication should be strongly encouraged, and the normal company channels should be used to invite specialists to attend meetings and offer advice. The experts may be considered to be 'consultants', allowing the quality circle/Kaizen team to retain responsibility for improving a process or solving the particular problem. The overriding purpose of quality circles or Kaizen teams is to establish the powerful motivator of allowing people to take some part in deciding their own actions and futures.

Kaizen Blitz events

Rapid benefits realization and effective employee engagement are clear features of the so-called 'Kaizen Blitz' approach. Kaizen means continuous improvement and Blitz means lightning fast. A Kaizen Blitz is an intense, accelerated, team improvement event which may take place over a number of days. Such an event will typically focus on a problem area where significant improvement is needed and the event may use a combination of process analysis and improvement tools such as Value Stream Mapping, 5S and so on. An effectively run Kaizen Blitz event can produce tangible benefits in a relatively short time frame. Events should be hands-on and have an analysis/problem-solving focus, using the insights and experience of cross-functional teams. The aim is to engage a team in the improvement process, to break it out of day-to-day, low morale fire-fighting so they actually see benefits being realized through their ideas.

REVIEW, CONTINUOUS IMPROVEMENT AND CONCLUSIONS

In organizations that achieve outstanding performance and deliver real improvement, processes for reviewing performance and continuous improvement exist at the individual, team, departmental/divisional and organizational levels. These include such processes as:

- annual staff surveys and subsequent actions which are viewed as the cornerstones of continuous improvement (the people surveys are also critically reviewed against data from other world-class organizations and benchmarks to determine best practice and feed into the continuous improvement processes);
- lean quality committees, the HR department and cross-functional teams drawn from sites, depots, regions, units and divisions review feedback from surveys as well as the format of the surveys;
- ongoing performance feedback and development through on the job coaching plus regular one-to-one individual and team reviews; and
- collecting feedback from internal and external customers to provide second- and third-party insight into performance and opportunities for performance improvement at the team level.

This chapter has highlighted the main people management activities that are currently being used in some world-class organizations. A general conclusion from the research supporting this is that successful organizations pay much more than lip service to the popular idiom 'people are our most important asset'. Indeed, successful organizations value and invest in their people in a never-ending quest for effective management and development of their employees. This involves rigorous planning of processes, skilful implementation, regular review of processes and continuous improvement practices.

From a theoretical viewpoint, these findings about people management activities in successful organizations are hardly surprising, since the management literature abounds with examples of the benefits of systematic planning, followed by strategic implementation, regular review and continuous improvement. Nonetheless, from a practical viewpoint, the real value of the findings is that they flesh out in some detail those people management activities that are being used to good effect in world-class organizations which are reaping the benefits of putting theory into practice.

ACKNOWLEDGEMENT

The authors are grateful to Dr Susan Oakland for the contribution she made to this chapter.

BIBLIOGRAPHY

Blanchard, K. and Herrsey, P. *Management of Organizational Behaviour: Utilizing Human Resources* (4th edn), Prentice Hall, Englewood Cliffs, NJ, 1982.

Boyatsis, R. *The Competent Manager: A Model for Effective Performance*, Wiley, New York, 1982.

Collins, J. and Porras, J. *Built to Last: Successful Habits of Visionary Companies*, Harper Collins, New York, NY, 1994.

Imai, M., *Kaizen*, McGraw-Hill, New York, 1986.

Imai, M., *Gemba Kaizen: A Common Sense, Low Cost Approach to Management*, Quality Press, Milwaukee, 1997.

Katzenbach, J.R. and Smith, D.K. *The Wisdom of Teams*, McGraw-Hill, Singapore, 1994.

Kotter, J.P. and Heskett, J.L. *Corporate Culture and Performance*, The Free Press, New York, 1992.

Larkin, T.J. and Larkin, S. *Communicating Change*, New York, McGraw-Hill, New York, 1994.

Maguire, S. Learning to change, *European Quality*, 2(6): 8, 1995.

Marchington, M. and Wilkinson, A. *Core Personnel and Development*, Institute of Personnel and Development, London, 1997.

Oakland, J.S. *Total Organizational Excellence*, Butterworth-Heinemann, Oxford, 1999.

Oakland, S. and Oakland, J.S. Current people management activities in world-class organizations, *Total Quality Management*, 12(6): 25–31, 2001.

CHAPTER HIGHLIGHTS

Strategic alignment of HRM policies

- In recent years the way people are managed has been recognized as a key to improving performance. Research on world-class, award-winning organizations has identified the main people management activities used in leading-edge organizations.
- World-class organizations value and invest in people through: strategic alignment of HRM policies, effective communications, employee empowerment and involvement, training and development, teams and teamwork, review and continuous improvement.
- Leading-edge organizations adopt a common approach to aligning HR policies with business strategy. Key elements of policy such as skills, recruitment and selection, training, health and safety, appraisal, employee benefits and remuneration are first

identified. The HR plan is then devised as part of the strategic planning process, following a plan, do, check, improve (PDCI) cycle.

- Review the case studies in relation to their HRM policies and practices, in particular how the Graniterock case study (CS5) details its HRM practices in some detail.

Effective communication

- Regular two-way communication, particularly face-to-face, is essential for success.
- Again the PDCI cycle provides a systematic process for ensuring effective communications, which uses benchmarking and self-assessment as part of the improvement effort.

Employee empowerment and involvement

- To encourage employee commitment and involvement, successful organizations place great importance on empowering employees. This can include people setting their own goals, judging their own performance, taking ownership of actions and identifying with the organization itself (perhaps as shareholders).
- Common initiatives include: employee suggestion schemes, culture change programmes and measurement of KPIs. Generally commitment is increased by involving more employees in planning and shaping the vision.
- Publicly recognizing excellent contributions by individuals and teams is an essential part of a HR programme.

Training and development

- Training and development has been highlighted by many initiatives as a critical success factor, although lack of effective training still predominates in many organizations.
- In successful organizations, training is planned through needs analysis, use of databases, training delivery at local levels and peer assessments for evaluation.

Teams and teamwork

- Leading-edge organizations place great value in people working in teams because this motivates and causes them to work better.
- Teams are most effective when their activities are clearly linked to the strategy, which in turn is communicated to influence direction. Cross-functional teams are particularly important to address end-to-end processes.
- World-class companies take the same planned and structured approach to social activity as they do to everything else they do.

Organizing people for lean quality

- The lean quality function should be the organization's focal point integrating the business interests of customers and suppliers into the internal dynamics of the organization.
- The role of the lean quality function is to encourage and facilitate lean quality and process improvement; monitor and evaluate progress; promote the lean quality chains; plan, manage, audit and review systems; plan and provide lean quality training, counselling and consultancy; and give advice to management.
- In larger organizations a lean quality director will contribute to the prevention strategy. Smaller organizations may appoint a member of the management team to this task on

a part-time basis. An external lean quality adviser is often beneficial, especially in the early stages of change.

- In devising and implementing lean quality for an organization, it may be useful to ask first if the managers have the authority, capability and time to carry it through.
- A disciplined and systematic approach to continuous improvement may be led by a steering committee/council whose members are the senior management team.
- Lean quality improvement or Kaizen teams and lean quality circles report to the local process management teams/site steering committees and they in turn report to the lean quality steering committee.

Quality circles and Kaizen teams

- Kaizen is a philosophy of small step, continuous improvement by all employees. In Kaizen teams, the suggestions and rewards are small but the implementation is rapid.
- A quality circle or Kaizen team is a group of people who do similar work meeting voluntarily, regularly, in normal working time, to identify, analyse and solve work-related problems under the leadership of their supervisor. They make recommendations to management. Alternative names may be given to the teams, other than 'quality circles'.
- Rapid benefits realization and effective employee engagement are clear features of the so-called 'Kaizen Blitz' approach which uses short duration events with multi-functional teams, focused on single problems or pieces of plant.

Review, continuous improvement and conclusions

- Effective organizations use processes for reviewing performance and continuous improvement at the individual, team, divisional/departmental and organizational levels. These include surveys of staff, committees/teams and ongoing performance feedback.

Culture change through teamwork

17

THE NEED FOR TEAMWORK

How often have you come out of a meeting frustrated, feeling it is a complete waste of time? People in the construction sector are extremely production-oriented, and this feeling of frustration only arises when meetings are poorly planned, poorly managed and at times poorly attended. Sadly, too many meetings fall into this category; yet the complexity of most of the processes that are operated in industry, commerce and the services places them beyond the control of any one individual. Furthermore, as any supply chain becomes more fragmented, the interface between organizations gives rise to many of the more significant problems and opportunities. This is typical of building and construction projects, and developing effective solutions requires the involvement of all the stakeholders in the problem area. The only really efficient way to tackle process management and improvement is through the use of some form of teamwork; this has many advantages over allowing individuals to work separately:

- A greater variety of complex processes and problems may be tackled – those beyond the capability of any one individual or even one department or organization – by the pooling of expertise and resources.
- Processes and problems are exposed to a greater diversity of knowledge, skill and experience, enabling more effective solutions to be developed more efficiently.
- The approach is more satisfying to team members, boosting morale and ownership through participation in process management, problem solving and decision-making.
- Processes and problems that cross departmental or functional boundaries can be dealt with more easily, and the potential/actual conflicts are more likely to be identified and solved.
- Team recommendations are more likely to be implemented than individual suggestions as the quality of decision-making in *good* teams is high.
- On construction projects, processes and problems that cross organizational boundaries can only be effectively dealt with by inter-organizational teams.

Most of these factors rely on the premise that people are willing to support any effort in which they have taken part or helped to develop.

When properly managed and developed, teams improve the process of problem solving, producing results quickly and economically. Teamwork throughout any organization is an essential component of the implementation of lean quality and process management: it builds trust, improves communications and develops interdependence. Much of what has been taught previously in management has led to a culture in the West of independence, with little sharing of ideas and information. Knowledge is very much

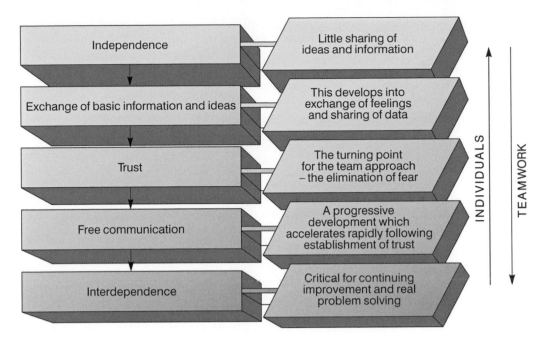

FIGURE 17.1 Independence to interdependence through teamwork

like organic manure – if it is spread around it will fertilize and encourage growth; if it is kept closed in it will eventually fester and rot.

Good teamwork changes independence into interdependence through improved communications, trust and the free exchange of ideas, knowledge, data and information (Figure 17.1). Interaction through face-to-face communication with a common goal develops a sense of interdependence over time. This forms a key part of any quality improvement process, and provides a method for employee recognition and participation through the active encouragement of group activities.

Teamwork provides an environment in which people can grow and use all the available resources effectively and efficiently to achieve continuous improvements. As individuals grow, organizations grow. It is worth pointing out, however, that employees will not be motivated towards continual improvement in the absence of:

- commitment from top management;
- the right organizational 'climate';
- a mechanism for enabling individual contributions to be effective; and
- recognition of individual contributions.

All these are focused essentially at enabling people to feel, accept and discharge responsibility. All world-class organizations have made this part of their strategy – to 'empower people to act'. If one hears from employees comments such as 'We know this is not the best way to do this job, but if that is the way management want us to do it, that is the way we will do it' then it is clear that the expertise existing at the point of operation has not been harnessed and the people do not feel responsible for the outcome of their actions. Responsibility and accountability foster pride, job satisfaction, and better work.

Empowerment to act is conceptually very easy to express but it requires real effort and commitment on the part of all managers and supervisors to put into practice. A good place to start is to recognize and applaud partially successful but good ideas and attempts

Culture change through teamwork   375

rather than criticizing them. Encouragement of ideas and suggestions from the workforce, particularly through their involvement in team or group activities, requires investment. The reward is total commitment, both inside the organization and outside through the supplier and customer chains.

Teamwork to support process management and improvement has several components. It is driven by a strategy, needs a structure and must be implemented thoughtfully and effectively. The strategy that drives the improvement comprises the:

- vision and mission of the organization;
- critical success factors; and
- core process framework.

These components have been dealt with in other chapters. The structural and implementation aspects of teamwork are the subject of the remainder of this chapter.

RUNNING PROCESS MANAGEMENT AND IMPROVEMENT TEAMS

The JB Henderson case study raises an interesting proposition. JBH as a part of their change strategy significantly increased the time their people spent in meetings at the work-crew level and yet they improved labour productivity by some 150 per cent in a year. They went from spending some 3 per cent of their day in meetings (a 15 minutes morning start-up) to 10 per cent of their day divided between a 40 minutes start-up meeting and a 15 minute wrap and review meeting each day. In their meetings, they plan work and review work outcomes; they study lessons from books; they share lessons learned within JBH; and they discuss opportunities for improvement.

Process management and improvement teams are groups of people, with the appropriate knowledge, skills and experience, who are brought together specifically by management to improve processes and/or tackle and solve particular problems, usually on a project basis. They are cross-functional and often multidisciplinary.

The 'task force' has long been a part of the culture of many organizations at the 'technical' and management levels. A task force tends to be used as a high-level review committee, and as a rule, task forces do not get into the detail of creating solutions, they stop at making recommendations. Process teams go a step further: they expand the traditional definition of 'process' to cover the entire end-to-end operating system. This includes technology, paperwork, communication, other units operating procedures and the process equipment itself. By taking this broader view, the teams can address new problems. However, the actual running of process teams calls the following factors into play:

- selection and leadership;
- objectives;
- meetings;
- assignments;
- dynamics; and
- results and reviews of the team.

Team selection and leadership

The most important element of a process team is its members. People with knowledge and experience relevant to the process or solving the problem are clearly required. However, there should be a limit of five to ten members to keep the team small enough to be manageable but still allow a good exchange of ideas. Membership should include appropriate people from groups outside the operational and technical areas that are directly 'responsible' for the process, if it is considered that they will add value. In the selection of team members, it is often useful to start with just one or two people who are concerned directly with the process. Once they need to draw maps or flowcharts (see Chapter 12) of the relevant processes, others who are needed to ensure the team fully understands the process should join the group. This approach will also ensure that all those who can make a significant contribution to the process and its improvement are represented. In construction settings, it will often be important to have representation from other companies in the supply chain – selecting the right individual is crucial.

The process owner has a primary responsibility for team leadership, management and maintenance; therefore, his/her selection and training is crucial to success. The leader need not be the highest ranking person in the team but must be concerned about accomplishing the team objectives (this is sometimes described as 'task concern') and the needs of the members (often termed 'people concern'). Weakness in either of these areas will lessen the effectiveness of the team in solving problems or making breakthroughs. The need for team leadership training is often overlooked; never assume that just because people have been elevated to supervisory or even project management roles, they necessarily know how to lead a team: many companies do not train in these basic skills. Skill development may be needed in areas such as facilitation, meeting management, negotiation and motivation. Needs should be identified, and training directed at correcting deficiencies in these crucial aspects.

A number of case study companies have recognized the need for this broader skill set among their supervisors, especially as they move from a command and control culture to one of collaboration. Construction supervisors increasingly need negotiation and meeting management skills. In addition to these normal meeting management skills, meeting leaders need to develop a performance tracking mentality. They need to learn to use visual dashboards to focus team attention on performance successes and weaknesses. They need to get into the habit of ensuring that clear commitments are sought and made, reservations are identified and dealt with, and performance tracking is effective in highlighting areas of poor performance.

The other feature of the case study companies that is noteworthy is that all have established active learning cultures which are evidenced through study action teams at the crew and leadership level. For example in the JB Henderson case study, this involves the field leadership group in an area meeting monthly to read, discuss and analyse process improvement books in order to learn from them. Key lessons are then fed down to the crew level.

Team objectives

At the beginning of any process improvement project, it is important that the objective should be clearly defined and agreed. This may be in problem or performance improvement terms and it may take some time to define – but agreement is important. Also at the start of every meeting, the objectives should be stated as clearly as possible by the leader. This can take a simple form:

> This meeting is to continue the discussion from last Tuesday on the development of our design manual and its trial and adoption throughout the company. Last week we agreed on the overall structure of the manual and today we will look in detail at the structure of the first section.

Project and/or meeting objectives enable the team members to focus thoughts and efforts on the aims, which may need to be restated if the team becomes distracted by other issues. The other important issue in the start phase of all process management meetings is a short review of the successes and failures in recent work, especially discussing causes of failures and strategies for performance improvement in the forthcoming period.

Team meetings

Meetings need to be seen as a part of a process working towards a longer-term goal – and hence, planning for each meeting and maintaining the continuity between meetings is important. Most important is the team leader's ability to maintain focus and momentum in the meetings, and structure meetings to make sure that important information is prepared and circulated in advance in a format that is clear and consistent. An agenda should be prepared by the leader and distributed to each team member before every meeting. It should include the following information:

- meeting place, time and how long it will be;
- list of members (and co-opted members) expected to attend;
- preparatory assignments for individual members or groups; and
- supporting material to be discussed at the meeting (this would include material from the previous meeting if it is an ongoing project).

Early in a project, the leader should orient the team members in terms of the approach, methods and techniques they will use to solve the problem. This may require a review of:

1 the systematic approach (Chapter 15);
2 the procedures and rules for using some of the basic tools (e.g. brainstorming – no judgement of initial ideas);
3 the role of the team in the continuous improvement process; and
4 the authority of the team.

To make sure that the meeting process is used to maximum advantage it is important that the team leader manages the meeting process; there are several important aspects to this. First of all, bear in mind the overall meeting plan (including an approximate time frame) as well as one for each topic that is addressed:

- maintain the participation of everyone;
- maintain focus on the topic being considered (visual information on a wall or screen is an effective way of supporting focus);
- maintain momentum, keep the process moving forward;
- achieve closure before moving on, capture where the group is up to and where and how it will proceed.

A team secretary should be appointed to take minutes of the meeting and distribute them to members as soon as possible after each meeting. The minutes should not be formal but reflect decisions and carry a clear statement of the action plans together with assignments of tasks. They may be handwritten initially, copied and given to team members at the end of the meeting, to be followed later by a more formal document that will be seen by any member of staff interested in knowing the outcome of the meeting. In this way the minutes form an important part of the communication system, supplying information to other teams or people needing to know what is going on.

Team assignments

To be effective, meetings must result in a series of action plans that assign specific tasks to team members. This is the responsibility of the team leader. Agreement must be reached regarding the responsibilities for individual assignments together with the timescale, and this must be made clear in the minutes. Task assignments must be decided while the team is together and not by separate individuals in after-meeting discussions. Make sure that task assignments are realistic to the time frame and resources available. This may need the allocation of additional resources and the team leader may need to negotiate for this with senior management.

Team dynamics

In any team activity the interactions between the members are vital to success. If solutions to problems are to be found, the meetings and ensuing assignments should assist and harness the creative thinking process. This is easier said than done because many people have either not learned or been encouraged to be innovative. Training in effective meeting leadership and negotiation skills is essential for the development of effective team leaders. The team leader clearly has a role here to:

- generate a 'climate' for creativity;
- encourage all team members to speak out and contribute their own ideas, or build on others;
- allow differing points of view and ideas to emerge;
- remove barriers to idea generation such as incorrect preconceptions that are usually destroyed by asking 'Why'; and
- support all team members in their attempts to become creative.

In addition to the team leader's responsibilities, the members should:

1   prepare themselves well for meetings by collecting appropriate data or information (*facts*) pertaining to a particular problem;
2   share ideas and opinions;
3   encourage other points of view;
4   listen 'openly' for alternative approaches to a problem or issue;
5   help the team determine the best solutions;
6   reserve judgement until all the arguments have been heard *and* fully understood;
7   accept individual responsibility for assignments and group responsibility for the efforts of the team.

Team results and reviews

A process approach to improvement and problem solving is most effective when the results of the work are communicated and acted upon. Regular feedback to the teams, via their leaders, will assist them to focus on objectives and review progress.

Reviews also help to deal with problems that may arise in teamwork. For example, some members may be more concerned with their own personal objectives than those of the team. This can result in manipulation of the problem-solving process towards different goals, and may result in the team splitting apart. If recognized, the review can correct this problem and demand greater openness and honesty.

A different type of problem is the failure of certain members to contribute and take their share of individual and group responsibility. Allowing other people to do their work results in an uneven distribution of effort and leads to bitterness. The review should make sure that all members have assigned and specific tasks, and perhaps lead to the documentation of duties in the minutes. A team roster may even help. If some members of a team are not contributing and cannot be induced to do so, consideration should be given to their replacement. However, this can become a more complex issue if the team leader does not manage team processes well and people believe they are wasting their time. There may be a high level of frustration that could lead to some members withdrawing support.

A third area of difficulty, which may be improved by reviewing progress, is the ready-fire-aim syndrome of action before analysis. This often results from team leaders being too anxious to deal with a problem before it is properly understood. A review should allow the problem to be redefined adequately exposing the real cause(s). This will release the team from the trap of doing something before they really know what should be done. The review will provide the opportunity to rehearse the steps in the systematic approach.

TEAMWORK AND ACTION-CENTRED LEADERSHIP

Over the years, there has been much academic work on the psychology of teams and on the leadership of teams. Three points on which all authors are in agreement are that teams develop a personality and culture of their own, respond to leadership and are motivated according to criteria usually applied to individuals.

Such key figures in the field of human relations as Douglas McGregor (Theories X and Y), Abraham Maslow (Hierarchy of Needs) and Fred Hertzberg (Motivators and Hygiene Factors) all changed their opinions on group dynamics, over time, as they came to realize that groups are not the democratic entity that everyone would like them to be. On the contrary, they respond to individual, strong, well-directed leadership, both from without and within the group, just like individuals (see also *Total Quality Management*, Oakland, 1993).

Action-centred leadership

During the 1960s John Adair, senior lecturer in Military History and the Leadership Training Adviser, the Military Academy, Sandhurst, and later Assistant Director of the Industrial Society, developed what he called the action-centred leadership model, based on his experiences at Sandhurst where he had the responsibility to ensure that results in the cadet training did not fall below a certain standard. He had observed that some instructors frequently achieved well above average results, owing to their natural ability with groups and their enthusiasm. He developed this further into a team model, which has become the basis of the approach of the authors and their colleagues to this subject.

In developing his model for teamwork and leadership, Adair clearly demonstrated that for any group or team (big or small) to respond to leadership, they needed a clearly defined *task*. The response and achievement of that task are interrelated to the needs of the *team* and the separate needs of the *individual members* of the team (Figure 17.2).

The value of the overlapping circles is that it emphasizes the unity of leadership and the interdependence and multifunctional reaction to single decisions affecting any of the three areas.

Leadership tasks

Drawing upon the discipline of social psychology, Adair developed and applied to training the functional view of leadership. The essence of this he distilled into the three interrelated but distinctive requirements of a leader. These are to define and achieve the job or task; to build up and coordinate a team to do this; and to develop and satisfy the individuals within the team (Figure 17.3).

FIGURE 17.2
Adair's model of action-centred leadership

1   *Task needs:* The difference between a team and a random crowd is that a team has some common purpose, goal or objective (e.g. a football team). If a work-team does not achieve the required results or meaningful results, it will become frustrated. Organizations have to make a profit, provide a service or even just survive; so anyone who manages others has to achieve results, whether it be in production, marketing, selling or whatever. Achieving objectives is a major criterion of success.

2   *Team needs:* To achieve these objectives, the group needs to be held together. People need to be working in a coordinated fashion in the same direction. Teamwork will ensure that the team's contribution is greater than the sum of its parts. Conflict within the team must be used effectively: arguments can lead to ideas or to tension and lack of co-operation.

3   *Individual needs:* Within working groups, individuals also have their own set of needs. They need to know what their responsibilities are; how they will be needed; and how well they are performing. They need an opportunity to show their potential, take on responsibility and receive recognition for good work.

The task, team and individual functions for the leader are as follows:

Task functions:
- Defining the task;
- Making a plan;
- Allocating work and resources;
- Controlling quality and tempo of work;
- Checking performance against the plan;
- Adjusting the plan.

Team functions:
- Setting standards;
- Maintaining discipline;
- Building team spirit;
- Encouraging, motivating, giving a sense of purpose;

- Appointing sub-leaders;
- Ensuring communication within the group;
- Training the group.

Individual functions:
- Attending to personal problems;
- Praising individuals;
- Giving status;
- Recognizing and using individual abilities;
- Training the individual.

The team leader's or facilitator's task is to concentrate on the small central area where all three circles overlap. In a business that is introducing lean quality this is the 'action to change' area where the leaders are attempting to manage the change from *business as usual* to lean quality: *Making lean quality business as usual,* using the cross-functional quality improvement teams at the strategic interface.

In the action area the facilitator's or leader's task is to try to satisfy all three areas of need by achieving the task, building the team and satisfying individual needs. If a leader concentrates on the task (e.g. in going all out for production schedules) while neglecting

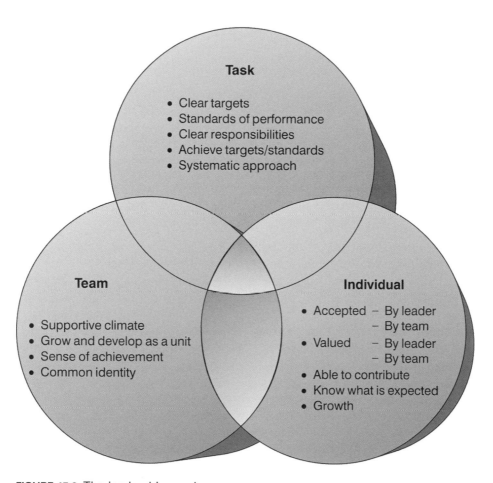

FIGURE 17.3 The leadership needs

the training, encouragement and motivation of the team and individuals, he/she may do very well in the short term. Eventually, however, the team members will give less effort than they are capable of. Similarly, a leader who concentrates only on creating team spirit, while neglecting the task and the individuals, will not receive maximum contribution from the people. They may enjoy working in the team but they will lack the real sense of achievement that comes from accomplishing a task to the utmost of the collective ability.

So the leader/facilitator must try to achieve a balance by acting in all three areas of overlapping need. It is always wise to work out a list of required functions within the context of any given situation, based on a general agreement as to the priorities. Here is Adair's original Sandhurst list which may be adapted to individual needs:

- *Planning* (e.g. seeking all available information):
  - defining group task, purpose or goal;
  - making a workable plan (in right decision-making framework).

- *Initiating* (e.g. group briefing on the aims and the plans):
  - explaining why aims or plans are necessary;
  - allocating tasks to group members.

- *Controlling* (e.g. maintaining group standard):
  - influencing tempo;
  - ensuring all actions are taken towards objectives;
  - keeping discussions relevant;
  - prodding group to action/decision.

- *Supporting* (e.g. expressing acceptance of persons and their contribution):
  - encouraging group/individuals;
  - disciplining group/individuals;
  - creating team spirit;
  - relieving tension with humour;
  - reconciling disagreements or getting others to explore them.

- *Informing* (e.g. clarifying task and plan):
  - giving new information to the group – keeping them 'in the picture';
  - receiving information from the group;
  - summarizing suggestions and ideas coherently.

- *Evaluating* (e.g. checking feasibility of an idea):
  - testing the consequences of a proposed solution;
  - evaluating group performance;
  - helping the group to evaluate its own performance against standards.

Situational leadership

In dealing with the task, the team and with any individual in the team a style of leadership appropriate to the situation must be adopted. The teams and the individuals within them will, to some extent, start 'cold' but will develop and grow in both strength and experience. With this change in the team, the interface with the leader must also change according to the Tannenbaum and Schmidt model (Figure 17.4).

Initially a very directive approach may be appropriate, giving clear instructions to meet agreed goals. Gradually as the teams become more experienced and have some success, the facilitating team leader will move from coaching and support to less directing and eventually a less supportive and less directive approach as a more interdependent style permeates the whole organization.

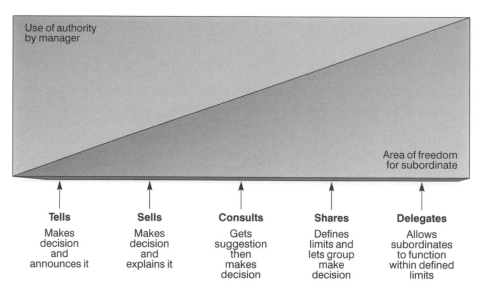

FIGURE 17.4 Continuum of leadership behaviour

The four leadership styles

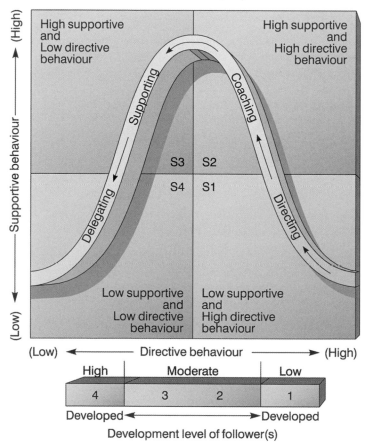

FIGURE 17.5 Situational leadership – progressive empowerment through lean quality

This equates to the modified Blanchard model in Figure 17.5 where directive behaviour moves from high to low as people develop and are more easily empowered. When this is coupled with the appropriate level of supportive behaviour, a directing style of leadership can move through coaching and supporting to a delegating style. It must be stressed, however, that effective delegation is only possible with developed and capable 'followers' who can be fully empowered.

One of the great mistakes in recent years has been the expectation by management that teams can be put together with virtually no training or development (S1 in Figure 17.5) and that they will perform as a mature team (S4). The Blanchard model emphasizes that there is no quick and easy 'tunnel' from S1 to S4. The only route is the laborious climb through S2 and S3.

STAGES OF TEAM DEVELOPMENT

Original work by Tuckman suggested that when teams are put together, there are four main stages of team development: the so-called forming (awareness), storming (conflict), norming (co-operation) and performing (productivity). The characteristics of each stage and some key aspects to look out for in the early stages are given below.

Forming – awareness

Characteristics
- Feelings, weaknesses and mistakes are covered up.
- People conform to established lines.
- Little care is shown for others' values and views.
- There is no shared understanding of what needs to be done.

Watch out for:

- increasing bureaucracy and paperwork;
- people confining themselves to defined jobs; and
- the 'boss' ruling with a firm hand.

Storming – conflict

Characteristics
- More risky, personal issues are opened up.
- The team becomes more inward-looking.
- There is more concern for the values, views and problems of others in the team.

Watch out for:

- The team becomes more open but lacks the capacity to act in a unified, economic and effective way.

Norming – co-operation

Characteristics
- Confidence and trust to look at how the team is operating.
- A more systematic and open approach, leading to a clearer and more methodical way of working.

- Greater valuing of people for their differences.
- Clarification of purpose and establishing of objectives.
- Systematic collection of information.
- Considering all options.
- Preparing detailed plans.
- Reviewing progress to make improvements.

Performing – productivity

Characteristics
- Flexibility.
- Leadership decided by situations, not protocols.
- Everyone's energies utilized.
- Basic principles and social aspects of the organization's decisions considered.

The team stages, the task outcomes and the relationship outcomes are shown together in Figure 17.6. This model, which has been modified from Kormonski, may be used as a framework for the assessment of team performance. The issues to look for are:

- How is leadership exercised in the team?
- How is decision-making accomplished?

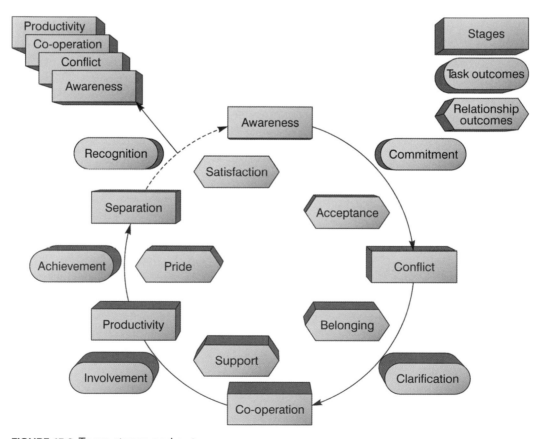

FIGURE 17.6 Team stages and outcomes

- How are team resources utilized?
- How are new members integrated into the team?

Teams that go through these stages successfully should become effective teams and display the following attributes.

Attributes of successful teams

Clear objectives and agreed goals
No group of people can be effective unless they know what they want to achieve, but this is more than knowing what the objectives are. People are only likely to be committed to their objectives if they can identify with and have ownership of them. In other words, objectives and goals are agreed to by team members.

Often this agreement is difficult to achieve but experience shows that it is an essential prerequisite for the effective group.

Openness and confrontation
If a team is to be effective, then the members of it need to be able to state their views, their differences of opinion, interests and problems without fear of ridicule or retaliation. No team works effectively within a cut-throat atmosphere where members do not feel safe to express themselves openly – as a result much energy, effort and creativity are lost.

Support and trust
Support naturally implies trust among team members. Where individual group members do not feel they have to protect their territory or job and feel able to talk straight to other members about both 'nice' and 'nasty' things, then there is an opportunity for trust to be shown. Based on this trust, people can talk freely about their fears and problems and receive from others the help they need to be more effective.

Co-operation and conflict
When there is an atmosphere of trust, members are more ready to participate and are committed. Information is shared rather than hidden. Individuals listen to the ideas of others and build on them. People find ways of being more helpful to each other and the group generally. Co-operation causes high morale. Individuals accept each other's strengths and weaknesses and contribute from their pool of knowledge of skill. All abilities, knowledge and experience are fully utilized by the group; individuals have no inhibitions about using other people's abilities to help solve their problems, which are shared problems.

Allied to this, conflicts are seen as a necessary and useful part of the organizational life. The effective team works through issues of conflict and uses the results to help objectives. Conflict prevents teams from becoming complacent and lazy, and often generates new ideas.

Good decision-making
As mentioned earlier, objectives need to be clearly and completely understood by all members before good decision-making can begin. In making decisions effective, teams develop the ability to collect information quickly and then discuss the alternatives openly. They become committed to their decisions and ensure quick action.

Appropriate leadership
Effective teams have a leader whose responsibility it is to achieve results through the efforts of a number of people. Power and authority can be applied in many ways, and team

members often differ on the style of leadership they prefer. Collectively, teams may come to different views of leadership but, whatever their view, the effective team usually sorts through the alternatives in an open and honest way.

Review of the team processes

Effective teams understand not only the group's character and its role in the organization but also how it makes decisions, deals with conflicts, etc. The team process allows the team to learn from experience and consciously to improve teamwork. There are numerous ways of reviewing team processes: using an observer, a team member giving feedback or the whole group discussing members' performance.

Sound inter-group relationships

No human being or group is an island; everyone needs the help of others. An organization will not achieve maximum benefit from a group of quality improvement teams that are effective within themselves but fight among each other.

Individual development opportunities

Effective teams seek to pool the skills of individuals, and it necessarily follows that they pay attention to development of individual skills and try to provide opportunities for individuals to grow and learn, and of course have FUN.

Once again, these ideas are not new but are very applicable and useful in the management of teams for quality improvements, just as Newton's theories on gravity still apply!

Personality types and the MBTI

No one person has a monopoly of good characteristics. Attempts to list the qualities of the ideal manager, for example, demonstrate why that paragon cannot exist. This is because many of the qualities are mutually exclusive; for example:

Highly intelligent	*v*	Not *too* clever
Forceful and driving	*v*	Sensitive to people's feelings
Dynamic	*v*	Patient
Fluent communicator	*v*	Good listener
Decisive	*v*	Reflective

Although no individual can possess all these and more desirable qualities, a team often does.

A powerful aid to team development is the use of the Myers-Briggs Type Indicator (MBTI). This is based on an individual's preferences on four scales for:

- giving and receiving 'energy';
- gathering information;
- making decisions; and
- handling the outer world.

Its aim is to help individuals understand and value themselves and others in terms of their differences as well as their similarities. It is well researched and non-threatening when used appropriately.

The four MBTI preference scales, which are based on Jung's theories of psychological types, represent two opposite preferences:

- Extroversion–Introversion – how we prefer to give/receive energy or focus our attention.
- Sensing–iNtuition – how we prefer to gather information.
- Thinking–Feeling – how we prefer to make decisions.
- Judgement–Perception – how we prefer to handle the outer world.

To understand what is meant by preferences, the analogy of left- and right-handedness is useful. Most people have a preference to write with either their left or their right hand. When using the preferred hand, they tend not to think about it, it is done naturally. When writing with the other hand, however, it takes longer, needs careful concentration, seems more difficult, but with practice would no doubt become easier. Most people *can* write with and use both hands, but tend to prefer one over the other. This is similar to the MBTI psychological preferences: most people are able to use both preferences at different times, but will indicate a preference on each of the scales.

In all, there are eight possible preferences – E or I, S or N, T or F, J or P, (i.e. two opposites for each of the four scales). An individual's *type* is the combination and interaction of the four preferences. It can be assessed initially by completion of a simple questionnaire. Hence, if each preference is represented by its letter, a person's type may be shown by a four letter code – there are sixteen in all. For example, ESTJ represents an *extrovert* (E) who prefers to gather information with *sensing* (S), prefers to make decisions by *thinking* (T) and has a *judging* (J) attitude towards the world – a person who prefers to make decisions rather than continue to collect information. The person with opposite preferences on all four scales would be an INFP, an introvert who prefers intuition for perceiving, feelings or values for making decisions, and likes to maintain a perceiving attitude towards the outer world.

The questionnaire, its analysis and feedback must be administered by a qualified MBTI practitioner, who may also act as external facilitator to the team in its forming and storming stages.

Type and teamwork

With regard to teamwork, the preference types and their interpretation are extremely powerful. The *extrovert* prefers action and the outer world, whilst the *introvert* prefers ideas and the inner world.

- *Sensing–thinking* types are interested in facts; they analyse facts impersonally and use a step-by-step process from cause to effect, premise to conclusion. The *sensing–feeling* combinations, however, are interested in facts; they analyse facts personally and are concerned about how things matter to themselves and others.
- *Intuition–thinking* types are interested in possibilities: they analyse possibilities impersonally and have theoretical, technical or executive abilities. On the other hand, the *intuition–feeling* combinations are interested in possibilities; they analyse possibilities personally and prefer new projects, new truths – things not yet apparent.
- *Judging* types are decisive and 'planful'; they live in orderly fashion and like to regulate and control. *Perceivers*, on the other hand, are flexible; they live spontaneously and understand and adapt readily.

As we have seen, an individual's type is the combination of four preferences on each of the scales. There are sixteen combinations of the preference scales and these may be displayed on a *type table* (see Figure 17.7). If the individuals within a team are prepared to share with each other their MBTI preferences, this can dramatically increase understanding, and frequently is of great assistance in team development and good team-

FIGURE 17.7
MBTI type table form

ISTJ	ISFJ	INFJ	INTJ
ISTP	ISFP	INFP	INTP
ESTP	ESFP	ENFP	ENTP
ESTJ	ESFJ	ENFJ	ENTJ

working. The similarities and differences in behaviour and personality can be identified. The assistance of a qualified MBTI practitioner is absolutely essential in the initial stages of this work.

INTERPERSONAL RELATIONS – FIRO-B AND THE ELEMENTS

The FIRO-B (Fundamental Interpersonal Relations Orientation-Behaviour) is a powerful psychological instrument that can be used to give valuable insights into the needs individuals bring to their relationships with other people. The instrument assesses needs for *inclusion*, *control* and *openness* and therefore offers a framework for understanding the dynamics of interpersonal relationships.

Use of the FIRO-B instrument helps individuals to be more aware of how they relate to others and how to become more flexible in this behaviour. Consequently, it enables people to build more productive teams through better working relationships.

Since its creation by William Schutz in the 1950s to predict how military personnel would work together in groups, the FIRO-B instrument has been used throughout the world by managers and professionals to look at management and decision-making styles. Through its ability to predict areas of probable tension and compatibility between individuals, the FIRO-B is a highly effective team-building tool that can aid in the creation of the positive environment in which people thrive and achieve improvements in performance.

The theory underlying the FIRO-B incorporates ideas from the work of Adomo, Fromm and Bion. It was first fully described in Schutz's book *FIRO: A Three-Dimensional Theory of Interpersonal Behaviour* (1958). In his more recent book *The Human Element* (1994), Schutz developed the instrument into a series of 'elements', B, F, S, etc. and offers strategies for heightening our awareness of ourselves and others.

The FIRO-B takes the form of a simple-to-complete questionnaire, the analysis of which provides scores that estimate the levels of behaviour with which the individual is comfortable, with regard to his/her needs for inclusion, control and openness. Schutz described these three dimensions in the form of the decision we make in our relationships regarding whether we want to be:

- 'in' or 'out' – inclusion;
- 'up' or 'down' – control; and
- 'close' or 'distant' – openness.

The FIRO-B estimates our unique level of needs for each of these dimensions of interpersonal interaction.

The instrument further divides each of these dimensions into:

1 the behaviour we feel most comfortable *exhibiting towards* other people – *expressed* behaviours; and
2 the behaviour *we want* from others – *wanted* behaviours.

Hence, the FIRO-B 'measures', on a scale of 0–9, each of the three interpersonal dimensions in two aspects (Table 17.1).

The *expressed* aspect of each dimension indicates the level of behaviour the individual is most comfortable with towards others, so high scores for the expressed dimensions would be associated with:

TABLE 17.1 The FIRO-B interpersonal dimensions and aspects

	Inclusion	Control	Openness
Expressed behaviour	Expressed inclusion	Expressed control	Expressed openness
Wanted behaviour	Wanted inclusion	Wanted control	Wanted openness

Modified from: W. Schutz (1978) *FIRO: Awareness Scales Manual*, Palo Alto, CA, Consulting Psychologists Press.

High scored expressed behaviours

Inclusion: Makes efforts to include other people in his/her activities – tries to belong to or join groups and to be with people as much as possible.
Control: Tries to exert control and influence over people and tell them what to do.
Openness: Makes efforts to become close to people – expresses friendly open feelings, tries to be personal and even intimate.

Low scores would be associated with the opposite expressed behaviour.

The *wanted* aspect of each dimension indicates the behaviour the individual prefers others to adopt towards him/her, so high scores for the wanted dimensions would be associated with:

High scored wanted behaviours

Inclusion: Wants other people to include him/her in their activities – to be invited to belong to or join groups (even if no effort is made by the individual to be included).
Control: Wants others to control and influence him/her and be told what to do.

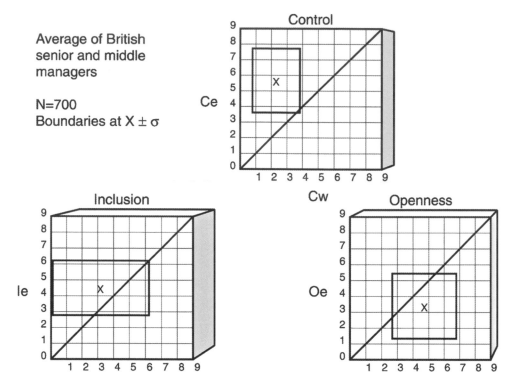

FIGURE 17.8 Typical manager profiles

Openness: Wants others to become close to him/her and express friendly, open, even affectionate feelings.

Low scores would be associated with the opposite wanted behaviours.

It is interesting to look at typical manager FIRO-B profiles based on their scores for the six dimensions/aspects in Table 17.1. Figure 17.8 shows the average of a sample of 700 middle and senior managers in the UK with boundaries at one sigma, plotted on expressed/wanted scales for the three dimensions.

On average the managers show a higher level of expressed inclusion – including people in his/her activities – than wanted inclusion. Similarly, and not surprisingly perhaps, expressed control – trying to exert influence and control over others – is higher in managers than wanted control. When it comes to openness, the managers tend to want others to be open rather than be open themselves.

It is even more interesting to contemplate these results when one considers the demands of some of the recent popular management programmes such as lean quality management, employment involvement and self-directed teams. These tend to require from managers certain behaviours, for example, lower levels of expressed control and higher levels of wanted control so that the people feel empowered. Similarly, managers are encouraged to be more open. These, however, are *opposite* to the apparent behaviours of the sample of managers shown graphically in Figure 17.8. It is not surprising then that lean quality has failed in some organizations where managers were being asked to empower employees and be more open – and who can argue against that – yet, their basic underlying needs caused them to behave in the opposite way.

Understanding what drives these behaviours is outside the scope of this book, but other FIRO and Elements instruments can help individuals to further develop understanding of themselves and others. FIRO and Schutz's Elements instruments for measuring *feelings* (F) and *self-concept* (S) can deepen the awareness of what lies behind our behaviours with respect to inclusion, control and openness. The reader is advised to undertake further reading and seek guidance from a trained administrator of these instruments, but the overall relationship between the B and F instruments is given below:

Behaviours related to:	*Feelings related to:*
Inclusion	Significance
Control	Competence
Openness	Likeability

Issues around control behaviour then may arise because of underlying feelings about competence. Similarly, underlying feelings concerning significance may lead to certain inclusion behaviours.

FIRO-B in the workplace

The inclusion, control and openness dimensions form a cycle (see Figure 17.9), which can help groups of people to understand how their individual and joint behaviour develops as teams are formed. Given in Table 17.2 are the considerations, questions and outcomes under each dimension. If inclusion issues are resolved first, it is possible to progress to dealing with the control issues, which in turn must be resolved if the openness issues are to be dealt with successfully. As a team develops, it travels around the inclusion, control and openness cycle time and time again. If the issues are not resolved in each dimension, further progress in the next dimension will be hindered. It is difficult to deal with issues

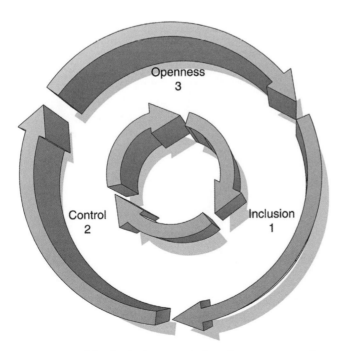

FIGURE 17.9 The inclusion, control and openness cycle

of control if unresolved inclusion issues are still around and people do not know whether they are 'in' or 'out' of the group. Similarly it is difficult to be open if it is not clear where the power base is in the group.

This I-C-O cycle has led to the development, by the authors and their colleagues, of an 'openness model', which is in three parts. Part 1 is based on the premise that to participate productively in a team, individuals must firstly be involved and then committed. Figure 17.10 shows some of the questions which need to be answered and the outcomes from this stage. Part 2 deals with the control aspects of empowerment and management. Figure 17.11 summarizes the questions and outcomes. Finally, Part 3 ensures openness through acknowledgement and trust (see Figure 17.12). The full openness cycle (see Figure 17.13) operates in a clockwise direction so that trust leads to more involvement, further commitment, increased empowerment, etc. Of course, if progress is not made round the cycle and trust is replaced by fear, it is possible to send the whole process into reverse – a negative cycle of suspicion, fault-finding, abdication and confusion (see Figure 17.14). Unfortunately this will be recognized as the culture in some organizations where the focus of enquiry is 'what has gone wrong', leading to 'whose fault was it?'

Fortunately, organizations and individuals seem keen to learn ways to change these negative communications that sour relationships, dampen personal satisfaction and reduce productivity. The inclusion, control and openness cycle is a useful framework for helping teams to pass successfully through the forming and storming stages of team development. Unfortunately, as teams are disbanded, for whatever reason, the process reverses and the first thing to go is the openness.

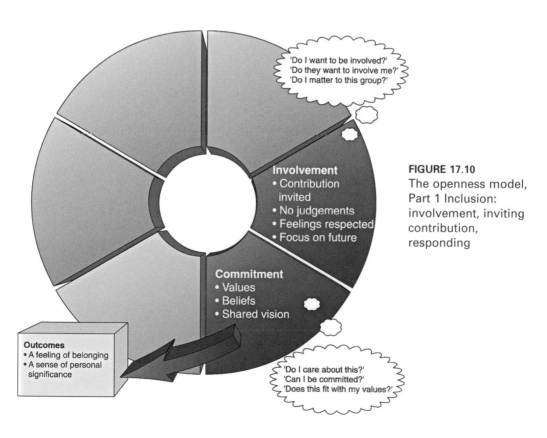

FIGURE 17.10
The openness model, Part 1 Inclusion: involvement, inviting contribution, responding

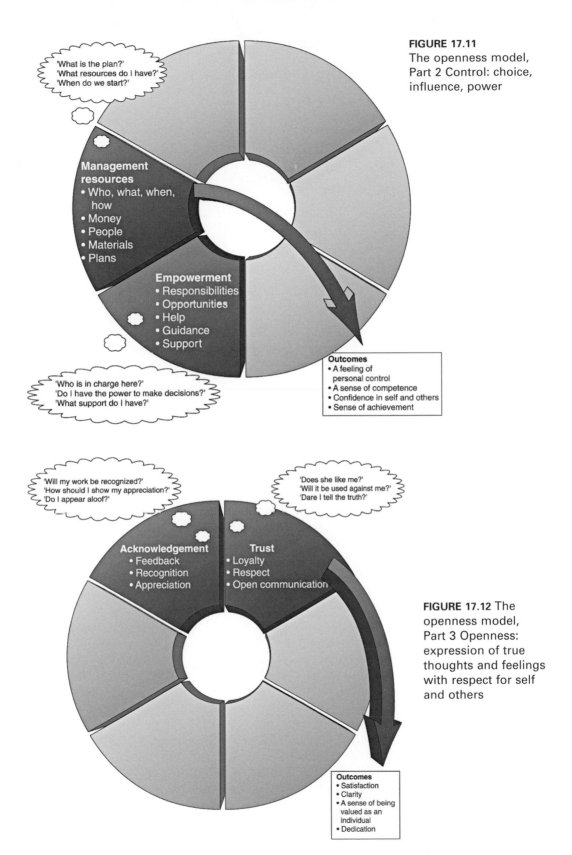

FIGURE 17.11
The openness model, Part 2 Control: choice, influence, power

FIGURE 17.12 The openness model, Part 3 Openness: expression of true thoughts and feelings with respect for self and others

Culture change through teamwork 395

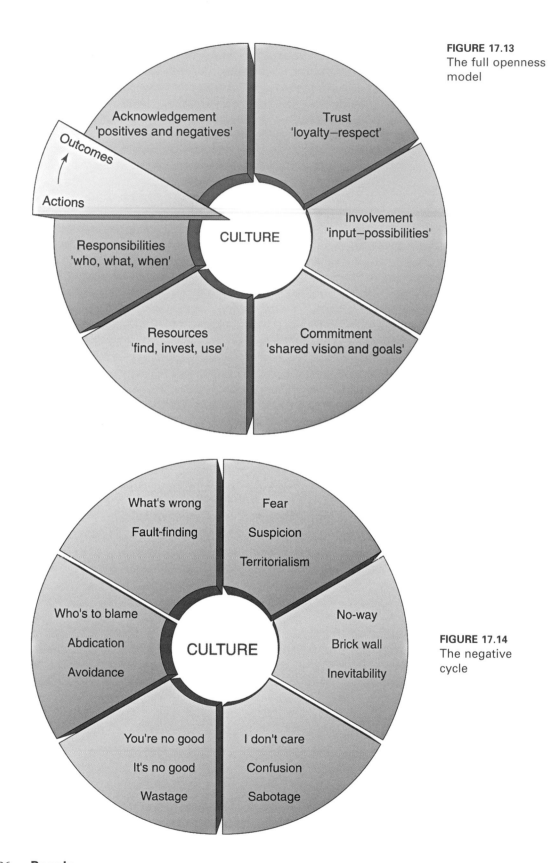

FIGURE 17.13
The full openness model

FIGURE 17.14
The negative cycle

TABLE 17.2 Considerations, questions and outcomes for the FIRO-B dimensions

Dimension	Considerations	Some Typical Questions	If resolved we get:	If not resolved we get:
Inclusion	Involvement – how much you want to include other people in your life and how much attention and recognition you want.	Do I care about this? Do I want to be involved? Does this fit with my values? Do I matter to this group? Can I be committed? ... leading to ... Am I 'in' or 'out'?	A feeling of belonging A sense of being recognized and valued Willingness to become committed	A feeling of alienation A sense of personal insignificance No desire for commitment or involvement
Control	Authority, responsibility, decision-making, influence.	Who is in charge here? Do I have power to make decisions? What is the plan? When do we start? What support do I have? What resources do I have? ... leading to ... Am I 'up' or 'down'?	Confidence in self and others Comfort with level of responsibility Willingness to belong	Lack of confidence in leadership Discomfort with level of responsibility – fear of too much – frustration with too little 'Griping' between individuals
Openness	How much are we prepared to express our true thoughts and feelings with other individuals.	Does she like me? Will my work be recognized? Is he being honest with me? How should I show appreciation? Do I appear aloof? ... leading to ... Am I 'open' or 'closed'?	Lively and relaxed atmosphere Good-humoured interactions Open and trusting relationships	Tense and suspicious atmosphere Flippant or malicious humour Individuals isolated

The five 'A' stages for teamwork

The awareness provided by the use of the MBTI and FIRO-B instruments helps people to appreciate their own uniqueness and the uniqueness of others – the foundation of mutual respect and for building positive, productive and high-performing teams.

For any of these models or theories to benefit a team, however, the individuals within it need to become **aware** of the theory, e.g. the MBTI or FIRO-B. They then need to **accept** the principles as valid, **adopt** them for themselves in order to **adapt** their behaviour accordingly. This will lead to individual and team **action** (Figure 17.15).

Particularly in the early stages of team development, the assistance of a skilled facilitator to aid progress through these stages is necessary. This is often neglected, causing failure in so many team initiatives. In such cases, the net output turns out to be lots of nice warm feelings about 'how good that team workshop was a year ago', with the nagging reality that no action was taken and nothing has really changed.

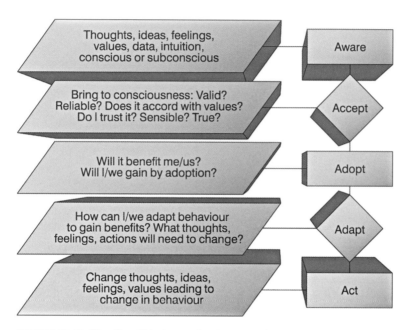

FIGURE 17.15 The five 'A' stages for teamwork

BIBLIOGRAPHY

Adair, J. *Effective Teambuilding* (2nd edn), Pan Books, London, 1987.

Atkinson, P. *Creating Culture Change: The Key to Successful Total Quality Management*, IFS, Bedford, 1990.

Blanchard, K. and Hersey, P. *Management of Organizational Behaviour* (4th edn), Prentice Hall, Eaglewood Cliffs, NJ, 1982.

Katzenbach, J.R. and Smith, D.K. *The Wisdom of Teams – Creating the High Performance Organization,* McGraw-Hill/Harvard Business School, Boston, MA, 1994.

Kormanski, C. A situational leadership approach to groups using the Tuckman Model of Group Development, *The 1985 Annual: Developing Human Resources,* University Associates, San Diego, CA, 1985.

Kormanski, C. and Mozenter, A. A new model of team building: a technology for today and tomorrow, *The 1987 Annual: Developing Human Resources,* University Associates, San Diego, CA, 1987.

Krebs Hirsh, S. *MBTI Team Building Program, Team Member's Guide,* Consulting Psychologists Press, Palo Alto, CA, 1992.

Krebs Hirsh, S. and Kummerow, J.M. *Introduction to Type in Organizational Settings,* Consulting Psychologists Press, Palo Alto, CA, 1987.

McCaulley, M.H. How individual differences affect health care teams, *Health Team News,* 1 (8): 1–4, 1975.

Myers-Briggs, I., *Introduction to Type: A Description of the Theory and Applications of the Myers-Briggs Type Indicator,* Consulting Psychologists Press, Palo Alto, 1987.

Oakland, J.S. *Total Quality Management* (2nd edn), Butterworth-Heinemann, Oxford, 1993.

Scholtes, P.R. *The Team Handbook,* Joiner Associates, Madison, NY, 1990.

Schutz, W. *FIRO: A Three-Dimensional Theory of Interpersonal Behaviour,* Mill Valley WSA, CA, 1958.

Schutz, W. *FIRO: Awareness Scales Manual,* Consulting Psychologists Press, Palo Alto, CA, 1978.

Schutz, W. *The Human Element – Productivity, Self-esteem and the Bottom Line,* Jossey-Bass, San Francisco, CA, 1994.

Tannenbaum, R. and Schmidt, W.H. How to choose a leadership pattern, *Harvard Business Review,* May-June, 1973.

Tuckman, B.W. and Jensen, M.A. States of small group development revisited, *Group and Organizational Studies,* 2 (4): 419–427, 1977.

Wilkinson, A. and Willmott, H. (eds). *Making Quality Critical – New Perspectives on Organizational Change,* Routledge, London, 1995.

CHAPTER HIGHLIGHTS

The need for teamwork

- The only efficient way to tackle process improvement or complex problems is through teamwork. The team approach allows individuals and organizations to grow.
- Within fragmented supply chains there is often a need for effective teams that cross organizational boundaries.
- Employees will not engage continual improvement without commitment from the top, a quality 'climate' and an effective mechanism for capturing individual contributions.
- Teamwork for quality improvement must be driven by a strategy, must have a structure and must be implemented thoughtfully and effectively.

Running process management and improvement teams

- Process management and improvement teams are groups brought together by management to improve a process or tackle a particular problem on a project basis. The running of these teams involves several factors: selection and leadership, objectives, meetings, assignments, dynamics, results and reviews.
- The need for training in the basic skills of team leadership should not be under-estimated if successful outcomes are sought.

Teamwork and action-centred leadership

- Early work in the field of human relations by McGregor, Maslow and Hertzberg was useful to John Adair in the development of his model for teamwork and action-centred leadership.
- Adair's model addresses the needs of the task, the team and the individuals in the team, in the form of three overlapping circles. There are specific task, team and individual functions for the leader, but he/she must concentrate on the small central overlap area of the three circles.

- The team process has inputs and outputs. Good teams have three main attributes: high task fulfilment, high team maintenance and low self-orientation.
- In dealing with the task, the team and its individuals, a situational style of leadership must be adopted. This may follow the Tannenbaum and Schmidt and Blanchard models through directing, coaching and supporting to delegating.

Stages of team development

- When teams are put together, they pass through Tuckman's forming (awareness), storming (conflict), norming (co-operation) and performing (productivity) stages of development.
- Teams that go through these stages successfully become effective and display clear objectives and agreed goals, openness and confrontation, support and trust, co-operation and conflict, good decision-making, appropriate leadership, review of the team processes, sound relationships and individual development opportunities.

Personality types and the MBTI

- A powerful aid to team development is provided by the Myers-Briggs Type Indicator (MBTI).
- The MBTI is based on an individual's preferences on four scales for giving and receiving 'energy' (extroversion-E or introversion-I), gathering information (sensing-S or intuition-N), making decisions (thinking-T or feeling-F) and handling the outer world (judging-J or perceiving-P).
- An individual's type is the combination and interaction of the four scales and can be assessed initially by completion of a simple questionnaire. There are sixteen types in all, which may be displayed for a team on a type table.

Interpersonal relations – FIRO-B and the Elements

- The FIRO-B (Fundamental Interpersonal Relations Orientation–Behaviour) instrument gives insights into the needs individuals bring to their relationships with other people.
- The FIRO-B questionnaire assesses needs for inclusion, control and openness in terms of expressed and wanted behaviour.
- Typical manager FIRO-B profiles conflict with some of the demands of lean quality and can, therefore, indicate where particular attention is needed to achieve successful lean quality implementation.
- The inclusion, control and openness dimensions form an 'openness' cycle that can help groups to understand how to develop their individual and joint behaviours as the team is formed. An alternative, negative cycle may develop if the understanding of some of these behaviours is absent.
- The five A's: for any of the teamwork models and theories, the individuals need to be aware, to accept, adopt and adapt in order to act. A skilled facilitator is always necessary.

Communication, innovation and learning 18

COMMUNICATING THE LEAN QUALITY STRATEGY

People's attitudes and behaviour clearly can be influenced by communications; one has to look only at the media or advertising to understand this. The essence of changing attitudes is to gain acceptance for the need to change and, for this to happen, it is essential to provide relevant information, convey good practices and generate interest, ideas and awareness through excellent communication processes.

This first step of convincing people of the need for change and setting the vision is possibly the most neglected part of many organizations' operations, yet failure to communicate effectively creates unnecessary problems, resulting in confusion, loss of interest and eventually in declining quality through apparent lack of guidance and stimulus.

To generate enthusiasm for change, leadership has to capture the imagination of its people. In situations where management has a 'burning platform', it is easier to create a sense of urgency. We see in the case of the multinational high-tech manufacturing company (CS4) the cost and speed of its construction programme is critical to maintaining competitiveness. In the JB Henderson case study (CS8), the finding that only 17 per cent of on-site labour was value-adding shocked the organization into action.

> When somebody told us (like we're telling you) that **'you're not as good as you think you are!'** we took it as a challenge to prove them wrong. Well, unfortunately, we found out that we weren't. If you are in a space where you feel like you have very little or no room for improvement, the future holds **no possibilities.**

In the construction industry, projects are generally delivered by highly fragmented supply chains, making the need for effective communication even greater. Managers and workers from myriad organizations work together to achieve a shared project outcome. The achievement of the vision in terms of product and process goals and the quality of those outcomes depend largely on the effectiveness of leadership and communication at the project level.

Lean quality management significantly changes the way organizations operate and 'do business'. The first step is for top management to convince all staff and employees of the need for change and the opportunities presented by lean quality. This communication should also explain the plan to focus on improvement at the process level. While initially this must be addressed to employees, in today's outsourced business model, this will have to be communicated to every subcontract designer or subcontract employee as well.

An excellent start is to already have identified some critical performance gaps, and possibly already have some data on current performance levels, and the cost for the business of poor performance in these areas. This should be accompanied by an outline implementation strategy that describes the steps of the change process and who is to lead the initiative. The message has to make clear management's commitment to driving the change and improvement process. This can be in the form of a quality policy (see Chapters 4 and 5) or a specifically designed statement about the organization's intention to integrate a lean quality-based approach into the business operations.

Such a statement might read something like the following example.

The Board of Directors (or appropriate title) have identified the following significant weaknesses in our performance, and we believe that to be successful as a market leader we have to close the following three performance gaps:

1   We have measured the cost of rework on our projects at 6 per cent, this is unsustainable and is undermining our profitability and that of our supply chain partners.
2   We have been benchmarking our design documentation; it is not error-free and far too often the late issue of design documents is compromising our site teams' capability to work efficiently.
3   We have measured the reliability of on-site work plans to be 60 per cent at the weekly level. We need to improve the reliability of on-site workflows so that our supply chain partners can become more productive, our pricing more competitive and our projects can be completed with fewer outstanding defects.

The senior management team is enthusiastic and committed to the lean quality approach; however, our success relies entirely on your support, your ideas and your innovations. Only you can improve our performance. We need every one of you to become as convinced as we are that lean quality-based business process improvement is critical for our survival and continued success.

We are putting in place a programme of education, training, teamwork development and resourcing built on lean quality-based business process improvement. We believe that we can move forward together to achieve our business goals.

The lean quality director/leader should then assist the senior management team to prepare a directive. This must be signed by all business unit, division or process leaders and distributed to everyone in the organization.

The directive should include the following:

* need for improvement;
* concept for lean quality;
* importance of understanding business processes;
* approach that will be taken and people's roles;
* individual and process group responsibilities;
* principles of process measurement;
* a commitment to the resources needed to support the change initiative.

The systems for disseminating the message should include all the conventional communication methods of seminars, departmental meetings, posters, newsletters, intranet, etc. First-line supervision will need to review the directive with all the staff, and a set of questions and answers may be suitably pre-prepared in support.

Once people understand the strategy, management must establish the infrastructure (see Chapter 16). However, the required level of individual commitment is only likely to be achieved if everyone is motivated, understands the aims and benefits of lean quality and the role each of them must play. They must understand how they can implement process improvements and have confidence that the necessary resource commitment to support them in their efforts will be made. For this understanding a constant flow of information is necessary, including:

1   when and how individuals will be involved;
2   what the process requires;
3   the successes and benefits achieved; and
4   regular updates on what is being achieved.

The most effective means of developing the personnel commitment required is to ensure people know what is going on. Otherwise, they will feel left out and begin to believe that lean quality is not for them, which will lead to resentment and the undermining of the whole process. The first line of supervision again has an important part to play in ensuring key messages are communicated, and in building teams by demonstrating everyone's participation and commitment.

Naturally the extent to which such policy is formally set out and formally communicated is influenced by the size of the organization and the stage of organizational development at which lean quality values are first introduced. For example, an interesting case in point is Mirvac, the Australian property group that started as a developer and builder of residential medium density housing in 1972. A case study of the company was included in the authors' 2006 edition of *Total Quality in the Construction Supply Chain*. From its inception, the company CEO and his co-founder saw the opportunity for their business as creating housing of high quality: not Rolls-Royce quality but dependable, well-designed, defect-free products. Their motto from day one was: 'If you wouldn't live in it yourself, you should not be building it'. This focus of senior management on quality in every decision created a company whose brand image for quality housing in the Australian market is next to none. It is acknowledged that their product holds its value better than any other in the market. They have achieved this through clear-sighted management focus, selecting suppliers and employees carefully and quickly learning from mistakes.

In the Larkins' excellent book, *Communicating Change* (1994), the authors refer to three 'facts' regarding the best ways to communicate change to employees:

1   Communicate directly to supervisors (first-line).
2   Use face-to-face communication.
3   Communicate relative performance of the local work area.

The language used at the 'coal face' is important. Avoiding complexity and jargon in written and spoken communications will help to get the message through. When written business communications cannot be easily understood, they receive only a cursory glance rather than the detailed study they require. *Simplify and shorten* must be the guiding principles. The communication model illustrated in Figure 18.1 indicates the potential for problems through environmental distractions, mis-matches between sender and receiver (or, more correctly, decoder) in terms of attitudes – towards the information and each other – vocabulary, time pressures, etc.

In relatively recent work with a major contractor, one of the authors encountered the use of safety and environmental planning documents that were up to 200 pages in length for a relatively straightforward construction operation in a high-risk rail environment.

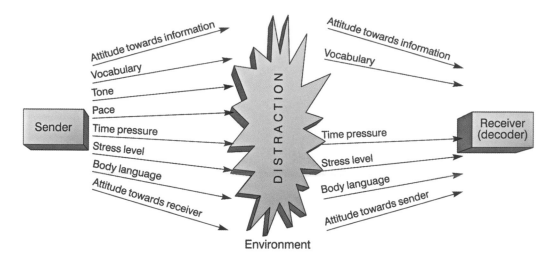

FIGURE 18.1 Communication model

No one at the workface will read a 200-page planning document; all the important messages are inevitably lost in the forest of words. Documents for use at the workface must be simple and succinct. Long wordy documents such as the ones described above are as far from lean quality thinking as it is possible to be. This documentation strategy was not designed for the purpose of communication; it would have been designed with input from lawyers as protection against prosecution should an accident occur. This is a very confused basis for developing documentation designed to outline plans and processes for use on the shop floor or on site.

All levels of management should introduce and stress 'open' methods of communication by maintaining open offices, being accessible to staff/employees and taking part in day-to-day interactions and detailed processes. Lean quality managers practise standard work, create routines that are visible to others around them, interact on a regular basis with their direct reports and undertake management tasks in a planned and regular manner. This regular rhythmic approach to management will lay the foundation for improved interactions *between* management and employees, which is essential for information flow and process improvement.

Opening lines of communication may lead to some confrontation, barriers might go up and some resistance may be encountered. Workforce training and the behaviour of supervisors and management should encourage and help people to accept responsibility for their own behaviour – this may be confronting and can raise barriers. To help overcome these barriers it is important to concentrate on improving the process outcomes rather than on 'departmental' issues.

Resistance to change will always occur and should be expected. First-line management should be trained to help people deal with it. This requires an understanding of the dynamics of change and the support necessary – not an obsession with forcing people to change. Opening up lines of communication through a previously closed system and publicizing people's efforts and achievements will aid the process. Change can be – even should be – exciting once employees start to share their development, growth, suggestions and questions. Management needs to encourage and participate in this by creating the most appropriate and effective communication and recognition systems.

In construction, the authors' experience has been that this resistance to change is likely to be significant, especially at the project director/manager level. Construction companies are generally led by former project directors who succeeded in a cost driven, exploitative culture. If cost and time goals are being met on a project, the authority and independence of the project manager is rarely questioned, except in relation to safety. The reason for this is that in this sector, all responsibility for schedule and cost success rests on the project manager's shoulders, and executive management will do nothing to diminish this. Yet, in this very thinking lie the seeds of the sector's inability to change. Any changes in business practices have to be sold to each and every project director/manager anew because executive management is reluctant to lead.

Furthermore, across the supply chain, managers' suggestions for change are likely to be met with cynicism from general contractors. The cost driven focus and exploitative management practices in this industry have undermined trust and the capacity for genuine collaboration. Subcontractors are well used to hearing lip-service to quality and process change, but seeing a short-sighted management focus that often accepts lowest cost in one or two work packages on a project, regardless of the penalty everyone pays for having one weak link on a project. The drive for speed at the expense of efficiency and the acceptance of the high cost of rework as normal have undermined the credibility of many contractors. In some construction businesses, expensive legal costs in litigation are accepted as a part of normal business practice. The implication of this situation for management is that as much effort has to be invested in driving the change in the attitudes of project directors/managers as in bringing the supply chain partners along on the lean quality journey. Strong leadership at the executive level, supported by training, is needed to drive change at the project directors/managers' level. A proven way to influence leadership of supply chain partners is through the joint training of key people; after all, they work next to each other on a daily basis.

COMMUNICATING THE LEAN QUALITY MESSAGE

The people in most organizations fall into one of four 'audience' groups, each with particular general attitudes towards lean quality:

1 *Senior managers,* who should see lean quality as an opportunity both for the organization and themselves.
2 *Middle managers,* who may see lean quality as another burden without any benefits and may perceive a vested interest in the status quo; they need to be convinced of the benefits.
3 *Supervisors* (first-line or junior managers), who may see lean quality as another 'flavour of the month' or current campaign, and who may respond by trying to keep their heads down so that it will pass over. They too need to be convinced of the benefits.
4 *Other employees,* who may be less interested as long as they still have jobs and get paid. These people are critical to the success of any change as they must be the custodians of the delivery of lean quality to the customer.

Senior management needs to ensure that each group sees lean quality as being beneficial for them. Lean quality training material and support (whether internal from a quality director and team or from external consultants) will be of real value only if everyone is motivated to respond positively. The implementation strategy must then be based on two mutually supporting aspects:

1 'marketing' any lean quality initiatives; and
2 a positive, logical process of communication designed to motivate.

There are of course a wide variety of approaches to lean quality implementation, as well as to the methods of lean quality and business improvement. Every organization's process improvement strategy must be designed to meet the needs of its own business and its people. It should reflect the structure of the organization and be sensitive to the group's capacity for continuous improvement activities. These days, very few organizations are starting from a green-field site. The key is that groups of people must feel able to 'join' the lean quality process at the most appropriate point for them. For middle managers to be convinced that they must participate, lean quality must be presented as the key to help them turn the people who work for them into 'lean quality employees'. At the senior management level, an early and critical task is to win the support of management from supply chain partners to a joint initiative. Without co-operation, the task of implementing a lean quality initiative is much greater.

The noisy, showy, hype-type activity is not appropriate to any aspect of lean quality. Lean quality 'events' should of course be fun, because this is often the best way to persuade and motivate, but the value of any event should be judged by its ability to contribute to understanding and to the change to lean quality. Key words in successful exercises include discovery, affirmation, participation and team-based learning. Middle and junior managers can easily undermine a change process; it is important that they understand that the change promises improvements for them, rather than being a threat to their position. In workshops designed for middle and junior managers and supervisors, they should be made to feel that the future is in their hands, rather than being blamed for the past. Training programmes should be delivered by specially trained people. The environment and conduct of the workshops must also demonstrate the organization's commitment to lean quality and the people who can make the changes being sought.

The key medium for motivating employees and gaining their commitment to lean quality is face-to-face communication and *visible* management commitment. Much is written and spoken about leadership, but success in a change process comes down to effective communication. If people are good leaders, they are invariably good communicators. Effective leadership relies on the communication between the leaders and the followers. It calls for skills that can be *learned* from education and training, but can only be *developed* through practice.

COMMUNICATION, LEARNING, EDUCATION AND TRAINING

It is the authors' belief that once there has been commitment to change, effective education, training and empowerment are the most important factors in actually improving lean quality and business performance. For it to be effective, and to provide the right sort of learning experience, education and training must be planned and delivered in a systematic and objective manner. The content and delivery methods must be continuously updated to take advantage of changes in technology and to reflect changes in the environment in which an organization operates, its changing structure and, perhaps most important of all, the people who work there.

An interesting paradox is that while industry in general tends to under-invest in training, world-class companies invest far above the industry average. Cost-focused organizations are concerned about the payback on training and see it as an overhead; managers tend to want to minimize investment. Value-focused companies see training as the key to maximizing the potential of their workforce; hence, they are active in identifying priorities and ensuring that training is effective. All the case study companies invest far above the industry average in training – how far above varies.

Training in the workplace can provide a powerful stimulus for personal development; it should be a productive improvement for the capacity of the organization as well. Value-focused companies realize that to attract and retain the best people, they have to help them to develop to reach their potential. It is interesting to review the Graniterock HR policy drivers (CS5); this company has been ranked in the top 10 places to work in the US economy by Fortune 500. The company policy is to encourage its people to use individual professional development plans to plan and track their own skill and knowledge development, and to pursue their personal development and career aspirations consistent with the goals and needs of the company.

People are motivated to learn for a range of reasons, some of which include:

- self-betterment;
- self-preservation;
- need for responsibility;
- saving time or effort;
- sense of achievement;
- pride of work;
- curiosity.

The case studies at the back of the book illustrate the many different ways that leading companies engage with their people, support their aspirations and development, because they realize that, ultimately, the only true asset they have is their people. The people are the company and shape its future based on their enthusiasm, capacity and energy. When planning the most appropriate method(s) of communication and workforce development, two objectives need to be considered: both the company's interests and skill requirements as well as providing support to the development of people's aspirations.

There is a range of different forms of communication, each with its strengths and limitations. An underlying truism is that few things are as effective as face-to-face communication. However, that is not always possible, nor is it always practical.

In lean practice, the information centre has been shown to be enormously effective, lean workplaces share plans on the wall, work status is indicated, causes of process failure are identified and problem and countermeasure boards not only focus the workgroup's attention, but people walking by are captured by the information and may be able to contribute to solutions.

It is fair to say that the wall of site offices on many construction sites around the world are largely wasted real estate. Companies tend to put up A4-sized sheets of the company's site-related policies (not as if anyone reads them there) for legal reasons, so that in the case of an accident, the company can point to fact that it did everything to let people know about its policies. If you want to focus people's attention on an aspect of safety practice, use an A2-sized poster like the one shown in the VNGC case study (CS2).

Companies have to realize that shaping their communication practices around poorly advised legal reasoning makes little sense. The walls are an empty canvas which can be used to focus people's attention on work plans, process outcomes, lessons learned, work practices, challenges to be met and the celebration of successes. The VNGC case study (CS2) shows a number of photos of visual information on the walls on the site office.

Depending on the goals to be achieved, and the circumstances, the principal forms of communication available are:

- *Verbal communication* either between individuals or groups, using direct or indirect methods such as public address or other broadcasting systems and recordings.

- *Written communication* in the form of notices, bulletins, information sheets, reports, email and recommendations.
- *Visual communication* such as posters, films, videos, internet/intranet, exhibitions, demonstrations, displays and other promotional features. Some of these may also call for verbal and written communication.
- *Examples* in the way people conduct themselves and adhere to established working codes and procedures through their effectiveness as communicators and ability to 'sell' good practices.

The characteristics of each of these methods should be carefully selected to achieve the desired result.

Education and training cycle of improvement

Education and training activities can be considered in the form of a cycle of improvement (Figure 18.2), the elements of which are the following.

Ensure education and training is part of the policy

Every organization should define its policy in relation to education and training. The policy should contain principles and goals to provide a framework within which learning experiences may be planned and operated. This policy should be communicated to all levels.

FIGURE 18.2 The quality training cycle

Establish objectives and responsibilities for education and training

When attempting to set education and training objectives, three essential requirements must be met:

1 Senior management must ensure that learning outcomes are clarified and priorities set.
2 The defined education and training objectives must be realizable and attainable.
3 The main objectives should be 'translated' for all functional areas in the organization. Large organizations may find it necessary to promote a phased plan to identify these.

The following questions are useful first steps when identifying education and training objectives:

* How are the customer requirements transmitted through the organization?
* Which areas need improved performance?
* What changes are planned for the future?
* What are the implications for the process framework?

Education and training must be the responsibility of line management, but there are also important roles for the individuals concerned.

Establish the platform for a learning organization

The overall responsibility of ensuring that education and training has been properly organized must be assumed by one or more designated senior executives. All managers have a responsibility for ensuring that personnel reporting to them are properly trained and competent in their jobs. This responsibility should be written into every manager's job description. The question of whether line management requires specialized help should be answered when objectives have been identified. It is often necessary to use specialists, who may be internal or external to the organization.

Construction organizations have a particular problem: site staff tend to be completely focused on the delivery process. Therefore, unless training is seen as a site responsibility led by senior project management, it will tend to be seen as a head office requirement and a distraction, creating a tension between production and personal development. One way of overcoming this tension is to design certain training to be workplace problem-based, using the challenges on the project as the setting for shared learning within production teams.

Specify education and training needs

The next step in the cycle is to assess and clarify specific education and training needs. The following questions need to be answered:

1 Who needs to be educated/trained?
2 What competencies are required?
3 How long will the education/training take?
4 What are the expected benefits?
5 Is the training need urgent?
6 How many people are to be educated/trained?
7 Who will undertake the actual education/training?
8 What resources are needed (e.g. money, people, equipment, accommodation, outside resources)?

Prepare education/training programmes and materials

Senior management should participate in the creation of overall programmes, although line managers should retain the final responsibility for what is implemented, and they will often need to create the training programmes themselves.

Training programmes should include:

* The training objectives expressed in terms of the desired behaviour.
* The actual training content.
* The methods to be adopted.
* Who is responsible for the various sections of the programme?

Implement and monitor education and training

The effective implementation of education and training programmes demands considerable commitment and adjustment by the trainers and trainees, alike. Training is a progressive process that must take into account any learning problems of the trainees. The authors believe that in construction, given the fragmentation of the supply chain and the consequent need to develop deep relationships within the supply chain, some joint training could be an important part of an effective organizational development strategy.

Assess the results

In order to determine whether further education or training is required, line management should themselves review performance when training is completed. However good the training may be, if it is not valued and built upon by managers and supervisors, its effect can be severely reduced.

Review effectiveness of education and training

Senior management will require a system whereby decisions are taken at regular fixed intervals on the:

* policy;
* education and training objectives;
* education/training organization; and
* progress towards a learning organization.

Even if the policy remains constant, there is a continuing need to ensure that new education and training objectives are set either to promote work changes or to raise the standards already achieved.

The purpose of management system audits or reviews is to assess the effectiveness of the management effort. Clearly, adequate and refresher training in these methods is essential if such checks are to be realistic and effective. Audits and reviews can provide useful information for the identification of changing quality-training needs.

The education/training organization should similarly be reviewed in the light of the new objectives and, here again, it is essential to aim at continuous improvement. Training must never be allowed to become static, and the effectiveness of the organization's education and training programmes and methods must be assessed systematically.

A SYSTEMATIC APPROACH TO EDUCATION AND TRAINING FOR LEAN QUALITY

Education and training for lean quality should have, as its first objective, an appreciation of the personal responsibility for meeting the 'customer' requirements by everyone from the most senior executive to the newest and most junior employee. Responsibility for the training of employees in quality rests with management at all levels, but in particular with the person nominated for the coordination of the organization's quality effort. Education and training will not be fully effective, however, unless responsibility for the deployment of the policy rests clearly with the chief executive. One objective of this policy should be to develop a *climate* in which everyone is quality conscious and acts with the needs of the customer in mind at all times.

The main elements of effective and systematic lean quality training may be considered under the following broad headings:

- error/defect/problem prevention;
- error/defect/problem reporting and analysis;
- error/defect/problem investigation, problem solving and innovation using approaches such as PreStart, Continuous Improvement including waste identification, and Value Stream Mapping; and
- training in lean quality tools and approaches, including Last Planner® System, 5S and Language in Action.
- Review.

The emphasis should be on error, defect or problem prevention, understanding customer satisfaction, problem solving and continuous improvement.

Error/defect/problem prevention

The following contribute to effective and systematic training for prevention of problems in the organization:

1   issued quality policy;
2   written management system;
3   job specifications that include quality requirements;
4   effective steering committees, including representatives of both management and employees;
5   efficient housekeeping standards;
6   preparation and display of maps, flow diagrams and charts for all processes;
7   simple transparent performance measures in all critical outcome areas.

Error/defect/problem reporting and analysis

It will be necessary for management to arrange the necessary reporting procedures and ensure that those concerned are adequately trained in these procedures. All errors, rejects, defects, defectives, problems, waste, etc. should be recorded and analysed in a way that is meaningful for each organization, bearing in mind the corrective action programmes that should be initiated at appropriate times.

This area is a critical challenge on construction projects simply because of the high degree of supply chain fragmentation and the independent way in which subcontractors build on each other's work. This means that effective error identification and reporting have to be done by the employees of subcontractors, who are driven by piece rate payment for their output and, hence, are generally not motivated to report a problem as it may delay their work. The biggest challenge is to develop motivation and procedures across the supply chain.

The Australian Centre for Construction Innovation conducted research into this very question on construction projects in Sydney. With a group of eighteen contractors, general contractors and subcontractors, the research consortium experimented with the areas of motivation, performance measurement and improved information capture, and information flow to identify error detection and improve management response.

That research generated the following insights:

- It measured the cost of rework, the correction of defective work within the contract period at just under 6 per cent of project value.
- It developed the performance metric of 'time from defect occurrence to avoidance of defect repetition' as a metric of management proactivity.
- It introduced the practice of *PreStart* meetings at work handover to identify the level of immediate customer satisfaction and find opportunities for improvement.
- It identified that the greatest motivator for subcontractor excellence was public recognition.

In Chapter 1 there is a section on the identification of waste; most people on construction sites regard waste as inevitable, and they do not see it. People working at all levels in construction have to be taught to see waste.

The core tools of lean quality practice need to be taught to managers, supervisors and work teams alike as they should be practised to create reliability in processes and actions by all parties. They are introduced in Chapter 15 and training could include:

- Process mapping in general and Value Stream Mapping, in particular The Rosendin case study (CS11), gives an excellent example of what can be achieved using VSM.
- Kaizen-based improvement workshops.
- Data recording and analysis.

Error/defect/problem investigation

The investigation of errors, defects and problems can provide valuable information that can be used in their prevention. Participating in their investigations offers an opportunity for training. The following information is useful for the investigation:

- nature of problem;
- date, time and place;
- product/service with problem;
- description of problem;
- causes and reasons behind causes; refer to the '5 Whys' techniques which are introduced in Chapter 15, Force field analysis, affinity diagrams and interrelationship diagraph, and Matrix diagrams;
- action advised;
- action taken to prevent recurrence.

Once again, this area is one of the construction industry's great challenges. Most project managers rely on error detection and rectification to get the job done and never take their project organizations into this critical phase of continuous improvement. Through their experience, the authors are aware of countless examples of quality and safety errors that have been allowed to happen again and again on projects (often detected and corrected), at times ultimately leading to very significant injury or cost impacts. Fragmentation and time pressure are the two main challenges to overcome.

Review of quality training

Review of the effectiveness of quality training programmes should be a continuous process. An assessment of perceived quality of training in the first instance is an important measure and all training courses should be assessed by attendees. However, the measurement of effectiveness is complex. One way of reviewing the benefits of a particular training course is to assess, through audit, how the practices and behaviours of people change before and after the training. This review can be taken a stage further by comparing employees' behaviour with the objectives of the particular quality-training programme. Every organization should routinely assess the quality of training and the benefits derived.

Once again the challenge on construction projects is to ensure that a similar approach to training is implemented across the entire project workforce. Most organizations consider only the training of their own people; however, when more than three-quarters of the workforce on a project are employed by subcontractors, limiting the training strategy to any one organization limits the potential benefits of the training programme.

Education and training records

All organizations should establish and maintain procedures for the identification of education and training needs and the provision of the actual training itself. These procedures should be designed (and documented) to include all personnel. In many situations it is necessary to employ professionally qualified people to carry out specific tasks (e.g. accountants, lawyers, engineers, chemists, etc.), but it must be recognized that all other employees, including managers, must have or receive from the company the appropriate education, training and/or experience to perform their jobs. This leads to the establishment of education and training records.

Once an organization has identified the special skills required for each task and developed suitable education and training programmes to provide competence for the tasks to be undertaken, it should prescribe how the competence is to be demonstrated. This can be by some form of examination, test or certification, which may be carried out in-house or by a recognized external body. In every case, records of personnel qualifications, education, training and experience should be developed and maintained. National vocational qualifications may have an important role to play here.

At the simplest level, this may be a record of training dates and outcomes for each employee for specific skills, including details of attendance on external short courses, in-house induction and training schemes. What must be clear and easily retrievable is each person's training record, including currency of licenses and certificates in areas that require renewal. For example, on a site, qualifications to operate specific equipment should be easy to check. Clearly, as the complexity of jobs increases and managerial activity replaces direct manual skill, it becomes more difficult to make decisions on the basis of such records alone. Nevertheless, they should document the basic competency requirements.

On a construction project, training of subcontract staff is as critical to project success as the training of the general contractor's staff. This includes critical areas such as safety and quality as well as areas of technical competency. However, a prerequisite to ensure that all key subcontractors do maintain training records is a close and durable relationship between the general contractor and the subcontract supply chain.

STARTING WHERE AND FOR WHOM?

Education and training needs to occur at four levels of an organization:

1 *very senior management* (strategic decision-makers);
2 *middle management* (tactical decision-makers or implementers of policy);
3 *first-level supervision* and quality team leaders (on-the-spot decision-makers); and
4 *all other employees* (the doers).

Neglect of education/training in any of these areas will, at best, delay the implementation of lean quality and the improvements in performance. It is also critical that organizations within the supply chain have the same commitment at all levels of their organizations. The provision of training for each group will be considered in turn, but it is important to realize that an integrated programme is required, one that includes follow-up learning-based activities and encourages exchange of ideas and experience.

All the case study companies place a heavy emphasis on developing their people. The Graniterock case study showcases a very solid approach to personal development consistent with the organization's needs using personal development plans for each employee at every level within the organization.

Very senior management

The chief executive and his team of strategic policy makers are of primary importance, and the role of education and training here is to provide awareness and instil commitment to lean quality. The importance of developing real commitment must be established, and often this can only be done by a free and frank exchange of views between trainers and trainees. This has implications for the choice of the trainers themselves; at this level trainers need to be senior and experienced. The fresh-faced graduate, sent by a *package consultancy* operator into the lion's den of a boardroom, will not make much of an impact. The authors recall thumping many a boardroom table, and using all their experience and whatever presentation skills they could muster, to convince senior managers that without a lean quality-based approach they would fail. It is a sobering fact that the pressure from competition and customers has a much greater record of success than enlightenment, though taking a group of senior executives down into the workplace has been successful on more than one occasion.

Executives responsible for marketing, sales, finance, design, operations, purchasing, personnel, distribution, etc. all need to understand lean quality. They must be shown how to define the policy and objectives; how to establish the appropriate organization; how to clarify authority; and generally, how to create the atmosphere in which lean quality will thrive. This is the only group of people in the organization that can ensure that adequate resources are provided and directed at:

1 meeting customer requirements – internally and externally;
2 setting standards to be achieved – zero failure;
3 monitoring of quality performance – quality-related costs;
4 introducing a good quality management system – prevention;
5 implementing process control methods – SPC, Six Sigma and Lean; and
6 spreading the idea of lean quality throughout the whole workforce.

The senior management of any principal contracting organization has the task of finding like-minded suppliers: suppliers who, because of their commitment to the same quality ideals, will deliver services and products that support the overall customer requirements.

Middle management

The basic objectives of management quality training should be to make managers conscious and keen to secure the benefits of the lean quality effort. One particular 'staff' manager will require special training – the lean quality manager – who will carry the responsibility for management of the lean quality management system, including its design, operation and review.

The middle managers should be provided with the technical skills required to design, implement, review and change the parts of the lean quality management system that will be under their direct operational control. It will be useful throughout the training programmes to ensure that the responsibilities for the various activities in each of the functional areas are clarified. The presence of a highly qualified and experienced lean quality manager in the organization should not undermine these responsibilities. An internal 'consultant' can easily weaken the leader's role and create 'not-invented-here' feelings by stepping in without adequate consultation with those charged with implementation.

Middle management should receive comprehensive training on the philosophy and concepts of teamwork, as well as the techniques and applications of data capture and analysis, lean tools and whatever other approaches are to be used. Without the right leadership, knowledge of tools and practices at the workplace level and empowerment, the lean quality management system will lie dormant and lifeless, and teams will continue with business as usual. Middle management must learn how to put the lean quality process together through the planning–process–people–performance value chain creating sustainable change.

A suggestion for organizations with fragmented supply chains is to develop some joint training in lean quality management with their suppliers' managers and supervisors, not only to ensure that values and goals are aligned, but also to build the capability of their supply chains.

First-level supervision

These are the people who have the capacity to change the organization. They are closest to the work teams and their knowledge, attitudes and example have the greatest potential to shape the attitudes and behaviours of the workforce. In a sense, all industries are in transition as younger, more IT-capable people are moving into supervisory roles. However, the experience and knowledge of the older generation of supervisors is valuable, and it is important to bring these people on the change journey. Effort invested in training and especially enthusing the older supervisors about the potential lean quality processes is essential.

One of the authors was in an on-site training situation where after a Last Planner® System training session, a seasoned older veteran of the industry was really engaged and asking questions, when suddenly a young and arrogant site engineer called him out for asking too many 'stupid' questions. Nothing could have been more damaging. On the other hand, any old supervisor who is not capable of coming on the lean quality journey will need to be replaced.

The first level of supervision is where the implementation of lean quality is actually 'managed'. Supervisors' training should include an explanation of the principles of lean quality, a convincing exposition on the commitment to lean quality of the senior management and an explanation of what the lean quality policy means for them. The remainder of their training needs to be devoted to explaining their role in the operation of the lean quality management system, teamwork, collecting and using data to drive focus, and to gaining *their* commitment to the concepts and techniques of lean quality.

It is often desirable to involve the middle managers in the training of front-line supervision in order to:

- ensure that the message they wish to convey is not distorted;
- convince the front-line supervision that the organization's entire management structure is serious about lean quality; and
- demonstrate that everyone is being suitably trained.

The change process requires patience and understanding: old habits are not easily changed; however, any display of arrogance towards the training of supervisors and the workforce can undo an investment in careful planning and will certainly undermine the educational effort.

All other employees

Awareness and commitment at the point of production or service delivery is just as vital as at the very senior level. If it is absent from the latter, the lean quality programme will not begin; if it is absent from the shop floor, lean quality will not be implemented. The training here should include the basics of lean quality and particular care should be given to using easy reference points for the explanation of the terms and concepts.

Most people see the waste and dysfunction in their workplace. Allowing them to talk about it and listening to their views on potential improvements is a certain way of unlocking the door to their interest and commitment. Most people can relate to lean quality and how it should be managed if they can think about its applications in their own lives and at home. Lean quality is really such common sense that, with sensitivity and the use of an approach that is suitable to the experience and prior knowledge of the group, little resistance should be experienced.

All employees should receive detailed training on the processes and procedures relevant to their own work. Obviously they must have appropriate technical or 'job' training, but they must also understand the concepts of waste and the requirements of their customers. A very successful programme has been used by several of the case study companies through what they call 'small wins' programmes. Through this approach, people are encouraged to come up with improvement ideas in their everyday work, to photograph them or create two-minute videos in their smartphones and these are shared within the work team and across entire businesses. Look at the examples in the VNGC (CS2), Southland Industries (CS6) and JB Henderson (CS8), which illustrate the creative ideas that come from the workface.

One problem on construction projects is that while subcontractors may be contractually bound to the general contractor or client in their supply relationships, they build on each other's work and, hence, they need to understand and work towards satisfying the needs of their immediate customer – the following trade. On-site training across the supply chain is one very effective way of breaking down these barriers.

TURNING EDUCATION AND TRAINING INTO LEARNING

For successful learning, training must be followed up. This can take many forms, but the managers need to provide the lead through the design of improvement projects and follow-up 'surgery' workshops.

The authors and their colleagues have found that a successful formula is the in-company training course plus follow-up within a few weeks by a 'surgery' workshop at which participants on the initial training course present the results of their efforts to improve processes and their use of the various methods learned. Specific implementation problems are discussed. A series of such workshops will build depth into people's experience, and may lead on to the creation of process or quality improvement teams which can spread the lessons learned across the organization.

Information and knowledge

Information and knowledge are two words used very frequently in organizations, often together in the context of 'Knowledge Management' and 'Information Technology'. But how well are they managed and what is their role in supporting lean quality?

Recent researchers and writers on knowledge management (e.g. Dawson, 2000) have drawn attention to the distinction between explicit knowledge – one we can express to others – and tacit knowledge, the rest of our knowledge – one we cannot easily communicate in words or symbols.

If much of our knowledge is tacit, meaning that we do not fully know what we know, we will find it very difficult to explain or communicate it. Explicit knowledge can be put into a form that we can communicate to others – words, figures and models in this book are an example of that. In many organizations, however, especially the service sector, much of people's valuable and useful knowledge is tacit rather than explicit.

The creation and expression of knowledge takes place through social interaction between tacit and explicit knowledge, and the matrix in Figure 18.3 shows this as four modes of knowledge conversion.

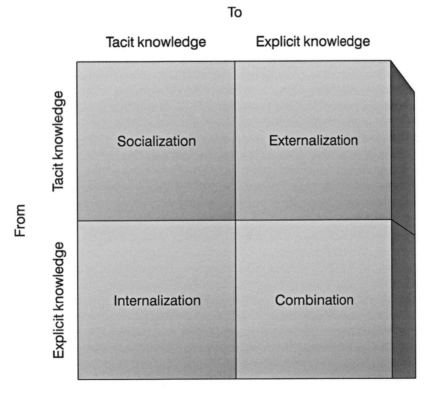

FIGURE 18.3 Modes of knowledge conversion

From *The Knowledge-Creating Company: How Japanese Companies Create the Dynamics of Innovation*, by Ikujiro Nonaka and Hirotaka Takeuchi © 1995, Oxford University Press, Inc. (Source: Dawson, 2000)

Socialization is the term used to describe the sharing of tacit knowledge between people; this takes place through the face-to-face sharing of experiences. This is how apprentices learn from their masters. The conversion of tacit knowledge to explicit knowledge is *externalization*; this involves codifying knowledge and experience so that it can be communicated to many others in an efficient though impersonal way. This may be in a written form or, for example, through a video or PowerPoint presentation. *Internalization* converts explicit knowledge to tacit knowledge by translating it into personal knowledge through reflection and by linking ideas gleaned explicitly with previous tacit knowledge – this is a form of learning. The re-organization or integration of different types of explicit knowledge, for example, creating new conceptual frameworks, is *combination*.

Explicit knowledge as information

When knowledge is made explicit by putting it into words, diagrams or other representations, it can then be typed, copied, stored and communicated electronically – it becomes *information*. Perhaps then a useful definition for information is something that is or can be made explicit. Information, which represents captured knowledge, has value as an input to human decision-making and capability development. Tacit knowledge remains intrinsic to individuals and only they have the capacity to act effectively with its use.

These ideas about information and knowledge enable us to substitute the word *information* for explicit knowledge, and simply use the word *knowledge* – in the business sense of capacity to act effectively – as tacit knowledge. This clarification of the distinction between *information* and *knowledge* makes the knowledge conversion framework more directly applicable to inter-organizational interaction.

In the same way, externalization is capturing people's knowledge – their capacity to act in their business roles – by making it explicit and turning it into information, as in the form of written documentation or structured business processes. This remains information until other people internalize it to become part of their own knowledge – or capacity to act effectively. Having a document on a server or bookshelf does not make individuals knowledgeable, nor does reading it. Knowledge comes from understanding the document by integrating the ideas into existing experience and knowledge, and thus providing the capacity to act usefully in new ways. In the case of written documents, language and diagrams are the media by which the knowledge is transferred. The information presented must be actively interpreted and internalized, however, before it becomes new knowledge to the reader.

The process of internalization is essentially that of knowledge acquisition, which is central to the whole idea of learning, knowledge management and knowledge transfer. Understanding the nature of this process is extremely valuable in implementing effective business improvements and in adding greater value to customers.

Socialization refers to the transfer of one person's knowledge to another person, without an intermediary or captured information in documents. It is a most powerful form of knowledge transfer. As we know from childhood, people learn from other people far more effectively than they learn from books and documents, in both obvious and subtle ways. Despite technological advances that allow people to telecommute and work in different locations, organizations function effectively mainly because people who work closely together have the opportunity for rich interaction and learning on an ongoing and often informal basis. This presents challenges, of course, in today's 'Virtual Organization'.

The learning–knowledge management cycle

One way of thinking about learning and knowledge management is as a dynamic cycle from tacit knowledge to explicit knowledge and back to tacit knowledge. In other words,

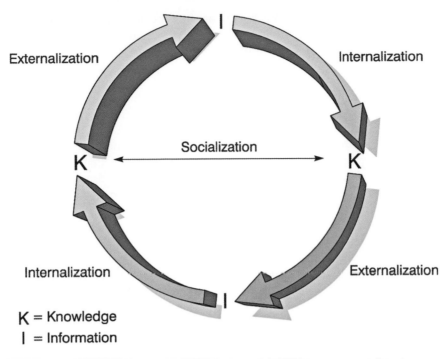

Externalization

Internalization

Socialization

Internalization

Externalization

K = Knowledge
| = Information

FIGURE 18.4 The knowledge management cycle

people's knowledge is externalized into information, which to be useful must then be internalized by others (learned) to become part of their knowledge, as illustrated in Figure 18.4. This flow from knowledge to information and back to knowledge constitutes the heart of organizational learning and knowledge management. Direct sharing of knowledge through socialization is also vital. In large organizations, however, capturing whatever is possible in the form of documents and other digitized representations means that information can be stored, duplicated, shared and made available to people on whatever scale desired.

The fields of learning and knowledge management encompass all the human issues of effective externalization, internalization and socialization of knowledge. As subsets of those fields, information management and document management address the middle part of the cycle, in which information is stored, disseminated and made easily available on demand. It is a misnomer to refer to information sharing technology, however advanced, as knowledge management. Effective implementation of those systems must address how people interact with technology in an organizational context, which only then is beginning to address the real issues of knowledge.

THE PRACTICALITIES OF SHARING KNOWLEDGE AND LEARNING

In world-class organizations there is clear evidence that knowledge is shared to maximize performance, with learning, innovation and improvement encouraged. In such establishments information (explicit knowledge) is collected, structured and managed in alignment with and in support of the organization's policies and strategies. These days this is often achieved in large organizations through intranet and/or common network mounted file servers. These can provide common access to online reports, training material and dash-boards of performance figures vs targets. The key is to provide appropriate access for both internal and external users to relevant information, with assurance required of its validity, integrity and security.

Another important element of knowledge sharing in large organizations and across the supply chain is the effective exchange of tacit information among people – creating networks within the organization and its integrated supply chain. This is probably best supported by an 'intranet'; however, the goal, rather than capturing the knowledge, is to identify knowledge sources within the expanded organization and putting people in touch with others who may be able to assist in solving a problem.

Cultivating, developing and protecting intellectual property can be the key in many sectors to maximizing customer value, for example in professional services, such as legal and consultancy. Firms in this sector can survive only if they are constantly seeking to acquire, increase, use and transfer knowledge effectively. This is the process of organizational learning and it underpins the capacity for innovation. The proper use and management of information and knowledge resources leads to the generation of creative thinking and innovative solutions. It is important to make information available as widely as possible on the most appropriate basis to improve general understanding and increase efficiency. It is important for management to know whether the knowledge system is being used and is generating value. Nowadays, with the ability to automatically capture data on the number of 'views' of information, and the number of ideas being posted, this data can be used to support conclusions as to the effectiveness of the information system. It is important, however, to recognize that as organizations get larger, a critical function of any knowledge management system is its ability to put a person on site somewhere in Australia in touch with a colleague on the other side of the globe who has tacit knowledge that can help to solve a burning problem down under. The ability to leverage tacit knowledge is a critical frontier in organizational knowledge management and development.

In the EFQM Excellence Model there are now clear feedback loops of 'Innovation and Learning', both from the results – on people, customers, society and key performance – to the enablers and within the enablers themselves. The challenge in the construc-tion industry is the capture and effective use of data. High-level comparisons of say customer satisfaction, branch margin and project margin can only be meaningfully analysed if there is detailed performance data at the process level. For example, to compare projects, Costain (CS12) have found that an early leading indicator of the ultimate performance of a project is the degree of compliance with the Costain Way assessed only six weeks into the project. The Costain Way defines 33 project requirements with detailed compliance criteria.

There may also be learning and innovation loops and opportunities within the enabler's criteria. For example, key performance indicators related to the management of processes, such as cycle time or conversion of bids to businesses won, can help to generate innovations in strategies, particularly if there is a good alignment between the two. Information on staff, customers and other sources of information can be readily kept on databases and their effective and efficient use can mean the difference between success and failure in many industries and sectors.

BIBLIOGRAPHY

Adair, J. *The Challenge of Innovation,* Talbot Adair Press, Guildford, 1990.
Dawson, R. *Developing Knowledge-Based Client Relationships,* Butterworth-Heinemann, Oxford, 2000.
Ettlie, J.E. *Managing Technological Innovation,* Wiley, London, 2000.
Francis, D. *Unblocking the Organizational Communication,* Gower, Aldershot, 1990.

Larkin, T.J. and Larkin, S. *Communicating Change – Winning Employee Support for New Business Goals*, McGraw-Hill, New York, 1994.

Purdie, M. *Communicating for Total Quality*, British Gas, London, 1994.

Tidd, J., Bessant, J. and Pavitt, K. *Managing Innovation*, Wiley, London, 2001.

Zairi, M. *Best Practice Process Innovation Management*, Butterworth-Heinemann, Oxford, 1999.

CHAPTER HIGHLIGHTS

Communicating the lean quality strategy and message

- People's attitudes and behaviour can be influenced by communication, and the essence of changing attitudes is to gain acceptance through excellent communication processes.
- The strategy and changes to be brought about through lean quality should be clearly and directly communicated from top management to all staff/employees both within the organization and across the supply chain. The first step is to issue a 'lean quality message', which should be followed by a signed directive.
- People must know when and how they will be brought into the lean quality process; what the process is; and the successes and benefits achieved. First-line supervision has an important role in communicating the key messages and overcoming resistance to change both within the organization and among the employees of supply chain partners.
- The complexity and jargon in the language used between functional groups needs to be reduced in many organizations. Simplify and shorten are the guiding principles.
- 'Open' methods of communication and participation should be used at all levels. Barriers may need to be broken down by concentrating on process rather than 'departmental' issues.

Communicating the lean quality message

- There are four audience groups in most organizations – senior managers, middle managers, supervisors and employees – each with different general attitudes towards lean quality and the senior management must ensure that each group sees it as being beneficial.
- Good leadership is mostly about good communications: skills that can be learned through training but must be acquired through practice.

Communication, learning, education and training

- There are four principle types of communication: verbal (direct and indirect), written, visual and by example. Each has its own requirements, strengths and weaknesses.
- Education and training is the single most important factor in improving quality and performance, once commitment is present. This must be objectively, systematically and continuously performed.
- All education and training should occur in an improvement cycle of ensuring it is part of policy, establishing objectives and responsibilities, establishing a platform for a learning organization, specifying needs, preparing programmes and materials, implementing and monitoring, assessing results and reviewing effectiveness.

A systematic approach to education and training for lean quality

- Responsibility for education and training of employees rests with management at all levels. The main elements should include error/defect/problem prevention, reporting and analysis, investigation and review.

- Education and training procedures and records should be established to show how job competence is demonstrated.

Starting where and for whom?

- Education and training needs occur at four levels of the organization: very senior management, middle management, first-level supervision and quality team leaders, and all other employees.

Turning education and training into learning

- For successful learning all quality training should be followed up with improvement projects and 'surgery' workshops.
- It is useful to draw the distinction between explicit knowledge (that which we can express to others) and tacit knowledge (the rest of our knowledge which cannot be communicated in words or symbols).
- The creation and expression of knowledge takes place through social interaction between tacit and explicit knowledge, which takes the form of socialization, externalization, internalization and combination.
- When knowledge is made explicit, it becomes 'information', which in turn has value as an input to human decision-making and capability. Tacit knowledge (simply 'knowledge') remains intrinsic to individuals who have the capacity to act effectively in it use.
- One way of thinking about learning and knowledge management is as a dynamic cycle from tacit knowledge to explicit knowledge (information) and back to tacit knowledge.

The practicalities of sharing knowledge and learning

- In world-class organizations there is clear evidence that knowledge is shared to maximize performance, with learning, innovation and improvement encouraged. This is often achieved through an intranet or common network mounted file servers, providing common online access to information.
- Managing intellectual property is key to success in many sectors and this has strong links with learning and innovation. Where information must be made available as widely as possible, excellent knowledge management is essential.
- The clear feedback loops of 'innovation and learning' in the EFQM Excellence Model drive increased understanding of the linkages between the results and the enablers, and between the results and the enablers, and between the enabler criteria themselves.

Part V Discussion questions

1 The so-called process approach has certain implications for organizational structures. Discuss the main organizational issues influencing the involvement of people in process improvement.

2 Various lean quality teamwork structures are advocated by many writers. Describe the role of the various 'lean quality teams' in the continuous improvement process. How can a construction organization ensure that the outcome of teamwork is consistent with its mission?

3 Describe the various types of lean quality teams which should be part of introducing a lean quality approach in an infrastructure organization. Explain the organizational requirements associated with these and give some indication of how the teams should operate.

4 A large residential homebuilder has decided that teamwork is to be the initial focus of its lean quality improvement programme. Describe the role of a Lean Quality Council or Steering Group and Process Improvement Team in managing teamwork initiatives in lean quality improvement.

5 Explain the difference between quality improvement teams and quality circles. What is their role in quality improvement activities?

6 Discuss some of the factors that may inhibit teamwork activities in a lean quality programme.

7 Suggest an organization for teamwork in a quality improvement programme and discuss how the important aspects must be managed, in order to achieve the best results from the use of teams. Describe briefly how the teams would proceed, including the tools they would use in their work on a large complex office block building site.

8 Describe in full the various types of process improvement teams which are necessary in a lean quality programme. Give some indication of how the teams operate at each level and, using the DRIVER model, discuss the problem-solving approach that may be adopted.

9 Discuss the various models for teamwork within a lean quality approach to business performance improvement. Explain through these models the role of the individual in lean quality, and what work can be carried out in this area to help cross-functional construction teams through the 'storming' stage of their development.

10 Teamwork is one of the key 'necessities' for lean quality improvement. John Adair's 'Action-centred Leadership' model is useful to explain the areas which require attention for successful teamwork. Explain the model in detail showing your understanding of each of the areas of 'needs'. Pay particular attention to the needs of the individual, showing how a psychometric instrument, such as the Myers-Briggs Type Indicator (MBTI) or FIRO-B, may be useful here.

Part VI

Implementation

Implementing lean quality **19**

LEAN QUALITY AND THE MANAGEMENT OF CHANGE

John Oakland recalls the managing director of a large support services group who decided that a major change was required in the way the company operated if serious competitive challenges were to be met. The Board of Directors went away for a weekend and developed a new vision for the company and its 'culture'. A human resources director was recruited and given the task of managing the change in the people and their attitudes. After several programmes aimed at achieving the required change, including a new structure for the organization, a staff appraisal system linked to pay and seminars to change attitudes, very little change in actual organizational behaviour had occurred.

Clearly something had gone wrong somewhere. But what, who, where? Everything was wrong, including what needed changing, who should lead the changes and, in particular, how the change should be brought about. This type of problem is very common in organizations that desire to change the way they operate to deal with increased competition, a changing marketplace, new technology or different business rules. In this situation, many organizations recognize the need to move away from an autocratic management style with formal rules, hierarchical procedures and narrow work demarcations. Some have tried to create teams to delegate (perhaps for the first time) and to improve communications.

Some of the senior managers in such organizations recognize the need for change to deal with the new realities of competitiveness, but they lack an understanding of how the change should be implemented. They often believe that changing the formal organizational structure, having 'culture change' programmes and new payment systems will, by themselves, make the transformations. In much research work carried out by The Oakland Institute (*previously ECforBE*) it has been shown that there is almost an inverse relationship between successful change and having formal organization-wide 'change programmes'. This is particularly true if one functional group, such as HR, 'owns' the programme.

In several large organizations in which lean quality management has been used successfully to effect change, the senior management did not focus on formal structures and systems but set up *process management* teams to solve real business or organization problems. The key to success in this area is to align the employees of the business, their roles and responsibilities with the organization and its *processes*. When an organization focuses on its key processes, that is, the core activities and tasks rather than on abstract issues such as 'culture' and 'participation' then the change process can begin in earnest.

An approach to change based on process alignment, and starting with the vision and mission statements, identifying the critical success factors *and* moving on to the core processes, is the most effective way to engage the staff in an enduring change process (see Chapter 5).

Although all the case studies give glimpses of how companies implement lean quality and through those we can see that they vary widely, two interesting and contrasting stories about implementation are found in the JB Henderson and Graniterock cases. The common thread is that both define goals through the outcomes that are to be achieved, and they focus on improvement at the process level. The JB Henderson case study describes the genesis of a lean quality journey on a single project; it starts with the realization of how much time is wasted in every worker's day. This led to a very sharp focus on improving labour productivity. The company created a framework for learning, continuous improvement and collaboration and empowered its workers to come up with ways of becoming more efficient, both on site and in the fabrication shop. In contrast the Graniterock case study describes the very detailed process that has driven this company's lean quality journey for 20 years. This is a process through which an entire company creates its plans for the forthcoming year, defining detailed targets, roles and responsibilities. It then decides on specific activities and tasks that are designed to achieve them. This case study describes how every part of the company is engaged in continuous process improvement driven by internal and external customer feedback, suggestions from its workforce and by continuously scanning technology for technology breakthroughs that create new opportunities for the business.

Many change programmes do not work because they begin trying to change the attitudes and beliefs of individuals. This is based on the theory that changes in attitudes and beliefs will lead to changes in behaviour throughout the organization. It relies on a form of religion spreading though the people in the business.

However, what often works best is almost the opposite. It is based on the belief that people's behaviour is determined largely by the roles they are given and the goals they are set. If they are given clear goals and responsibilities, and team goals are clearly defined within a process-driven environment, a very different situation develops: one that focuses everyone's attention and energy on the improvement of work processes. This in turn changes the culture. *Teamwork* is an especially important part of the lean quality model in terms of bringing about change.

The Last Planner® System is an interesting example of just this principle. LPS® is a team-based planning process designed to improve the quality and reliability of weekly work plans. To that end, entire project teams engage with the challenges of improving foresight, gaining explicit commitments from team members for their forthcoming work, and with the clearer use of language they learn to formulate sound requests and make reliable promises. If all these practical work-focused behaviours are done with the right intentions, a collaborative team culture is created without the word culture ever being mentioned in any meeting.

At the level of the corporation, if changes are to be made in quality, costs, market, product or service development close coordination among the marketing, design, production/operations and distribution groups is essential. This can best be brought about by tasking multifunctional teams to work on a company's key business processes and understanding their interrelationships. In addition to the knowledge of the business as a whole (which will be brought about by an understanding of the mission, CSFs and process breakdown links), certain *tools, techniques* and *interpersonal skills* are required for good *communication* around the processes. These are essential if people are to identify and solve problems as teams.

Finally, senior level *commitment* is a key element of support for the high levels of co-operation, initiative and effort that are required to change the processes existing in most organizations. This is often a stumbling block in construction organizations as executive management generally struggles with the development of an effective implementation strategy. Senior management usually delegates implementation and funding of initiatives to the project level, and often does not take a personal interest in the change project. Without proper senior level leadership, resourcing or commitment change initiatives are likely to fail.

If any of these elements, *closely coordinated multifunctional teams, excellent communication* and *senior management commitment,* are missing, the total change process will collapse.

Two case studies, those of the VNGC Hospital project (CS2) and the Multinational High-Technology Manufacturer (CS4), are particularly interesting here. In each case the client mandated the adoption of a lean quality-based framework on its projects by its entire supply chain. Both clients had become convinced that a lean quality-focused collaborative supply chain would yield the best outcomes for them on their major projects. In such cases, the client (if sufficiently significant for the suppliers' business) can be the initial driver behind a lean quality transformation in the entire supply chain. Both clients in these case studies are shaping the business process development focus of their regular suppliers. The transformation is starting at the local level and then spreading back up through the supplier corporation.

Organizations will avoid the risk of 'failed change programmes' by concentrating on creating alignment around the need to improve core processes – recognizing that people's goals, roles and responsibilities must be related to the processes in which they work. Senior managers should begin the process of alignment by developing clearly articulated improvement goals for core processes, defining milestones and metrics that can be used to guide and assess progress in the change process.

In the introduction of a lean quality-based approach for managing change, timing can be critical and an appropriate starting point is often a broad review of the organization and the changes required by the top management team. This is the starting point of the annual strategic review process undertaken by the Graniterock executive team. By gaining this shared diagnosis of what changes are required, what the 'business' problems are and/or what must be improved, the most senior executive mobilizes the initial commitment that is vital to begin the change process.

In getting the top team working as a team and in creating effective teams to drive process change, techniques such as MBTI and/or FIRO-B will play an important part (see Chapter 17).

PLANNING THE IMPLEMENTATION OF LEAN QUALITY IMPROVEMENT

The task of implementing lean quality can be daunting, and the chief executive faced with this may draw little comfort from the 'quality gurus'. The first decision is where to begin, and this can be so difficult that many organizations never get started. This has been called TQP – total quality paralysis!

The chapters of this book have been arranged in an order which should help senior management bring total quality into existence. The preliminary stages of understanding the challenges and committing to change are vital first steps that also form the foundation and direction of the whole lean quality strategy. Too many organizations skip these phases, believing that they already have the right attitude and awareness, when, in fact, there are some fundamental gaps in their 'lean quality credibility'. These will soon lead to insurmountable difficulties and the collapse of the process.

While an intellectual understanding of quality and efficiency provides a basis for lean quality, it is clearly only the planting of the seed. The understanding must be translated into commitment, policies, plans and actions for lean quality to germinate. Making this happen requires not only commitment but a competence in leadership and in making changes and a trust in the capacity of the workforce to come up with the necessary changes. Without a strategy to implement lean quality through process management, capability and control, the expended effort will lead to frustration. Poor quality management can become like poor gardening: a few weed leaves are pulled off only for others to appear in their place days later, plus additional weeds elsewhere. Problem solving is very much like weeding – tackling the root causes and digging deep is essential for better control.

Individuals working on their own, even with a plan, can only ever generate limited results. This is like a single construction project attempting to initiate changes locally. The results will be limited to that project, unless they are a part of a larger strategy to develop lead practices and to disseminate lessons learned across the organization. Individual efforts are required for improvement to take place, but they must be coordinated and become integrated with the efforts of others to be truly effective.

The implementation begins with the drawing up of a lean quality policy statement, and the establishment of the appropriate organizational structure, both for managing and encouraging involvement in process efficiency and quality improvement through teamwork. The planning stage will involve all managers. Collecting information on process inputs and outcomes, including the reliability of workflows and the costs of quality helps to identify the prime areas in which improvements will have the largest impact on performance. A crucial early stage involves putting lean quality systems in place to monitor that quality standards are maintained and to make sure that any problems encountered remain solved forever, using structured corrective action procedures. It is also important to monitor customer satisfaction. At the same time, the organization must actively search for waste and opportunities for continuous process improvement.

Once the plans and systems have been put into place, the need for continued education, training, communication and evaluation becomes paramount. Organizations that try to change the culture, operating systems, procedures or control methods without effective, honest two-way communication will experience the frustration of being a 'cloned' type of organization that can function but inspires no confidence in being able to survive the changing environment in which it lives.

An organization may, of course, have already taken several steps on the road to lean quality. If a good understanding of lean quality and how it should be managed already exists, then there is top management commitment, a written quality policy and a satisfactory organizational structure, and the planning stage is a routine annual event, as illustrated in the Graniterock case study. Whenever lean quality improvements are contemplated, in either a new implementation or an ongoing one, priorities amongst various projects must be identified. For example, a lean quality system which conforms to the requirements of ISO 9001 may already exist, and the systems step will not be a major task, but introducing a quality costing system may well be. It is important to remember, however, that a review of current performance in all key areas of a business, even when well established, should be part of normal operations to ensure continuous improvement.

The major steps, illustrated in Figure 19.1, may be used as an overall planning aid for the introduction of lean quality, and individual steps and major projects should appear on a planning or Gantt chart. Major projects should be time-phased to suit each individual unit's requirement, but this may be influenced by outside factors such as the need to meet the requirements of a standard or a specific client requirement. Major projects may need to be split into smaller sub-projects as better control and review can be exercised on smaller discrete work plans with their own tasks and milestones. This is certainly true of management system work, the introduction of statistical processes and the establishment of improvement teams.

The education and training component will be continuous and draw together the requirements of all the steps into a cohesive programme of instruction. The timing of the training inputs, follow-up sessions and advisory work should be coordinated and reviewed on a regular basis.

It is desirable at regular stages of the implementation to undertake checks and reviews. For example, before seeking top management commitment, objective evidence should be obtained to show that the next stage is justified. Commitment should be demonstrated by the publication of a signed quality policy; there may be the formation of a Council, and/or Steering Committee(s). Delays here will prevent real progress being made towards lean quality through teamwork activities.

The launch of process improvement requires a balanced approach, and the three major components must be 'fired' in the right order to lift the campaign off the ground.

1   A good system of management must be established first of all to define leadership, commitment, and to provide a framework for teams to follow.
2   Goals, milestones and performance measures must be in place before teams can start their work, otherwise support for the initiatives will weaken with time.
3   Effective improvement teams with clear goals can then get on with the job to make the initiatives come 'alive', and a regular reporting structure is required to ensure the initiative stays on track.

An effective coordination of these three components will result in lean quality improvements through increased capability. This should in turn lead to consistently satisfied customers and improved outcomes in terms of efficiency and product and process quality.

Lean quality management may be integrated into the strategy of any organization through an understanding of the core business processes and involvement of the people. In Figure 19.1, the recommended framework shows that it all starts with the vision and high-level goals set by the executive. The mission, strategies and detailed plans should be fully thought through, agreed and shared within the business. What then follows determines whether these are achieved. The factors which are critical to success, the CSFs – the building blocks of the mission – are then identified. The key performance indicators (KPIs) and metrics, the measures associated with the CSFs, tell us whether we are moving towards or away from the mission or just standing still.

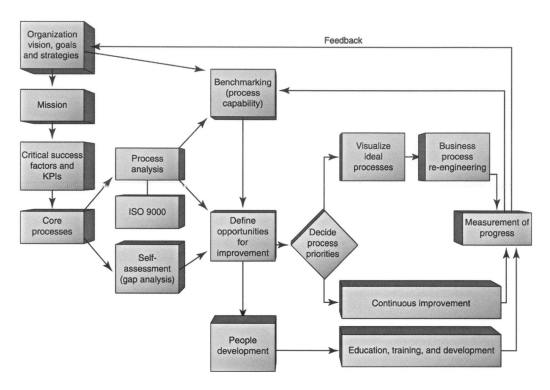

FIGURE 19.1 The framework for implementation of lean quality

Having identified the CSFs and KPIs, the organization should know its *core processes*. This is an area of potential bottleneck for many organizations because, if the core processes are not understood, the rest of the framework is difficult to implement. If the processes are known, we can carry out process analysis, sometimes called mapping and flowcharting, to fully understand our business and identify opportunities for improvement. Once this process is well underway, rather than duplicating the ISO 9001 compliance effort, the lean quality process improvement initiative and the ISO compliance effort should be integrated.

The identified processes should be prioritized into those that require continuous improvement, those which require re-engineering or redesign and those which lead to a complete rethink or visioning of the business.

Self-assessment to the European (EFQM) Excellence Model or Baldrige Quality model and benchmarking will identify further improvement opportunities. This will create a very long list of things to attend to, many of which will require people development, training and education. What is clearly needed next is prioritization – to identify those processes which are running pretty well. They may be marketing and the estimating processes; they should be subjected to a continuous improvement regime. Those processes which are identified as being poorly executed – perhaps workflow reliability, aspects of on-site quality or even financial management – should be completely re-visioned and redesigned. That is where process mapping and BPR come in.

Performance-based measurement of all processes and people development activities is necessary to determine progress so that the vision, goals, mission and critical success factors may be examined and reconstituted if necessary to meet new requirements for the organization and its customers, internal and external.

World-class organizations, of which there need to be more in most countries, are doing all of these things. They have implemented their version of the framework and are achieving world-class performance and results. What this requires first, of course, is world-class leadership and commitment.

In many successful companies lean quality is not the very narrow set of tools and techniques often associated with failed 'programmes' in organizations in various parts of the world. It is part of a broad-based approach used by world-class companies such as those presented in the case studies in this book, organizations that have achieved organizational excellence, based on customer results: the highest weighted category of all the quality and excellence awards. Lean quality embraces *all* of these areas. If used properly and fully integrated into the business, lean quality will help any organization (including private sector clients and public sector agencies) deliver its goals, targets and strategy. This is because it is about people and their identifying, understanding, managing and improving processes – the things any organization has to do particularly well to achieve its objectives.

CHANGE CURVES AND STAGES

A useful device in thinking about any implementation programme is the change curve (Figure 19.2). This represents a journey on which employees of both the company and its key supply chain partners need to be taken if change is to lead to actions and be successful. The five stages that can be expected on such a well-managed journey are noted below.

- *Unaware*: employees have a sense that something needs to be done but not what, how or why.
- *Awareness*: employees starting to become aware of what changes are needed, where the business wants to be and how to get there.

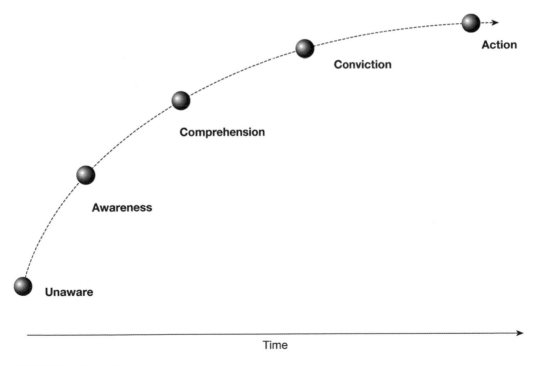

FIGURE 19.2 The change curve

- *Comprehension*: employees understand the desired environment and have a more detailed understanding about how the organization intends to reach its goals; they will be concerned about how the change is likely to affect them.
- *Conviction*: employees understand the change and are reaching their own conclusions with respect to the necessity and feasibility of the change and the effect upon them.
- *Action*: employees will be ready to take part in actions and plans that will deliver the change.

The change curve is useful to plan the various changes of involvement and engagement. For example, Figure 19.3 shows three stages (there may be more, depending on the organization's size and complexity) in which the change curve repeats, in this case, through different layers of management and staff. It shows how the impetus for change develops to a point where it becomes 'irresistible'.

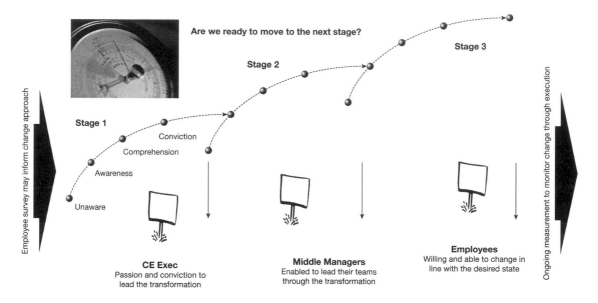

FIGURE 19.3 The stages of change

Stage 1 – **CE Exec** – passion and conviction to lead the transformation:

- clear and compelling driver for change;
- sense of urgency for the change;
- clear vision and goals; and
- power to make the change happen.

Stage 2 – **Middle Managers** – enabled to lead their teams through the transformation:

- understand what and why change is happening;
- support the need to change;
- have confidence that the CE Exec can successfully lead the transformation;
- understand how the change will affect their team and those with whom they interact;
- understand the steps to be taken to effect the change;
- have confidence that the 'system' is being considered;
- have sufficient information to guide their teams;

- have appropriate 'tools' to help guide change;
- feel that they can rely on support from their leaders;
- have a clear idea of their contribution to the change and clarity over the effect on them.

Stage 3 – **Employees** – willing and able to change in line with the desired state:

- understand what and why change is happening;
- support the need to change;
- have confidence that their line managers can help them through the change;
- understand how the change will affect them and those with whom they work;
- understand the steps to be taken;
- have access to the information they need to help them understand/monitor change;
- have confidence that they will be equipped to deal with changes that affect them; and
- have a clear idea of their contribution to the change.

Overcoming resistance to change

There can be many reasons why people resist change and, as far as possible, they need to be identified and thought about if that resistance is to be overcome. Common reasons include:

1. *Perceived negative outcomes* – change can unleash a multitude of fears: of the unknown, loss of status or position, loss of freedom, loss of responsibility and/or authority, and loss of good working conditions and, of course, money.
2. *Fear of more work* and less opportunity for rewards.
3. *Habits must be broken* – dozens of interrelated habits lead to a style of management.
4. *Lack of communication* – the organization does not effectively communicate the what, why and how of change and does not clearly spell out expectations for future performance.
5. *Failure to align with the organization as a whole* – the organization's structure, business systems, technology, core competencies, employee knowledge and skills, and culture are not aligned and/or integrated with the change effort.
6. *Lack of clear leadership commitment* – in many construction organizations, leaders do not give a clear lead. They show interest and then delegate implementation to project managers without demanding following-up reporting. This is a subset of the previous point but is worth mentioning as it is a particular failing in the sector.
7. *Employee rebellion* – there is a view that people do not resist the intrusion of something new into their lives as much as they resist the resulting loss of control; this can be a huge yet 'hidden' factor.
8. *Lack of trust* – in construction, as some 85 per cent of work on site is undertaken by subcontract labour, site management has to empower subcontractor leaders and trust that their input will help to create the best work plans.

A complementary change curve can be helpful in overcoming resistance to change (Figure 19.4). The stages likely to be encountered include:

1. *Pre-awareness*: a sense exists that something needs to be done but not what, how or why.
2. *Awareness*: thoughts about what changes are needed, where we want to be and how to get there are coming into focus but are not yet defined.
3. *Self-concern*: the desired environment and possibly some elements of the projects are now seen in detail, whereupon the concern 'How will this affect me?' becomes primary. At some point between self-concern and mental tryout the transition to acceptance begins.

4 *Mental tryout*: changes are beginning to be viewed as inevitable, attitudes shift to 'How do I make this work for me?'
5 *Hands-on*: simulation of the new environment in the form of pilot projects, prototypes or training is formalized. The point of no return is reached somewhere between hands-on and acceptance when the momentum for the change and near-acceptance of change has increased to the point from where turning back is unlikely.
6 *Acceptance*: the changed order of things is achieved, and the new environment becomes status quo.

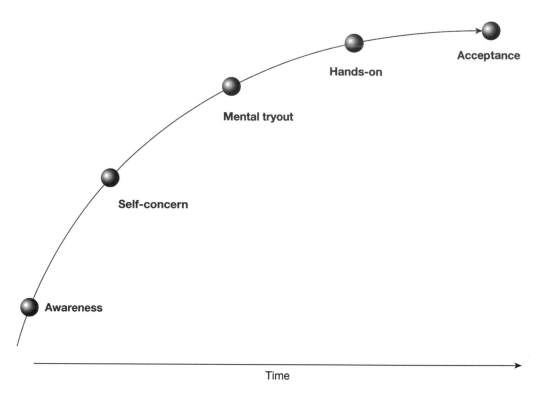

FIGURE 19.4 Overcoming resistance to change

Managing change effectively

In Chapter 11 on Benchmarking and Change Management, the Figure of 8 framework for the effective management of change was introduced (Figure 19.5). The essence of achieving successful change during the implementation of approaches such as TQM, Lean Six Sigma, Continuous Improvement involves:

- *Establish a need to change*; if you want people to change don't give them a choice.
- *Create a clear and compelling vision* that shows people how their lives will be better; without this, any transformation effort can easily degrade into a long list of incompatible and time-consuming, often confusing, projects that can take people in the wrong direction.
- *Go for true performance results and create early wins*; successful change programmes begin with results – clear, tangible, bottom-line results – and the earlier they occur the better.

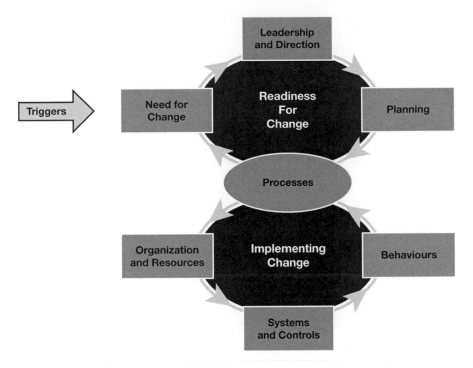

FIGURE 19.5 The organizational change framework

- *Communicate, communicate, communicate;* poor and/or inadequate communication is one of the chief reasons for failed change efforts.
- *Build a strong, committed, guiding coalition* that includes top management.
- *Redesign the processes, systems and structures;* if we fail to look at the processes that are needed for the 'new order of things' together with the structures and systems around them a lot of old behaviour can just be reinforced – the new desired behaviours need to be rewarded.
- *AND FINALLY – People do not resist their own ideas;* people who participate in deciding *what* and *how* things will change not only are more likely to support the change but are actually changed themselves by the act of participation.

USING CONSULTANTS TO SUPPORT CHANGE AND IMPLEMENTATION

Sometimes viewed as a costly necessity by business owners, the role of a consultant can be perceived as intrusive and disruptive. Used wisely, however, consultants can provide specialist skills, offer sound advice, suggest practical solutions and inject new life into a business without costing the earth or upsetting staff.

However well a business is managed, problems exist in every organization, caused either by external or internal events. Good leaders recognize this fact and tackle problems head on, plan changes to address them and implement solutions. In some problem-solving or change management situations, the best strategy is to combine internal and external resources in client/consultant teams. This can enhance the skills and competence of internal people, significantly reduce the costs of solving the problem or making the changes effectively, and avoid the internal resistance that can be associated with consultants.

Performance problems typically manifest themselves in gaps between expected and actual performance, whether this is the quality of a product (through the eyes of a customer), failure to deliver something on time, excessive costs or poor profitability. So the first step in identifying problems is to provide accurate performance data. Sometimes performance gaps require specialist skills that may not exist within an organization, and it is not feasible to develop the in-house capability. If the problems are non-recurring and there will be no ongoing use for the skills, the use of a consultant is appropriate. They can also fill a competence gap until management can obtain the necessary skills or find time to close the gaps in performance.

If the decision is made to employ a consultant, users have to be clear about who should be involved during the assignment and ensure that staff have the time required to do the work that the organization has to undertake. They should manage each stage of the consultancy and ensure that all stakeholders are signed up to the diagnosis, the recommendations, the planning and implementation.

The brief for potential consultancies should always be carefully defined. It should include a short description of the organization and of the problem or situation that has led to requiring a consultant. An outline of what the consultancy is expected to achieve, the starting date and expected duration of the work, some idea of how the consultancy is expected to proceed and who might be involved should also be in the initial briefing. It will be helpful to think through whether the consultant will be asked to:

1   perform a specific task (e.g. draw up a quality plan, develop a strategy);
2   help people think and talk through what needs to be done regarding a particular task;
3   help improve the way people communicate with and relate to each other;
4   help resolve a conflict; and
5   conduct training / coaching to help people learn specific skills or increase awareness.

How to choose a consultant

The professional and ethical standards that guide the consultant chosen should complement the organization's values and philosophy. It will then be necessary to determine how specifically the consultant's knowledge, competence and past experience relates to the client's own particular issues and needs. It will be important to find someone who seems to offer an effective mix of thought leadership, expertise and experience. They should also be seen to work closely with clients to address their needs, both in the short and long term.

The ethos driving the consultant's approach should be to ensure sustained improvement as they transfer their knowledge to the client organization. They should also provide customer feedback that confirms an understanding of the real issues and should be adept at applying a set of basic consulting competences such as diagnosis, strategic planning, change management and execution in addition to specific technical skills such as Lean Six Sigma.

During the consultancy

Consultancy can be a sensitive matter, so it is vital that the consultant has the appropriate interpersonal competences, including problem root-cause analysis and confrontation skills, risk management, collaboration, conflict management and relationship building.

They *should also be recogniz*ed by a trade body such as the Management Consultancies Association (www.mca.org.uk). Consultants are there to assist, so clients should not be defensive – they need to describe and explain all the issues surrounding the project. Clients should not be afraid of providing too much information; consultants much prefer this to a lack of detail. Make sure you have identified the potential 'life of project' (the time the project will take, including implementation), the expected benefits as well as the initial investment required.

A consultancy project usually commences with a diagnostic of the situation, hopefully with an internal team, and employs a variety of techniques to induce high levels of internal commitment in the change process, building support for actions necessary for its success. A final and very important step at the end of the project or programme of work is to undertake a joint post-programme/project completion review with the consultancy team to see what has been gained and learned from the project or programme, and to give feedback to the consultant–client team. This should take place within six months of completion.

Getting the best from an ongoing consultancy project:

- closely liaise with the consultancy team;
- define, with the consultant, what information will be available and when;
- identify and communicate to the consultancy team any problems so remedial action can be taken promptly;
- obtain routine progress reports on the project from the consultant, with interim results;
- hold regular progress reviews with actions agreed in writing;
- where necessary and where agreed, provide staff, facilities and information promptly;
- be open to innovative approaches and methods proposed by the consultancy team, but seek supporting evidence for these;
- establish an agreed protocol for dealing with proposed changes in the client/consultant team;
- ensure client staff and the consultancy team are briefed on the confidentiality state of information; and
- undertake a joint post-project completion review with the consultancy team within six months of completion to build on the experience.

SUSTAINED IMPROVEMENT

Never-ending or continuous improvement is probably the most powerful concept to guide management. It is a term not well understood in many organizations, although that must begin to change if those organizations are to survive. To maintain a wave of sustained improvement, it is necessary to develop generations of managers who not only understand but are dedicated to the pursuit of never-ending improvement in meeting external and internal customer needs.

The concept requires a systematic approach to lean quality management that has the following components:

1. *planning* the processes and their inputs;
2. *providing* the inputs;
3. *operating* the processes;
4. *evaluating* the outputs;
5. *examining* the performance of the processes;
6. *modifying* the processes and their inputs.

This system must be firmly tied to a continuous assessment of customer needs and satisfaction. It depends on a flow of ideas on how to make improvements, reduce variation and generate greater customer value. It also requires a high level of commitment, and a sense of personal responsibility from those operating the processes.

The never-ending improvement cycle ensures that the organization learns from results, standardizes what it does well in a documented process management system and improves operations and outputs from what it learns. But the emphasis must be that this is done in a planned, systematic and conscientious way to create a climate – a way of life – that permeates the whole organization.

There are four basic principles of sustained improvement:

1 focusing on the *customer* and understanding what the customer values;
2 understanding the *process*;
3 all *employees* committed to lean quality; and
4 all *suppliers* committed to lean quality.

1 Focusing on the customer

An organization must recognize throughout its ranks that the purpose of all work and all efforts to make improvements is to serve the customers better. This means that the organization must always know how well its outputs are creating value for the customer through measurement and feedback. The most important customers are the external ones, but the supply chains can break down at any point in the flow of work. Internal customers must also be well served if the external ones are to be satisfied.

2 Understanding the process

In the successful operation of any process, it is essential to understand what determines its performance and outputs. This means intense focus on the design and control of the inputs, working closely with suppliers and understanding process flows to eliminate bottlenecks and reduce waste. If there is one difference between management/supervision in the Far East and the West, it is that in the former, management is closer to and more involved in the processes. To be truly effective, lean quality managers must be close to and understand the work. It is not possible to stand aside and manage in never-ending improvement. In a lean quality organization everyone has the determination and ability to use their detailed knowledge of processes and to make improvements, to use data and all appropriate analytical methods to understand the root cause of problems and to create action plans to improve outcomes.

3 All employees committed to lean quality

Everyone in the organization, from top to bottom, from offices to technical service, from headquarters to local sites must play their part. People are the source of ideas and innovation and their expertise, experience, knowledge and co-operation have to be harnessed to get those ideas implemented.

When people are treated like machines, work becomes uninteresting and unsatisfying. Under such conditions it is not possible to expect quality services and reliable products. The rates of absenteeism and of staff turnover are measures that can be used in determining the strengths and weaknesses, management style and people's morale in any company.

The first step is to convince everyone of their own role in the creation of lean quality products and services. Employers and managers must of course take the lead, and the most senior executive has a personal responsibility. The degree of management's enthusiasm and drive will determine the ease with which the whole workforce is motivated.

Most of the work in any organization is done away from the immediate view of management and supervision, and often with individual discretion. If the co-operation of some or all of the people is absent, there is no way that managers will be able to cope with the chaos that will result. This principle is extremely important at the points where the processes 'touch' the outside customer. Every phase of these operations must be subject to continuous improvement and for that everyone's co-operation is required.

Never-ending improvement is the process by which greater customer satisfaction is achieved. Its adoption recognizes that improving lean quality is a moving target.

4 All suppliers committed to lean quality

Every supplier, both in the planning and design phase as well as the construction and maintenance phase, is a key member of your supply chain; every one of them must share your commitment to lean quality.

They are a critical part of your team and your ability to provide your end customers with quality services and products depends on your entire team's commitment to lean quality at every step. Furthermore, your ability to solve problems and improve relies on your entire team's ability to innovate.

Your quality, safety and productivity risks are in the hands of your supplier's managers and employees; they are the people doing most of the actual work. You must develop close relationships with each of them to ensure that they share your values and that their employees have the skills you need.

Here too you must employ all the strategies for continuous improvement to ensure that the entire team is performing as well as possible.

A model for lean quality management

The concept of lean quality management is basically very simple. Each part of an organization has customers, whether within or without, and the need to identify customer requirements, and then set about meeting them, forms the core of a lean quality approach. Good *performance* requires the three hard management necessities: *planning* (including the right policies and strategies), *processes* (and supporting management systems and improvement tools) and *people* (with the right knowledge, skills and training, see Figure 19.6). These are complementary in many ways, and they share the same requirement for an uncompromising top-level commitment, the right culture and good communications. This must start with the most senior management and flow down through the organization.

Having said this, teamwork, performance analysis based on data, or a process-oriented management system may be used as a spearhead to drive lean quality through an organization. The attention to the many aspects of a company's operations – from purchasing through to distribution, from data recording to control chart plotting – that is required for the successful introduction of a good lean quality management system, or the implementation of SPC will have a 'Hawthorne effect': concentrating everyone's attention on the customer–supplier interface, both inside and outside the organization.

Lean quality management calls for consideration of processes in all the major areas: marketing, design, procurement, operations, distribution, etc. Clearly, these each require considerable expansion and thought, but if attention is given to all areas using the concepts of lean quality, then very little will be left to chance. Much of industry, commerce and the public sector would benefit from the continuous improvement cycle approach represented in Figure 19.7, which also shows the 'danger gaps' to be avoided. This approach will ensure

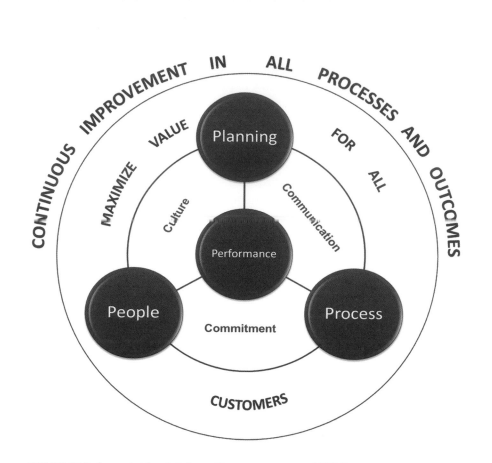

FIGURE 19.6 A model for total quality management (TQM)

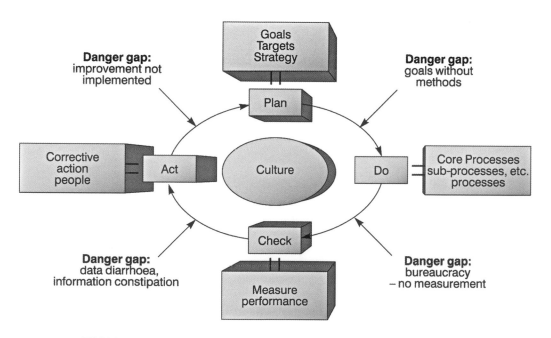

FIGURE 19.7 TQM implementation: all done with the Deming continuous improvement cycle

the implementation of the management commitment represented in the quality policy, and provide the environment and information base on which teamwork thrives.

By implementing the concepts of lean quality and the approach of managing, controlling and improving business processes many organizations have improved their day-to-day operations. Established lean quality management systems, statistical process control methods and an enhanced quality consciousness have allowed these organizations to provide customers with quality products and services that match customer requirements and organization policies, and are properly planned, developed, designed, produced and deployed. As a result, there have been reductions in development and design problems, defects, errors, installation problems, service failures and market claims and complaints. Excellent product and service quality has been achieved to generate satisfied and loyal customers. Results also include improved reliability, safety and environmental outcomes that meet society's needs. Together with positive economic outcomes these improvements have enabled many organizations to acquire a world-class reputation.

BIBLIOGRAPHY

Albin, J.M. *Quality Improvement in Employment and Other Human Services – Managing for Quality through Change*, Paul Brookes Pub., Baltimore, MD, 1992.

Antony, J. and Preece, D. *Understanding, Managing and Implementing Quality*, Routledge, London, 2002.

Cook, S. *Customer Care – Implementing Total Quality in Today's Service Driven Organization*, Kogan Page, London, 1992.

Fox, R. *Six Steps to Total Quality Management*, McGraw-Hill, NSW, 1994.

Kanji, G.P. and Asher, M. *100 Methods for Total Quality Management*, Sage, London, 1996.

Morgan, C. and Murgatroyd, S. *Total Quality Management in the Public Sector*, Open University Press, Milton Keynes, 1994.

Munro-Faure, L. and Munro-Faure, M. *Implementing Total Quality Management*, Pitman, London, 1992.

Oakland, J.S. *Oakland on Quality Management*, Butterworth-Heinemann, Oxford, 2006.

Oakland, J.S. *Total Organizational Excellence*, Butterworth-Heinemann, Oxford, 2006.

Senge, P.M., Kleiner, A., Roberts, C., Ross, R., Roth, G. and Smith, B. *A Fifth Discipline Resource – the Dance of Change*, Nicholas Brearley, London, 2000.

Strickland, F. *The Dynamics of Change*, Routledge, London, 1998.

CHAPTER HIGHLIGHTS

Lean quality and the management of change

- Senior managers in some organizations recognize the need for change to deal with increasing competitiveness, but lack an understanding of how to implement the changes.
- Successful change is effected not by focusing on formal structures and systems, but by aligning process management teams. This starts with writing the mission statement, analysing the critical success factors (CSFs) and understanding the critical or key processes.
- Senior managers may begin the task of process alignment through a self-reinforcing cycle of commitment, communication and culture change.

Planning the implementation of lean quality improvement

- Making quality happen requires not only commitment but competence in the mechanics of lean quality. Crucial early stages will comprise establishment of the appropriate organization structure; collecting information, including quality costs; teamwork; quality systems; and training.

- The launch of quality improvement requires a balanced approach through systems, teams and tools.
- A new implementation framework allows the integration of lean quality into the strategy of an organization through an understanding of the core business processes and involvement of people. This leads through process analysis, self-assessment and benchmarking to identifying opportunities for improvement, including people development.
- The process opportunities should be prioritized into continuous improvement and re-engineering/redesign. Performance-based measurement determines progress and feeds back to the strategic framework.

Change curves and stages

- A useful device in thinking about any implementation programme is the change curve, which represents a journey on which employees need to be taken if change is to lead to actions and be successful. The stages that can be expected on such a well-managed journey are: unaware, awareness, comprehension, conviction and action. Use of the curve leads to the planning of stages of involvement and engagement for executives, middle management and employees.
- There can be many reasons why people resist change, and these need to be identified and thought about if that resistance is to be overcome. A complementary change curve can be helpful in overcoming resistance to change.
- The essence of achieving successful change during the implementation of approaches such as TQM, Lean Six Sigma, Continuous Improvement involves establishing a need to change; creating a clear and compelling vision; going for true performance results and creating early wins; communicating well; building a strong, committed, guiding coalition; and redesigning the processes, systems and structures.

Using consultants to support change and implementation

- Used wisely, consultants can provide specialist skills, offer sound advice, suggest practical solutions, and inject new life into a business without costing the earth or upsetting staff. The best strategy is to combine internal and external resources in client/consultant teams.
- The brief for potential consultancies should always be carefully defined and include a short description of the organization, the problem or situation, an outline of what the consultancy is expected to achieve, the starting date and expected duration of the work, some idea of how the consultancy is expected to proceed and who might be involved.
- The professional and ethical standards that guide the consultant chosen should complement the client organization's values and philosophy, and the ethos driving the consultant's approach should be to ensure sustained improvement as they transfer their knowledge to the client.

Sustained improvement

- Managers need to understand and pursue never-ending improvement. This should cover planning and operating processes, providing inputs, evaluating outputs, examining performance, and modifying processes and their inputs.

- There are four basic principles of continuous improvement: focusing on the customer, understanding the process, ensuring that all employees are committed to quality and working with suppliers similarly committed to quality.
- In the model for lean quality the customer–supplier chains form the core, which is surrounded by the hard management necessities of planning, processes and people. These are complementary and share the same needs for top-level commitment, the right culture and good communications, all wrapped round with continuous improvement!

Part VI Discussion questions

1   You have just joined a construction company as the 'Lean Quality Executive.' The method of quality control of supplied material is based on the use of inspectors who return about 15 per cent of all goods inspected for modification, rework or repair. The monthly cost accounts suggest that the scrap rate of materials is equivalent to about 10 per cent of the company's turnover and that the total cost of employing the inspectors is equal to about 15 per cent of the direct labour costs. Outline your plan of action to address the situation over the first 12 months.

2   You have recently been appointed as equipment manager of the construction division of an expanding company and have been alarmed to find that maintenance costs seem to be higher than you would have expected in an efficient organization. Outline some of the measures that you would take to bring the situation under control.

3   TQM has been referred to as 'a rain dance to make people feel good without impacting on bottom-line results'. It was also described as 'flawed logic that confuses ends with means, processes with outcomes'. The arguments on whether to focus on budget control through financial management or quality improvement through process management clearly will continue in the future. Discuss the problems associated with taking a financial management approach which has been the traditional method used in many organizations. How could implementing 'lean quality' improve things?

4   You are a management consultant who has been invited to make a presentation on lean quality to the board of directors of a construction company employing around 500 people. As they hope to win contracts in the UK rail industry and achieving their supplier registration status is vital, the board have asked you to stress the role of lean and quality management systems. Prepare your presentation, including references to appropriate frameworks and approaches.

5   Describe the key stages in integrating lean quality into the strategy of an infrastructure construction organization and illustrate your answer by reference to one organization of your choice.

6   What are the critical elements of integrating lean quality management or business improvement into the strategy of a house building programme? Illustrate your approach with reference to an organization with which you are familiar, or which you have heard about and studied.

7   You are the new Lean Quality Director of part of a large electrical contractor and service company. Some members of the top management team have had some brief exposure to Six Sigma and Lean, and you have been appointed to consider plans for implementation. Set down your arguments and plans for the process which you might initiate to deal with this situation. Your plans should include reference to any training needs, outside help and additional internal appointments required, with timescales.

8   You have been appointed as an external personal adviser to the Chief Executive of ONE of the following:

   –   A national bank or
   –   State Department of Health or
   –   A major multi-site university.

The members of the top management have had some brief exposure to 'Business Excellence' and you have been appointed to help the Chief Executive lay down plans for its implementation in relation to the company's procurement of construction services.

Choose any of the above organizations and set down plans for the process which you would initiate to help the Chief Executive achieve this. Your plans should be as fully developed as possible and include reference to any training needs, further outside help and any internal appointments required, with a realistic timescale.

Case studies

READING, USING AND ANALYSING THE CASES

The cases in this book provide a description of what occurred in 14 quite different organizations operating with the construction industry. They give insight into various aspects of their lean quality and operational performance improvement efforts. They may each be used as a learning vehicle, as well as providing information and descriptions which demonstrate the application of the concepts and techniques of lean quality and business excellence in the sector.

The objective of writing the cases has been to offer a resource through which the student of lean quality (including the practising manager) may understand how organizations that adopt the approach operate. It is hoped that the cases provide a further useful and distinct contribution to lean quality education and training.

The case material is suitable for practising managers, students on undergraduate and postgraduate courses and all teachers of the various aspects of business management and lean quality implementation. The cases have been written so that they may be used in three ways:

1   as orthodox cases for student preparation and discussion;
2   as illustrations, which teachers may also use as support for their other methods of training and education;
3   as supporting/background reading on lean quality and TQM/business excellence.

If used in the orthodox way, it is recommended that first, the case is read to gain an understanding of the issues and to raise questions which may lead to a collective and more complete understanding of the organization, lean quality implementation and the issues in the particular case. Second, case discussion or presentations in groups will give practice in putting forward thoughts and ideas persuasively.

The greater the effort put into case study preparation, analysis and discussions in groups, the greater will be the individual benefit. There are, of course, no perfect or tidy cases in any subject area. What the directors and managers of an organization actually did is not necessarily the best way forward. One object of the cases is to make the reader think about the situation, the problems and the progress made, and what improvements or developments are possible.

Each case emphasizes particular challenges or issues which were apparent for the organization. This may have obscured other more important ones. Imagination, innovation and intuition should be as much a part of the study of a case as observation and analysis of the facts and any data available.

Lean quality cases, by their nature, can be very complicated and, to render the cases in this book useful for study, some of the complexity has been omitted. This simplification is accompanied by the danger of making the implementation seem clear-cut and obvious, but that is never the case with lean quality!

The main objective of each description is to enable the reader to understand the situation and its implications, and to learn from the particular experiences. The cases are not, in the main, offering specific problems to be solved. In using the cases, the reader/student should try to:

- *Recognize or imagine* the situation in the organization.
- *Understand* the context and objectives of the approaches adopted.
- *Analyse* the different parts of the case (including any data) and their interrelationships.
- *Determine* the overall structure of the situation/problem/case.
- *Consider* the different options facing the organization.
- *Evaluate* the options and the course of action chosen, using any results stated.
- *Consider any recommendations* which should be made to the organization for further work, action or implementation.

The set of cases has been chosen to provide a good coverage across different types of industry and organization. The value of illustrative cases in an area such as lean quality is that they inject reality into the conceptual frameworks developed by authors on the subject. The cases are all based on real situations and are designed to illustrate certain aspects of managing change in organizations, rather than to reflect good or poor management practice. The cases may be used for analysis, discussion, evaluation and even decision-making within groups without risk to the individuals, groups or organization(s) involved. In this way, students of lean quality and business excellence may become 'involved' in many different organizations and their approaches to implementation, over a short period and in an efficient and effective way.

The organizations described here have faced translating lean quality theory into practice, and the description of their experiences should provide an opportunity for the reader of lean quality literature to test his/her preconceptions and understanding of this topic. All the cases describe real lean quality processes in real organizations and we are grateful to the people involved for their contribution to this book.

John Oakland
Marton Marosszeky

BIBLIOGRAPHY

Easton, G. *Learning from Case Studies* (2nd edn) Prentice Hall, London, 1992.

CASE STUDY ORDER

Each of the case studies in this section may be used as illustrations and the basis of discussions on a number of parts of the text. The cases are listed in the order given below.

Case study

1. Boulder Associates – implementing lean thinking in design
2. VNGC delivered through IPD (IFOA) contract for Sutter Health
3. ConXtech re-engineers the structural steel frame using lean thinking
4. Multinational high-tech manufacturing company deploys lean
5. Continuous improvement and growth at Graniterock
6. BIM and lean practices drive lean transformation at Southland Industries
7. Crossrail: elevated focus on quality to match safety
8. Worker empowerment transforms operations at JB Henderson
9. Safety, quality and BIM drive lean transformation at DPR
10. Quality and operational excellence in Heathrow Development
11. Lean deployment at Rosendin Electric, Inc.
12. The development of the Costain Way
13. Re-engineering timber floors in the Australian housing sector: an example of process innovation
14. Highways England

Boulder Associates – implementing lean thinking in design

INTRODUCTION

Boulder Associates has specialized exclusively in healthcare and senior living design since being founded in 1983. The company's hands-on approach focuses on one goal: to become a leader in design for health and aging by establishing itself as trusted adviser to its clients. An indicator of the success the practice has achieved is that 93 per cent of all its work is from repeat clients.

The business is focused on the core issues that concern its clients; from broad economic and regulatory pressures to day-to-day operational challenges Boulder Associates take these challenges on as if they were their own. The practice meets these by combining beautiful, innovative design with high-quality, efficient and cost-effective solutions.

Among its clients Boulder Associates has some of the most progressive healthcare and senior living organizations in the United States. Boulder Associates helps its clients set new standards for healing environments by aligning their facility investments with business strategies and goals. Boulder Associates is known for its innovation, design and technical excellence, and a collaborative process that forms lasting partnerships.

With offices in Colorado, California and Texas, Boulder Associates maintains a staff of architects, interior designers and graphic designers who all share a belief in the power of design to enrich lives. The leadership team believes that good design directly serves the needs of its clients, their patients and residents, and the surrounding communities.

Primary products and services include all phases of planning and design delivery from existing facility evaluation, master planning and briefing, to detailed design, as well as interior, furnishing and graphic design, budgeting and scheduling, wayfinding and LEED certification coordination.

THE BOULDER WAY

Boulder Associates is passionate about developing The Boulder Way, the firm's unique approach to its business:

- To demonstrate how a professional design organization can honour creativity while seeking to drive efficiency and effectiveness in its business using lean thinking.
- To engage more closely with clients and end users through the People–Planning– Process (3P) process, developing design solutions through full-scale 3D prototyping and post-occupancy evaluation.
- To encourage employee development and support the disadvantaged in local communities where the company operates.

LEAN PRINCIPLES IN DESIGN

A better way to design

3P refers to **People Preparation Process**, and is an adaptation of a concept of the Toyota Production System, called Production Preparation Process. It is a very effective design tool for healthcare because of the emphasis on efficient operating processes.

FIGURE CS1.1 Team simulating a medical interview room

The 3P approach develops the ideal process first, followed by the design of the layout that best supports that process. The design team works alongside facility employees, who have in-depth knowledge of their own workflow processes. This approach comprises of a 3P event, typically an intensive week-long, hands-on workshop conducted with representatives from all of the key stakeholder groups. This will include everyone in the trenches, from providers and administrators to receptionists and materials managers.

During this event, the users work with the design team to develop an ideal work process that suits how they work while eliminating waste and inefficiencies. The users and the architects build full-scale mock-ups out of cardboard to test, simulate and finalize design ideas. This allows for:

- turning ideas into physical models that people can visualize and relate to;
- concentration on flow and process improvement;
- simulating designs to test what works and modify what doesn't; and
- providing an affordable and highly visible way to do a 'Plan, Do, Check, Act' (PDCA) process that will result in a product that truly meets the users' needs.

DAY 1	DAY 2	DAY 3	DAY 4	DAY 5
Establish objectives	Build mock-ups to simulate flow	Simulate	Simulate	Finalize mock-ups
Document and explore current processes (Go to Gemba)	Simulate the process	Kaizen	Kaizen	Finalize schematic design
Review current metrics	Evaluate alternatives, choose best three	Simulate	Simulate 3D process	Report out
Document current state		Reflect and repeat	Lay out material flow	
Create alternatives			Simulation review	
Start mock-ups				

FIGURE CS1.2 Programme for 5-day mock-up construction

FIGURE CS1.3 Design concepts are developed and modified on site during the workshop based on the team's design development and evaluation

The 3P event was by far the most valuable exercise that I have every participated in, and I believe the efforts of the 3P will save us months of design time. We modelled the clinic based entirely on the patient experience, and made significant improvements in this process. It was also a wonderful team-building experience.

(Sharon Booker, University of Iowa Health Center)

Such full-scale modelling is undertaken with user participation for rooms (up to 200 sqf) and in some cases patients and other users may be involved in the planning workshop.

Currently for planning of room assemblies, layouts are designed on the basis of functional relationships between spaces, and travel patterns and travel distances of the staff between activities. Currently data on travel behaviour are collected by asking staff to describe their work patterns; however Boulder Associates is currently working with a start-up to implement technology that can capture staff travel patterns in real-time and model travel times for different functional arrangements.

Standard work processes to improve efficiency and reduce error

Based on the understanding that a number of different benefits can be realized through standardization, the practice has embarked on developing standard components and solutions wherever such an approach is beneficial. Benefits to be obtained include:

- better resolved designs on the basis that more effort can be put into a design that is to be repeated than one which will only be used once;
- designs can be error proofed because they are standardized and optimized;
- constructability can be optimized;
- client engagement can be increased as the client and/or end users can review design alternatives in a library of alternate solutions; and
- efficiencies can be realized in the documentation process because of the use of standard design elements.

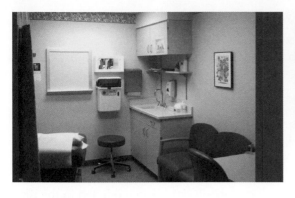

Current exam room

- Not zoned – provider has to cross exam room and squeeze by.
- No accommodation for family or bariatric patients.
- Computer set-up does not promote eye contact with patients.
- Design was never tested – equipment does not reach.

Future exam room

- Zoned with patient side and provide side.
- Design accommodation for family or bariatric patients.
- Computer set-up for patient engagement.
- Mock-up tested for equipment accessability.

FIGURE CS1.4 Comparison of present and future exam room

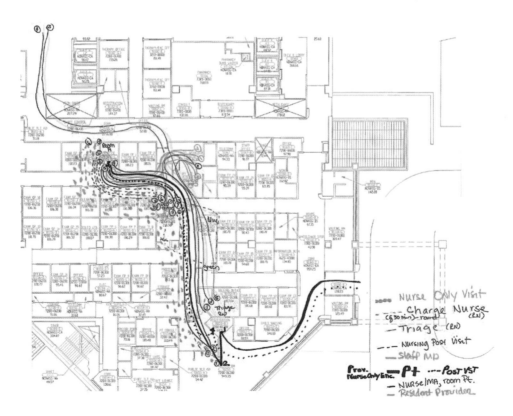

FIGURE CS1.5 The current process for illustrating travel patterns and distances is with the use of spaghetti diagrams such as this which can indicate travel distances by various users

Design solutions can be set up in a virtual showroom for clients and users to view and select preferred solutions:

1 Setting up single point models which are then used to populate the design:
   a Joinery (casework);
   b Partitions;
   c Doors;
   d Furniture.
2 Controlling variety by starting with a library of standard objects and then deploying them throughout the model.
3 Making actual process in the documentation of projects standardized.

The virtual showroom is a digital construct that involves a room modelled in the BIM and populated with design geometry. The environment of the room is set up such that all geometry is clearly identified and organized in a rational manner. The room is also set up as an environment that can be rendered photo realistically, including lighting, materials and environmental effects. The rendering pre-set allows for a quick review of finishes, materials, fabrics and hardware.

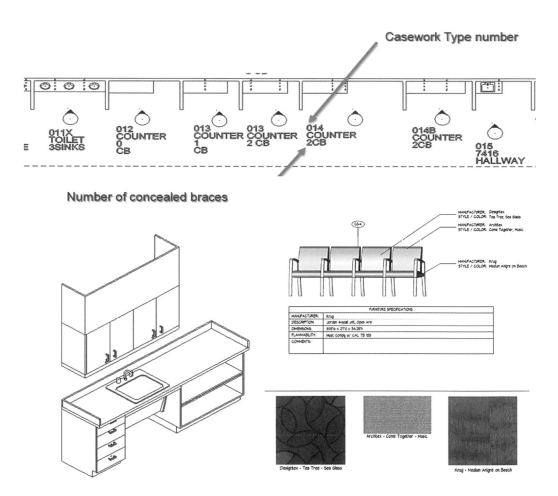

FIGURE CS1.6 Elements from the virtual showroom

Choosing by advantages (CBA)

Choosing-By-Advantages (CBA) arose from decades of research performed by the United States Forest Service, in collaboration with Utah State University. Styled as a Sound Decision-Making System, CBA seeks to illustrate beneficial differences, advantages, between two or more alternatives. As healthcare architects, Boulder Associates staff have found CBA helpful in making decisions on projects. This includes architectural and building systems decisions (mechanical unit types, steel vs concrete structural system, type of building skin, etc.), as well as planning decisions (number of exams per provider, number of operating rooms, etc.).

Beyond the output of a decision, Boulder staff also leverage the process of working through a decision using CBA. The rigour required to build alternatives, identify factors and determine the importance of advantages carries with it a strong team-building component as interdisciplinary decision-makers work together to achieve consensus on a path forward.

Figure CS1.7 illustrates the use of this method in the decision-making in relation to the design of a birthing and bed tower configuration. This process allows the client to become involved in the design decision-making in an informed way.

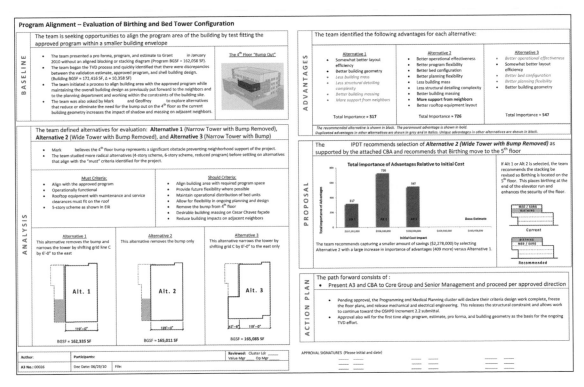

FIGURE CS1.7 Example of choosing by advantages

ORGANIZATIONAL INNOVATION

Last Planner adoption

The principals of Boulder Associates convened a planning retreat in October of 2010. At this retreat, the Board of Directors adopted a strategic plan for 2011 that included a lean goal for individual work planning. The initiative centred on improving:

1 reliability;
2 lookahead planning;
3 continuous improvement through variance capture; and
4 application of the Deming Cycle through trend monitoring, applied at a team and an individual level.

Boulder Associates has created a Kaizen team to drive its lean initiative; the membership consists of the company-wide lean champion, lean champions from each office and a member of the executive.

The individual work planning goal challenged 90 per cent of staff to submit weekly work planning numbers 90 per cent of the time. The goal stated that a complete work plan must include at least two weeks of lookahead planning, reporting of planned per cent complete (PPC), tasks anticipated (TA – the percentage of completed tasks that were planned more than a week in advance) and variance tracking with the use of 5-Whys. The goal also stated that a complete work plan should include 15–20 planned tasks per week, and that staff should utilize trend monitoring tools as a means for seeking continuous improvement. To date BA does not mandate a specific work planning tool, but rather allows individuals to adopt the technique that suits them best. This is to drive innovation and maintain the interest of experienced staff.

The firm-wide lean champion devised a rollout plan that would re-energize voluntary planners, bring back into the fold the lapsed planners and teach the programme to the new planners. A team of eight lean mentors were identified to act as trainers to a group of four to six new planners. A standard work planning process was developed; this was used by the lean mentors to train staff. The process identified eight key elements for Individual Weekly Work Planning (IWWP):

1 lay out the work;
2 review the past week;
3 ask why, five times;
4 re-plan unfinished work;
5 plan upcoming work;
6 plan with the team;
7 confirm the work plan daily; and
8 check and adjust.

The training materials also included detailed descriptions of what each step entailed, as well as a glossary of firm-wide lean resources and lean terms and definitions.

The groups also served as beta testers for a new automated reporting system. Staff logged their PPC, TA and variance in a PDF form and sent the XML file from the form via automatic email to a central work planning email account. The XML attachments were stripped from the emails and uploaded into an Excel spreadsheet designed for data synthesis and monitoring. This data allowed the programme administrator to monitor the programme and make adjustments as needed.

Each of the mentors (selected on the basis of enthusiasm and experience) met with their groups to teach them to develop a work plan, use the work plan during the work

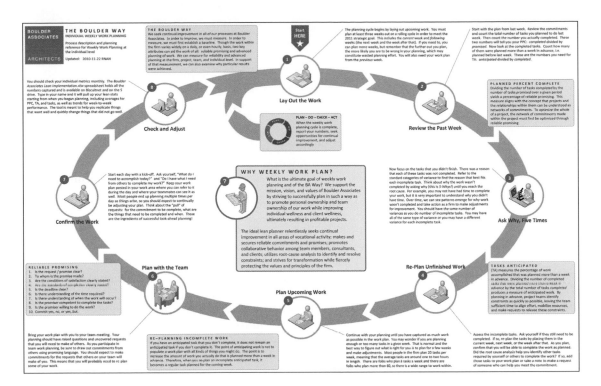

FIGURE CS1.8 Individual weekly work plan

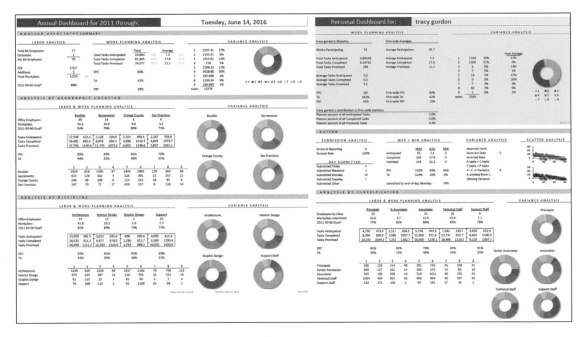

FIGURE CS1.9 Firm-wide reporting dashboard

week, review the work plan at week end, tabulate numbers and re-plan work, and submit numbers using the PDF form. Mentors engaged in one-on-one sessions with staff to answer questions.

The firm-wide lean champion also developed metrics to measure and track the success of the programme implementation. These metrics included:

1 the number of forms submitted each day;
2 the number of forms submitted by end-of-business Monday; and
3 variances for submission errors, including incorrect form use, incorrect date, inverted data, errors in tabulating tasks and missing variances.

WHAT HAS BEEN ACHIEVED AND SOME CONCLUSIONS

Boulder Associates staff found achieving reliability to be easier than lookahead planning. However, people who practised lookahead planning achieved higher TA percentages over time, indicating that lookahead planning can be taught and improved. Analysis of the statistics indicated areas where focused improvement efforts should be targeted. Next steps will be to take the improvement opportunities, particularly around lookahead planning, and work to bring all staff to a higher level of lookahead planning.

Lookahead planning is developing the ability to look further ahead, anticipate impediments to work being completed as planned and make firm plans further in advance.

Where to next – Using 'big data' to drive process improvement

Boulder Associates recognizes an opportunity to leverage existing data. By indexing and filtering data from existing internal sources, the firm is working to create various dashboards to review, monitor and project various aspects of the practice.

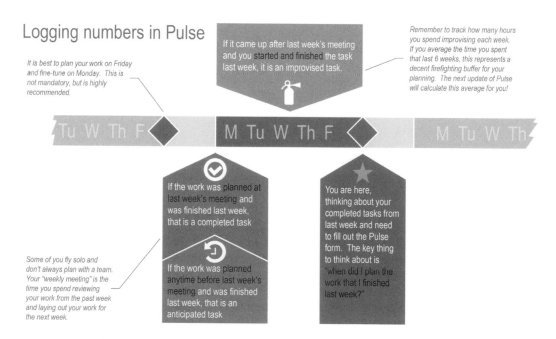

FIGURE CS1.10 Work planning guidelines for designers

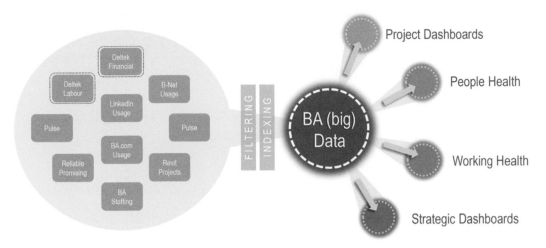

FIGURE CS1.11 Big data management dashboard

Goals of this effort include:

- Better visibility of project work by equipping older generations with tools to peer inside electronic models and drawings.
- Transforming trailing indicators into leading indicators, allowing project managers to accurately triage work and apply the right assistance as it is needed.
- Developing the ability to take a snapshot of firm-wide health that is in real-time, or as close as possible to real-time, without having to manually gather or accumulate data.

Investing in people

- BA encourages employees in the team to find creative ideas they are passionate about and to pursue them to resolution. These ideas need not be based in architecture nor be immediately profitable. The firm leadership believes that enabling the passion of employees will reap dividends by increasing employee loyalty, promoting individual investment in the business of the firm and fostering the growth of the whole person. The firm maintains an annual budget to allow for direct costs of ideas to be studied and developed. Employees can also submit project ideas for pro bono work.
- To support local communities where Boulder Associates works, the firm also offers a benefit that pays staff to take time each year to volunteer at a charity of their choosing. This allows employees to build connections with their local community apart from their professional work.

VNGC delivered through IPD (IFOA) contract for Sutter Health

FIGURE CS2.1
Artists impression of
VNGC campus

SUTTER'S PROCUREMENT JOURNEY

Sutter Health is a not-for-profit, community-based healthcare and hospital system head-quartered in Sacramento, California. It is one of the largest healthcare providers in Northern California, caring for more patients than any other network. After the 1994 Northridge earthquake caused significant damage to healthcare facilities, as well as other infrastructure in Southern California, the Hospital Facilities Seismic Safety Act known as SB 1953 was enacted. This act required healthcare facilities to be seismically retrofitted or replaced by 1 January 2013. In Sutter's case this led to a $5.5 billion design and construction programme over a period of some seven years; this included acute healthcare facilities, non-acute outpatient facilities, medical office buildings and parking structures. The first of these categories is permitted by California's Office of Statewide Health Planning and Development (OSHPD), a process that can take up to 20 months.

To meet the challenges of this large programme, Sutter's Facility Planning and Development Department (FPD) set about developing systems to manage this task. The department's focus was to increase the reliability of projects, including some that would take up to five years to design, permit and construct. The organization was looking for

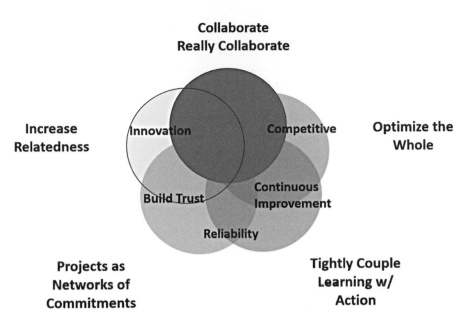

FIGURE CS2.2 Sutter Health's five big ideas

reliability on project duration and budget without being subjected to claims, safety violations or burnout of critical departmental staff. The team developed internal standards of practice that included the implementation of lean project delivery. With the support of its consultants, they developed an approach to more effectively drive collaborative work and a learning culture on their projects, the basic framework was called the Five Big Ideas.

Sutter had its attorneys develop a collaborative contract which came to be known as an Integrated Form of Agreement (IFOA). It also mandated that its supply chains would use a common BIM platform and adopt lean construction practices. In making these changes Sutter found that some of its existing suppliers were reluctant to adopt the new procurement regime, values and methods and this led to some suppliers being replaced. Sutter became a leader in the lean construction community. It spoke openly about its programme, principles and methods and shared the results that were achieved on early projects such as the Fairfield MOB, Eden Medical Centre, Temecula Valley, Castro Medical Office Building (MOB) and El Camino MOB.

Results achieved on these projects were reported in detail in terms of metrics such as cost vs budget, schedule against plan, project on-site time, labour productivity, rework, BIM model accuracy, RFIs, change orders and failed inspections.

In 2013, on the Castro MOB, Sutter used an 11-party IPD agreement. It was decided by the client that this proved to be a bit clumsy so for the VNGC project, in an attempt to simplify the management of the agreement, a three-party IPD agreement was used. Under this Sutter had an agreement with the general contractor, the HerreroBoldt partnership between Herrero Builders and The Boldt Company, and the project Architect SmithGroupJJR. These parties then had back-to-back agreements with their main suppliers. This essentially meant that Sutter had the same agreement with its 17 main suppliers. Some key features of the agreement are:

- Overhead is fixed and not variable.
- The same painshare/gainshare agreement ties all 16 parties together, costs are recoverable but profit is at risk if target costs are not achieved. On the upside, cost gains

beyond the target cost are shared between Sutter and the supply chain. As the savings increase the share that the supply chain receives increases.

- The Sutter Project Executive is on site and involved in day-to-day discussions about risk and opportunity; Sutter's role is that of team member and at times facilitator to drive team engagement.

INTRODUCTION TO VNGC

In response to the Hospital Facilities Seismic Safety Act (SB 1953) Sutter Health realized that it required a new replacement hospital in its San Francisco market. It selected a site on the corner of Geary Boulevard and Van Ness Avenue in downtown San Francisco. This project came to be named the Van Ness and Geary Campus or VNGC. Initially Sutter went through a traditional design process; however it found itself with a project that was significantly over budget even before construction began. Two issues were identified as causing the budget blowout; the hospital was larger than required and finishes were beyond Sutter's requirements.

Sutter decided to adopt an IPD agreement with its architect SmithGroupJJR and general contractor HerreroBoldt, underpinned by back-to-back agreements with the main designers, suppliers and contractors in the supply chain. The risk/reward members of the team numbered 17 parties (their logos are shown in Figure CS2.19 at the end of the case study). They included the key designers, fabricators and constructors of the project. The team immediately turned its attention to more clearly identifying Sutter's needs and modifying the design to better meet them.

Figure CS2.3 illustrates the results of the target value design process over this period. The City of San Francisco entitlement process delayed the project start by some 18 months. During this time the team undertook value engineering work to drive the design cost down

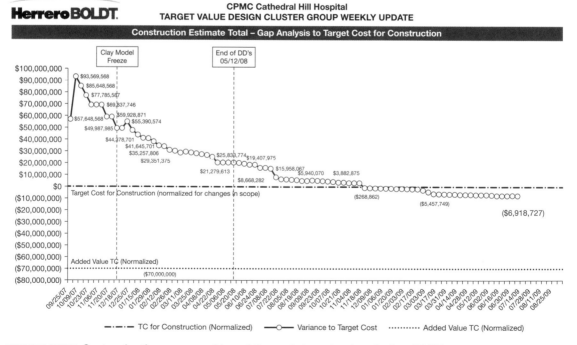

FIGURE CS2.3 Cost reduction curve achieved through target value design (TVD)

to achieve Sutter's target value and to improve the constructability of the design. They cut the budget for the hospital by 20 per cent, reduced its size and increased its functionality.

The final project has 12 stories above grade, an area of 74,000 square metres and a total project cost, including equipment and fit-out, of $2.1 billion.

Significant savings were achieved by prefabricating bathrooms, utilizing shared overhead utility racks for services, displacement ventilation and improved scheduling for the structural steel frame.

'We were looking for two categories when it came to innovation from lean processes and IPD,' said Paul Reiser, project executive for VNGC during the pre-construction stage. 'Small, incremental savings from lean alone and, this is the trickier one, a push for innovation to look at other projects all over the world and incorporate technologies that aren't commonly used.'

Demolition commenced in September 2013 and the hospital will be open for patients in the first quarter of 2019.

COLLABORATION AND CULTURAL INTEGRATION

Sutter invited the initial members of the team: general contractor HerreroBoldt, concrete trade partner Pankow, and the architect SmithgroupJJR. From there, the team members went through a rigorous review and selection process to add additional trade partners until the team was complete. The selection was focused on finding the trades with a collaborative and innovative culture that would fit well into the integrated project atmosphere.

When asked what was unique about working on the VNGC site, everyone gave the same answer, the high level of collaboration. As a part of routine team processes, each week, all team members, including Sutter Health, meet to review the status of the project and to review and evaluate risks and opportunities.

The VNGC site office has the core project teams from all the participants co-located in one office. In addition to proximity, which encourages conversation and joint problem solving, the VNGC office has numerous communication boards throughout. These are designed to improve communication, to standardize processes and to create a common culture. The boards fall into a number of categories, among them:

- policy statements;
- process instructions;
- plans;
- progress reports against plans;
- problem and countermeasure boards;
- statements to define the culture.

Figures CS2.4 and CS2.5 illustrate some of these boards.

An outstanding feature of the VNGC project team was its commitment to shared goals, the potential to change and improve, and to continuous improvement. On a project such as this, there is constant change and this is managed through a deep commitment to planning and continuously learning from experience on a daily basis.

CONTINUOUS IMPROVEMENT

Work experiences are reviewed by each team daily and lessons learned are identified and implemented. For example, the team leader from Pankow Constructions, the contractor pouring concrete on site, described how after every concrete pour, a debriefing meeting

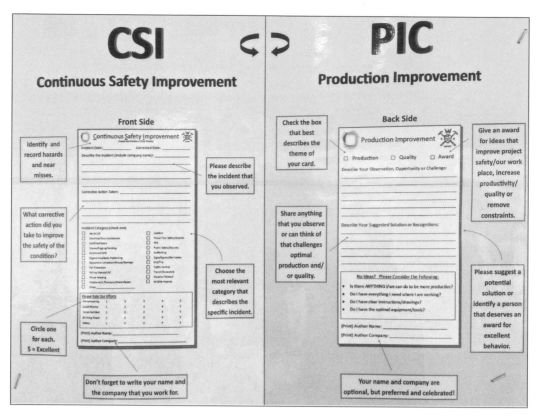

FIGURE CS2.4 Explanation of safety and production improvement cards carried by all personnel

FIGURE CS2.5 Daily site start-up meeting, planning boards showing the delineation of areas and the daily work plans for a 3-week window. Foremen and site engineers meet in front of these boards each morning for a 15-minute planning session

was held to review how the work went and identify any potential improvements. Up till the time of our conversation, these meetings had been held on the morning following the concrete pour. However, plans were in place for these meetings to be held on the slab immediately after the pour was completed so that problems and ideas were fresh in everyone's mind.

Each discipline has what the team calls a PDCA (Plan, do, check, adjust) weekly huddle where they meet with project leaders to discuss lessons learned in regards to safety, quality and production. The structural team meets up on the elevated decks so they can tie the learnings to hands-on observations. The framing field leadership team comes in early on Wednesdays to have breakfast and discuss their learnings before the workday starts. The process looks a little different for each trade but the importance is that each discipline does meet each week and the ideas and learnings are being addressed.

THE VNGC PRODUCTION SYSTEM

The overall production planning and control system on the project is designed to ensure predictable and streamlined workflow through a simple construction process.

In each of these areas, the delivery of plans and resources and information is structured in five steps. These steps are laid out in detail providing all team members with detailed instructions regarding the requirements of the system. Process steps and outcomes are specified in detail.

The project emphasizes the importance of thinking about the project as a system, where all the resources are tied to the same physical output. The system starts with the production

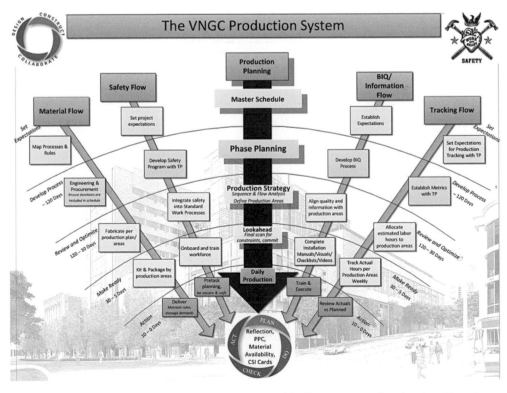

FIGURE CS2.6 Planning and reporting is structured in five streams; Production Planning; Material Flow; Safety Flow; the BIQ Information Flow; and Tracking Flow

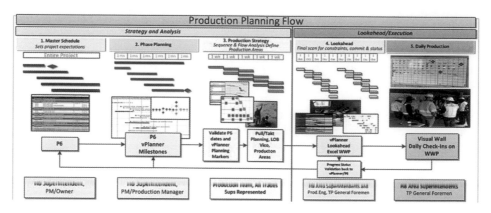

FIGURE CS2.7 Production flow planning on VNGC

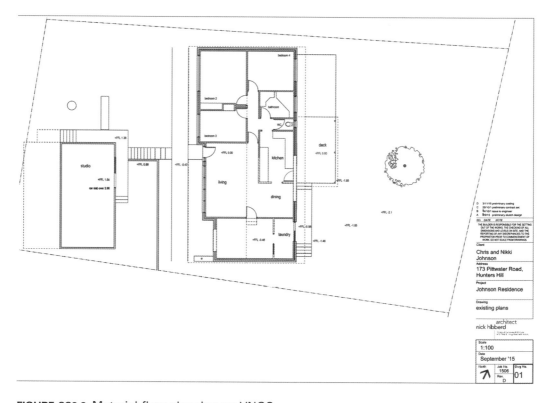

FIGURE CS2.8 Material flow planning on VNGC

planning efforts. The production planning efforts are divided into five levels of detail, from high-level milestones, to the daily execution in the field.

Once the production plan is agreed to, all trades align their supply chain with the identified work areas on the deck and flow. It is important that the needs of the project define the order of fabrication and deliveries to support a just-in-time approach.

The team also aligns the labour tracking to the same production plans and work areas on site to make sure that the project is tracked to reflect the actual sequence in which it is built.

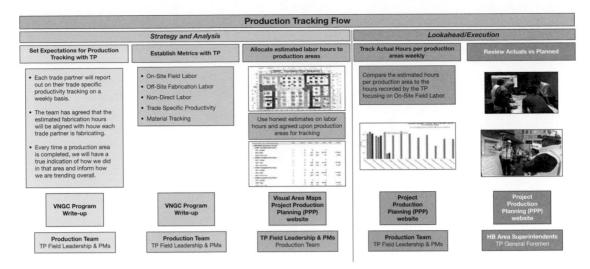

FIGURE CS2.9 Production tracking flow on VNGC

Once the overall flow is established, trade contractors map their own installation process and identify all the important steps. Once mapped, each step is reviewed to make sure that all quality expectations specified in the project documents are built into the installation process. The team also tries to anticipate issues from past failures and develop countermeasures within the trade installation process, equipment usage, sequence and flow to make sure that it is built right the first time. Once the work starts, the team adjusts the processes to ensure continuous improvement, and to suit field conditions.

Safety is the most important aspect of the production planning effort and is included in every step of the process. The team additionally undertakes a final detailed review of each trade's installation process to make sure that each step is contributing to a safe working environment.

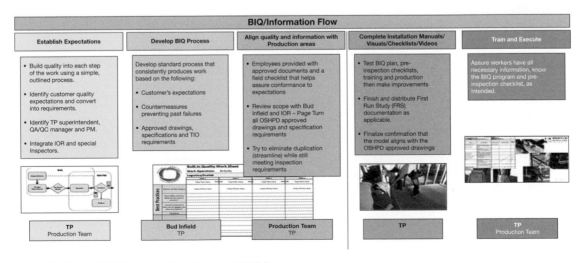

FIGURE CS2.10 BIQ information flow on VNGC

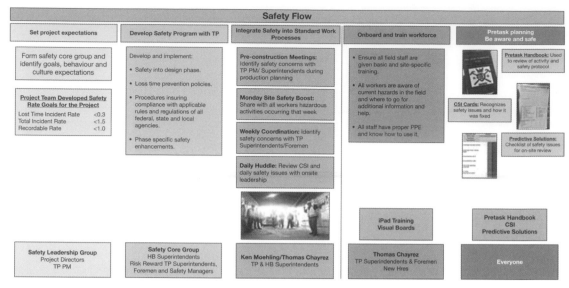

FIGURE CS2.11 Safety flow on VNGC

FIGURE CS2.12
Weekly planning is undertaken by the team using the Last Planner® System in front of the project planning board in the Bigroom. The project uses the software vPlanner to keep track of the commitments and weekly work plans

SMALL WINS PROGRAMME

Essentially a small wins programme is the active encouragement of a continuous improvement culture in everyday work. The purpose of this programme is to encourage workers to capture simple improvement opportunities that add value and simplify their work, and to share these lessons with fellow workers. The programme empowers everyone to fix what bugs them and helps to create a community of support and recognition.

Ideas are presented weekly at the Last Planner meeting in the Bigroom where all stakeholders are represented, where these are evaluated, sharing wins. There is no approval process for small wins; the programme encourages a culture of experimentation and the sharing of ideas.

Two examples of small wins on VNGC are illustrated in Figures CS2.13 and CS2.14.

Limited water suction capacity with
original shop vac attachment

Used top of 5 gal water jug to increase
suction capacity and a broom to push
water to vacuum

Limited Capacity!

Sucks up water faster and more efficiently

Mike Rangel   Field   1/20/2016

FIGURE CS2.13 Small wins water suction idea

Removal of Diagonal Bars of Columns

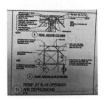

Diagonal bars were installed around
columns same way as reinforcement
around slab openings. Caused
complicated rebar installation in the slab.

Team confirmed with inquiry that the
diagonal rebar is not needed at
columns less than 2' x 2', were
removed.

Easier rebar installation to maintain clearances, saves time, increases productivity

Chuck/Carrie   Field   1/20/2016

FIGURE CS2.14 Small wins rebar simplification

BUILDING INFORMATION MODELLING

The VNGC project leverages Building Information Models (BIM) developed from its co-located trade partners. This consists of a team of over 100 people working to design, engineer and coordinate the project. Approximately 600 BIMs are assembled and the team coordinates on a daily basis. The coordination team works to resolve over 1,000 clashes a week. In addition to the clash coordination the team ties data from the schedule to help prioritize work through the use of 4D Sequencing. Laser scanning is being used to check the as-built conditions in the field. The point clouds from the laser scans are combined with the planned models. These scans help answer the question 'did we build what we intended and if not, how do we adjust?'

The work from the modellers and detailers in the coordination phase of the project is made visible to the field through the use of an iPad application called BIManywhere. Each level of the building is available on the iPad and makes the model easily accessible to anyone.

FIGURE CS2.15
Project BIM 'war room'

One of the challenges the team faced early in the project pertained to treating the model as a construction document before it was released for construction. The model is a flexible tool, which can be used to prototype potential solutions for coordination and to understand constructability. BIM should be used to move through versions of possible solutions to flush out issues. Over the course of the project, the maturity of BIM, and its ease of implementation, has helped the team understand the value of developing versions of complex conditions and details in order to help the construction teams install work easily and quickly in the field.

TEAM AGILITY

The co-location of the entire team, and a shared purpose to improve efficiency and reduce cost, has created opportunities for improvement on a regular basis.

One example was the decision to cast the basement slab on ground before steelwork erection. It had been planned that the steelwork would be erected first and the basement slab cast afterwards as it was thought that this would be faster. However, it was realized in time that casting the slab first and erecting the steel on top of the slab would save time in the overall programme and would make the casting of the slab simpler and more efficient. This change was able to be implemented because of the IPD agreement and the co-location of the parties on site.

Another example was that the team decided to change their plan and delay erection of some of the structural steel members of the central utility plantroom roof. While this meant that both the steel erector and the concrete crew would have to return to finish their work in the area, the mechanical trades benefited from improved access and an ability to set many of their pieces of equipment right down on their pads with the tower crane. This resulted in overall net savings to the project in both labour hours and schedule, both of which had a positive impact on cost.

Another example of overall savings brought about through collaboration is the changing of work sequence to help improve the productivity and schedule. In this instance the fireproofing spray crew agreed with the plumbing trade partner to adjust their installation sequence. This change sped up the plumber and created net savings to the team. Since the production tracking is tied to the same areas as the schedule, the team can make adjustments at short notice to steer production in the most efficient sequence for the overall project.

CAPACITY TO INNOVATE

One of the biggest innovations introduced was the use of viscous wall dampers that use less than 20lbs/ft of steel to brace VNGC's structural frame for seismic compliance. Viscous wall dampers are hydraulic devices that dissipate energy when a fluid is forced to pass through an orifice. Their original use was pioneered as shock absorbers in long-range military weapons; however they have been used to dampen earthquake shocks in hundreds of buildings in Japan since 1990. VNGC is the first hospital project in California to use viscous wall dampers for seismic design and one of around 60 buildings in the US to use viscous wall dampers. The challenge of introducing the viscous dampers could only be taken up because of the IPD arrangement. A great deal of work was required to get OSHPD to approve the use of this technology, and this would not have happened under a conventional contractual arrangement.

Another key area of innovation was in the acceleration of the OSHPD review and approval process. Normally getting full approval for the detailed design of a project of this size and complexity would take several years. By assisting OSHPD to adopt electronic submissions through the provision of hardware, software and technical support, communications were radically improved and the project was able to submit designs electronically. Designs were able to be reviewed with OSHPD staff progressively and this accelerated the approval process significantly. The approval of the structural and architectural design, a process which would normally take up to two years for a project such as this, was achieved in six months.

Once the integrated team transitioned into design, the design cluster teams shifted their focus to construction. To ensure that clashes were avoided, the 3D model included everything larger than 2.5cm. The 3D model was also used for constructability planning to simplify and standardize work, and to increase the opportunity for prefabrication. 4D sequencing and scheduling using Revit and Tekla models were used to improve workflow and hence increase construction speed and reduce construction cost.

One example of this approach was in the collaboration between the structural designers and the concrete contractor Pankow Constructions. To optimize the reinforcement detailing at column junctions to improve constructability, the parties sat together and developed an optimum solution jointly working over a 3D model.

TAKT TIME PLANNING AT VNGC

The purpose of Takt time planning is to align resource requirements with demand in real-time. It is a management approach which dictates that all production units should progress through the building at the same rate, in a tight sequence. It also readily indicates whether a production unit is ahead or behind the production rate of the overall system. While it is relatively easy to understand the concept that all the work stations in the process should operate at the same pace in a manufacturing environment, in construction, the application of this concept is less obvious.

In a high-rise building where production moves forward on a regular floor-to-floor cycle, a rhythm is relatively easy to create. Every trade balances its teams to move up the building at the same rate one behind the other. In a hospital this is not so easy. The floor plate is large and progress does not have such a readily recognizable pattern. Different areas have different intensity of services and, as different trade contractors move at different rates, this causes disruptions to the overall flow of work. This variability creates the need for trade groups to 'jump around' to keep individual crews working and avoid running into each other.

At VNGC, the team is working to optimize flow by introducing a schedule that is based on Takt (rhythm) principles from the manufacturing industry. All trades move through the building at the beat of the same drum, or set the 'cruise control' to the same speed. The team uses clearly marked production areas and standard weekly durations to identify progress expectations. Similar to an assembly line, the tangible indication of production flow makes it easy for all trades to identify and align their capacity to the overall flow of the project.

The team use the analogy of a train to explain their approach to new team members:

- the railway track outlines the area sequence the team is set to follow;
- each train car is one discipline;
- the speed of the train is set by the lead car;
- all cars are connected and move together in the required sequence;
- all cars moving through the areas at the same pace in a finish to start order; and
- if done right, the result will be a balanced workflow and no crashes!

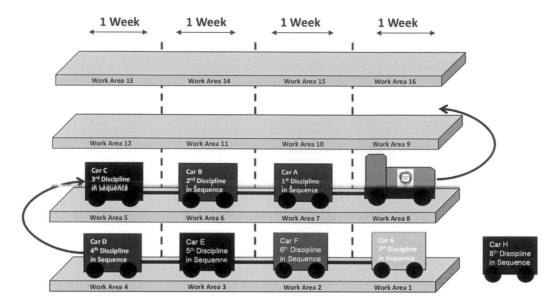

FIGURE CS2.16 Illustration of tightly coupled workflow

The focus on Takt helps the team to create alignment and establish an overall flow of production.

- The focus is to make sure that everyone moves at the same speed through the building.
- The trades optimize their resources to fit the plan; they benefit from stable workflow and uniform resource demand.

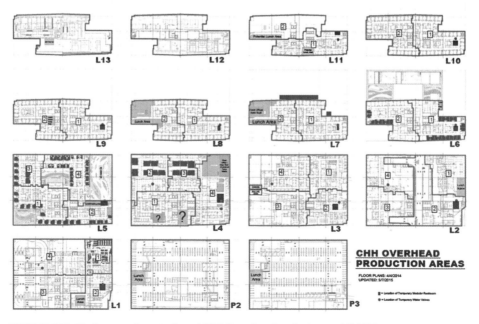

FIGURE CS2.17 Areas allocated on VNGC as having balanced SOW

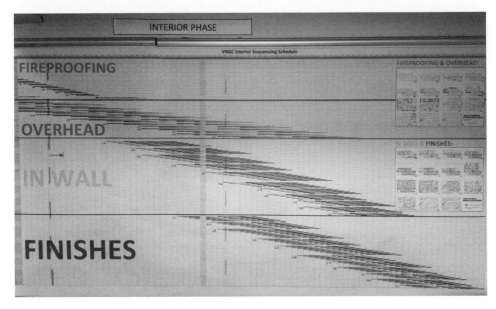

FIGURE CS2.18 Continuous flow in fit-out and finishing trades on VNGC

- The predictability allows the trades to plan and focus on one area at a time, achieving the needed throughput with fewer resources. The project team stresses the importance of exploring prefabrication opportunities, waste elimination and production process improvements before relying on an increase in manpower.

The faster trades are challenged to reduce resources to avoid overproducing, while the slower trades, or bottlenecks, need to be carefully analysed to discover any opportunities to increase their output.

FIGURE CS2.19
The risk/reward members of the project team

The approach demands an increased level of planning upfront, and the system is continuously challenged by traditional behaviour and a strong culture of trades wanting to 'jump ahead'. Overall, the project is experiencing very strong production for the phases where Takt is used to steer production and supply chain.

CONCLUSION

VNGC is on track to be the best large project Sutter Health has procured. Its success is attributed to the open collaboration and the client's close involvement in discussions on risk and opportunity on a weekly basis.

The risk/reward members of the project team are shown in Figure CS2.19.

CASE STUDY 3

ConXtech re-engineers the structural steel frame using lean thinking

PREAMBLE

The opening chapter of the book describes innovation in complex products as taking place at the overall product, subsystem and process levels. Large sophisticated multipurpose buildings are among the most complex products made by mankind.

This case study is designed to showcase ConXtech's innovation in developing a rapid assembly structural 'chassis' for buildings. ConXtech's structural innovation has increased labour productivity in steel structure assembly approximately five-fold.

This company has shown what can be achieved through radical innovation and sets an example for other subsystem manufacturers to follow. From an overall lean delivery perspective, there is limited value in a single subsystem being delivered significantly faster if the following trades do not take advantage of early delivery and modularity. As buildings are assembled on line, one component on top of the other, optimum productivity improvement is achieved if all the subsystems can be erected at the same rate.

CONXTECH

The ConX system

In late 2000, inventor Robert J. Simmons conceived the ConXR 200™ connection to address the need for an efficient structural framing system for the then robust high-density residential and mixed-use markets. By 2003 the system had been developed, peer reviewed, tested and refined.

ConXtech was founded in 2004, and the team began to build manufacturing, fabrication and world-class structural testing capabilities in a 12,200 sqm fabrication facility in Hayward, California. This facility is capable of producing 600,000 sqm of structural steel frame annually.

ConXtech offers a complete structural solution for owners and developers who specify ConX as the structural 'chassis' for their projects. ConXtech's present capabilities include

FIGURE CS3.1
ConXtech's ConXR 200 Lower &
Locking™ connection

FIGURE CS3.2
ConX® field assembly
is conducted from the
safety of personnel lifts

education on the ConX System for all stakeholders and trades, design assistance for designers and engineers, structural engineering and detailing services, connector manufacturing, fabrication and field installation. Initially these capabilities were delivered as a turn-key solution, but now the company supports outside fulfilment teams (including engineering and design teams, fabricators and erection contractors) in optimizing and delivering ConXtech's disruptive technology.

ConX steel structural components are quickly and easily assembled on site by ironworkers. Typically, 400 sqm of gross floor area of ConXR and 1,000 sqm of gross floor area of ConXL can be assembled per day. The ConX System is designed to create not only a more efficient workflow for the actual assembly of the structure, but also to make vertical sections of the building accessible to follow-on trades (mechanical, electrical and plumbing, etc.) very early in the construction schedule. For example, sections of a three-storey commercial building in the 10,000 sqm range can be ready for follow-on trades in the second week of construction.

FIGURE CS3.3
ConX® column weld fixture enables all welds to be performed ergonomically in a horizontal position

Development phases

Through its developmental years ConXtech delivered 50+ structures into various markets, totalling approximately 930,000 sqm of built space. The company has invested in two major development phases.

Phase I: From 2004 through 2007 ConXtech achieved 100 per cent year-on-year growth in the high-density residential market working for some of the largest multi-family housing developers in the US. During Phase I, ConXtech experienced consistent, rapid growth until the collapse of the residential market in 2007.

Phase II: The second development phase ran from 2007 through 2011. In mid-2007, the company began development of a new connection to broaden ConXtech's offering and to position the company to sell structural frames into other market segments which required longer spans between columns. This innovative new connection design, ConXL, was adjustable, accommodating not only longer spans but also different beam depths. By April 2008 (one year from the inception of the design) the factory was reconfigured and had an operating ConXL production line in addition to the ConXR line. At this time a five-storey office building totalling 15,600 sqm was expedited through fabrication and erected in the field in 14 working days (1,200 sqm/day).

During this phase the company was able to establish itself in the commercial office, industrial/energy, healthcare, institutional, data centre, military and temporary structures sectors. ConXtech began to enable a Certified Network of fabricators and erectors to deliver ConXtech's structural solution.

Additionally, ConXtech developed a modular turn-key fabrication facility, which could be replicated and deployed into existing fabrication operations. This approach was necessary to overcome the scepticism of the fabrication community, and to prove that:

- outside fabricators could achieve the tolerances necessary to allow the ConX System to be predictably assembled in the field; and
- those fabricators could produce ConX at a rate of 2x the industry average by leveraging automation, Mating Surface Simulating Fixtures™ and a process-driven production line that, when combined, predictably delivered dimensional precision consistent quality.

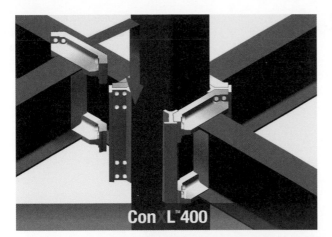

FIGURE CS3.4
ConXtech's ConXL 400 Lower &
Locking™ connection

FIGURE CS3.5 ConXtech's ConXL™ System utilized in a Silicon Valley commercial office
building

Phase III: Over the last four years ConXtech has expanded into a broader set of markets
and geographies; grown and built redundancies in their supply chain and fulfilment;
developed software and other tools that further simplify design integration; refined existing
products; and documented processes in a set of simple visual tutorials called ConX Process
Instructions (CPIs). This has enabled outside stakeholders in the ecosystem – including
design teams, detailers, fabricators and trades – to extract further efficiencies, despite the
inherent ripple created by disruptive technology. These efforts resulted in a 4x expansion
of fabrication capacity through a growing network of ConX certified fabricators and
erectors located domestically and abroad.

A concerted effort to understand customer pain points beyond the frame and further
investment in product development resulted in a modular Kit-of-Parts for pipe rack
structures that can be assembled in hours vs a week or more if built conventionally. In
addition to standard framing members, modular accessories such as handrails, ladders,
stairs, modular access platforms and innovative safety-related devices to assist in hoisting

FIGURE CS3.6
ConXtech's column weld fixtures ready for shipment to an outside fabricator

and fall-protection are now offered. Other Kit-of-Parts systems for healthcare and data centre facilities have also been productized. ConXtech's growth and innovation continues with the ultimate mission to 'Improve Lives by Changing the Way the World Builds'.

OPTIONAL SYSTEM ELEMENTS

By this time other elements of the system had also been developed.

FIGURE CS3.7
The ConX® modular stair system is fast and simple to assemble

ConX stair – Modular permanent stair system

Within the second or third day of construction, ConX stairs are added to the structure. The ConX stair system provides permanent vertical access and circulation to each floor of the building very early in the construction process. There are both construction efficiencies and safety benefits to having stable vertical access to each floor as well as the roof of the building so early in the building process.

ConX floor – Corrugated metal deck with lightweight concrete fill

From almost the onset of erection, installation of the structural floor system that spans between the horizontal wide-flange beams can begin. Comprised of a corrugated metal deck which eventually is filled with concrete, this system, when complete, delivers the required separation between floors to meet a 1-hour fire rating. Ceilings can be added, moved or finished without affecting the required horizontal area separation.

FIGURE CS3.8
Decking is efficiently installed in a ConX data centre due to standardized columns

ConX System perfectly suited for exterior wall panel systems

Because of the inherent precise geometry of the frame system, modular components such as exterior wall panels can be efficiently pre-manufactured to fit right on to the frame with a minimal need for adjustment. This results in the building being rapidly enclosed, offering perimeter fall protection (where panels are installed) early in the assembly process.

FIGURE CS3.9
ConX® beams are pre-cut or drilled to the MEP contractors' specifications

MEP COORDINATION/ INTEGRATION

Mechanical, electrical and plumbing (MEP) can be installed below floors through pre-cut (factory) penetrations in the wide-flange beams. ConXtech coordinates with the MEP trades throughout their design process in order to automate and pre-drill or cut ConX beams to accommodate their on-site work. The trades not only commence their work on the site earlier in the building process, but can take advantage of a simplified, clean and efficient process.

While from a lean construction perspective, the new system has obvious benefits, in the case of this technology, the work of the design architects and design engineers is changed and hence there can be natural resistance to the adoption of this new system. However, for design teams willing to embrace a more systemic approach, the benefits can far outweigh the anticipated impacts of something 'new'. Architects and engineers are more productive due to standardized connection details and components which minimize iterations between these stakeholders. Additionally, the architect is the beneficiary of a brace and shear wall free structure, simplifying layout and leaving room for future programmability.

MANUFACTURING

The demand to build high-quality, sustainable buildings is evident. ConXR and ConXL System components are precision manufactured in a highly automated manufacturing facility where technology minimizes waste and carbon emissions while enhancing quality and cost efficiencies. Electronic CAD/CAM files from Building Information Models (BIM) feed data to CNC cut and drill lines, increasing precision and reducing the risk of human error. Robotic weld cells, CNC machine centres, Mating, Surface Simulating and Fixturing streamline the flow of materials into a near Just-In-Time (JIT) delivery system. The result? Superior quality and unprecedented efficiencies in the use of materials, time and energy.

Precision, repeatable quality and scalability of ConX structures are achieved by using robotic welding, CNC-driven machines, modular jigs and fixture stands.

PIPE RACK CASE STUDY

An adaptation of the ConXtech steel chassis system to the construction of pipe racks in the oil and gas sector was a logical application of the system.

FIGURE CS3.11
ConXtech robotic beam weld cell efficiently welds ConXR™ connectors to beam ends

FIGURE CS3.10
ConXtech's ConXR™ robotic beam weld cell takes welding out of the field and ensures quality

FIGURE CS3.12 Mazak machine centres mill ConX® connection components to precise dimensions

Safety and speed with a modular solution

The structurally robust ConX® System is an ideal solution for the efficient design, fabrication and construction of pipe rack and process structures. ConXtech and its growing network of certified fabricators leverage standardized connectors and employ automated manufacturing technologies to deliver modular and typically brace-free structures. This technology-enabled approach accelerates design through delivery allowing industrial facilities to come online months or years earlier compared to more conventional structural options.

ConXtech's innovative connectors are manufactured and factory welded onto wide-flange beams and square columns (HSS or box), resulting in prefabricated assemblies. This full-scale erector set comprised of ConX assemblies is easily transported and safely and rapidly 'lowered and locked™' together in the mod-yard (as modules) or field (stick-built). ConX is intuitive to assemble and ideal where there is a shortage of qualified labour.

ConXtech has also developed foundation systems, layout tools, templates and other accessories which ensure unprecedented accuracy and predictability in the field.

Central California pipe rack productivity

- 1,411 assemblies, 567 tons, 823 metres of rack.
- ConX Shop Man Hours Per Ton: 7.2 (conventional 15 to 50).
- ConX Field Man Hours Per Ton: 4.23 (conventional 15 to 25+).
- ConX Detailing and Design: 2 weeks with less than 20 shop drawings (conventional = many months/iterations/hundreds of drawings).
- Owner credits ConX for taking 9 months off a 3-year schedule.

Photo represents:

- 61 metres of rack.
- Less than 5 working hours.
- 4 Crew Members + Superintendent.
- 22 Columns and 38 Beams = 60 total pcs.
- No crane required.

Specific benefits in pipe rack construction

Brace free, flexible structure: simplifies pipe layout and routing, and often reduces rack levels required. The structural frames can be fully stick-built on site or transported in prefabricated modules which are simply joined on site. The system is much faster and more efficient than traditional approaches. Either way, the frame components are fully prefabricated as a precision product in a controlled factory environment.

Site maintenance is easier because the absence of bracing improves access for inspection and repair. Future modifications are safer and easier to implement due to the system's 'lower and locking' connections and standardized parts.

FIGURE CS3.15
ConXtech's Modular Pipe
Rack System is brace
free which vastly
simplifies pipe and
process design and
installation

Modular components and accessories

ConXtech has developed modular components to streamline pipe rack design and installation, as well as offer improved access and configurability. The Modular Assembly System offers alignment tools to simplify erection in the mod-yard or field. Modular Access Platforms (including safety rail and toe kicks) and standardized ladders are designed at a consistent width so they can be easily installed and relocated anywhere along the structure. Cable tray attachments, pipe shoes and other accessories further simplify the ConX design compared to conventional custom pipe rack structures.

GENERAL CONSTRUCTION BENEFITS

Safer, faster and more productive assembly processes

ConXtech has taken a holistic approach to developing their systems. Not only are the frames faster and easier to erect, the simple moment bearing connection markedly improves labour productivity. Other elements of the overall system such as the foundations, integrated stairs, flooring and facade accelerate not only the frame assembly, but the entire building completion increasing return on investment. The improved safety is reflected in lower insurance premiums during construction.

ConXtech's bolted bi-axial moment frame connections are structurally robust and perform comparably to field welded moment frames without the need for field welding. The elimination of field welding is significant in that it reduces risk-related hot and high work on a structure and it also vastly reduces the requirement for highly skilled labour on site.

The ConXtech frame is achieved with higher levels of accuracy than timber and concrete construction can offer, creating the potential for efficient prefabrication in the other building subsystems. Furthermore the system meets the most demanding seismic and blast resistance requirements.

FIGURE CS3.16
ConXtech's ConXR™ System is simple and brace free

FIGURE CS3.17
ConXL™ is a bi-axial moment frame connection system

FIGURE CS3.18
ConXtech's ConXL™ System simply lowers and locks together

The optimization of materials often results in lighter and greener structures. Both ConXR and ConXL are Cradle to Cradle Certified™ Silver, making ConX one of the most sustainable/green structural options available. Additionally, the system is unique in that it can be safely disassembled and repurposed into entirely new structures extending its lifecycle far beyond that of other more conventional structural systems.

IMPLEMENTATION

Over the last decade, implementation of the ConX System has been both challenging, (particularly during the great recession of 2008–2012) as well as highly encouraging. The company boasts 100 per cent reference-able clients, and several clients have made ConX structures their standard, citing shorter design cycle; a condensed, more predictable construction cycle; and reduced Total Installed Cost (TIC).

The challenge and opportunity now is for follow-on trades to develop complementary systems that can keep up with the ConX frame.

FIGURE CS3.19 ConX® Digital Chassis from model to reality

Multinational high-tech manufacturing company deploys lean

A MAJOR CUSTOMER FOR CONSTRUCTION

This confidential multinational technology company is one of the largest semiconductor chip makers in the world. Like many other owners, it has been dissatisfied with the flat productivity gains of the construction industry over the past half century and is proactively partnering with other owners to bring fundamental change and improvement to the productivity and service delivery capability of the construction sector.

In manufacturing semiconductors, silicon is purified, melted and cooled to form an ingot that is then sliced into discs called wafers. Chips are built simultaneously in a grid formation on the wafer surface. A chip is a complex component that forms the brains of every computing device. While chips look flat, they are three-dimensional structures and may include as many as 30 layers of complex circuitry. Chips are fabricated in batches of wafers in 'clean rooms' that are thousands of times cleaner than hospital operating theatres.

Unarguably, high-tech manufacturing is the world's fastest changing industry. In 1971, a computer chip much, much smaller than the size of the nail on your little finger (12 sq mm), contained 2,300 transistors; by 2013 a computer chip with an area of 160 sq mm, roughly the size of your thumbnail, contained some 1,400,000,000 transistors. During this period, the world has seen not only the introduction of smaller and ever more powerful computers, but also the proliferation of new mobile and wearable devices, pervasive electronics in automobiles and massive growth in cloud networking data centres – all requiring chips with greater capacities for performance and reductions in size.

The process of making semiconductor chips is called fabrication. The factory where chips are made is called a fabrication facility, or also commonly called a fab and sometimes foundry. Such facilities are among the most technically advanced manufacturing facilities in the world. Every year as this company develops new products for its rapidly developing and changing markets, it requires either modified or additional new manufacturing and distribution facilities.

This makes this business one of the most significant global clients for construction services as it develops and builds new buildings and installs new manufacturing tools each year. In a given year, facilities renovation, construction and equipment installation can represent a very sizeable portion of the business' capital expenditures.

PRESSURES OF OPERATING IN A RAPIDLY CHANGING GLOBAL MARKETPLACE

In the high-technology sector, international competition is fierce; smartphones, tablets and wearable devices have altered the market and have created demand for new and different products. Entirely new technologies are emerging, and miniaturization as well as new products and applications are constantly altering demand in a nuanced way. As the market is rapidly expanding and changing, all the key players in the sector are seeing their products, applications and global market share constantly under pressure. For any business to remain competitive in this sector, it has to continuously reduce the cost of its products, as well as shorten time-to-market for new products.

Given the significance of construction costs in its business, and the fact that the cost of construction had not reduced in decades, company management realized that, as a part of its overall business strategy, construction costs would have to come down at a rate that would match its own competitive improvement curve. The company also recognized that it had a key role as the client to take the lead in driving changes within its supply chains.

The corporation's internal construction group was set the challenge of reducing its internal procurement costs and improving the productivity of its construction supply chains. The team identified three primary goals in relation to its construction programme. It had to:

1   reduce its structural costs by 30 per cent, including its indirect costs;
2   improve the productivity of its key external supply chain business partners; and
3   integrate better both internally and with its key business partners to increase the level of collaboration.

To accomplish these three things, early on, the team identified the need for culture change. Past success had shaped the business' construction procurement practices. As in the case of many powerful and confident client organizations in both the public and private sector, the business specified its requirements in minute detail and employed contract supervisors to ensure its requirements were met, to the letter of the contract. The internal change team realized that if it was to drive change, it had to attract the very best suppliers, and it needed to work more closely with its business partners to establish better ways of working.

After some internal and external research and reflection, the change team recognized that parties across the construction sector had a tendency to blame each other for the industry's flat productivity curve. The team identified that what is required is a culture of shared ownership of problems and a sector-wide collaborative approach to the development of solutions. The business decided that, in addition to driving change on its own projects, it would take a lead role in championing the movement for change in the construction sector globally.

PLANNING THE CHANGE JOURNEY

As a starting point, for its own projects the business identified the need for behavioural change both internally and across its supply chain; it targeted the following areas:

1   collaborative learning;
2   100 per cent accountability for all interactions; and
3   mapping and analysis to drive continuous learning and innovation throughout the entire process.

The business also recognized the need to change its procurement model from one of transferring risk to one of shared ownership of the problems and their solutions. This meant going to a procurement model under which the entire team is responsible for the outcomes, and risks and reward are shared.

Figure CS4.1 describes the key steps in the business transformation journey that the company has undertaken.

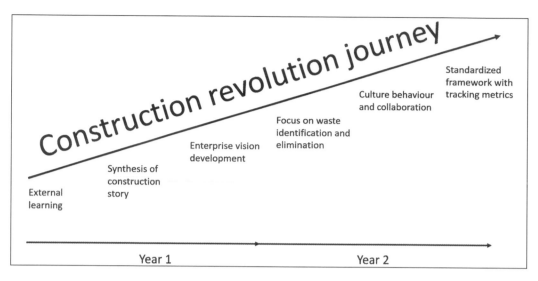

FIGURE CS4.1 Construction change process plan

MAPPING THE CURRENT STATE AND DEFINING THE FUTURE STATE

Once the business had identified the need to improve the construction efficiency of its projects, it set about developing a strategy that drew on internal lessons from its core business as well as lessons from early adopters of lean construction and key supply chain partners.

From its own experience in running its core business, the change team was familiar with a basic six-step process that it had used to drive change throughout its business. The steps adopted are described in Figure CS4.2 and set out in the following steps:

1 Cross-functional teams were established that include leaders and individuals who work closely together to the value-added work.
2 A current state value stream map was created and analysed by asking:
   – What process steps are non-value-adding?
   – What steps can be eliminated or replaced?
   – Where people might be obtaining inaccurate, incomplete or delayed information?
3 The ideal state was defined considering the following questions:
   – Where can we create continuous flow?
   – How should we respond to problems?
   – How will we clearly connect every customer–supplier pair?
   – What would a map the idea process look like?
4 Barriers were identified and ranked.
   – What is currently standing between the current reality and the ideal state?
   – What is the barrier's impact on overall performance or progress towards the ideal state?
   – How quickly, cheaply or easily can the barrier be removed or reduced?
5 An action plan was developed.
   – Name improvement projects.
   – Define expected outcomes.
   – Develop master schedule.

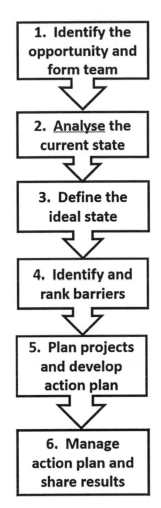

1. **Identify the opportunity and form team**

2. **Analyse the current state**

3. **Define the ideal state**

4. **Identify and rank barriers**

5. **Plan projects and develop action plan**

6. **Manage action plan and share results**

FIGURE CS4.2
Mapping the change process

6   The plan was managed and results shared.
    – Schedule a new event to recreate the current reality based on project completion.
    – Choose next set of barriers to work on.

One of the early tasks was to scan the construction industry in order to identify best practices. This started with conversations with lead owners and with contractors who were seen to be leaders in innovation and practice. Many of these organizations were active members of the Lean Construction Institute. The change team also spoke to lead consultants in the areas of lean construction and productivity improvement, and linked up with relevant industry associations.

With support from its supply chain partners, the business set about mapping the current state of its value streams and defining its desired future state. It established targets and a 'language' to begin collaborating with its supply chain partners, and together they established clear measurable project goals and developed a path to achieve competitive market pricing.

The collaboration with supply chain partners also led to the development of owner/supplier commitments for each work package. This involved defining what the owner and its suppliers needed from each other to achieve the project goals that had been identified.

The next part of the journey focused on the risks associated with the programme; whereas, previously, the company had negotiated a lump sum price and transferred risk to its suppliers, a new collaborative approach that set out to define a win/win relationship from the outset was developed. A cost loaded risk matrix was developed jointly with suppliers, and an incentive programme that was jointly managed by the owner and its suppliers was established. As a result of this approach, it was found that the majority of the project risk could be mitigated.

EARLY EXPERIENCE ON EXPANSION PROJECT

In June 2013, prior to introducing the lean construction changes, a pilot project was started in Arizona to start to drive efficiency into the construction process on a new expansion project.

As a lead-up to reducing waste, work study was used to see how much value-added time was being spent by trade workers on site. The amount and type of non-value-added time was also identified to guide further improvement initiatives.

A benchmark study undertaken in late 2013 showed that 46 per cent of total time was being spent on:

1   moving (13 per cent);
2   waiting (15 per cent);
3   set-up (8 per cent); and
4   retrieving materials (10 per cent).

Parallel work study measurement indicated that, on similar projects elsewhere, productivity was similar.

The change team realized that its practices and delivery business model had to fundamentally change. The window for making the change was eight weeks, Thanksgiving to Christmas – as all changes had to be implemented by January 2014. The group started with an intense one-week session, with just the internal team, to develop and share the vision of the new business model and to discuss the cultural changes needed by the entire project team and its supply chain partners.

The change team recognized that there were too many contractors involved so they decided to reduce the number of the process piping and electrical contractors. The team and its advisers wanted an Integrated Form of Agreement (IFOA agreement) for the whole project. This type of collaborative contract, which included painshare/gainshare provisions, had been developed by Sutter Health for its lean construction project delivery strategy in Sacramento almost a decade earlier. However, at this stage, the piping contractors were not willing to sign such an agreement with their competitors in the risk pool, so the project proceeded during the next stage with four IFOA agreements centred on each of the four process piping contractors.

The project continued under enormous time pressure. The owner's project team worked hard at improving the transparency that it provided around cost and schedule. Although costs stopped increasing, they were still too high to meet business needs.

At this point, the contractors came forward with a proposal. They wanted the owner to change its practices even further. They offered to sign a single IFOA agreement for the balance of the project; however, they identified several areas where they considered that the owner was still unnecessarily adding to cost. They argued that the current contracting model was still too wasteful and they advocated shifting greater responsibility to the supply chain partners with a commensurate reduction in oversight. They wanted to see a shift to a more open, trusting and collaborative culture.

The owner had its own supervisory and quantity surveying teams and it had engaged a construction management firm to oversee operations; this company had significant project personnel on site. The supply chain partners questioned whether these were all needed. The owner took up the challenge and further reduced the size of its quantity surveying and supervisory teams and implemented a dramatic reduction in the size of its construction management team. Furthermore, the owner changed the role of its safety and probity auditors, retraining them to become problem solvers and coaches.

The project team focused on high impact items such as:

1 cable pulls requiring equipment vendor support;
2 field welding output;
3 improving 3D modelling quality and accuracy; and
4 wireless connectivity in the field.

To drive improved workflow and productivity, the trade contractors on the team started to really collaborate in their planning and work practices. The owner was involved in regular pull planning sessions and weekly work planning sessions as well. Wherever appropriate, the contractors used community cut-stations, they shared resources and they shared equipment and lay down spaces in the field. Everyone's 3D modellers were co-located so that they could collaborate more effectively. Costs saved were shared with all supply chain partners in the IFOA.

As a result of the changes made by all project team partners, the project was finished on time; the cost reduction goals were achieved; and the supply chain partners improved their profit. Additionally, no disputed claims were issued for the IFOA portions of the work. The business' supply chain partners reported that the project culture had been transformed from an adversarial to one of open collaborative teamwork.

THE ROLE OF DATA

Throughout this period, within the owner's organization a debate had been underway about which way to go, some arguing that design and commissioning had to be significantly accelerated, while others focused on labour productivity. Early on, based on work study measurement, the owner's team had identified that approximately 50 per cent of on-site craft labour activity was non-value-adding. The owner's change team also recognized that any lump sum price would automatically include this 50 per cent of non-value-adding time because it was a real cost within the existing procurement model under which the company had been operating.

Once the Arizona team changed to working on the basis of IFOA agreements, the change team continued to use the work study measurement to further highlight continuous improvement opportunities. There had been two rounds of work study conducted in the second half of 2013 and an additional round in early 2014. The early 2014 measures suggested that labour productivity had improved by 30 per cent within an eight-month period.

Within 18 months, by early 2015, the following gains had been identified:

- productivity of field welders doubled;
- labour productivity measured as earned hours was 20 per cent above previous best case estimates;
- labour productivity in electrical work improved after the site was working on a single IFOA agreement;
- rework reduced from 25 per cent to 10 per cent between June 2013 and December 2013; however, within 6 months of the introduction of the single IFOA agreement, this had dropped to zero.

By early 2015, less than 18 months after the start of the pilot and a little over 12 months post the change to working under a single IFOA agreement, value-adding time for trades on site had improved from 17 per cent to 57 per cent. Non-value-adding time had reduced to 32 per cent in total. The pilot project had well and truly turned the corner, and the company has taken the learnings from the Arizona project to other projects globally and has experienced similar results.

THE BROADER STRATEGY – INTERNAL WASTE REDUCTION

While the Arizona pilot project was underway, the change team had set about establishing a measurable baseline to define what is possible. Benchmarking standards were established, improvements measured and an education programme across the supply chain initiated. This benchmarking spread to a global effort to identify standards and drive best practices. The team also started to identify waste and focused its efforts on reducing waste in its own operations.

With the input from its supply chain partners, the owner looked at its own impact on costs and identified that it provided excessive requirements and documentation to its contractors. The existing document count was benchmarked and the owner decided to go on a 'Doc Diet' targeting a 75 per cent reduction as a goal. By early 2015, a year later, it had reduced its documents by 44 per cent while maintaining its stringent core requirements.

Another area identified for improvement was the early delivery of equipment. Equipment delivered to the expansion project before it was needed was increasing the amount of work in progress; in addition, it was creating logistical and stock control challenges, including the need for greater care against damage during handling and re-handling.

Finally the owner also realized that in its traditional design–bid–build process, the time lost during tender and contractor selection delayed the commencement of construction. The removal of this wasted time was recognized as a further benefit of the IFOA approach to procurement.

CHANGING HABITS AND THE ACHIEVEMENTS OF THE FIRST 18 MONTHS

The initial focus area was to train trade workers to recognize waste. Recognizing that in traditional organizations operators do not see waste, it becomes invisible, as it is simply a part of business as usual, the change team and its supply chain partners adopted a standard classification of waste and created a laminated card that everyone on site carried in his/her pocket.

A second focus was in the area of cultural change. This was achieved by building trust and commitment across the supply chain and investing in the development of the project team as well as in technology. The owner also wanted to see ongoing continual improvement. One important element of this strategy was to introduce transparency across all data – especially, cost and time-related data. This helped to build partnership and trust.

The third focus was to align goals across the project team. To accomplish this, individual goals were aligned with group goals. The company wanted to establish a culture of continuous improvement as a way of life.

This broad-based engagement has led the owner and its team of suppliers towards adopting more prefabrication, Kaizen or team-based improvement events, utilization of Target Value Design (refer to Chapter 8) and the implementation of various management tools to support continuous improvement, including the incentivizing of innovation.

In the initial period described by this case study, the owner was able to bring down its costs to equal or better than market rates and significantly reduced the cost of its projects. The company is committed to becoming the client of choice for suppliers of construction services and to reducing its costs on every project by a further 10 per cent year on year.

When reviewing the sources of the gains that have been made, no one strategy stands out as being the primary cause; rather, the business leadership credits all the efforts put into changing the culture and processes on its projects. Numerous changes have occurred driven by the change in the contract form, which has allowed a much more open and collaborative culture to develop. The information gathered through work study has helped to identify areas of inefficiency and joint opportunity. Finally, the attention to cultural change to drive collaboration, transparency and team alignment broke down the barriers that prevented innovation and helped to integrate improvements.

Under the new IFOA contracts, the owner's people are a part of the team, rather than the overseers. Staff are involved in weekly risk and opportunity review meetings with their supply chain partners. Additionally, the supply chain partners have also radically improved the way they operate.

THE ROAD AHEAD

The internal challenges faced by the owner's construction procurement team are not insignificant: the construction procurement team is some 400 strong worldwide and works across three business groups. Taking any change across a global business of this size has its own challenges.

Now several new projects have been contracted under an Integrated Form of Agreement (IFOA), which includes the following elements:

1 a combination of fixed and reimbursable compensation;
2 key trades and builder on board before design commences;
3 target costs based on best-in-class;
4 innovation plans to improve on costs;
5 jointly managed risk so everyone has skin in the game;
6 shared savings for cost savings beyond target and for risk mitigated;
7 transparent goal setting, budget and schedule;
8 aligned method to track progress and define 'great';
9 defined Conditions of Satisfaction, 360 degrees;
10 indirect cost target includes owner's costs; and
11 shared decision-making.

This new approach sees the owner meeting with its suppliers on each major project weekly to jointly review progress, risks and opportunities. This represents a radical shift from the past where project reviews were often one-sided.

This owner is not only seeking to change its own practices and those of its immediate suppliers but it has also taken a leadership role in helping leading private and government construction clients to better understand the impacts of their policies and practices on the behaviours and efficiency of the construction supply chain and the industry as a whole. The business believes, that by creating informed and engaged owners/clients, it can help the construction sector improve its efficiency at a rate that is consistent with the rate of improvement in other sectors of the economy.

CASE STUDY 5

Continuous improvement and growth at Graniterock

PREAMBLE

This is a company that has for decades focused on the challenges of building internal capability, enhancing customer quality and at the same time improving the efficiency of all its operations. While traditionally Graniterock has not used the term *Lean Production* to describe its business approach, its practices which are deeply rooted in the philosophy of total quality and its striving for continual improvement in efficiency define it as an outstanding example of a lean enterprise.

COMPANY BACKGROUND

On 14 February 1900, in the days of manual labour and horse drawn wagons, the Granite Rock Company was incorporated as a quarrying business. By 1924, both the quarry and transport system had become increasingly automated and the business had expanded to include the Granite Construction Company (spun off in the early 1900s) and the Central Supply Company. In the 1930s, Central Supply opened California's first asphaltic concrete plant, which then continued to automate and expand through to 1960.

The tremendous development of the Monterey and San Francisco Bay areas during the 1960s and 1970s contributed to the continuing growth of Graniterock. Central Supply merged with Graniterock to form one company for construction materials production and sales while numerous new plants were opened for the sand, concrete, asphaltic concrete and building materials operations.

Prior to 1987, construction material needs were served by a number of medium-sized, family-owned businesses; most of these were locally based and served local demand. However, in 1987, competitive conditions changed dramatically: large multinational corporations acquired and consolidated smaller suppliers into large businesses. Although Graniterock remained a California business, it now had to compete with the world.

The company was under pressure when Bruce W. Woolpert took the helm in 1987; under his guidance Graniterock increased its focus on its people and on its innovation in every area of its business, resulting in increased competitiveness and higher than industry performance levels: Graniterock won *the National Baldrige Award for Quality* in 1992. Despite price and market share driven competitors, Graniterock has continued to grow (today it has over 800 employees) and increase its share of the market. Due to its values, the quality of its people, products and services, and its innovation, Graniterock has become a commercially successful enterprise that is repeatedly listed in the top 100 employers in the USA.

In 2012 Bruce W. Woolpert, the president of Graniterock, died suddenly in a boating accident; he had been a very generous and wise leader. Graniterock had prospered with him at the helm, the business had been his life and he had been a very hands-on CEO. He was deeply involved in every aspect of the running of the company and one of his primary commitments had been to his people, a legacy that would serve the company well in its transition to a new period in the life of the company. Since 2000 Graniterock has repeatedly been voted among the top 20 most preferred employers in the US economy according to Fortune 500.

Leadership challenges

'Bruce was larger than life, and as CEO he filled many roles'. Tom Squeri, the new President and CEO immediately identified that he could not do all of what Bruce W. Woolpert had done. He did not have the capacity or knowledge to fill all the roles that the former CEO had taken upon himself. One of his first challenges in the transition period was to recast the roles and responsibilities of leaders throughout the business to more fully share in the task of leading and shaping the business. He needed to build the capability of the entire leadership team to share and take forward the legacy that Bruce W. Woolpert had created.

Because of Bruce W. Woolpert's total commitment to his people, the company had excellent resources at all levels in the business who could step up; the challenge has been to redistribute some of the roles and share the responsibilities of leadership more broadly across the entire leadership team.

In addition, some of the management systems that were well suited to Bruce W. Woolpert's leadership approach did not work as well in his absence. Bruce fostered a 'Make it Better' core value in the company, and in his absence the new leadership team set about changing processes and systems to deal head on with the leadership transition. The key to this work was maintaining and enhancing the company's culture and shared values, centred around care and respect for people, while improving management systems to respond to the change in leadership.

One company

The second major challenge that Tom Squeri and his team identified early in his leadership was the need to more closely integrate the construction arm and the materials supply arm of the business to operate as 'one company'. The two branches of the business were set on somewhat divergent paths, each focused on its own issues without thinking of the other. This culture of separation meant that opportunities for growth and capacity to enhance service quality in parts of the business were being missed.

Dual objectives were identified to better leverage the potential within the two businesses:

- to grow the overall business more effectively; and
- to improve the ways that the parts of the business worked together to better serve customers.

The theme of collaboration within the company has been taken through every aspect of the company's cultural strategy and processes, and is reflected in the accountabilities and performance review process of every staff member within the entire business. It has also been worked into the language used to talk about the company's Core Values and Core Purpose.

Basis of achievements

The achievements of this exceptional company are the result of the collective efforts of individuals – each person having the freedom to do the job in the best way he or she knows; however, few achievements are derived purely by the effort of individuals. For a company to translate such individual achievements into maximum value, teamwork is essential. Therefore, when financial resources, human talent and energy are directed by *core values, a core purpose, and clear role accountability*, the co-operative effort and team-work result in both a greater individual and company achievement.

CORE VALUES

Graniterock staff and the company are expected to adhere to the following core values.

Safety before all else

Graniterock people and the company place safety ahead of achieving any of its other plans and goals. Our commitment to safety extends to Graniterock people, our customers and the public. Graniterock people are expected to implement work plans which achieve safety as the primary goal and to improve work site conditions to eliminate potential safety risks. The company will provide effective safety training, including health and wellness education and support, and finance plant and equipment safety innovations and improvements.

Dedication to customer service excellence

Customer service excellence requires Graniterock people to anticipate and exceed customer needs and expectations. We work hard to build long-term valuable business partnerships with our customers, and to live by our 'Yes, we will' standard of responding positively and creatively to every ethical request for special products and services.

Building great lives

Graniterock provides an opportunity for personal growth and professional achievement through support of 'lifelong learning,' through trust in people to do a good job and thereby creates an environment of freedom for individuals to direct their own work and work improvement. The company strives to develop job responsibilities that are rewarding, worthwhile and challenging. Graniterock people and the company demonstrate respect and caring for all members of the Graniterock team and their families, and for our neighbours, customers and suppliers.

Honesty and integrity

Graniterock and Graniterock people shall conduct their activities with uncompromising honesty and integrity. People in every job are expected to adhere to the highest standards of business ethics and fairness in all of our dealings with customers, suppliers, communities, government officials and agencies, and other Graniterock people. We expect the same high standards from the people and organizations with whom we work.

Make it better

Graniterock people are achievement-oriented, unsatisfied with 'things the way they are' when improvements are possible. Rather than preserving or protecting the work of the past, Graniterock people encourage and support both incremental and sweeping change and recognize that risk-taking and honest mistakes are unavoidable parts of doing this. In everything we do, we will compare ourselves with the results achieved by role-model high performance companies in any industry. We are not satisfied with simply being 'good', we must be among the 'best' in every important thing we do. Graniterock people consistently provide a 'can do' commitment to individual – and team – driven improvements, and provide support for improvement ideas in other areas within the company.

CORE PURPOSE

Graniterock thrives when our people thrive. We exist to provide a place where inspired people can do their best work – building great projects, producing quality materials, and developing enduring customer relationships.

CORPORATE OBJECTIVES

The company categorizes its business and management systems under nine corporate objectives. The corporate objectives are presented in a circle format in Figure CS5.1 – rather than in a list because each of the corporate objectives is important to the company's long-term success, with safety being the first of equals. Each corporate objective is equally weighted in business performance assessments, and each receives prime consideration in business investment and other decisions made by all Graniterock people. In addition, the circle format helps to convey the supportive and synergistic relationship between the objectives.

FIGURE CS5.1 Graniterock corporate objectives

Safety

To operate all Graniterock facilities and jobsites with safety as the primary goal. Meeting production schedules or customer commitments are secondary. Every person will put safety first – before all else.

Customer satisfaction and service

To earn the respect of our customers by partnering with them and providing them with products and services in a timely manner that anticipate and exceed their needs, and by providing them product and service advantages which contribute to the customer's business success.

Financial performance and growth

The commitment of Graniterock people to knowledge development, coupled with available financial resources, will drive business growth strategies and an expanding new product offering. Diligently plan and implement very long-term mining and resource plans needed to ensure a supply of construction materials for the next 100 or more years.

People

Graniterock people believe they can make a difference in the success of the company and make their jobs more effective. Every person in the organization is given an opportunity to attain a sense of satisfaction and accomplishment from work achievements, to be recognized for individual and team accomplishments, and to be rewarded based upon demonstrated skills and job ownership performance.

Product quality assurance

To supply products and services which provide lasting value to our customers and end-point users (e.g. owners and architects), while leading industry standards and practices to 'zero defects'.

Community contribution and responsibility

To be exemplary citizens of each of the communities in which we operate by actively participating in achieving community goals. The environment is important to us; our actions shall be consistent with environmental responsibility.

Efficiency

To produce and deliver our services and products more efficiently than anyone in the world.

Management

To foster initiative, creativity and commitment by expecting each individual to desire personal and professional growth. To provide each person with the freedom of job ownership to establish work improvement objectives and the freedom of action to implement individual work improvement methods.

Profit

To earn a fair profit to provide resources needed to fund the achievement of our other objectives.

QUALITY IN MANAGEMENT

Leadership

Information gained from its customers is the foundation upon which objectives, goals and quality values are built. While the executive leadership defines and redefines quality values and expectations through the articulation of the company core values and core purpose, the nine corporate objectives are critical to the future of the company as they provide the focus areas for quality improvement.

Annually, the *executive committee* establishes goals and initiatives within these corporate objectives, thereby focusing even more on the specific goals through associated and measurable performance criteria. These goals are implemented through annual written role descriptions for employees throughout the company, with care taken by managers to ensure 'line of sight' from the corporate objectives to individual accountabilities and measured goal-specific tasks. Performance of the general accountabilities and goal-specific tasks are monitored throughout the year, and at the end of the year through formal Performance Evaluation Reviews (PERs). The PERs start with an employee self-assessment that includes both qualitative and quantitative ratings, followed by a manager's assessment. A final check is conducted by the 'Manager once Removed' (the manager's manager). Executive remunerations/compensation and promotional opportunities are directly affected by performance. All PERs are reviewed by the Executive Committee as part of the compensation review process, to ensure consistency, fairness and internal equity. The Executive Committee determines individual compensation for all employees not covered by a collective bargaining agreement, and any performance-based pay enhancements for those who are.

Senior management are encouraged to attend appropriate courses and to be involved in the detail of company training: assisting to design courses and teaching in some instances. This reinforces the company's strong lifelong learning focus. Annual Recognition Day celebrations are held in every branch, creating the opportunity for senior management to acknowledge, face-to-face, the successes of individuals and teams throughout the company.

The leadership also makes it clear that Graniterock is a company of impeccable integrity; companies that do not meet Graniterock standards do not qualify as customers or suppliers. Graniterock's quality commitment to the end-point customer is communicated throughout the market and widely acknowledged.

Strategic quality planning

Quality and business plans start from customer needs. These are established through regular surveys, which are developed and aligned with information gathered from other relevant sources both from the supply side and from the perspective of the market trends. Inputs into the process include feedback from customers, employees and suppliers. Strategic quality goals with measurable performance outcomes are set and they form the basis of performance reporting.

All processes are evaluated regularly through the process shown in Figure CS5.2. Information is gathered early to help establish goals that are most often set to exceed current industry capability. Benchmarks, supplier capability and customer aspirations are routinely sought to assist in the systematic planning that will ensure superiority of performance.

Statistical process control and R&D through the Research Technical Services department underpin continuous improvement and growth. A commitment to continuous improvement

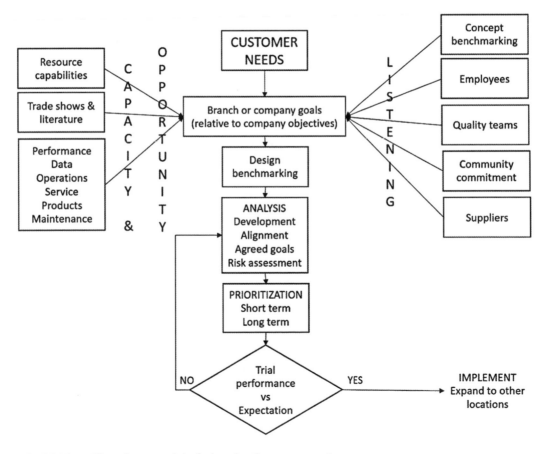

FIGURE CS5.2 Flowchart model of planning/improvement process

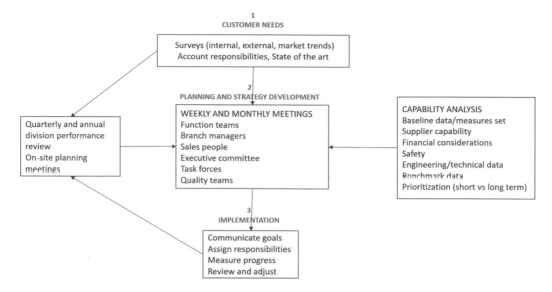

FIGURE CS5.3 Model of plan deployment

through technological advancement ensures timely service improvements in areas such as product load out and on-time delivery, billing speed and accuracy, and problem resolution.

Plans are deployed as outlined in the flowchart in Figure CS5.3.

Once customer needs are established the planning and strategy formulation stages begin. After annual improvement targets are set, branches and divisions develop their own implementation plans. Cross-functional teams developed to resolve specific Critical Issues, identified in the strategic planning process, oversee and help align improvement efforts across the entire organization and foster coordination across divisions. Although senior executives chair committees, members include managers, salaried professional and technical workers, and hourly union employees.

Information and analysis

A comprehensive measurement system is in place to enable management to make decisions based on facts. The system identifies the key areas for which information is to be gathered and measured, and allows the comparison between results and goals and benchmarks.

Criteria for selecting key data are:

- *Customer needs* – established through surveys and regular focus groups.
- *Product quality assurance* – measures of process quality to ensure downstream excellence.
- *Production efficiency* to ensure lowest cost production.
- *Supplier performance* to ensure the highest quality from suppliers.
- *Financial performance* to maintain business efficiency in terms of resource utilization.
- *People development* to ensure the development, health, welfare and safety of team members.

The use of key data in the management system is critical to achieving Graniterock's corporate objectives. They provide the information needed to guide the development of the company and to deploy its resources wisely and effectively. Benchmarking with the world's best companies is also undertaken to provide an external yardstick to evaluate progress.

QUALITY PRODUCTS AND SERVICES

Customer focus and satisfaction

Graniterock's key driver is value creation for the end customer, not just the immediate customer (e.g. the contractor), for the products or services. Focusing on both of these customer groups (each with its own perception of quality) gives the company clear direction for the establishment of marketplace plans which will support current and evolving customer needs.

Customer survey destinations are determined through customer files: trade association membership, project bids and building permit lists as well as supplier input and telephone directories. While the company surveys its principal customer groups at least once a year, it also solicits data from all potential customers, annually. Customer product and service factors are prioritized and customers are asked to rank Graniterock performance against that of market competitors.

Validation of survey accuracy and assessment of customer service levels are accomplished through two mechanisms: a *short pay* system – any customer can omit payment on any product or service that does not meet his or her quality expectations – which is a service and product guarantee to customers, allowing them to be the judge as well as ensuring a rapid response from the company to any complaint; a *product/service discrepancy* (PSD) system – any complaint is root-cause analysed, rapidly resolved and a preventative action plan generated to eliminate recurrence – is used for overall product and process improvement as well as for immediate problem solving.

Statistical process control (SPC) and *numerical performance measures* are used to monitor performance and product quality. These are also the basis for ensuring continuous improvement in all the targeted product and quality service factors. Internal customer/ supplier relationships receive the same attention as external ones do.

Being a leader in new product and process development has also helped to build the company's competitive edge. For example, they developed GraniteXpress 2, an automatic customer card controlled loading system that enables regular customers to drive up to the loading facility 24 hours a day, 7 days per week and accurately fill their truck with the materials required. This reduces waiting time for trucks and increases flexibility for customers.

Quality and operational results

Quality performance is assessed both in terms of physical product quality and service quality. Both areas are represented by hard measures that can be compared to other products and organizations. In the quarries, SPC is used to control variables such as aggregate gradation and cleanliness. Graniterock has reached quality levels that are not generally attained by industry competitors. Variability in some key measures has been reduced to six sigma levels – a standard that is not generally seen in such a 'low tech' industry.

One of the key measures of overall quality in concrete is low drying shrinkage. Although there will always be some level of shrinkage, understanding and then reducing can improve long-term concrete performance. Shrinkage creates cracking and concrete

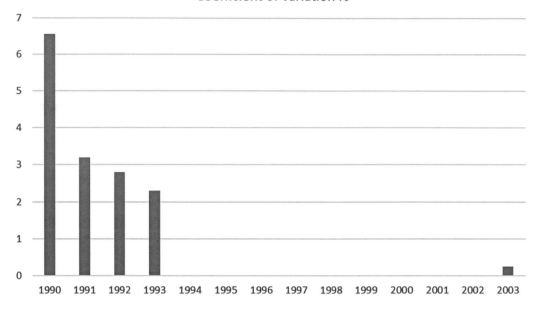

FIGURE CS5.4 Concrete batching accuracy

'curl' problems – both an aesthetic and a durability concern. Crack elimination or at least minimization is in the interests of all customers and users, present and future. To achieve this there has been a focus on improving the quality of the ingredients, the control of the mix proportions and improving mix design.

Graniterock led the industry with shrinkage values that were less than half of those of their competitors for a long time. More recently, as the demand for low shrink concrete has risen, Graniterock is among the industry leaders in terms of this measure; however it is no longer the differentiator that it once was. A recent investment has been made in in-line cameras to monitor particle size and shape in real-time during production. This enables unwanted variability to be identified as soon as it appears, and preventative action can be initiated early to keep the plant working to tight tolerances.

Improvements in service quality receive the same attention as improvements in product quality. Graniterock's on-time-delivery performance record illustrates this (see Figure CS5.5). This on-time-delivery performance – getting concrete into mixers on time, delivered to the jobsite and unloaded quickly – has been listed as the number one concern of its contractor customers: a concern that is even above price and concrete quality. In contrast, owners' and architects' (representing the end-use customers) interests are most concerned with concrete quality. Recently GPS has been fitted to concrete trucks in major markets to further improve reliability and predictability of deliveries.

Similar measures are used in each business area. In the manufacture of asphaltic concrete, SPC has been used to improve batching reliability. Whereas, previously, it was rare to complete an asphaltic concrete project without some rejected material, product quality has now improved to the extent that it is rare to have a single specification

Concrete On-Time Delivery
Per cent on time

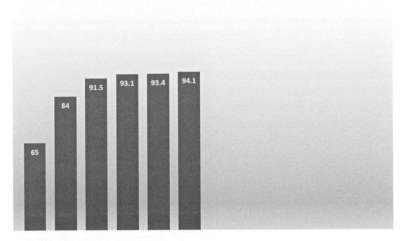

FIGURE CS5.5 Concrete on-time delivery

discrepancy on any project. The company regularly earns maximum bonuses for exceeding quality benchmarks applicable to certain public works projects that reward quality.

Plant production and operation measures such as variable costs and production efficiency show generally improving trends. Increasing market share is a measure which many companies seek to improve. Between 1986 and 2004 Graniterock saw its market share steadily rising, however, since the GFC price competition has intensified and the market has fragmented. While Graniterock's market share in the premium end of the market, where quality and service are valued, has continued to increase, the company has lost market share to lower cost suppliers in the market segments where lower quality and service levels are accepted.

A key function of product quality is the quality of suppliers' products and services. Partnering arrangements with key suppliers have been put into place and are working well. Three of Graniterock's outstanding relationships are with its primary cement supplier, asphalt supplier and railroad company. These relationships have created real benefits for all involved.

To assure effective vendor contact, the number of suppliers has been reduced. For example, the company only has one cement supplier and two petroleum product suppliers, and ten building materials suppliers provide 80 per cent of the products purchased for resale.

Continuous improvement

To offset some of the rising costs of materials, equipment and labour, Graniterock is constantly improving the productivity of every business process and job function. Every business area and department is expected to demonstrate higher productivity and safety performance each year. Graniterock strives to increases productivity by changing and improving work practices, investigating and implementing new technologies, and eliminating errors and waste. A critical aspect of this endeavour is to find and define suitable SMART performance measures that will drive the company's quest for continuous improvement in each area of its business.

At Graniterock, individuals are responsible for championing new ideas to improve their job efficiency, by implementing improvement processes which compare current performance with goals. The company also expects people to work with teams to improve company-wide practices. As Graniterock implements improved productivity practices in each of its business entities, management expects cost performance ratios to decline, making the company one of the leaders in cost efficiency and market competitiveness.

QUALITY PEOPLE

Human resource development and management

The company has a policy of favouring internal applicants for senior positions. Therefore, already at the entry-level, it seeks to hire people who demonstrate the capacity for senior management. The company seeks to hire lifelong employees.

Graniterock encourages all its team members to use *individual professional development plans* (IPDPs) to plan and track their own skill and knowledge development. This is a comprehensive process for integrating the company's human resource needs and quality objectives with the individual's aspirations and abilities through education and training. The IPDP is a forward-looking system because individuals can use it in alignment with company needs as a tool to plan their professional development goals. The IPDP is used in conjunction with the written Role Description and PER to ascertain knowledge and skill development needs as well as long-term career aspirations. It is a contrast to the *past performance assessment* system, in that it represents a proactive approach to personnel development.

The backbone of the IPDP is a simple one-page plan drafted by individual team members and their immediate supervisors. Short- and long-term goals are selected and a training/development plan is created to help the individual achieve these goals.

Apart from helping to set Graniterock repeatedly in the top 100 employers in the US (*Fortune* magazine), the process has benefited both managers and team members. It has improved the coaching and personnel management skills of managers; it has helped to identify and highlight the talent within the company; and it has helped to avoid the underutilization of people throughout the company.

Other performance measures used to track the company's human resource development practices are:

- number attending the Graniterock university and other seminars;
- number of cross-trained people;
- number requesting special education and training;
- number of absentees;
- number of turnover rates;
- analysis of exit interviews and employee surveys;
- satisfaction with benefit surveys;
- speed of processing accident and health insurance claims.

Teamwork is widely used within the quality improvement processes of the company. The types of teams include quality teams, function teams and Critical Issues task force teams. There are normally dozens of teams active within the company at any one time, and a high level of involvement reflects the company's team culture.

Recognition Day, held annually at each branch and department, plays the most significant role in the all-important recognition process. It allows face-to-face presentations to recognize all team member achievements in the past year. Senior management attends these days to send a clear message of support. This process is just one part of the overall communication and support network that makes up the company's employment culture. Other activities include company awards, recognition of achievements in the company's twice-monthly newsletter to all employees and occasional letters from the president of the company to all team members.

Employee morale

The company arranges many recreational opportunities for its team members, encouraging them to play together as well as work together. Events range from concerts, parties at elegant hotels with famous speakers to parties for the whole family. In this way Graniterock is committed to building a community of excellence among its team members.

QUALITY IN THE COMMUNITY

Graniterock is an exemplary corporate citizen. Company team members provided record financial support for United Way agencies, in 2014, with a total contribution of nearly $200,000. Graniterock matched this contribution dollar-for-dollar, resulting in a doubling of overall community financial support.

The team members demonstrate the values of their employer by investing their own time and money in building the communities in which they live and work. This is exhibited by the more than 600 volunteers from the Graniterock team who have volunteered their personal talents and shared their financial support with more than 700 organizations in Alameda, Monterey, San Benito, San Mateo, Santa Cruz and Santa Clara Counties in 2014. Graniterock matches employee donations to charities dollar for dollar.

Graniterock people respond to a broad spectrum of needs within the community with real-life benefits: they coach little league teams, provide senior care, staff domestic abuse hotlines and support children's hospitals. They raise funds for community recreational programmes and deliver food, companionship and good cheer to people and families in need. They heighten awareness for safety programmes and volunteer in fire-fighting. They apply their skills to the advancement of women in construction, the protection and cleaning of our coastlines, to emergency flood repairs and relief work and to the recycling effort. Graniterock people apply the total quality management approaches that they have learned in business to their dealings with government, educational and non-profit organizations.

Recurring events

Graniterock is involved in many great events through the year, including these regular occasions:

- A.R. Wilson Quarry Open House;
- Contractor's Expo;
- Masonry and Stone Expo;
- Pavement Expo;
- Rock & Roll Poker Run;
- Remodelers' Day;
- Rock & Run;
- Share Your Holiday with KSBW;
- Silicon Valley Expo; and
- Wag 'n' Walk.

RECOGNITION FOR EXCELLENCE

Graniterock has received more than 50 awards in recognition of the group's performance. In the category of business awards these have come from Californian and national organizations for:

- Accuracy in company accounts;
- Excellence in business;
- Outstanding business leadership;
- Excellence in engineering management;
- Excellence in personnel management;
- Outstanding customer service;
- Outstanding website;
- Producer of the year;
- Sixteenth best employer in the USA.

Since 1993 the company has won numerous awards for excellence for its products and services, from Californian and national organizations for:

- Award for construction innovation;
- Extremely high standards of safety;
- High-quality products and services;
- High-quality plant operations;
- Information executive top 100 list;
- Winner of the US Baldrige award for overall quality.

In the area of community awards the company has gained recognition for a broad range of outstanding performance achievements:

- Environmental responsibility through waste reduction;
- Excellence in community service;
- Outstanding community relations;
- Partners in educational excellence;
- Support for charities;
- Resource conservation for the use of water.

REVIEW QUESTIONS

1 Consider how Graniterock uses performance measurement to drive improvement and innovation.
2 In the construction supply chain it is often said that one of the impediments to quality is the fact that everyone focuses on the immediate customer and only the builder is focused on the end customer. Explain and evaluate how Graniterock uses end-customer satisfaction as the main driver for its improvement processes.
3 How does the Planning/Improvement process compare with 'theory' – identify the strengths and areas for improvement in the Graniterock approach.

ACKNOWLEDGEMENTS

The authors gratefully acknowledge the assistance of the Graniterock team led by president and CEO Tom Squeri. Some of the text and illustrations have been taken from company information. Further information on Graniterock may be obtained from its website: www.graniterock.com.

CASE STUDY 6

BIM and lean practices drive lean transformation at Southland Industries

INTRODUCTION

Founded in 1949, Southland Industries (SI) provides innovative engineering, construction, service and energy service solutions through a holistic approach to building performance. Advocating a design–build–maintain model, Southland believes in offering customers the option of optimizing each stage of the building lifecycle through an integrated, customized project or by selecting any of its services and capabilities to be implemented individually. For jobs large and small, SI's in-house experts remain connected, sharing knowledge and information in order to produce the innovative, practical solutions that have earned Southland its unmatched reputation as one of the top design-build firms in the nation.

Utilizing a variety of progressive tools such as building information modelling (BIM) and lean methods, Southland specializes in the design, construction and service of mechanical, plumbing, fire protection, process piping, automation and controls systems as well as comprehensive energy service needs. As a company that has always prided itself on innovation and collaboration, Southland, as an industry leader in sustainability and energy efficiency, continues to pave the way towards improved building design, build and maintenance.

Beginning as a Southern California-based supplier of residential heating solutions, SI has organically grown and exponentially expanded its services and capabilities over the years to serve a wide variety of markets and industries. Recognized as one of the nation's largest building systems experts, today Southland delivers superior results for commercial, data-centre, education, healthcare, government, hospitality, industrial, life sciences, entertainment, and mixed-use buildings and clients.

This case study focuses on the lean construction aspects of SI engineering, fabrication and construction activities.

ENGINEERING

Southland's Engineering group specializes in the planning, design, construction and operation of mechanical, electrical, plumbing, process piping and fire protection systems for a wide range of facility types and projects of varying in complexity. SI has a unique ability to integrate knowledge of engineering, construction and facility operations into a single technical solution, enabling it to achieve the performance, energy efficiency, sustainability, constructability, maintainability and cost goals its clients seek.

Southland's engineers are respected thought leaders in the architectural, engineering and construction community. Combined with Southland's experience in construction and facility operations, its engineers understand actual system performance, total costs and opportunities to improve construction and maintenance services. Believing deeply in the power of collaboration, Southland's people recognize that the best solutions are inspired when people with many differing perspectives work together to drive innovation.

With extensive industry documentation of the benefits that accompany a collaborative and an integrated approach, Southland is well versed in collaborative project delivery methods such as integrated project delivery (IPD), design-build (DB) and design-assist.

Additionally, process and technology tools such as lean principles, modular methods, and construction processes streamlined for efficiency, maximize value and improve team interaction. Through ongoing research and partnering with a wide range of industry experts, Southland Industries continues to innovate and drive next generation solutions.

CONSTRUCTION

In the crucial stage when designs become reality, Southland approaches construction with proven procedures to execute objectives efficiently and, ultimately, deliver the best project value. By utilizing available resources and the wealth of lessons learned, Southland Industries is able to consistently achieve outstanding results.

With personnel that are adept at handling the manpower planning required for large, complex projects, Southland installs each new project with the goal of making it SI's best project yet. Experienced foremen work with the construction manager and trade superintendent to develop a plan of bringing on the appropriate number of skilled craftsmen with the specific specialties a project requires.

One of Southland's biggest opportunities for improving construction quality and controlling a project schedule is through implementing high levels of prefabrication and modular construction. With fabrication facilities located near each of its divisional offices, SI's goal is to manufacture and pre-assemble as much material as possible in order to provide more efficient logistics, materials handling, superior installation and site work that is safer overall.

Utilizing process and technology tools like BIM, SI streamlines its efficiency, maximizes value and improves team interaction. Taking a holistic approach to all projects, SI continues to innovate and drive next generation solutions that elevate building operations, lower energy consumption and reduce overall owner costs.

BUILDING INFORMATION MODELLING

Southland's building information modelling (BIM) strategy is fully integrated with its engineering, construction and maintenance services. The company utilizes a streamlined process that leverages technology to share information and improve collaboration during the design, construction and operation phase for any facility. To achieve these goals and maximize the benefits, Southland focuses on the following three key components of its successful BIM strategy:

- **People** – Southland's commitment to BIM is evident through the key resources, including an engineering lead, a constructability lead and a BIM lead, who are assigned to every project supporting its BIM goals. The engineering lead provides technical leadership, while the constructability lead provides constructability knowledge to the design team. The BIM leader is responsible for managing the technology that supports the process. Ongoing BIM training is also provided to employees in order to keep Southland in step with process and technology changes as well as improve its use of BIM.
- **Process** – By championing the use of a single model to achieve the requirements of both design and fabrication, Southland has established a single source for project information. To support this approach, Southland has integrated its design-coordination and spooling resources as well as adjusted its internal workflow. This approach has led to exceptional solutions, highly collaborative and well-informed teams, fully coordinated facilities, fabrication-ready models and a streamlined workflow that produces real and actionable intelligence for the entire team.

FIGURE CS6.1
Complex prefabricates enabled by BIM

The degree of prefabrication illustrated on the adjacent components can only be completed with confidence that there will be no rework through the use of advanced 3D BIM technologies.

- **Technology** – The strategic use of technology is a critical component of Southland's BIM strategy. Southland's modelling, coordination and fabrication software is based on several Autodesk products. In addition, the business has developed many tools and standards to improve interoperability with other modelling solutions as well. Leveraging a proprietary database, Southland engineers are able to manage a wide array of information associated with the model on their projects. Technical solutions also incorporate the use of radio frequency identification (RFID) and mobile devices to support its fabrication, construction and facility service activities. Expanding its use of data sharing models and cloud computing has allowed SI to leverage its talented workforce across the country.

As its clients' needs change and technology evolves, so will Southland's BIM capabilities. As leaders in the industry, SI will assist others in recognizing the value of BIM, expanding its adoption and influencing key players and software developers to create products that support the ideals of a BIM-enabled world.

MANUFACTURED AND MODULAR CONSTRUCTION

At Southland, staff go to great lengths to keep pace with the most recent manufactured and modular construction trends in this industry. Manufactured construction provides benefits in all key cost-driving aspects of a project. With a focus on creating repeatable units and a highly standardized approach to design and construction, regardless of the project type, costs are reduced. By keeping a large portion of the labour that is required to build a project within an on-site (or off-site) manufacturing facility, significant improvements to labour productivity, quality, safety and project duration are achieved.

Figure CS6.2 illustrates a large prefabricated module with multiple services incorporated into the modular assembly.

Southland has also invested significant time and resources into understanding and producing multi-trade assemblies that allow the efficiencies already identified to occur across an entire project. Multi-trade assemblies like overhead pipe/conduit racks and bathroom pods allow for multiple design and construction entities to benefit from a manufactured construction mindset. Southland is experienced in collaborating with architects and structural engineers for solutions to integration challenges that present themselves early in the project lifecycle. Southland has implemented several examples of multi-trade construction, overhead racking (mechanical, plumbing, electrical, framing and fire protection), and equipment skids (mechanical and electrical) on a number of projects and actively looks for the opportunity to do more during the design phase of a project.

Southland has travelled the globe in search of the best and most challenging modular implementations. Its research has improved SI's ability to develop ideas and solutions which improve the efficiency of construction through manufactured strategies. SI believes that there is much to be gained through the better alignment of construction techniques and the installation of bigger, more complete elements.

LEAN DESIGN AND CONSTRUCTION

In 2004 Southland Industries was introduced to the ideas of lean construction on the Sutter Health El Camino medical office building (MOB). On this project, Last Planner® System was implemented and 3D BIM was used. Since that time, lean thinking has permeated the entire business. It is used in the fabrication shop, on site and to integrate shop and site operations. More effort is applied to gain a fuller understanding of the client's needs as early as possible.

An early lesson about the need to better understand the client's expectations was received on an office building where SI installed the HVAC system. When the client walked the site with the SI job foreman, he asked, 'What can you see that is wrong with that ductwork?' The foreman replied, 'not much'. The owner then went on to complain that the standard of finish on the ductwork was far too good, he did not want to pay for the unnecessarily high standard of work that had been delivered. Southland recognized that there was a need to better understand client needs so that the right level of quality was provided and the work is not 'gold plated' in the eyes of the customer.

Since the implementation of lean, labour productivity in the fabrication shop has increased by 15 per cent and on certain activities on site it has improved by 20 per cent. Southland site foremen implement LPS® regardless of the owner or general contractor. The benefit that the company receives from this practice is that improved scheduling on site gives reliable advance notice of demand to the fabrication shop and allows the shop to run more efficiently. Each general foreman on site has a schedule with dates for detailing, shop fabrication and delivery, as well as a reference to the relevant drawing.

Southland foremen get training in 5S and in LPS®. They are taught to lead and promote innovation, identify waste and find opportunities for improvement. The company has developed 5S tools to support its processes in the shop and in the field. It has also developed a stock re-order process to improve the reliability of ordering, and it has implemented the use of shadow boards both in field toolboxes and in the fabrication shop.

SI continually invests in upgrading its fabrication facilities. In one case, research students from the UC Berkeley Project Production Systems Lab (P2SL) helped to design workflow within one of its factories. Throughout Southland, there is a focus on identifying waste and maximizing customer value. Metrics are used to measure productivity both in the fabrication shop and in the field. Measures such as panels/hour, weld inches/hour and fabricated hangers/hour are kept so that improvements in productivity can be measured and productivity can be compared between fabrication shops.

There is a real focus on innovation that will simplify and error proof site operations. Tradesmen throughout the organization are encouraged to come up with new ideas and to record them in a two-minute video, on their smartphones. These are posted on the company intranet and evaluated, and the best ideas are rewarded.

Innovations are spread throughout the shop and field operations: one theme is around prefabrication, packaging and kitting for the field. Another is to improve efficiency, error proofing and improving quality. Yet another is improving production processes in the shop. In all cases, ideas come from the shop floor or site. Innovations include:

- a motorized lift for lifting duct in the field;
- clamps used to attach wheels to sectioned ductwork enable it to be easily moved on the jobsite, and yet are reusable for future duct sections;
- a trolley for moving toilet pans and cisterns to position on-site functions in the following way: all packaging is removed so that it does not have to be collected and removed from site; up to eight pans are seated in padded saddles on a single wheeled trolley together with all the required fittings; they are loaded onto a truck for delivery to the site; and, wheeled to the location where they are to be installed.

A further innovation was the development of an adjustable trolley in which a single pan can be positioned ready for mounting to the wall. This enables the tradesman to adjust the pan to the right height, wheel it into position and simply fix it to the wall. There is no strain on the worker to hold the pan in position while it is being fixed. It makes a two man operation into a one-man operation and installation is much more efficient.

Southland's Piping Supports (hangers) have had many improvements made to them over the past 5–6 years. Some of these improvements were due to BIM modelling processes. BIM allowed the coordinatation of supports in the above ceiling spaces of a building and the confident fabricatation of hangers without fear of rework. Using BIM enabled the

use of Trimble technology and resulted in a much faster layout time than our conventional way.

Reports detailing components could be extracted from models and then broken out by cut length, area, size, and floor section, or system type. Identifiers (labels) are also extracted from these reports and used to identify each individual hanger which correlates to an install plan. Reports lead to fabricating 'like' supports to optimize repetition in the shop and eliminating wasted steps and wasted materials.

The next step was to see what innovations could be made in shipping and in field installation. New pipe support carts were created allowing 40 per cent more hangers per cart than in previous renditions to be shipped but maintaining the same overall size cart. For field install, a scissor lift attachment was created that would receive a row of hangers from the shipping cart and reduce the trips on and off the lifts when installing hangers. Due to these changes the hanger install has been improved by roughly 40 per cent and in the true spirit of lean practices, much less wasted hardware and threaded rod has been seen.

Shop site integration

Since 2012, all designs have been modelled in 3D BIM to assess accessibility for workmen, to identify spatial constraints during installation and operational maintenance, to assess safety as well as ergonomic issues for workers. 3D BIM is also used to drive an increase in prefabrication and the standardization work. There has been a move to more complex and larger prefabricated elements including multiple services. Such components may take a week to prefabricate, but as little as 15 minutes to install on site.

There is always a tension on site around the demand for crane time for the assembly of large and heavy services components, as this can add a few weeks to the crane schedule. However, at the end of a project, prefabriaction can help to bring a project to completion up to six months early. Furthermore, field labour can be reduced by up to 50 per cent. To capture the benefits of prefabrication, it is important to start planning early, using virtual modelling, to identify opportunities for standardization and for a modular approach as it is critical to guide the design in this direction as early as possible. HVAC tends to have the greatest influence on the design within SI because it dominates in terms of complexity and size. Hence there is a greater need to plan ahead.

3D BIM is also being used to design clash avoidance into the building rather than for clash detection after the design is completed. Laser scanning is used in the field to check the accuracy of construction against the BIM design. There are still challenges in the use of this technology as many operators do not have sufficient experience in setting targets correctly to create an accurate scalable model. However, this is an area of rapid development and is seen as beneficial because it enables rapid assessment of existing services in an existing installation and provides a measure of the consistency between construction and design.

In the design process A3 analyses are used to set out alternative design solutions. This process helps SI engage with the owner and opens up the decision-making process within the whole design team, helping them to make more informed comparisons between different design solutions.

Transferring lean practices across the business

As a part of the overall lean initiative, there has been an increase in management coordination and review. In each region, there is a quarterly meeting of all foremen in the office to share lessons, and to drive best practices and innovation across all projects. At this time, outstanding contributions are identified and recognized within the business. In every region within SI, there is a lean board where lessons derived locally as well as from other regions are shared, benchmarking is displayed and challenges to be solved are listed. Finally, superintendents from SI fabrication shops around the country meet twice a year to compare improvement initiatives and operational challenges.

Each year there is an internal development programme in which six people are given a problem to study: in 2014 focus was on leadership, in 2015 it was on lean and the challenges for the organization to transition globally towards lean practices.

Overview of lean achievements

As a founding member of the Lean Construction Institute, Southland has long embraced the lean principles as part of its core operating philosophy. SI continuously seeks the systematic removal of waste in its processes and the efficient use of materials, resources and personnel to improve performance. By applying lean principles over the past 15 years, Southland's efforts have directly translated to the following benefits for Southland, its business partners and clients:

- **Process Improvement** – as a design–build–maintain provider, Southland has the opportunity to leverage lean principles throughout each stage of the building lifecycle in order to fully maximize the lean approach. A centrepiece of this effort lies within the BIM strategy, which focuses on a single model that addresses the needs of engineering, fabrication, construction and maintenance. This approach reduces waste, aligns project goals and improves communication amongst the entire team while providing clients with a faster and more flexible delivery model that reduces cost and improves schedule.
- **Efficient use of materials** – working to reduce the use of raw materials, Southland starts with the appropriate selection and design of innovative HVAC, plumbing and fire protection systems and utilizes prefabrication and modular construction to eliminate

redundant support systems and improve coordination. Through its extensive use of virtual construction, SI ensures systems are coordinated while mitigating potential rework and waste on the jobsite. With state-of-the-art fabrication facilities, Southland also eliminates scrap and raw material waste along with its associated energy, labour and ongoing operational costs.

- **Integrated schedules** – Southland recognizes that integrated schedules improve communication and maximize the efficient use of resources for all phases of a project. SI personnel's deep understanding of the design, coordination, fabrication and construction process coupled with the use of technology and lean tools create a unique opportunity to plan, communicate and track progress across the entire project. In doing so, SI is able to further optimize the project schedule, improve communication and assist the entire team to meet the schedule commitments made to clients.
- **Productive workforce** – essential to remaining competitive and cost-effective, Southland invests heavily in the continual training of employees to keep pace with rapidly changing technology, means and methods, and tools. SI's lean approach naturally focuses attention on prefabrication and modular construction to leverage the potential for improvements in quality, safety, productivity and scheduling. The efficient scheduling techniques adopted by SI allow Southland to retain talented employees through varying work-cycles.
- **Cost savings** – the culmination of SI's process improvements, efficient use of materials, increased productivity and efficient scheduling tools not only directly reduces costs for Southland but also for other project team members and the owner.

INTEGRATED PROJECT DELIVERY

Southland has been one of the core group of early adopters of lean construction practices in the Sutter Health supply chain. It is a partner in the Van Ness and Geary Campus (VNGC) project, a major Sutter Health project in downtown San Francisco, and the subject of another case study in this book. Southland has participated in many IPD projects, particularly for Sutter Health.

Southland has a history of both integrating its in-house engineering, construction and maintenance expertise and collaborating to develop innovative and cost responsible solutions. SI sees integrated project delivery (IPD) as a natural extension of that philosophy. Essentially expanding its natural approach by aligning the entire project team with those same principles at the start of the project, the use of IPD and the multi-party integrated form of agreement offers significant benefits to the owner:

- **Optimized solutions** – Forming an alliance between the owner, architect, engineers and contractors at the start of an IPD project, project decisions are made with valuable input from all team members to ensure solutions are optimized project-wide. By leveraging its own internal engineering, construction and maintenance resources, Southland is recognized for its ability to support IPD teams explore and discover new solutions.
- **Shared risk and reward** – A key tenet of the IPD approach is to use relational contracts that reinforce and reward team collaboration as well as distribute risk amongst the team to those in the best position to manage that risk. Proactively taking accountability for the risks that it is best suited to control, Southland utilizes a variety of tools and technologies to assist in risk management and keeping our team partners informed of progress.

- **Cost savings** – Using target value design to establish scope and project costs, IPD teams pursue solutions that maximize the value for the owner within the funding limits. Given its experience as a design–build–maintain firm, Southland is very skilled at conceptual design and cost estimating and is able to help the team establish realistic solutions and goals. Throughout the design and construction process, Southland engineers utilize their engineering and construction experience to track costs and ensure the client receives maximum value for money.

CASE STUDY 7

Crossrail: elevated focus on quality to match safety

INTRODUCTION

Opening from 2018 Crossrail is a new fast and convenient railway for London and the South East of England. It will take the pressure off London Underground ('The Tube') trains and buses – see Figure CS7.1 for the route map. It is also the biggest construction project in Europe – digging 21 kilometres of twin-bore tunnels under London to connect the City with Reading in the west to Shenfield and Abbey Wood in the east. The building of the new railway has made good progress safely, on-time, on-quality and on-budget, with a spend of ca £100m per month.

Crossrail have developed a comprehensive approach to managing quality which includes a quality vision and model with links to integrity and behaviours, a list of factors critical to success, a construction quality plan, a quality dashboard with a 'right-first-time' metrics procedure, a performance and risk assurance quality framework and procedures – an assessment process applied to all Tier 1 suppliers (ranked using a Quality Performance Index, QPI), supervisors responsible for quality under NEC3,[1] and an emphasis on quality/improvement-related training. There is a quality function with a team vision.

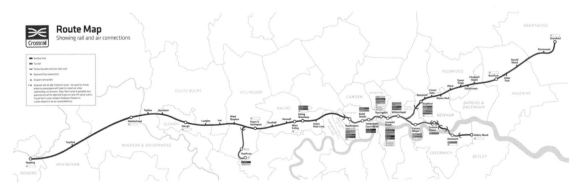

FIGURE CS7.1 Crossrail route map

CURRENT APPROACH

Original context

Crossrail recognized that, in terms of quality, the construction industry is generally many years behind other sectors and often does not take quality seriously enough. This is in contrast with other sectors which are reaping huge benefits from the implementation of formalized lean-quality management programmes, including increased efficiency, productivity and staff morale.

Crossrail based their thinking on the fact that the construction industry has the ability to adopt improvement programmes, such as in safety management, but that focusing only on safety does not go far enough. Hence, Crossrail undertook a mission to make quality part of the culture in the same way that safety was. In support of campaigns focused on 'right-first-time' and reducing waste, they developed a quality training programme as part of their commitment to 'change the game'.

Significant steps in the evolution

Crossrail recognized that they needed certain components in the development of the quality culture. Key steps in this were the creation of a Quality Vision and a Quality Model with Quality and Integrity tied together in clear simple messages for the staff to get the right behaviours. This was followed by the creation of a Quality Dashboard, with Quality and Right-First-Time Metrics, together with a Quality Plan and a Performance Assurance Quality Framework and Procedure (supplier assessment and QPI). They created a Quality Function with a clear team vision. Some of the thinking around the quality management approach was initially informed from the Channel Tunnel Rail Link but it has developed significantly since then. Key components were the supervisors' role under NEC3, the quality-related training and self-certification. There is also an independent verification team that monitors operations using a risk-based approach in carrying out surveillance, as an extra check, to verify compliance with systems and check workmanship is to the required standards.

CURRENT MODEL/APPROACH

Quality vision

The focus on quality is Right First Time – Every Time, the tenets of which are:

- a World-class Railway with efficient delivery and certification of compliant works;
- all defects are preventable;
- quality should be built in not inspected in.

The Sponsor and Infrastructure Management requirements are spelled out as:

- Crossrail's Client and Programme Management responsibilities, audit, surveillance and inspections;
- Review of certification.

The requirements are reviewed with the contractors' own programmes, targets, customer satisfaction and internal audits. Self-certification by contractors of their own work is a key part of the approach. Contractors' site-specific controls, procedures, Inspect and Test Plans (ITP), records and certification are spelled out in the PM Quality Programmes.

Client/project-wide contract requirements, initiatives and best practice are part of the approach to Contracts, Quality Awards, Bulletins, lessons learned, ITP guidelines and information sharing. Specifically and very importantly:

- Internal quality team of approx 25 people.
- 120 Field Engineers, integral to the team, to validate work out in the field (working into the quality supervisor within an NEC3 framework).
- Quality requirements on contractors include them having qualified quality managers in post (Chartered Quality Institute qualified) and the Quality Performance Index (QPI).
- 'Inspection and Test Plans' (ITPs) are a key element in the quality approach; Crossrail do not allow contractors to progress work that is not already covered by an ITP (few other clients use ITPs in this way); ITPs are written by the people who do the work, not quality people. Retrospective ITPs are not permitted.
- 'Progressive Assurance' is used – producing the paperwork as they go along, not just pulling it all together at the end – Crossrail require contractors to submit documents as they go along.
- NCRs (Non-Conformance Reports) – incentives are put on contractors to raise an NCR and the Field Engineers supervise the NCR process. Despite apparently high figures of NCRs raised, the overall cost of rework is less than 1 per cent. The Crossrail team believe that the transparent nature of NCRs means that long-term quality and reliability is being built in.
- As a result of the above, Crossrail have an excellent data set on failure modes and correction requirements. There is a high level of transparency with much of the data being shared within Crossrail and with the supply chain.

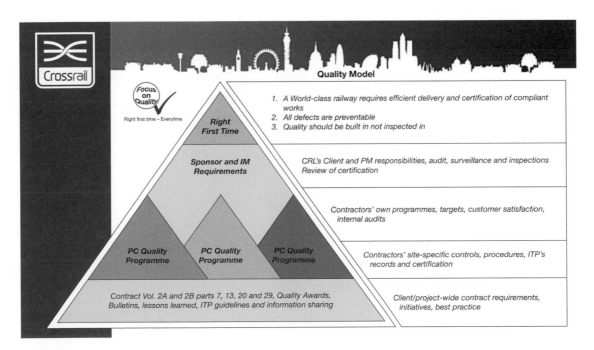

FIGURE CS7.2 Quality model

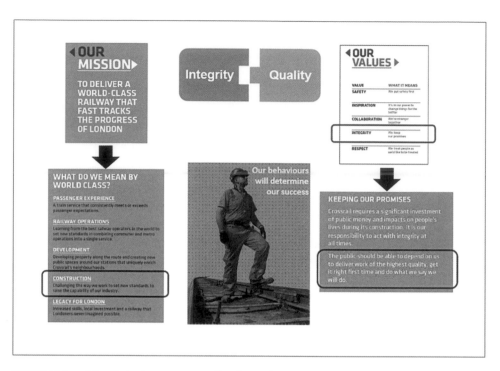

FIGURE CS7.3 The links between quality, integrity and values

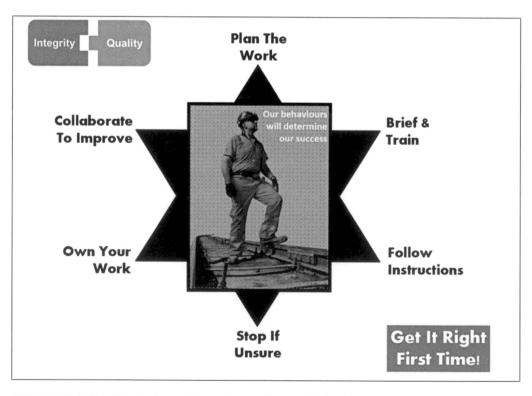

FIGURE CS7.4 The links between integrity, quality and behaviours

QUALITY AND INTEGRITY

Crossrail ensure everyone understands the links between quality, integrity, values and behaviours – see Figures CS7.3 and CS7.4.

QUALITY FUNCTION AND TEAM VISION

The above aspects of the approach to managing quality at Crossrail are key – everyone has a copy of the pocket book containing the key messages and people do 'walk-the-talk'. This is also driven down into the Tier 1 contractors and pushed further down into the supply chain. Quality has a major weighting in evaluating all contracts and a lot of time and effort is put into the tendering process to ensure requirements are met. Training is a major thrust.

CRITICAL SUCCESS FACTORS

Crossrail have identified five CSFs that are key to delivering its Quality Vision:

1   **Quality is Everyone's responsibility** – Getting the quality right on Crossrail is everyone's responsibility – 'after all it's why we are all here!'
2   **Building Crossrail to last 120 years** – When Crossrail is all done, they need to leave for the generations ahead a world-class railway that is built to last for a minimum of 120 years.
3   **The right people and systems** – Quality is all about having the right people and systems from the client throughout the supply chain to ensure that all design and construction work is adequately planned, controlled and certified.
4   **Getting it 'right first time'** – Quality falls under one of the five Crossrail core values – **Integrity** – 'The public should be able to depend on us to deliver work of the highest quality and do what we say we will do: – design and construct Crossrail right first time.'
5   **Gaining acceptance from IMs, Operators and ORR** – Quality is also about how Crossrail can be successfully accepted and handed over to the Infrastructure Managers, Operators and the Office of Rail Regulation.

CONSTRUCTION QUALITY PLAN

This document describes the organizational structure, responsibilities, processes and tools for the management of construction quality in the implementation of construction works. It describes the roles and responsibilities of the Delivery Team (DT) and the contractor in the delivery of construction quality. It does not cover the delivery of the system-wide contracts.

The document is applicable to procurement, manufacturing, design and construction field-based activities of the Central Section Project included within the Crossrail Programme managed by the Crossrail Delivery Team. This is part of the Quality Management System of the DT supplementing the Delivery Management Plan. It complements the Technical Assurance Plan and the Design Management Process.

The Construction Quality Plan and associated requirements, procedures and instructions are applicable to activities during the construction phase of the project from the time of site mobilization until each completed structure, system or component is tested, commissioned and accepted for turnover.

QUALITY DASHBOARD

Crossrail have established a Quality Dashboard which is reviewed monthly by senior management. The main headings reported against are:

Right First Time

RFT1 – Inspection and Test Plans.
RFT2 – Partial Rework Rate per Period.
RFT3 – Observations.
RFT4 – Non-Conformance Reports (plus summary by sector and top NCR categories).

KPIs

- Contract Rework Costs.
- Certification of Works – percentage cumulative completed on time.
- Corrective Action Requests (CARs) – percentage cumulative closed within due date.
- Status of Request for Information (RFIs).
- Status of Field Change Documents (FCDs).
- Overall Performance Summary:
  - Key accomplishments in period;
  - Key issues in period;
  - Key actives planned for next period.

RIGHT-FIRST-TIME METRICS PROCEDURE

The purpose of this procedure is to describe the metrics, targets and process for the collection, analysis and reporting of data associated with the Corporate Quality and Right-First-Time metrics.

The responsibilities of the Delivery Team are outlined in support of collecting the Quality and Right-First-Time metrics. Additionally, the Delivery Team are responsible for collecting and analysing the number of certification packages of works completed on time.

The Audit Team are responsible for collecting and analysing the number of Corrective Actions Requests that are closed within their due date.

The Quality Team owns the procedure and is responsible for compiling, analysing and reporting the data referenced in this procedure.

PERFORMANCE ASSURANCE VISION

Crossrail's Performance Assurance model was designed following consultation with key members of the programme and supply chain. Its objectives are to:

1 support contractor performance to enable world-class levels to be achieved;
2 provide an objective demonstration (a platform) for contractors' performance;
3 improve integration between functions and key contractors;
4 provide a mechanism for high performers to obtain wider industry opportunity;
5 facilitate good practice movement across the programme.

In summary Crossrail's vision for this process is simple:

World-class Performance will be assured across six functional areas shown in Figure CS7.5.

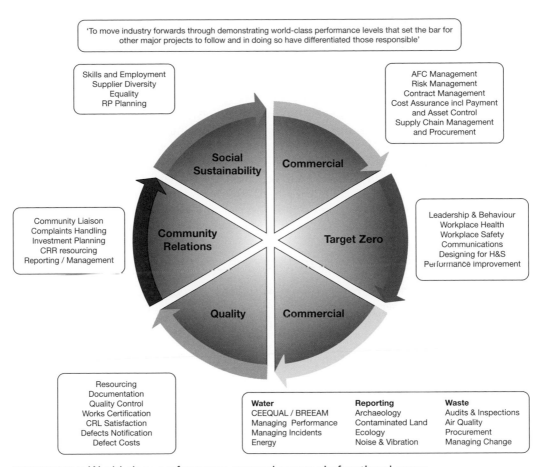

'To move industry forwards through demonstrating world-class performance levels that set the bar for other major projects to follow and in doing so have differentiated those responsible'

Skills and Employment
Supplier Diversity
Equality
RP Planning

AFC Management
Risk Management
Contract Management
Cost Assurance incl Payment
and Asset Control
Supply Chain Management
and Procurement

Social Sustainability

Commercial

Community Liaison
Complaints Handling
Investment Planning
CRR resourcing
Reporting / Management

Community Relations

Target Zero

Leadership & Behaviour
Workplace Health
Workplace Safety
Communications
Designing for H&S
Performance Improvement

Quality

Commercial

Resourcing
Documentation
Quality Control
Works Certification
CRL Satisfaction
Defects Notification
Defect Costs

Water	**Reporting**	**Waste**
CEEQUAL / BREEAM	Archaeology	Audits & Inspections
Managing Performance	Contaminated Land	Air Quality
Managing Incidents	Ecology	Procurement
Energy	Noise & Vibration	Managing Change

FIGURE CS7.5 World-class performance assured across six functional areas

PERFORMANCE ASSURANCE QUALITY FRAMEWORK

This document sets out the agreed process for operating Crossrail's Performance Assurance Function. The purpose is to provide a clear and consistent overview of:

1   the **vision** for Performance Assurance;
2   the **process** of operating Performance Assurance;
3   **responsibilities** for the implementation of Performance Assurance;
4   the **reporting** of the Performance Assurance assessments.

The performance assurance quality framework covers several areas:

• Positive action taken in response to Performance Assurance assessments;
• Continual Improvement;
• Quality Resource;
• Certification Resource;
• Quality Documentation;
• Design Coordination, Design Submission, Design Production and Construction Management, Technical Information Handover;
• Document Quality and Control;
• CRL satisfaction with works, Site Tour, Defects identified by contractor;

- Cost of defects against earned value;
- Quality Performance Index (QPI).

The following aspects are key to assuring quality in Crossrail:

- FDCs – Framework Design Contractors for engineering;
- Design Contractors for civil work;
- PFEs – Project Field Engineers – *all* they do is Quality (*not* time and cost);
- Quality Managers with Quality Engineers working for them;
- Dashboard performance subject to Management Review with all Directors (ExCom) looking at the KPIs;
- There is a Quality Sub-Committee of the above, chaired by the CEO;
- Independent verification team to monitor operations using a risk-based approach;
- Key leading indicators include RFT – ITPs reporting on where work started without one; Cost of Rework – every NCR in Crossrail is monitored in terms of CoR.

PERFORMANCE ASSURANCE PROCESS

There is an objective mechanism by which the performance of selected members of the supply chain (Tier 1) are measured. A Quality Performance Index (QPI) has been introduced which ranks Tier 1 suppliers in a publicized league table. The QPI is calculated using a formula that combines scores from some of Crossrail's leading and lagging KPIs. Each of the function headings are displayed and the measures are scored on a 0 to 3 scale for each sub-heading:

3 **World Class** – Exceptional performance likely to be industry frontier.
2 **Value-added Compliance** – Performance beyond the level set by CRL.
1 **Basic Compliance** – Performance compliant with CRL processes/procedures/contract.
0 **Non-Compliant** – Performance is non-compliant with CRL processes/procedures/contract.

Performance is measured on two axes:

- **Inputs** (Qualitative): approach to the delivery, quality of systems and processes.
- **Outputs** (Quantitative): results of the process such as number of programmes accepted, apprentices employed.

By correlating the performance of inputs against outputs, Crossrail are able to measure if improved performance of one metric directly affects another, and thus derive targets for each of the delivery teams to aim for, by their next assessment. An example of the framework assessment is shown in Table CS7.1.

TABLE CS7.1 An example of the framework assessment

Core Coverage Areas		Measure	Weighting	Pillar Weighting	1 Compliant	2 Beyond Expectations	3 Performance broadly recognizable as World Class
Tours	Input	Senior Management Leadership Tours (Involving Project Director or above)	0.3		1 hour per period	≥2 hrs per period	≥4 hrs per period
RIVO NM Reporting	Input	Near Miss Reporting rate in RIVO, normalized by 100,000 hours	0.15		210 Reports/100,000 hrs	≥20 Reports.100,000 hrs	≥40 Reports.100,000 hrs
Unsafe Observation Reporting	Input	Unsafe Observation Reporting rate in PC own system as reported on KPI, normalized by 100,000 hours	0.15	1	210 Reports/100,000 hrs	≥20 Reports.100,000 hrs	210 Reports/100,000 hrs
Training	Input	Supervisor SSSTS Training (This should include Tier 2 Supervisors)	0.25		≥75% trained	≥90% trained	≥100% trained
	Input	Manager SMSTS Training (This should include Tier 2 Managers)	0.15		≥75% trained	≥90% trained	≥100% trained

KEY PARTICIPANTS

In order to successfully integrate the Performance Assurance framework into each of the delivery team's working practices, Crossrail have a defined group of participants and process stages. There are four key groups of participants that have been identified – Table CS7.2.

TABLE CS7.2 Four key groups of participants

Participant	Definition
Performance Assurance Team	Facilitate the reviews and produce the final reports
Functional Teams	Undertake the reviews and produce the functional sections of the reports, consisting of the 6 PA Framework function headings (Target Zero, Quality, Commercial, Community Relations, Social Sustainability and Environment)
Crossrail Project Managers	The key interface between the contractors and the functions
Contractors	The group that is being assessed, required to evidence their performance

The Performance Assurance team is the ultimate owner of the framework and is responsible for keeping it up to date and driving its use. However, this can only be achieved through a concerted effort from everyone that is involved in the assessment. Each assessment consists of three main stages, the flowcharts in Figures CS7.6 to CS7.8 display the core tasks carried out in them.

PROACTIVE PROCESS

Crossrail adopt an intelligent approach to performance assurance and have ensured internal collaboration with 'Internal Audit' to incorporate a proactive mechanism into the procedure. This is designed to drive self-assurance into Tier 1 suppliers in a structured manner potentially resulting in the opportunity to reduce the amount of independent reviews/audits. The focus is on operating the model in an efficient manner whilst preserving the objectives that performance assurance was designed to achieve.

This enables the Tier 1 Contractors and Delivery Teams to self-certify performance and set their proactive improvement actions.

An example of performance controls and measures from the Target Zero section of the framework that would be expected to be in place on contracts is as follows:

- Senior Management Leadership **Tours** (involving Project Director or above).
- Never Miss **Reporting** rate in **RIVO** normalized by 100,000 hours.
- **Unsafe Observation Reporting** rate in PC own system as reported on KPI normalized by 100,000 hours.
- Supervisor SSSTS **Training** (This should include Tier 2 Supervisors).
- Manager SMSTS **Training** (This should include Tier 2 Managers).

As part of the performance assurance process Tier 1 suppliers may be required to self certify that these controls are present and effective. Subsequently the need to undertake contract reviews may be reduced significantly or even removed and would be discussed with Crossrail's Internal Audit function. The over-arching philosophy is for Performance Assurance to support the operation of a self-assured model reducing the need for independent central reviews.

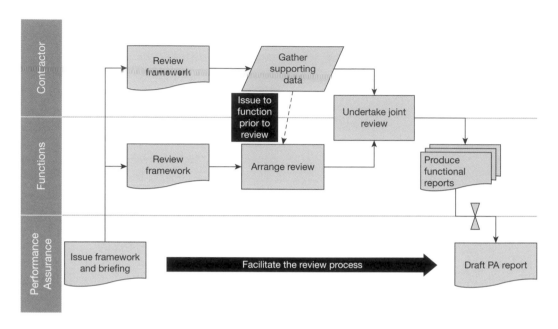

FIGURE CS7.6 Stage 1: Undertaking the assessments

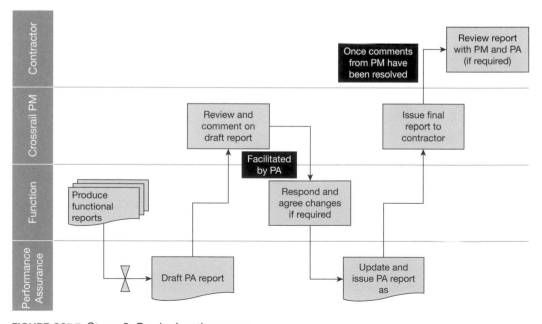

FIGURE CS7.7 Stage 2: Producing the reports

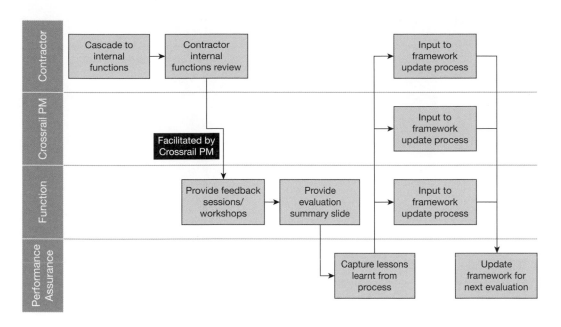

FIGURE CS7.8 Stage 3: Issuing the reports and obtaining feedback

Pillar	Core Coverage Areas		Measure	Weighting	Pillar Weighting	1 Compliant	2 Beyond Expectations	3 Performance broadly recognizable as World Class
Pillar 1: Leadership and Behaviour	Tours	Input	Senior Management Leadership Tours (Involving Project Director or above)	0.3		1 hour per period	≥2 hrs per period	≥4 hrs per period
	RIVO NM Reporting	Input	Near Miss Reporting rate in RIVO, normalized by 100,000 hours	0.15		210 Reports/100,000 hrs	≥20 Reports.100,000 hrs	≥40 Reports.100,000 hrs
	Unsafe Observation Reporting	Input	Unsafe Observation Reporting rate in PC own system as reported on KPI, normalized by 100,000 hours	0.15	1	210 Reports/100,000 hrs	≥20 Reports.100,000 hrs	210 Reports/100,000 hrs
	Training	Input	Supervisor SSSTS Training (This should include Tier 2 Supervisors)	0.25		≥75% trained	≥90% trained	≥100% trained
		Input	Manager SMSTS Training (This should include Tier 2 Managers)	0.15		≥75% trained	≥90% trained	≥100% trained

Senior Management Leadership Tours (involving Project Director or above)
Never Miss Reporting rate in RIVO normalized by 100,000 hours
Unsafe Observation Reporting rate in PC own system as reported on KPI normalized by 100,000 hours
Supervisor SSSTS Training (This should include Tier 2 Supervisors)
Manager SMSTS Training (This should include Tier 2 Managers)

FIGURE CS7.9 Example of four performance controls

The Quality Performance Index (QPI) has been developed to build healthy competition between contractors and to identify which ones are in need of improvement. The QPI is akin to the Healthy and Safety Performance Index (HSPI) and calculated utilizing the scores from existing leading and lagging indicators.

The average of the following indicators is taken: RFT1 ITP's, RFT 3 Satisfactory Observations, RFT 4 Contractor raised NCR's, WC1 – Certification on time, WC2 – CAR closure and Overdue NCR's. This score is then averaged with the latest Performance Assurance Framework (PAF) Assessment Score to create the QPI. The score from the Performance Assurance Assessment of Contractors process makes up half the QPI score (see Figure CS7.10 for a QPI summary example).

Period 12 / Contract	Work Starting without ITP	% Satisfactory Observation	% NCRs raised by Contractor	% Certification Completed on time	%CARs Closed on time	No. of NCRs Overdue	% of NCRs Overdue	PAF5 Score	
CXXX - <CONTRACT NAME>	0	N/A	100%	100%	N/A	0	0%	2.21	2.61
CXXX - <CONTRACT NAME>	0	100%	100%	N/A	N/A	0	0%	2.14	2.57
CXXX - <CONTRACT NAME>	0	92%	100%	100%	N/A	0	0%	1.97	2.39
CXXX - <CONTRACT NAME>	0	N/A	100%	100%	N/A	1	33%	2.33	2.29
CXXX - <CONTRACT NAME>	0	100%	100%	0%	100%	1	3%	2.17	2.25
CXXX - <CONTRACT NAME>	0	N/A	100%	N/A	N/A	0	0%	1.41	2.21
CXXX - <CONTRACT NAME>	1	76%	96%	N/A	100%	0	0%	2.29	2.05
CXXX - <CONTRACT NAME>	0	90%	83%	100%	N/A	1	3%	1.88	2.04
CXXX - <CONTRACT NAME>	0	100%	94%	67%	N/A	13	9%	2.25	2.03
CXXX - <CONTRACT NAME>	0	70%	97%	N/A	N/A	1	1%	2.19	1.97
CXXX - <CONTRACT NAME>	0	90%	100%	N/A	N/A	6	25%	1.87	1.94
CXXX - <CONTRACT NAME>	0	71%	88%	N/A	100%	1	2%	1.92	1.86
CXXX - <CONTRACT NAME>	0	82%	100%	N/A	N/A	0	0%	0.97	1.74
CXXX - <CONTRACT NAME>	0	67%	64%	N/A	100%	0	0%	1.66	1.73
CXXX - <CONTRACT NAME>	0	67%	94%	N/A	N/A	8	16%	1.99	1.50
CXXX - <CONTRACT NAME>	0	88%	77%	100%	N/A	23	22%	1.41	1.41
CXXX - <CONTRACT NAME>	0	75%	88%	46%	N/A	25	38%	1.33	1.07
CXXX - <CONTRACT NAME>	2	97%	92%	0%	N/A	11	30%	0.88	0.74

FIGURE CS7.10 QPI summary example

BENEFITS – WHAT IT HAS DELIVERED

The overall major benefit is that Crossrail is making good progress safely, on-time, on-quality and on-budget. This is essential when this infrastructure will be in place for over a century.

As a client, Crossrail now have a set of input/output metrics and they are not aware of others in the sector that have a complementary approach.

The ITP process is increasing the chance of getting it Right First Time. Rework costs are less than 1 per cent. If the approach was not in place, it is estimated the CoR would be 5–6 per cent. Three per cent CoR is equivalent to £300–400m which is about the cost of a new station.

So, there are in place cost of rework measures, together with some leading indicators; there is an Inspection Test Plan in place which provides percentage NCRs reports raised by contractors (95 per cent target); certifications closed on time, with a novel KPI that requires work done in 30 days; Crossrail are as far ahead as anyone on digital BIM (Building Information Modelling).

AREAS FOR FUTURE DEVELOPMENT

Rather than run maintenance programmes based on failures or timetables, Crossrail are able to interrogate the assets and decide the maintenance regime – this will be a key area of development.

The training and development of people is a key parameter and there are five major areas in which this is taking place:

1 Three groups of quality professionals taking CQI courses as major building blocks.
2 A programme for senior leaders on 'how your management system helps you run your business'.
3 Supervisors (8–10 people on a site) – to help them recognize quality and how to deliver it.
4 Training in Root-Cause Analysis, with the intent to reduce costs.

5   Pushing people development into the supply chain; e.g. the intent of getting Chartered Quality Professionals (CQI) in place throughout the contractors.

World Quality Day/Week events have been held in recent years showcased under 'A Quality Job is a Safe Iob' and 'Let's Talk About Quality.'

KEY TAKEAWAYS FROM THIS CASE STUDY FOR THE CONSTRUCTION SUPPLY CHAIN

- NEC3 Contract with self-certification:
    - supervisor is 'Head of Quality';
    - project field engineers (PFE) are supervisors' representatives.
- As the client, Crossrail have:
    - set KPIs (metrics) – RFT and Corporate KPIs at Board level;
    - surveillance (coordination between Central Team and Supervisors' reps);
    - organized training (CQI, Root-Cause Analysis, etc.);
    - carried out six monthly deep dives – Performance Assurance;
    - driven quality culture – World Quality Week – CEO;
    - owned Integrated Management System;
    - linked Quality to Assurance – IMS, quality is embedded through an institutional systemic function.
- Quality as a sub-committee to the Executive Committee (chaired by CEO) – now a main part of the Executive Committee.
- Quality is built in to the system:
    - accountability remains with the project manager and the contractor (self-certification);
    - work does not start if ITPs are not ready;
    - significance of PFEs and supervisors' ITPs – validating work done but not inspecting in quality.
- Suppliers encouraged to raise NCRs:
    - financial incentives if raised by supplier;
    - supply tends to foot the bill otherwise, has led to massive lift in quality;
    - policed by the PFEs and supervisors.
- Modified NEC contract:
    - quality was ca 20 per cent of contract award criteria;
    - stipulates quality upfront (on a robust design);
    - all suppliers will self-certify;
    - progressive certification;
    - stipulates contractor must have a quality manager (certified by CQI);
    - 6–8 performance assessments of contractors.
- The organization 'believes' its quality values.
- Robust quality management information:
    - transparency across the business and within the supply chain;
    - cost of rework currently ca 1 per cent (very low) – despite high numbers of NCRs.

REVIEW QUESTIONS

1   Crossrail have developed a comprehensive approach to managing quality which involves the entire supply chain. Evaluate in detail this approach and explain how it could be adapted for house or hospital building programmes and projects.

2   An important aspect of the approach is key performance indicator monitoring in the supply chain – what is special/different about this in Crossrail and how do you think this might be received in the supply chain of other construction sectors?

3   Key to success has been the training, development and qualification of people throughout the supply chain. How does this compare in general with the construction sector and how could the Crossrail approach to this be adapted and adopted?

ACKNOWLEDGEMENTS

The authors gratefully acknowledge the involvement and assistance of Chris Sexton, Technical Director, and Chris Titterton, Head of Quality, Crossrail.

NOTE

1   NEC (New Engineering Contract) is a family of contracts that facilitates the implementation of sound project management principles and practices as well as defining legal relationships. It is suitable for procuring a diverse range of works, services and supply, spanning major framework projects through to minor works and purchasing of supplies and goods. The implementation of NEC3 contracts has resulted in major benefits for projects both nationally and internationally in terms of time, cost savings and improved quality.

CASE STUDY 8

Worker empowerment transforms operations at JB Henderson

INTRODUCTION

Preamble

This case study is about the implementation of lean thinking and lean practices in the JBH (JB Henderson Construction) Arizona business unit, under the leadership of Tracy Lucero, JBH's Arizona Area Manager, and Kyle Price, Area Superintendent. The case study describes how this team set about their lean journey and gives some examples of what they have achieved. Throughout the case study we include some quotes from some email newsletters Tracy has published to share their experience.

History of the company

Jack B. Henderson established JBH in 1959, providing general contracting services. In 1967, JBH branched out from performing strictly as a general contractor and added a mechanical division. In 1990 after the death of Jack B. Henderson, his son Mark G. Henderson took over as President/CEO of the company and in 2012, long-time employee John Stroud was named President of the company. Mark and his wife Linda have been personally involved in running the company since 1990, maintaining the core values of Safety, Quality, Honesty and Integrity. In 1993, JBH built a pipe fabrication facility for the fabrication of carbon steel, stainless steel and other piping systems and components at the main physical plant. Two years later, a Class 100 clean room was built to enhance their services to the local microelectronics industry. In 1996, after operating in Albuquerque for 37 years, JBH opened an office in Los Alamos, New Mexico primarily to serve Los Alamos National Laboratories. In 2003, JBH completed work on their Rio Rancho, New Mexico office, dedicating staff to address the construction needs of one of their core customers in that area. At the invitation of the same customer, JBH expanded to the state of Colorado in 2004, Utah in 2005 and Arizona in 2008. Currently, JBH maintains operations in New Mexico and Arizona.

MISSION AND VISION

Vision

JB Henderson Construction Co. is committed to achieving the highest level of quality and performance, to being known as the supplier of choice, and to building the future of families through jobs, training and community involvement.

Mission

JB Henderson Construction Co. is a General/Mechanical Contractor. We deliver the highest quality and best value to our customers by integrating teamwork, dedicated project management and outstanding craftsmanship. We create a desirable workplace, an injury-free environment and good jobs through equal opportunity.

Core values

Safety, Quality, Honesty and Integrity.

THE LEAN ADVENTURE

How it all started

A confidential fabrication client sparked JBH's curiosity through their involvement in an experiment. JBH was invited to participate in an experiment that challenged them to figure out a way to incorporate the PDCA cycle in construction, not in the 'office' but in the field. The team charged with the experiment set out to implement a High Precision Maintenance Process (a process used in their customer's operation) to improve daily operations. Their PDCA is described below.

- PLAN: The Day's Operations;
- DO: Execute Work;
- CHECK: Status of the plan at the end of the day's work;
- ACT: Conduct an after action review:
  - What was supposed to happen?
  - What went well?
  - What was unexpected?
  - What did we learn?
  - What improvements can we make to help us in the future?

The experiment lasted for three weeks; it included twice a day field meetings including the crew, their foreman and JBH area management as well as direct observations. The direct observations measured how JBH workers spent their time. At the beginning of the experiment, the client spiked JBH management team's interest by sharing with them that they had studied a number of other contractors' labour practices and found that on average only 18 per cent of time on the job was spent in the direct installation of work, in other words adding value. The JBH team were surprised at how low the value-added time was and felt that their people would be significantly better. They readily agreed to a comparative work study on their crew. You can imagine the chagrin of the JBH team when they found that their value-adding time was 17 per cent, slightly below the team average of the other trades.

Figures CS8.1 and CS8.2 compare the Team average and JBH in week 1 of the JBH lean journey.

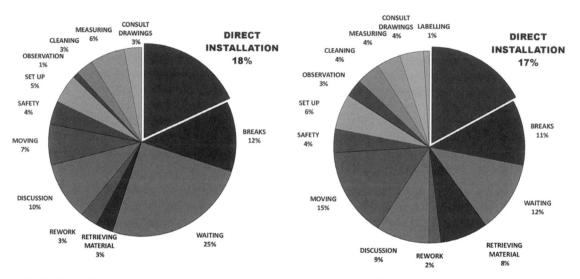

FIGURE CS8.1 Other contractors' summary   **FIGURE CS8.2** JBH 'HPM' experiment Week 1

Other key learnings from this experience involved the development of an employee learning and development programme, which included the following elements (some of which will be discussed in further detail throughout this case study):

1   Tip of the Month Learning;
2   Kaizen Crew Learning;
3   Field Leadership Learning;
4   Study Action Teams;
5   JBH Video Learning Library;
6   BIM Process Improvement.

In a recently published *Continuous Improvement* newsletter Tracy reflected, and I quote:

So step one is a reality check. It was pretty harsh for us. We thought we were 'wrapped tight' with what we had going on. We had a successful 56 year track record to prove it. When somebody told us (like we're telling you) that *'you're not as good as you think you are,'* we took it as a challenge to prove them wrong. Well unfortunately we found out that we weren't. If you are in a space where you feel like you have very little or no room for improvement, the future holds *no possibilities*. It's been said, if you're not moving forward, you're falling behind. We sincerely believe that. If others are improving and you don't think you need to, one day you will find you've been left behind.

The JBH team's interest was aroused, in Tracy's words 'we became extremely dissatisfied and we wanted to make a change'. JBH immediately embarked on its lean adventure (as they refer to it because it has been fun and exciting) working closely with their work crews, and support from their customer. They have not looked back since.

Standard Crew Routine

The daily meeting that was started during the experiment was so enlightening that they couldn't stop. Since that time they have developed what they call their Standard Crew Routine (which also includes the Daily Crew Plan, which will be covered in more detail in the following section). The Standard Crew Routine takes a total of 50 minutes of the crew's time each day. This time would be considered 'non-value-add' activities. It seems counterintuitive to take this much time of a workday to plan and educate employees but this is not seen as an additional 'cost', it's an investment in adding value. JBH contends that if not intentionally dedicated to planning, this time (if not more) is wasted due to lack of communication and direction.

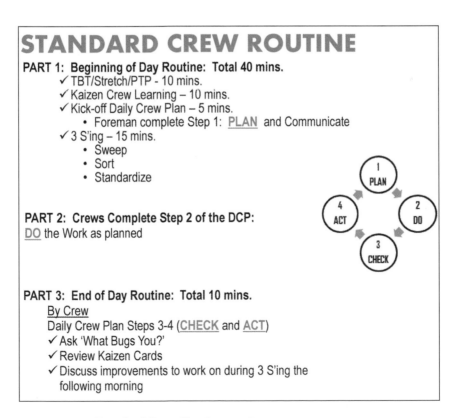

FIGURE CS8.3 Standard Crew Routine card

It starts out with a 40 minute 'Beginning of Day' routine where the entire crew meets together first thing in the morning. They start by discussing and reviewing their safety tool box topic, perform their stretch and flex as well as completing their pre-task plan for the day. The following ten minutes are spent investing in the crew. They call it their Kaizen (small improvement) Learning time. During this time, the foreman either reads a book (on lean, leadership or other industry interests) or they watch a video from the JBH Video Learning Library. The reading/video is followed up with a short crew discussion.

The next step involves the kick-off of the Daily Crew Plan (DCP). A further 15 minutes is spent 3-S'ing (Sweep, Sort and Standardize) which involves ensuring their work area is clean (sweep), gather all the tools, information, materials and equipment for the work outlined for them in their DCP (sort) and work on ways to standardize their resources and information (standardize). This 15 minutes invested in 3-S'ing is intended to help avoid other more time-consuming 'non-value'-added activities that the crews face throughout the day. After 3-S'ing, 'value'-add activities can begin.

Each day ends with a ten minute wrap up and review of the day. The plan laid out by the foreman at the beginning of the day is statused (it gets checked and actual performance is noted). After checking in on the plan the crew talks about 'what bugged them?' (adopted from 2 Second Lean by Paul Akers (2012)). The items get documented and potential improvement opportunities are discussed as well as any Kaizen Cards (a card where employees document any of the eight deadly wastes they experienced throughout the day as well as identifying how they can eliminate that waste, discussed in further detail in an upcoming section). During this time they may recommend some 3-S'ing activities to improve their operation as well. JBH crews spend almost 10 per cent of their day in planning and review, and their labour productivity continues to improve.

Daily Crew Plan

As mentioned previously the Daily Crew Plan is embedded within the Standard Crew Routine. The Daily Crew Plan was adapted directly from the PDCA 'experiment' that JBH participated in. The daily meeting that was started during the experiment was so enlightening that they couldn't stop. Their daily crew routine starts with the foreman communicating the PLAN to the crew (see sections 1, 1A and 1B of the DCP in Figure CS8.4). It includes the goals for the day as well as any workable backlog that is available to the crew if they should complete the plan or run into some unexpected roadblocks which can eliminate potential waiting time.

After the plan is kicked off (documented and discussed) the crew moves on to the DO part of the cycle (identified on the plan in Figure CS8.4 with the number 2). They go out and execute the plan that was communicated. At the end of the day, the crew comes together to perform the CHECK step. They check in on the status of the plan (sections identified as 3 in Figure CS8.4). The crew reports to the foreman what they completed on the plan and the workable backlog. They dig further into the CHECK step by asking 'What Went Well?' This is then followed by discussing 'What was Unexpected'? To facilitate communication, they utilize prompting questions focused on identifying one of the eight wastes. The final step of ACT includes how the crew will approach fixing the items that were identified during the CHECK step. If improvement opportunities are acted upon they can help eliminate waste in the future. The DCPs have served them with a great source of items to continuously improve their operations.

DAILY CREW PLAN

• Adopt-a –Crew Training by Tracy, Kyle and PM

CREW MEMBER	FAB (F) / SUBFAB (S)	TOOL	SYSTEM(S)	Demo	Hangers	Weld	Install	Label	Test	Punchlist	Quantity / How much?	Completed? / Status

1 ☆ PLAN AND COMMUNICATE   **3 ☆ CHECK**

1A ☆ Plan Tomorrow's Work (using format below) and Communicate at end of shift the today.
Be Specific - Where, Who, How Much? Critical Items Needed to Complete Work as well as Workable Backlog

2 ☆

1B ☆ CRITICAL ITEMS

1C ☆ WORKABLE BACKLOG When there are obstacles to completing the planned work above, look here for opportunities!

3 ☆ CHECK

ITEMS ESSENTIAL TO EXECUTING TOMORROW'S PLAN	FAB (F) / SUBFA	TOOL	SYSTEM(S)	Demo	Hangers	Weld	Install	Label	Test	Punchlist	Quantity / How much?	Completed? / Status

FIGURE CS8.4 Daily Crew Plan morning

DAILY CREW PLAN

KAIZEN LEADER: _____

3B ☆ MORE CHECKING...
WHAT WENT WELL? (Did we have everything we needed to complete our plan?)

_____

_____

_____

WHAT WAS UNEXPECTED? (What prevented us from completing our plan'

Did you spend time...

Working on fixing something (Defects)?
Handling fabrication or materials that weren't needed or used today (Overproduction)?
Waiting on anything (Waiting)?
Handling the same material multiple times (Transportation)?
Looking for tools or materials (Inventory)?
Making unnecessary movements (Motion)?
Doing unnecessary steps today (Extra Processing)?

WHAT BUGGED YOU?

_____

_____

_____

_____

D.O.W.N.T.I.M.E. Questions

4 ★ ACT
THE KEY 3-5 IMPROVEMENT OPPORTUNITIES FOR TOMORROW'S WORK OR OTHER FUTURE WORK (What did we learn? / Improvements--Big or Small!)

IMPROVEMENT -Tap into your Crew's Potential (Non-Utilized Talent):   ACTION BY:

FIGURE CS8.5 Daily Crew Plan end of shift

Field leadership learning

All field foremen, general foremen, superintendents and the area manager meet every two weeks for 90 minutes to invest in learning.

Meetings alternate between two different formats. One format is run as a Study Action Team (SAT) where they read a book and come prepared to discuss and learn together. As a team they agree to actions that would deepen the culture they are building which focuses on respect for people and continuous improvement.

The books that have been studied include:

- *The Five Minute Foreman* by Mark Breslin (2013);
- *2 Second Lean* by Paul Akers (2012);
- *Lean in Construction* by Ade Asefeso (2014); and
- *The Servant Leader* by James A. Autry (2004).

The alternating format includes conducting work improvement meetings at which the team leaders review improvement opportunities, develop and update their improvement strategy. They generate their learning by reviewing daily crew plans that include pluses (good things to continue) and deltas (things they should either improve or stop doing).

At these meetings, the field leaders review a range of issues that influence productivity. These include lessons learned by the work teams, problems that are bugging people and need action and opportunities for improvement identified from the shop floor. Kyle Price, area superintendent, emphasized that 'the magic is that all the ideas come from the field'. This leads to buy-in during implementation instead of the usual resistance that accompanies change.

The team manages their improvement ideas through a Kanban board (visual tool indicating status) as shown in Figure CS8.6.

FIGURE CS8.6 Foremen's planning Kanban board

The Kaizen card

The Kaizen card concept was 'hatched' in the field leaders group. The card provides a place for employees to document any waste they identify. Employees are empowered to eliminate the identified waste if it is within their realm of influence. Each card is logged and reviewed by the Kaizen Card Committee that meets on a weekly basis (consisting of the JBH area manager, area and project superintendents, safety manager and the design manager) to review the cards. The committee members document any actions that are warranted to address eliminating the waste identified by employees.

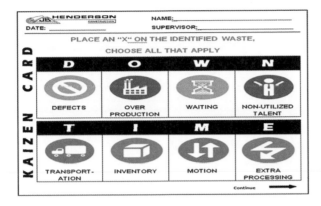
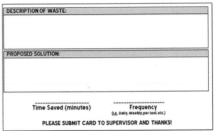

FIGURE CS8.7 Employee Kaizen card to log wastes

The cards are laminated and every employee carries one in their pocket at all times. On the front, shown in Figure CS8.7, the type of waste is identified, and on the back there is room for a proposed solution to be suggested along with an area to note the anticipated time savings if the waste was eliminated. Every month the minutes saved indicated on each Kaizen card are tallied and the top three employees are recognized at an all-employee meeting and bestowed the title of a Kaizen Warrior. Kaizen Warriors receive a custom designed T-shirt or cap that proudly displays their title.

OUTCOMES ACHIEVED FROM THE SHOP FLOOR

The following are some examples of achievements through 3-S'ing. The first examples are of a few before and after improvements.

Standardized General Foreman's Gang Box

The General Foreman's Gang Box was one of JBH's first standardization efforts. The intent of these boxes is to secure the tools that are shared resources and utilized by the entire crew. As Figures CS8.8 and CS8.9 (before pictures) show, the boxes were typically a mess with both needed and unneeded items, many of which were not visible because they were kept in unlabelled cardboard boxes or hiding in the back of the box.

After reviewing these photos and wanting to make improvements JBH engaged the Field Leadership Team. They engaged their expertise to develop the standardized list of tools needed by the crew. Essentially, it was developed by the experts. This has been a

FIGURE CS8.8 Foreman Box 1 Before 5S

FIGURE CS8.9 Foreman Box 2 Before 5S

FIGURE CS8.10 Foreman's toolbox after 5S

key to the successful implementation ensuring that it worked for crews. From their initial rollout, the boxes have been kept in the condition shown in Figure CS8.10. While building their lean culture, respect for each other has also grown. The employees keep the boxes in pristine order. The previously common occurrence of a lost or stolen tools has become a thing of the past for JBH.

Standardized pipe hangers

Pipe hangers are utilized everyday by JBH. Prior to their standardization efforts, cutting and assembly of these hangers was completed in the field. Often times meant gathering materials from carts similar to those shown in Figure CS8.11.

JBH has now put in place hangers cut to various standardized lengths and pre-capped. These hangers are manufactured at their fabrication shop. They are delivered to the site and placed in a 'shark' cage in the field ready for use. They are kept stocked to appropriate levels and when restocking is needed a (standardized) easy to use field requisition form is completed and transmitted to their fab shop (Figure CS8.12).

FIGURE CS8.11 Traditional pipe hanger trolley

FIGURE CS8.12 Stocked standardized pipe hanger trolley after 5S

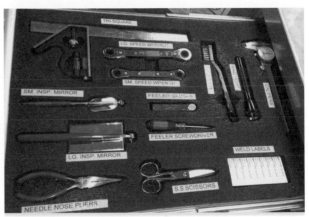

FIGURE CS8.13 Orbital welder boxes after 5S

Those were just a few of JBH's before and after examples. The following are examples that have come from 3-S'ing the workplace. Numerous ideas have been introduced, the following serve to illustrate a few of these.

Standardized orbital welding boxes

JBH regularly performs high purity welding that involves a very precise process. There are several components and steps involved with the joining of piping and the standardized box (Figure CS8.13) highlights just one of the drawers of the box. This design of this box was led by one of JBH's top welder with input from other company welders each considered an expert in the process. Again, the engagement of the 'experts' was key to the successful implementation and maintenance of this box. Empower the experts to tell you what they need then let them execute it!

Standardized materials boxes

Material storage was another opportunity to eliminate many of the eight wastes. Figure CS8.14 shows just one example of JBH's standardized material storage. The box contains

FIGURE CS8.14
Parts tool box after 5S, everything in clearly labelled drawers

hundreds of fittings and components, each sorted and easy to find. The key to this successful implementation was to clearly label the contents of each organized tray which eliminated the time spent trying to find the right parts. Another key was to leave the space in front of the tray open so nothing is blocking the labels from reading them and the space is available to pull out the tray open it up and pull what you need. The right material (identifiable and organized), available when it's needed, kept in the right amounts (not too little or in excess) contributes to increasing time spent on value-add activities.

CONTINUOUS IMPROVEMENTS

A key lean principle is continuous growth and improvement. In that effort, JBH continues to drive their progress through the development of new initiatives.

Union hall training

In an endeavour to improve the quality and work readiness of union labour working on JBH jobs, they worked with the local union hall as well as other signatory contractors in the development of a training course on lean thinking and lean practices. It is an eight-hour certified course titled 'Securing our Future with Lean Construction'. All JBH AZ employees have completed this course. To ensure this foundation of lean understanding continues for all new employees, there are nine leaders that have been certified to train this course.

Gemba walks

In 2015, JBH introduced the practice of Gemba walks as a step to formally check-in on the overall lean culture that has infected the operation over the previous year. The team has developed a standard card used by managers to guide their attention in a standard manner during the walks. Coined Gemba 'Go-See' walks, managers complete two each month, taking them out from behind their desks to where value is added. The walks include a general review of the work areas confirming successful implementation of the continuous improvement culture.

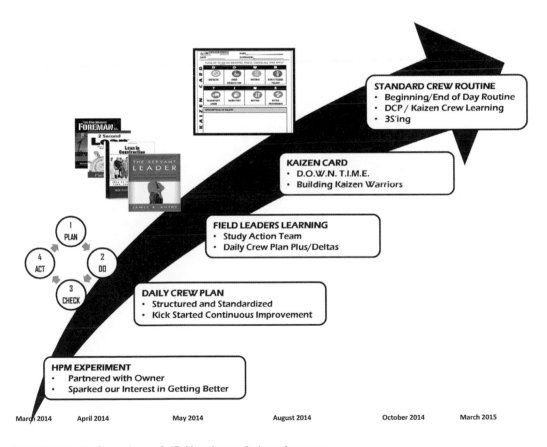

GENERAL CHECK-IN:

Yes No
☐ ☐ Is the Daily Crew Plan completed?

☐ ☐ Is 3S'ing being sustained or utilized?

☐ ☐ Are there Visual Management/Controls in place?

☐ ☐ Is the Work Status (Tracker) Posted?

☐ ☐ Is the **DOWNTIME** poster visible?

☐ ☐ Is the equipment in good working order?

EMPLOYEE CHECK-IN:

Yes No
☐ ☐ Do they know the 8 Wastes?

☐ ☐ Do they know what 3 S'ing stands for?

☐ ☐ Can they describe 3 S'ing?

GENERAL HOUSEKEEPING:

Yes No
☐ ☐ Clutter Free?
☐ ☐ Adequate Lighting?
☐ ☐ No Excess Waste Present?

RECOGNITION:
Who and Why?

Rev. 3/30/15

FIGURE CS8.15 Gemba walk Go-See support card

STANDARD CREW ROUTINE
- Beginning/End of Day Routine
- DCP / Kaizen Crew Learning
- 3S'ing

KAIZEN CARD
- D.O.W.N. T.I.M.E.
- Building Kaizen Warriors

FIELD LEADERS LEARNING
- Study Action Team
- Daily Crew Plan Plus/Deltas

DAILY CREW PLAN
- Structured and Standardized
- Kick Started Continuous Improvement

HPM EXPERIMENT
- Partnered with Owner
- Sparked our Interest in Getting Better

1 PLAN 2 DO 3 CHECK 4 ACT

March 2014   April 2014   May 2014   August 2014   October 2014   March 2015

FIGURE CS8.16 Overview of JB Henderson's lean journey

It also involves managers interacting with field employees in regard to their knowledge of the eight deadly wastes as well as 3-S'ing. General housekeeping is also reviewed and documented. Field crews enjoy this time and frequently 'show-off' their lean improvements. JBH also utilizes this as an opportunity to recognize their employees. At mass employee meetings each month, managers recognize employees that particularly impressed them during the walks. Employees take pride in knowing they 'did a job well done'. Additionally, all gemba walk results are summarized and shared with the management team and actions identified to address any opportunities for improvement in the field. The walks continue to reinforce employee engagement and continuous improvement, a true win–win for JBH.

OVERVIEW

In Tracy Lucero's words 'JB Henderson has experienced a culture shift, which I think we can all agree is not a small feat'. Some of the milestones on their adventure are identified in Figure CS8.16. Although it appears that this 'grand' plan was all mapped out for them to follow, it was not. It all started by recognizing that they were not as good as they thought they were (and they realize more and more each day how much this is true) and follow the path of respecting their employees by empowering and recognizing them for their genius. The employee-focused culture they are building ensures that these experts continue to share improvement opportunities and continue to make JBH better and better.

BIBLIOGRAPHY

Akers, P. *2 Second Lean: How to Grow People and Build a Fun Lean Culture*, FastCap Press, USA, 2012.

Asefeso, A. *Lean in Construction: Key to Improvements in Time, Cost and Quality*, Create Space, Amazon, USA, 2014.

Autry, J.A. *The Servant Leader; How to Build Creative Team, Develop Great Morale, and Improve Bottom-Line*, Random House, New York, 2004.

Breslin, M. *The Five Minute Foreman; Mastering the People Side of Construction*, McAlly International Press, Alamo, CA, 2013.

CASE STUDY 9

Safety, quality and BIM drive lean transformation at DPR

INTRODUCTION

DPR (named after the three founders Doug, Peter and Ron), just 25 years old, was set up to be different in an industry that traditionally resists change. The ambition of the founders was to build a truly great construction business.

Within 9 years, DPR was ranked in the top 50 US contractors and in 2013 T/O reached 2.5bn. Today DPR has offices in almost every technology hub in the USA, and many senior managers have grown with the organization after having started out as tradesmen; DPR's commitment to its people is second to none.

A significant point in the history of DPR was a meeting in the spring of 1992 with Jim Collins, who was a Stanford professor at the time and now is a management consultant and best-selling author of *Built to Last* and *Good to Great*. Collins helped to identify and articulate the purpose and core values that continue to drive the company today.

'We defined our purpose and established our first company mission during that meeting,' said Peter. Doug added that, 'To say we wanted to become a truly great construction company by the year 2000 was like a three-year-old saying that I want to graduate from college by the time I'm 10. We do set high goals, but with the people we have in place, we know we can reach every one of them.'

Ron felt positive that they could build a truly great company due to the network that each one of them had built and the individual skill sets they had developed. 'In the very beginning, we were just three senior guys in one organization with a passion for building,' said Ron. Customer focused from the outset, DPR has maintained long-term customer relationships as a core strength.

'It was Ron's relationships with the banks and insurers that really got us through the initial phase, providing us with the financial base and bonding capacity we needed to compete against larger contractors,' said Peter.

From Doug comes the 'audacious vision'. According to Peter and Ron, Doug is the one with the 'believe it and it will happen' attitude, and Doug demands more from himself than anyone.

DPR CONSTRUCTION EXISTS TO BUILD GREAT THINGS

We are a company of builders building great projects, great teams, great relationships and great value.

Our purpose and core values are a starting point that help clearly define who we are and what we stand for as a company. They underlie the passion that drives us to be better and different; they allow us the freedoms of our entrepreneurial organization, where people can make a difference with their ideas and hard work.

Purpose

WE EXIST TO BUILD GREAT THINGS.

Core values

DPR's fundamental, inviolable values and beliefs:

Integrity
We conduct all business with the highest standards of honesty and fairness; we can be trusted.

Enjoyment
We believe work should be fun and intrinsically satisfying; if we are not enjoying ourselves, we are doing something wrong.

Uniqueness
We must be different from and more progressive than all other construction companies; we stand for something.

Ever-forward

We believe in continual self-initiated change, improvement, learning and the advancement of standards for their own sake.

Mission

To be one of the most admired companies by the year 2030.

Vivid descriptions

Over the next 30 years our **people practices** will be recognized as being as progressive and influential as Hewlett Packard's were over the last 50 years.

When it comes to **quality and innovation**, we will do for the Construction Industry what Toyota did for the Auto Industry.

We will have created a **brand image** that is as positive and consistent as Disney's.

Like Microsoft and Sun Microsystems, our people and company will be known for being **aggressive and 'Bullet Smart'.**

We will be as **integral and indispensable** to the communities we operate in as are The Boys and Girls Clubs and The Red Cross.

We will consistently produce truly **great results and financial returns** comparable to the Most Admired Companies.

Committed to being at the forefront

- Industry leader in safety before safety was recognized as being as critical as it is today.
- Leader in the adoption of BIM for design and construction planning.
- Leader in sustainable design.
- Industry leader in the adoption of lean planning and lean thinking and the effective use of the bigroom.
- Now driving the lean transformation through the lever of quality.

COLLABORATIVE VIRTUAL BUILDING AND BIM

Providing measurably more value is DPR's guiding principle in using technology to enhance project delivery. As a long-established leader in Virtual Design and Construction (VDC) and Building Information Modelling (BIM), Andrew Fischer of DPR writes:

> To name just a few benefits, when VDC apps are used correctly, project teams can improve productivity, reduce field clashes, enhance building documentation, and feed into management systems, which can result in lower long-term costs.

In light of these potential benefits, it can be tempting to think of software as the 'silver bullet' to the many complex hurdles that project teams face in the AEC industry. But even the very best software is simply a tool and is only useful if the person (or team) who wields it, does so effectively.

DPR's experience has proven that it is a combination of technology and a collaborative ideology, such as integrated project delivery (IPD) or design-build that yields the greatest returns.

DPR employs more than 300 individuals trained in advanced BIM and virtual building across the country. DPR offers a wide range of virtual building services:

1 VDC/BIM Consulting;
2 VDC Execution Planning;

3   Model-Based Quantity Take-off;
4   Model-Based Estimating;
5   Self-Perform Work Detailing and Tracking;
6   MEP Coordination;
7   4D Sequencing;
8   BIM-Enabled Constructability Analysis;
9   Laser Scanning;
10 Total Station Integration;
11 Site Logistics Planning.

Modelling of metal stud framing

Over several years, DPR has been working with software specialists to develop the modelling of metal stud framing. As this is an area in which the company self-performs, this work brings tremendous value to DPR projects.

DPR modellers create a composite model with stick-built walls, wall panels and MEP systems. Multiple combinations of heights, lengths and sizes are tested using modelled conditions as a guide prior to erecting a material mock-up.

Using the framing model for coordination with mechanical, electrical, plumbing and fire protection (MEP/F) contractors allows DPR to save space for critical studs, corner studs and king studs around doors and large openings. Further coordination with overhead kickers, shaft-bottoms, soffits, leg-overs and backing makes the model more comprehensive, thus providing spatial and visual cues that indicate potential conflicts to anyone viewing the model.

This gives the project team a clearer picture of what will be required for construction in the field, allowing the team to decide what makes the most sense for the project: whether it's more efficient to place built-up headers or re-route MEP systems, for example. Framing comes to the site either in prefabricated or precision cut for on-site assembly. Not only are estimates much more accurate, detailing, workflow and hence costs are optimized.

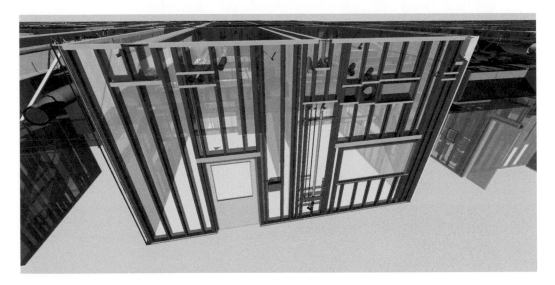

FIGURE CS9.1 Modelling of metal stud framing

Sutter Medical Center Castro Valley

DPR was the lead contractor for this $320 million Sutter Heath project. It was delivered under an integrated project delivery (IPD) contract with 11 partners. VDC was deployed on the project and Chris Murray, supervisor for health-facilities review with the Office of Statewide Health Planning and Development (OSHPD), is on record stating, 'We have seen fewer change orders of less substance and less rework than we typically see'. This was the result of VDC deployed within an IPD framework.

Deployment of BIM 360 on San Francisco Chinese Hospital

The Chinese Hospital Replacement project team is currently working on an extremely tight urban site in the middle of Chinatown to deliver a 100,000sqft, ground-up replacement hospital. The new project is connected to the 1979 hospital addition, which is fully operational and treating patients.

With 15,000 residents living within a 20-square-block area, Chinatown is a vibrant urban backdrop for this technically complex and logistically challenging project. The street in front of the hospital is home to fish and produce markets, which require several deliveries a day. Impeccable planning, logistics and communication are critical to this project.

The project was fully modelled in BIM360 prior to commencing construction. Each foreman starts each day by updating his model of the building. This means that not only has he the full documents with him on site, but everyone on site has the same, up-to-date documents all the time.

FIGURE CS9.2
Foreman accessing real-time data on site

THE SUSTAINABLE ADVANTAGE

DPR believes that being 'green' is no longer novel. Building sustainable structures is simply the right thing to do – right for the environment and right for business. Sustainable building projects routinely produce greater marketability, higher employee productivity and improved building efficiencies. To assist owners in developing and implementing the best strategies, DPR combines experienced people, a collaborative methodology and custom tools to help customers address the triple bottom line: environmental, social and economic.

A proven team for different shades of green

From inception to completion, DPR has repeatedly demonstrated the value of having knowledgeable resources in-house and readily available to fit specific customers' needs. Ranging from light to dark green, DPR's portfolio of sustainable projects consists of nearly 200+ LEED® certified projects and green projects worth more than $5 billion. DPR's LEED®-accredited professionals, located in every office across the country, not only have practical green project experience but also continuously research and implement the latest green building practices and techniques to further improve sustainability.

Custom tools to enhance the process

Understanding and evaluating cost premiums for various sustainable strategies are essential to any green building project. Imagine the advantage of having current first costs and lifecycle data at a very conceptual phase of a project – even before start of design. DPR's greenBook, a custom application, helps teams determine with a higher level of certainty which sustainability measures make the most sense for a specific project's goals. Using greenBook, DPR provides detailed cost data and identifies the 'break-even' point based on estimated design and construction costs and anticipated operational savings. greenBook is also used to analyse various LEED rating systems and levels of certifications from the earliest stages of pre-construction through to project completion.

DPR living labs

One of DPR's goals is to have 'a proven track record of being environmentally responsible in the way it does business'. DPR's passion for sustainability extends to what it builds for itself and how it runs its regional offices and jobsites. DPR's Sacramento location was the first privately owned office building in California's Central Valley to achieve LEED® certification in 2003. Also, in 2011, DPR's San Diego office became the first and only commercial office in San Diego to achieve both net-zero energy status and LEED for New Construction (NC) Platinum certification. In 2013, DPR's LEED®-NC Platinum Phoenix office became the first building in Arizona to achieve net-zero energy certification from the ILFI. In 2014, DPR built its own office in San Francisco, which is on track to become the first net-zero energy building in the city. It won ENR's Northern California's Project of the Year and Best Green Project.

You can see the up-to-the-minute energy and water usage for several of DPR offices online on the DPR website. Along with treating its offices as living labs for sustainability, DPR has instituted many other in-house initiatives, including:

1   *Carbon reduction*: DPR joined the U.S. Environmental Protection Agency (EPA) Climate Leaders and committed to reduce overall employee greenhouse gas emissions by 25 per cent by 2015. Eight years later, DPR met and exceeded that goal by reducing emissions by 30 per cent. In recognition, DPR was awarded the *EPA Climate Leadership Award* in the category of Excellence in Greenhouse Gas Management (Goal Achievement). To achieve this, DPR implemented a combination of solutions to reduce its carbon footprint, including improving building efficiency in its offices and providing employee education. Efficiency features range from simple energy audits to replaced lighting systems and appliances to complex net-zero energy designed offices.
2   *Green offices and jobsites*: DPR tracks energy use at offices and jobsites through the EPA's Energy Star program, implementing green purchasing and cleaning practices.

PRECISION PLANNING FOR PROJECT SUCCESS

Overall project success relies heavily upon the decisions made at the initial stages of the job. Combining skilled professionals with an open and collaborative approach, DPR's comprehensive pre-construction services offer advantages on projects of various sizes and complexities, including:

- providing the assessments and recommendations necessary to make timely, informed and data-driven business decisions;
- better serving the interests of the project through relevant suggestions and hands-on partnership with owners, designers and construction teams;
- helping to manage the overall process by tracking details and holding team members accountable.

DPR's pre-construction services go well beyond traditional open-book cost estimating and value engineering to create more reliable schedules and estimates. DPR uses the most relevant tools, technologies and processes, and gets the right people involved at the appropriate time to drive down project costs while maintaining design intent and the customer's overall goals and objectives.

Tracking performance

Pre-construction is one of six critical success factors DPR tracks on every project to ensure that its cost estimates are accurate and reliable. DPR collaborates with the delivery team to identify key performance metrics that are reported on regularly, helping to create greater alignment around project goals, which ultimately leads to a much smoother, more successful project.

Services

1. Target Value Design;
2. Cost Control;
3. Cost Estimating;
4. Constructability Reviews;
5. Last Planner®/Short-Interval Planning;
6. Life Cycle Cost Analysis;
7. Scheduling;
8. Subcontractor Selection;
9. Material Procurement.

Planning using the LPS® and managing quality

The DPR business unit in Phoenix, Arizona is one of the company's lead teams in achieving exceptional planning, quality and safety outcomes. Over a series of projects this team has developed approaches that have yielded zero defects combined with on-time or early completion.

The first project in the series was a laboratory building for the University of Arizona in Phoenix. On this project the team led by Peter Berg started implementing LPS® including pull planning in which the client participated. One of the team had come up with the idea of developing 'pump primers' for each trade to focus the entire team on areas of risk where things often go wrong with the aim of reducing defects. Three months prior to completion, during a routine site walk the client asked, 'where are all the tradesmen?' In fact the job was well advanced and finished early with only 50 defects remaining at practical completion (PC).

The next project the team took on was a cancer clinic. The project was a complex brown field project with between eight to ten zones per floor, and the challenge was to create workflow on the project. The client wanted to have a relational contract which the DPR team described as IPD 'light'. The project was fully modelled in BIM prior to the start in the field, and the team deployed a tablet/phone-based punch list technology for tracking the close-out of defects. The client started the project with a three-day team-building and project goal-setting workshop. Throughout the project, the entire project team was co-located in a 10-foot-wide trailer. The LPS® including pull planning was deployed, and each day, the team, including all the subcontractor superintendents, met for a 15-minute huddle to review plans for the day. From early on, the DPR team pushed the subcontractors to achieve the commitments they had made at the previous weekly planning meeting, to the extent that the SC workers had to work back on Fridays to get their work finished. At first this was a stretch for the SC teams; however, they soon developed a more disciplined approach to commitment making and plan achievement, and soon enough, they were completing their work on time without resorting to overtime.

On this project, using the combination of LPS® to support short range planning, the trade area 'Pump Primers' to focus on areas of quality risk, combined with the use of tablet and phone-based defect punch list control, a discipline approach to making and delivering on commitments, the use of 5 Whys to get to the root cause of failures and a focus on safety led to a defect-free completion and a safety outcome that was one of the best in the company. The project outcomes were so exceptional that the client awarded its next major contract to the team without going to tender.

SAFETY

Historically DPR has had a focus on safety, and the company's commitment to safety has long been a core value. Rodney Spencley, the company safety 'thought' leader, has a vision that 'the safety professional will no longer be needed, safety will be so integrated into the management system and culture as to how we do work and who we are that we won't need safety professionals'. He believes that 'If you are team leader, it's not what you say it's what you do, every day you have to approach your work in the same way, you have to live safety every minute of every day'.

The DPR belief is that you do not have to have any accidents on any jobs; it's not in the safety manual. To avoid accidents DPR managers have to be world-class planners. Safety at DPR isn't about money, it's not about schedule, it's about people, it's about everyone looking out for each other, making sure they go home safely to their family at night.

The DPR approach to safety centers on training, instilling and reinforcing safe behaviours throughout the company and the entire project team, including owners, architects and subcontractors. The company provides the tools and information people need to make informed decisions through detailed root-cause analysis and pre-task planning.

The outcome is a robust, value-based safety programme that looks upstream at leading indicators to produce exceptional downstream results:

- Over the last three years, DPR's OSHA recordable incident rate averaged 1.07. This is based on an average of more than 5 million total hours worked – making it one of the safest contractors in the US.
- Based on nearly 800 customer surveys, DPR ranked significantly higher in safety when customers were asked to compare the company with their best previous contractor experience.

At DPR everyone has a role in safety

Customer participation is a vital component of DPR's safety culture. To help achieve an Injury Free Environment (IFE), they ask owners to:

- pre-qualify contractors with safety as a key factor in the process;
- ask questions, not only about injury rates and trends but also about how safety standards and techniques are communicated between the office and field;
- recognize that, from a business standpoint, safety is not mutually exclusive and goes hand-in-hand with quality, schedule and budget;
- meet regularly with project teams and actively participate in a way that holds all members accountable.

QUALITY

Rodney Spencley moved on from being the company safety thought leader to developing an extremely effective quality focus using similar philosophies to those that drove the safety initiative in DPR under his guidance. His major criticism of the existing quality approach was that conversations about quality were ambiguous and that quality was checked after the work was completed, creating only lagging metrics around quality.

Rodney Spencley sees that a major challenge in achieving quality is the absence of a common language among the parties. He characterizes this in the following manner. The owner's language is about costs and key operational and aesthetic outcomes; the architect's language is primarily focused on the aesthetic, and end-user functionality; the fire engineer's language is about flame spread; the structural engineer's language is about structural integrity, etc. There are so many different languages being spoken that inevitably everyone is having a different conversation, even though they may be in the one room. He is concerned that some of the owner's key functional concerns are lost in the babble.

A key aspect of the new approach has been to work more closely with owners to better understand their requirements, defined as 'Key Distinguishing Features' for each project. DPR then introduces a detailed framework of quality alignment activities to ensure that DPR and its supply chain meet the client's requirements. These are described below. These alignment activities can then be used to ensure that quality requirements for each project are met, and they can also be monitored and used as leading indicators to demonstrate the focused effort that is going into quality achievement.

DPR's core values in relation to quality

- Build it right the first time is the essence of our QUALITY programme.
- Zero rework, zero errors, zero defects is the goal of every project. Every team member – from the corporate level through to the apprentice carpenter at the jobsite – is accountable and committed to continuously improving performance to deliver higher quality to our customers. Every project is an opportunity to learn and build smarter.
- DPR's QUALITY programme centres on the needs of the project and the customer with each jobsite developing and maintaining a site-specific quality plan. Using web-based tools and resources, our teams customize programmes to meet the unique challenges of the project, putting systems in place to better plan and coordinate work.
- As an ever forward thinking organization, DPR is dedicated to setting new standards in quality and have established a comprehensive programme that not only addresses

a specific error or defect but also closely analyses the processes across all projects. Quality at DPR goes hand in hand with improvement, driving greater efficiency and delivering more value.

Spencley believes,

> Build it right the first time is the essence of our QUALITY programme. There are two parts to this approach to quality: prescriptive (the plans and specs) and descriptive. The first is basic 'blocking and tackling' and cannot be overlooked. The second compels us to understand our client's intentions and expectations through work with responsible parties within the client's organization and the project team!

Producing quality begins with building the right project team. This involves two components:

1   Write instructions to bidders (ITB) in which DPR clearly identifies the distinguishing features of work using its 'pump primers', which are idea starters based on lessons learned on past projects. These are written with trade partners early in design.
2   Identify measurements with the trade contractors so that every work activity is a measurement activity and every day is a measurement day.

The DPR quality toolkit

To elevate DPR's quality performance, Spencley and his team have developed a web-based quality toolkit for use by project teams.

The approach involves four steps:

1   Really understanding the client's needs – this is developed into a documented set of stakeholder expectations in relation to quality for every work package, with goals and metrics to track the achievement of quality expectations.
2   Develop a site-specific quality plan – this is built on DPR's experience and knowledge from previous projects. The quality plan includes a clear description of the client's expectations and quality tools to be used. Key distinguishing features are identified for each trade package; the need for mock-ups; and BIM modelling.
3   Creating a culture of quality through team alignment and planning – this, first of all, involves selecting the right team; then, working with subcontractors to identify the distinguishing features of their work; followed by their response to the owner's expectations and DPR's key distinguishing features. This preparation and planning is consolidated through a series of pre-installation meetings.
4   The final step is measurement – it is essential to effectively track quality and make the outcome transparent for everyone. Technology such as the BIM 360 field, PlanGrid, Excel or CMiC are used to track quality issues. Root-cause analysis is conducted on all more significant quality issues.

DPR's quality system is an integral part of its approach to lean construction, driving continuous improvement both in terms of better meeting clients' expectations and in eliminating waste through avoiding quality defects and driving continuous improvement.

SELF-PERFORMED WORK

DPR also delivers value through its commitment and ability to self-perform critical trades. By executing fundamental scopes of work such as foundations, drywall and carpentry, the company is able to offer greater control and set the tone and pace for each project.

The DPR self-performed advantage

Self-performed work has the following benefits:

- Schedule enhancements: work starts earlier in the field, rather than waiting for bid-dable documents.
- Budget improvements: due to enhanced productivity and contractual arrangements, such as cost of work basis and return savings to owners.
- Superior quality: delivered by their own, highly skilled and continuously trained craftspeople.
- Increased safety: DPR crews working on site provide a clear picture of their high expectations and set a good example for other subcontractors.

Each DPR region may self-perform some or all of the following trades as this gives the company more control over the deployment of resources and the quality and safety of work practices:

- concrete, including foundations, slabs and horizontal/vertical concrete as well as tilt-up construction;
- drywall and taping;
- doors, frames and hardware;
- rough carpentry;
- acoustical ceiling work;
- light demolition and clean up;
- firestop (passive fire protection systems at penetrations, interior joints and slab edge conditions) and insulation;
- strut for overhead supports as well as ground-up MEP utility racks;
- Division 10 Specialties – office and restroom partition accessories.

RECOGNITION

DPR wins numerous awards each year. In 2014, DPR was ranked tenth in the Fortune 500 'Best Companies to Work For' rankings. Several state offices have been ranked at the top and near the top of Best Company to Work For rankings in their state. Also many of its projects receive industry recognition for being best-in-class.

SOCIAL RESPONSIBILITY

DPR's vision is to be integral and indispensable to our communities. It does this through local initiatives that match the needs of local communities with the skills and passions of local DPR employees. DPR regards its 'local' area as extending beyond its regional offices to its jobsites, where often DPR teams are called on to share their talents with the community. Beyond building buildings, DPR offers its expertise in organizational leadership, project planning, soft-skills training, marketing and other professional areas.

'The best way to truly make a positive impact is to understand the needs and match those with our strengths as a company,' said Peter Nosler, DPR's co-founder, who serves in leadership roles for the company's Community Initiatives effort and on the DPR Foundation board. 'We want to focus our energies in targeted areas, and go deeper to make a greater impact.'

Meaningful results

This company-wide vision is implemented on the local level through volunteer action spearheaded by individual jobsites, regional offices and grants from the DPR Foundation. DPR staff professional skills are the most valuable assets the company has to share. On a local level, employees find the intersection between community needs and their skills so that volunteer actions and resources can focus on helping organizations move forward to achieve their mission.

DPR works with many organizations nationwide including:

- Rebuilding together;
- YouthBuild;
- ACE mentoring;
- Boys and girls clubs;
- Habitat for Humanity.

The DPR Foundation

A major component of DPR's regional Community Initiatives is the DPR Foundation. Founded in 2008, the 501(c)(3) non-profit's mission is to support organizations that assist economically disadvantaged youth. Each year the Foundation awards grants and commits volunteer time to organizations local to its regional offices. As of 2015, the DPR Foundation had awarded more than $4.2 million to organizations across the country, with an average grant size of $45,000.

BIBLIOGRAPHY

Collins, J.C. *Good to Great*, Random House, New York, 2001.

Collins, J.C. and Porras, J.I. *Built to Last: Successful Habits of Visionary Companies*, Random House, New York, 2005.

CASE STUDY 10

Quality and operational excellence in Heathrow Development

GENERAL INTRODUCTION

Since 1955 Heathrow Airport has been connecting people from around the globe, making it the world's busiest international airport. At a size of over 1,200 hectares, the airport has two runways handling nearly half a million air transport movements in order to carry nearly 75 million passengers and 1.50 million metric tonnes of cargo for 80 different airlines to 185 destinations in over 80 countries. To service this volume of passengers and cargo, there are five terminals representing over 650,000 square metres of floor space and over 175 aircraft stands. The age of the infrastructure ranges from the 1960s (Terminal 3 1961/ Terminal 1 1968) to the 2000s (Terminal 5 2008 / Terminal 2 2014).

Since 2003, £11bn has been invested in improving Heathrow, which represents one of the largest private sector investments in UK infrastructure over this period. Heathrow has opened new Terminals 2 and 5, extensively refurbished the other terminals and made improvements right across the airport. Behind the scenes, there have been improvements to the management of baggage by investing £900 million to create the world's largest integrated baggage system that ensures passengers and their baggage 'take off' and are reunited on time. While Heathrow is developing the infrastructure to improve the passengers' journey experience, the airport is committed to minimizing the disruption to ensure that the airport continues to operate as normal.

The case study was written at a time when Heathrow Airport was embarking on a new quinquennial business plan (Q6) containing a significant investment up to approximately £3bn in assets. Whilst the spend value was reduced from Q5 due to the impact of T2, the scale and complexity of the demands on the delivery team, Heathrow Development, were no less and complicated by the interdependencies between the projects and, critically, working in a 24/7 live operation.

There was pressure on Heathrow to deliver improved shareholder returns in Q6, due to the impact of lower than forecast passenger numbers. Furthermore, how Heathrow Development would deliver Q6 investment is significantly different from Q5 and reflects some of the changes in regulation and organizational structures. Hence, the asset investment programme needed to deliver development projects 'better' and faster, with least cost. Projects also needed to deliver the potential to further reduce operating costs and improve the passenger experience. The Value for Money (VFM) aspect is critical as it allows Heathrow to demonstrate to the airline stakeholders that disciplined project management leads to the delivery of maximum benefits at lower costs.

For Heathrow, as a regulated business, consultation with stakeholders is absolutely critical. It allows their needs to be incorporated into the project requirements and allows the transparency for Heathrow to demonstrate the VFM promise.

HEATHROW VISION AND BUSINESS PLAN

Heathrow Airport Ltd (HAL) developed the vision to be 'the UK's direct connection to the world and Europe's hub of choice by making every journey better' with its airline customers. Together with the airline community, Heathrow has also defined a set of passenger principles, which capture what drives a positive experience throughout the passenger journey at Heathrow, and a set of service propositions. These were developed collaboratively with the airlines, based on a gap analysis of the current passenger experience.

Taking the vision and passenger principles, Heathrow developed four priorities for Q6:

- Continuous Improvement to enhance the passenger experience.
- Winning the Third Runway to deliver hub capacity and operational resilience.
- Beating the Regulatory Settlement to deliver a competitive cost of operation.
- Making Heathrow a great place to work.

These were agreed with the airline community and these priorities shaped and guided Heathrow's thinking, particularly when considering solutions for Q6. In this way passenger interests are embedded at the very heart of future developments at Heathrow. The framework in Figure CS10.1 sets out the relationships between the vision, priorities and service proposition for Q6 and how they are guided by the passenger principles.

Vision	Priorities	Service Propositions

Vision

The UK's direct connection to the world and Europe's hub of choice by making every journey better

Priorities

Deliver a noticeably better 'hub of choice' **passenger experience** through Heathrow, evaluating and delivering improvements in areas that matter for our passengers

Ensure a continued focus on **improved resilience** and availability of sufficient **hub capacity** for forecast aircraft and passengers

Ensure Heathrow Airport investments to support **efficient airline operations** have defined and realizable benefits to airlines and passengers

Ensure Heathrow Airport's **total cost of operation** is efficient and competitive relative to the passenger mix, service and facilities

Service Propositions

1 High-quality terminals
2 Courteous service
15 Minimize passenger stress
3 Improved Security
2 Wayfinding
10 Provides memorable moments
4 Efficient immigration
7 An end-to-end passenger experience

5 Enable punctuality and resilience
8 Provide airline facilities and services
6 Improved connections
14 Robust operations and collaboration

11 Increased Operational Efficiency
12 Increased non-aeronautical income
13 Efficient capital delivery

Guided by **Passenger Principles** and leading towards the **Masterplan;** enabled by **collaboration** between airport and airlines

FIGURE CS10.1 Visions, priorities, service propositions

HEATHROW DEVELOPMENT

In order to deliver the priorities and service propositions, Heathrow needs to deliver its capital investment programme on time and with a focus on realizing the benefits from the solutions delivered. Heathrow's Development directorate leads the infrastructure programmes and plays a leading role in the transformation of the airport with projects ranging from refurbishment projects to entire new terminal complexes. Heathrow Development delivers the capital investment programme in consultation with airport users ensuring that airlines and other key stakeholders are engaged so that project objectives and benefits are achieved.

Heathrow Development's role is not to construct buildings as that is the responsibility of the Tier 1 partner contractors. Their role is as a 'thin client', the airport's experts in defining the needs for the infrastructure development. This role is deployed through consultation with all the stakeholders and then competitively selecting the right sources, and working with the partner contractors and stakeholders to commission and bring the infrastructure into operational readiness. The portfolio and programme effectiveness is assessed through the realization of project benefits.

Within Heathrow Development there are four strategic programmes and three delivery areas around which the teams are structured. These are shown in Table CS10.1.

The department is a programme management structured organization with teams focused on programmes and delivery areas supported by a Programme Management Office (PMO) – see Figure CS10.2.

TABLE CS10.1 Strategic programmes and delivery areas

Strategic programmes	Delivery areas
• Airport Resilience	• T3/T5
• Asset Management	• T1/T2/T4
• Baggage	• Airside and Landside
• Passenger Experience	

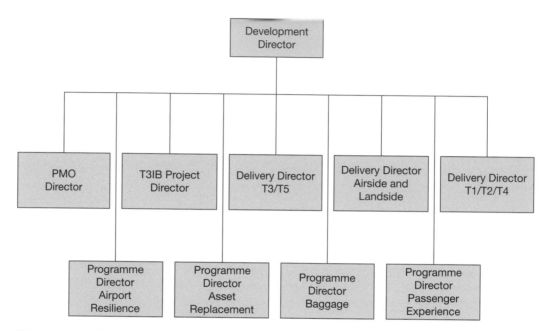

FIGURE CS10.2 Development leadership team

As the Heathrow team's role is that of a thin client, the programme's delivery is through partnered Tier 1 contractors. These are a group of framework contractors that have been selected to support the individual Development teams because of their particular capability to operate in multi-tiered partnering models. They bring technical expertise ranging from design and construction to cost control, as well as effective management of the Tier 2 contractors, where the majority of the build work is carried out.

The commercial procurement strategy is based on the principle of selecting Tier 1 contractors via a competitive process to operate within a framework agreement that bundles work into portfolios of sufficient value to attract high-quality teams. Specific project requirements and timescales can flex within the contract although the overall portfolio value is not expected to change significantly. This changed from the model adopted in the previous quinquennial period, where projects were tendered individually. This was found to be less efficient and encouraged the allocation of ad-hoc teams rather than building a high-quality Heathrow-focused team.

The partner contractors and their linkage to programmes and delivery are noted in Figure CS10.3.

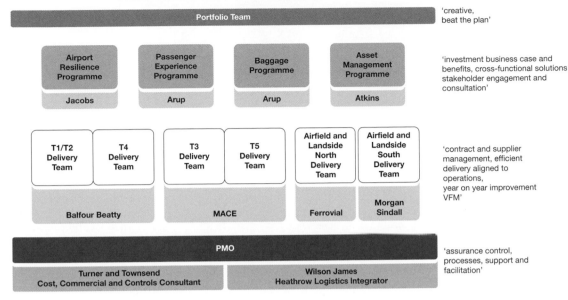

FIGURE CS10.3 Partner contractors and their linkages to programmes

Programme Management Office

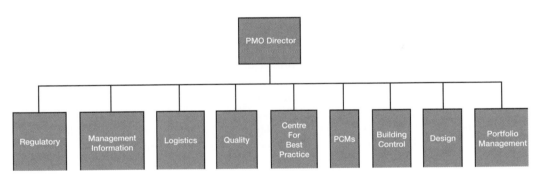

FIGURE CS10.4 Programme management office

QUALITY STRUCTURE

In Heathrow Development, quality management is a Development-wide approach to ensure that there is a collective understanding of the customers' needs and a consistent delivery of the agreed benefits safely, within budget, on time, every time. To support this approach, the quality organization structure is embedded in the Programme Management Office, whose role is to support the programme delivery teams – see Figure CS10.4.

To ensure that quality management is embedded into delivery, the quality team are allocated to support the strategic programmes and delivery areas – see Figure CS10.5.

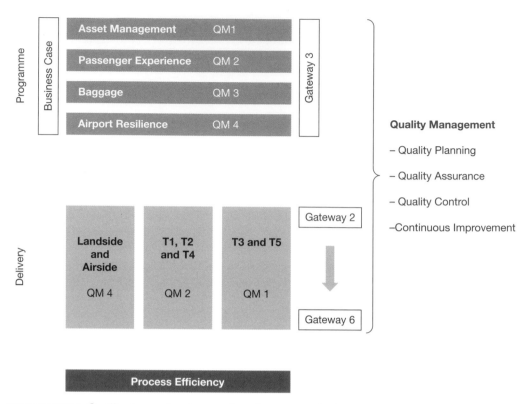

FIGURE CS10.5 Quality team development and resource deployment

PROGRAMME DELIVERY – HEATHROW GATEWAY LIFECYCLE

In terms of a quality management system, the development programme delivery is managed by the Heathrow Gateway Lifecycle. This is the product realization process for the programme with each of the programme's projects having to deploy the lifecycle. The Heathrow Gateway Lifecycle (HGL) is a series of project activities controlled by a set of Gateways (G0 to G8 – see Figure CS10.6). Each gateway involves short, intensive reviews at critical points in the project's lifecycle. These can be internal reviews on the smaller projects through to major reviews with the Board and major stakeholders on the larger projects. This provides an assessment of the project against its specified objectives, and an early identification of areas requiring corrective action. The Gateway is designed to assist the sponsoring team to deliver its project objectives. It is therefore, a project quality assurance methodology.

The Heathrow Gateway Lifecycle is jointly delivered by both Heathrow Development and their partner contractors. This approach needs to be managed effectively to ensure that the roles and responsibilities in the management of quality are clearly understood and controlled. The key controls for the partner contractors are the contract and the Employers Requirements and Standards. These documents are the approaches that communicate the quality process and standards needed to deliver the project. Heathrow Development predominately uses the NEC3 contract framework to support the contractors' delivery. This contract vehicle is collaborative in nature and describes how the project interfaces are to be managed and quality controlled. The high-level roles and responsibilities in a thin client contractor relationship are shown in Figure CS10.7.

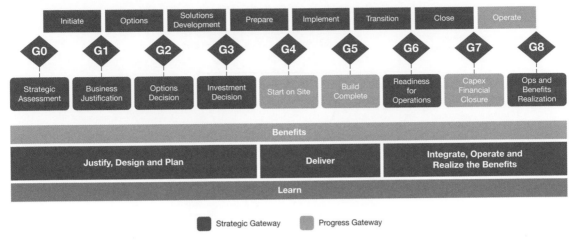

FIGURE CS10.6 The Gateway Lifecycle and its decisions

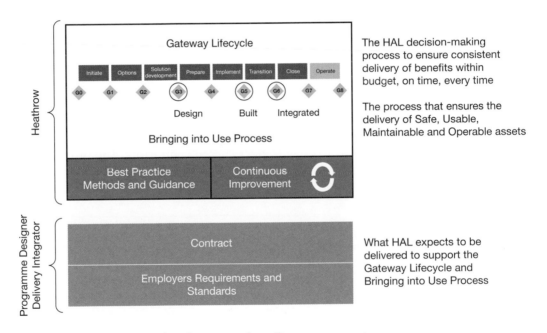

FIGURE CS10.7 Background and context of quality management

In terms of managing quality in the lifecycle, the critical phase is between Gateway 2 and Gateway 6. This is where the solution is developed, constructed and brought into operational use and thus requires careful control of quality. Early involvement of Tier 1 partners (and through them, Tier 2 contractors) is crucial in order to build quality into the solution. To support quality management throughout the project lifecycle there is a V framework to define the management of quality from Gateway 2 to Gateway 6. The framework is shown in Figure CS10.8.

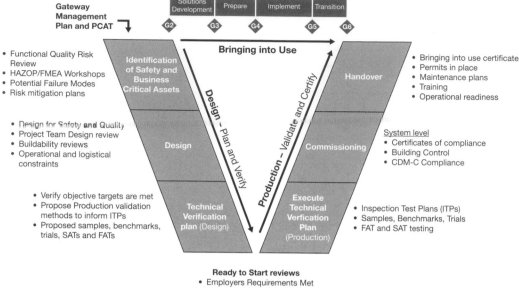

FIGURE CS10.8 Building quality into the solution – G2 to G6

QUALITY ROLES AND RESPONSIBILITIES

The responsibility for project delivery and thus the quality of the solution in the Heathrow Gateway Lifecycle is the project manager. To discharge this responsibility in a thin client, the project manager has to work with their chosen contractors. The project manager, in this respect, takes on the responsibility of quality assurance whilst the contractor has a quality control role. To support the project manager in this role; there in a dedicated quality

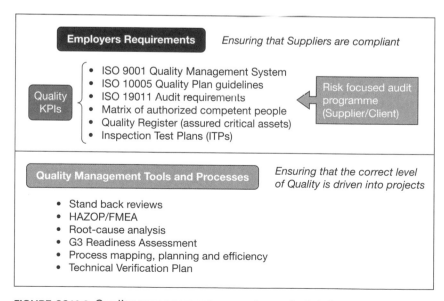

FIGURE CS10.9 Quality management support – project delivery

Core Management	Process Management
Audit & Review • Quality ER Compliance (QP, MACP etc.) • Risk focused audit programme • Quality focused site visits Reporting and investigation • Quality incident investigation & reporting • Quality KPI reporting for MPR Training • Training & behavioural change	Process Ownership • Heathrow Gateway Lifecycle Process Improvement • Continuous improvement activities
Programme Management – Pre-Gateway 3	**Delivery Management – Post-Gateway 3**
Lead • Planning of verification activities • Functional risk reviews • HAZOP / FMEA Workshops • Buildability Reviews Facilitate • Quality assurance activities • G3 Readiness Assessment Verify • Technical Verification Plan • Samples, benchmarks and trials • Review supplier readiness to start	Lead • Samples, benchmarks, trials • Attend FATs and SATs Facilitate • Response to identified quality issues Verify • Execution of Technical Verification Plan • Completion of Inspection Test Plans • Bringing into use process • Stand back reviews (high-risk events)

FIGURE CS10.10 Overview of the quality manager's role

manager to support the management of the V framework. The approach to support has two perspectives – an internal support for the project manager and an external compliance support to the supply chain – see Figure CS10.9.

From the internal perspective, quality managers use the tools shown to support the project managers in ensuring that the correct level of quality is driven into their projects. From the external perspective, the quality managers support the project managers in the delivery of projects through ensuring that the partner contractors are compliant with the employers' requirements. This is through an audit programme based on risk linked to the contractor's performance.

An overview of the quality manager's role is in four primary areas as shown in Figure CS10.10.

QUALITY KEY PERFORMANCE INDICATORS AND PERFORMANCE REPORTING

The delivery of the project's quality solution is linked to the demonstration that the associated solution's documentation has been delivered and meets the necessary standards. These outputs are linked to the quality framework embedded in the design and production phases of the Heathrow Gateway Lifecycle (G2 to G6) – see again Figure CS10.8.

The Key Performance Indicator is calculated to measure both the quality of delivery of outputs and that they are delivered to plan. Delivery to plan is a key indicator to successful asset handover and integration within the Heathrow Operation. The KPIs for the design and production phases are noted in Figure CS10.11.

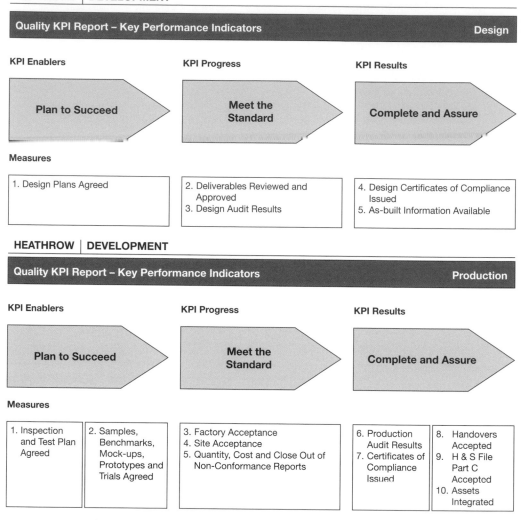

Quality KPI Report – Key Performance Indicators Design

KPI Enablers

Plan to Succeed

KPI Progress

Meet the Standard

KPI Results

Complete and Assure

Measures

1. Design Plans Agreed	2. Deliverables Reviewed and Approved 3. Design Audit Results

4. Design Certificates of Compliance Issued
5. As-built Information Available

Quality KPI Report – Key Performance Indicators Production

KPI Enablers

Plan to Succeed

KPI Progress

Meet the Standard

KPI Results

Complete and Assure

Measures

1. Inspection and Test Plan Agreed

2. Samples, Benchmarks, Mock-ups, Prototypes and Trials Agreed

3. Factory Acceptance
4. Site Acceptance
5. Quantity, Cost and Close Out of Non-Conformance Reports

6. Production Audit Results
7. Certificates of Compliance Issued

8. Handovers Accepted
9. H & S File Part C Accepted
10. Assets Integrated

FIGURE CS10.11 Quality KPI report – key performance indicators

These KPIs are used by the partner contractor to develop their reports to the Heathrow Development Quality Team. These contractor project-based reports are collated to identify the overall quality score for the project. Design and production on the report examples are shown in Figure CS10.12.

The collation process for the KPI report and associated report commentaries is noted in Figure CS10.13.

The overall portfolio view for quality is noted in the Monthly Programme Review (MPR) report, which notes the quality indicator for all the projects. This is a combination of the individual project reports and a consolidation into a programme/portfolio view. This high-level portfolio view is noted by three KPIs:

Quality: Measure of quality of outputs' delivery and delivery to plan.
Earned Value Post Gateway 3: Key projects cumulative Cost Performance Index (CPI) and Schedule Performance Index (SPI).
Spend and Forecast: Overall portfolio spend vs forecast.

Quality KPI Report - Design		Date	Jan-15

Supplier	
Design Review & Approval Plans Agreed	
Design Certificates of Compliance Due	
'As-Built Information' Due	

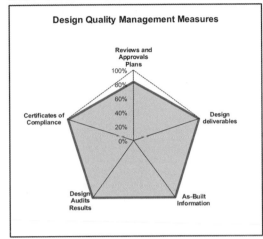

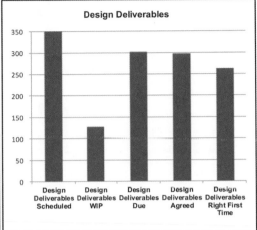

Quality KPI Report - Production		Date:	Jan-15

Supplier	Ferrovial Agroman UK
NCRs Raised	12
NCRs Awaiting Cost Analysis	1
Total Estimated Cost of NCRs	£95,000

Quality Management Measures

Meet the Standard

FIGURE CS10.12 Quality KPI report – design and production

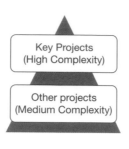

FIGURE CS10.13 Quality KPIs

QUALITY GOVERNANCE

The quality team engage with both the strategic programmes and delivery teams with quality management support – see Figure CS10.14.

This approach is mirrored in the quality governance structure, which integrates the KPI reporting into the Quality Leadership Forums – see Figure CS10.15.

A key element of the quality role is to understand the performance and efficiency of the Heathrow Development processes. This will be from both process compliance and process measurement perspectives.

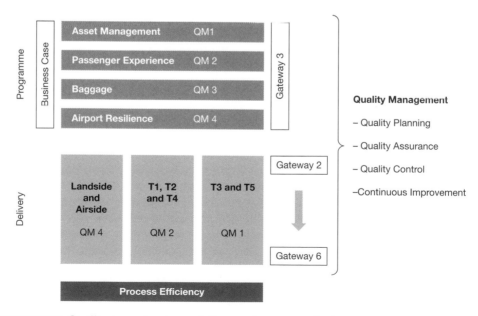

FIGURE CS10.14 Quality team implementation and resource deployment

Quality Leadership Meetings

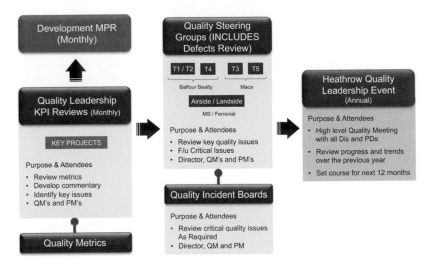

FIGURE CS10.15 Quality leadership meetings

QUALITY IMPROVEMENT

Heathrow Development has invested in creating a 'state-of-the-art' project management process, Heathrow Gateway Lifecycle, as proven by reference to external sources of good practice. This provides a good high-level platform for success; however, to realize its full potential more must be done to deploy it more efficiently, consistently and reliably throughout the project teams. The key element for quality improvement is the Heathrow Gateway Lifecycle and ensuring that the process delivers effective solutions that support the delivery of the Heathrow Priorities and Service Propositions.

In Q6, there was a change to the roles and responsibilities within the Heathrow Gateway Lifecycle, which can be deployed flexibly to ensure efficient delivery. There is a much greater importance placed on self-certification by the Programme Designers and Delivery Integrator contractors (Tier 1 contractors) along with professional collaboration between suppliers in Tiers 1 and 2. With the new Programme and Delivery model adopted, as with any significant change, embedding new ways of working brings grey areas around roles and responsibilities.

In order to minimize the variation in delivery of the Heathrow Gateway Lifecycle, the Quality Management Initiative (QMI) was developed by the quality team. This has the aim of bringing all the process definition work that was happening at Heathrow under one roof. The first step is to create fuller descriptions of the process to the level at which the project teams believe that they can deliver the levels of efficiency, consistency and reliability that the business requires. Previous attempts to rollout Gateway Lifecycle process and guidance have primarily focused on entry/exit at each Gateway rather than the activities and methods used between the Gateways. So there is a lack of guidance and information of approach, activities, roles and responsibilities between Gateways.

The QMI has prioritized the process development work around the key areas of interest from a programme perspective ensuring that the correct decision-making on the selected option and solution is critical. Thus the priority was the Options and Solution Development

Expert definition topics
- Preparing for a successful Gateway 3
- Gateway Management Plan and Project Management Documents
- Funding Process
- PCAT and Scalability
- Approach for low complexity projects
- Bringing into use

FIGURE CS10.16 QMI scope

phases, which are Gateway 1 to Gateway 3. The phases from G3 to G6 will be developed as a second phase of the QMI – Figure CS10.16.

The QMI was designed to deliver benefits based on improving the levels of efficiency, consistency and reliability through improved process management. The benefits were identified as:

- Process definition
  - Clear process definitions will increase the consistency with which projects are delivered.
- Process deployment
  - Project teams will gain a greater understanding of the workflow from G1 to G3.
  - This will improve integration and future performance.
- Risk reduction
  - Better upstream performance will reduce the risk of downstream 'failure cost'.
  - Reduce possible impact on other related projects in the portfolio.
  - Teams will be better prepared for Gateways.

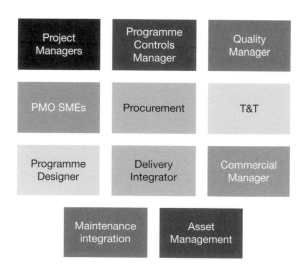

FIGURE CS10.17
QMI – workshop attendees

- Process improvement
  - Process improvement opportunities identified.
  - Continuous improvement approach implemented.

The approach to developing the process definitions and guidance was to deliver collaborative workshops to map the process and identify the key guidance. The workshops were a joint team comprising of both Heathrow and contractors representatives as noted in Figure CS10.17.

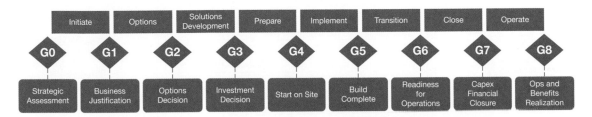

Level 1 – End-to-end process flows
- Key inputs, outputs, controls and resources
- Key outcome indicators
- Fully interlinked
- Describes whole gate-to-gate flow

Level 2 Sub-Process Method Guidance – for every process step
- Method Flow Description – Activity lists, tools & templates – What should get done
- Role Responsibility Matrix – Who is responsible for each activity
- Key Decision Matrix – Who makes the key decisions
- Success Criteria – Critical Success Factors – what must happen
- Exemplars – Good practice examples, hints and tips
- Key Performance Indicators for the Level 2 processes

FIGURE CS10.18 What has been delivered

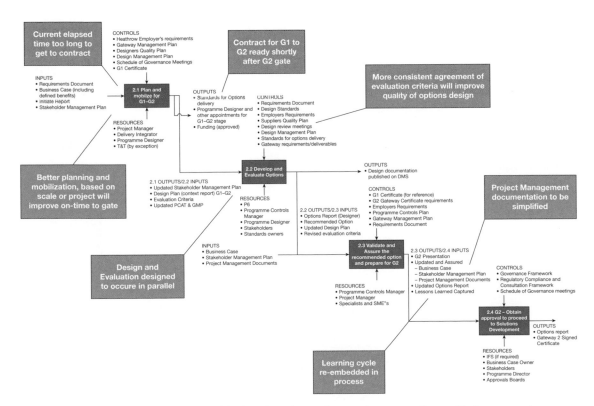

FIGURE CS10.19 Example Level 1 ICOR showing the end-to-end process flow with issues and opportunities

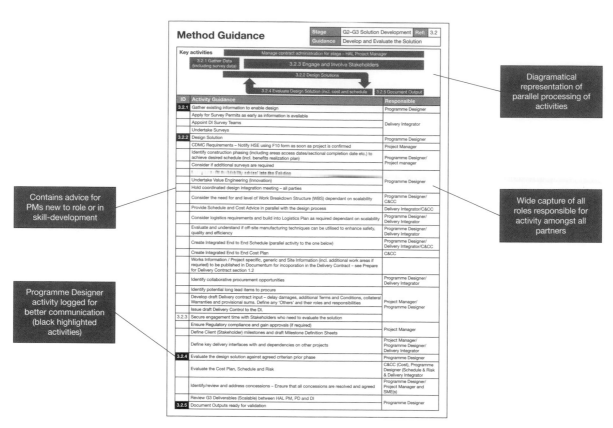

FIGURE CS10.20 Example of method guidance – activity listing

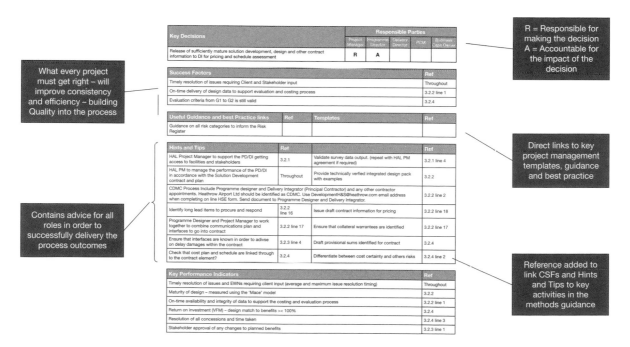

FIGURE CS10.21 Example of method guidance – supporting information

The 'Big Picture' for the Heathrow Gateway Lifecycle was to define the process down to three levels, Level 0 to Level 2 – see Figure CS10.18.

Through the workshops process, the joint teams identified the key process steps needed to take the outputs from the Gateway and convert them into the inputs for the next Gateway. This was in the form of an ICOR process (Input Control Output Resource). An example Level 1 ICOR showing the end-to-end process flow with issues and opportunities is shown in Figure CS10.19.

The workshop also identified the Method guidance necessary for each of the ICOR elements to be delivered. The Method Guidance was in two parts noting the Activities and the Supporting Information – see Figures CS10.20 and CS21.

REVIEW QUESTIONS

1 Heathrow Development has invested in creating a 'state-of-the-art' project management process; Heathrow Gateway Lifecycle. Discuss how this approach may be developed and deployed in another organization in a different area of the construction sector and its supply chain.
2 Develop an implementation plan for the deployment of that approach.
3 Explain how the benefits of the approach should be realized over periods of one, three and five years.

ACKNOWLEDGEMENTS

The authors gratefully acknowledge the involvement and assistance of David Myers, Head of Quality Development, Heathrow, and Mike Turner and Graham Dobell of Oakland Consulting LLP.

CASE STUDY 11

Lean deployment at Rosendin Electric, Inc.

INTRODUCTION

In 1919 Woodrow Wilson was President of the United States, the average house sold for under $4,000 and Moses Rosendin began Rosendin Electric Motor Works. The company serviced the Santa Clara/San Jose agricultural community by wiring homes and shops, rewinding motors for wells and pumps, and extending pole line facilities for utility companies. The company expanded its business to Central California through the 1920s, the Depression and World War II.

The end of World War II brought rapid commercial and industrial growth to the Santa Clara Valley and enormously expanded work for Rosendin. The company grew from eight employees in the late 1930s to ninety employees in the 1950s. It was at this time, in 1953, that Rosendin Electric was incorporated as an electrical contractor in the State of California.

Employee ownership

On 1 January 2000, the Rosendin Electric employees completed their buyout of the Rosendin family to become the largest employee-owned electrical contractor in the US. Employee ownership gave the employees a stake in the future success of the company. Motivated and empowered to produce the best value and service for their customers, the new owners have been focused on keeping customers for life. By providing the highest levels of productivity, exceptional customer service and a quality product, Rosendin Electric has been able to expand to an international contractor, with offices nationwide.

CULTURE

Mission

To set the standards of performance in the electrical construction industry by delivering unparalleled service and value to our customers.

Core values

Teamwork

We believe what we do as an individual, as a company or as members of a project team, affects others. Therefore, we strive for an environment that challenges and empowers people to be their best, to work safely, and to respect one another.

Quality

We believe the quality of our work will represent us for years to come. We take pride in what we build.

Service

We believe that people will do business with companies that meet their needs and provide solid, long-term customer service. The profitability of our organization and the return to shareholders will come from this relationship-based business.

Vision

We believe our innovation separates us from our competitors. Entrepreneurial ideas are encouraged, promoted and implemented. We set the standards.

Integrity

We believe that we cannot achieve our mission, nor the company's other core values, if we do not act with the utmost honesty and integrity in our dealings with our customers, vendors and each other.

Code of ethics

Rosendin uses its code of ethics to memorialize the highest ethical standards that have been the hallmark of the operation of the business for over 90 years and to help staff maintain these standards so that the company will continue to thrive and prosper.

The business has put in place a Business and Ethics Compliance Programme, including an Ethics Hotline to prevent violations of the Code, and to detect violations early and to take appropriate corrective action immediately.

LEAN CONSTRUCTION AT ROSENDIN

Rosendin Electric is one of the early adopters of the concepts of lean construction and an early member of the Lean Construction Institute in the US. Three main strategies have helped Rosendin and its partners to drive innovation:

1   Integrated Form of Agreement (IFOA);
2   Value Stream Mapping and analysis; and
3   Set-based Design.

Looking at projects as networks of commitments, lean construction concepts were found to work extremely well under IFOA agreements in which an integrated project team consisting of the owner, design team members and the major trade partners work together. These contracts incentivize all parties to collaboratively innovate and to drive process and productivity improvements. Benefits created by each party are shared through a consortium approach to painshare and gainshare.

Value Stream Mapping has been used to study workflow on specific activities in order to modify or eliminate non-value-adding tasks to improve workflow and save cost for the project.

Under IFOA agreements, Rosendin and the MEP trades are engaged early in the design–development phase. After gaining a detailed understanding of the client's expectations, the design/construction team use Set-based Design to concurrently develop several alternate design solutions, pricing and evaluating each in order to determine which creates the best value for the owner.

Utilizing the many tools of lean construction, Rosendin Electric, as a key trade partner, has delivered many extremely complicated projects on time and under budget. The company is proud of not only meeting its customers' expectations but going beyond them to add extra value without exceeding the customers' financial constraints.

The Cardiovascular Research Institute at UCSF

From the pre-construction of this $190 million project to the completion of the construction phase, general contractor Rudolph and Sletten constructed a 'bigroom' for representatives from all disciplines to work together in a collaborative setting. A 1,000 square metre trailer was situated only steps away from the construction site, providing space for more than 50 engineers, architects, contractors and consultants who were responsible for the project's design and coordination.

The bigroom environment created close access between all parties, creating streamlined workflow processes, improving communication and cutting response time dramatically. A quarter of the project RFIs were answered the same day, and 95 per cent in less than 15 days.

Utilizing the Last Planner® System scheduling technique (using software accessible by the entire team) provided transparency and held people accountable for each task, allowing any issue with deliverables and schedule to be promptly addressed.

Construction began in May 2008 and achieved substantial completion on 13 August 2010: ten weeks ahead of schedule.

Implementation strategy

Rosendin Electric has long held continuous improvement, standardization of its processes, innovation, creating customer value and respect for employees as its core beliefs. Lean/IPD takes those beliefs just a step further, committing the business to continually do this within a multifunctional team. This requires commitment and willingness from everyone in the organization to identify waste and look for ways to improve processes in order to respond to the needs of both internal and external customers.

The business established a steering committee to lead its development and practice in lean construction. The main goals of this group are to:

1 share information between the various project teams, divisions and departments;
2 participate in improvement projects, educate, expose employees to lean construction and share lessons learned.

This was done as a part of a business improvement initiative throughout the business to ensure that all of the company's teams were exposed to the same heightened level of planning, innovation, business improvement and team integration.

Study action

Throughout the business 'study action' teams meet regularly to drive improvement in processes and outcomes within their department. These groups study books on lean production and discuss lessons that can be applied to the problems in their work. Two books that are popular in the organization are *2 Second Lean* by Paul Akers (2012), and *Switch: How to Change Things When Change is Hard* by Chip and Dan Heath (2010). Rosendin also recognizes that there is much to be learned from other companies who have a lean construction culture and therefore groups tour other companies as part of their improvement journey.

Use of 3- to 6-week Lookahead within the Rosendin team

Rosendin Electric has made use of 3-week and 6-week Lookahead schedules for over 20 years. While these plans have been used with success on individual projects, the process did not provide the opportunity for planners to improve the reliability of their plans week on week. The introduction of the Last Planner® System (LPS) has allowed planning teams to leverage the metric of PPC (planned per cent complete) and improve plan quality from week to week. This has led to improved productivity and a closer evaluation of work activities from project to project, creating a basis for continuous improvement in work practices.

To date most of Rosendin's experience with LPS® and pull-planning has been on projects where implementation has been contractually supported by the general contractor (GC) or project owner. However, Rosendin is committed to training the entire organization in the use of LPS®, so that the benefits of pull-planning and PPC can be leveraged on projects regardless of the practices of the GC or owner.

Current lead practice is being demonstrated on a project in Fort Worth where the GC is using BIM360 Plan for managing LPS® updates, reporting PPC and charting the data. The team enter their activities into BIM360 Plan off-line, activities are tracked daily for completion and the plan is reviewed weekly at the regular planning meeting which is attended by all trades. The team also uses graphics and drawings to show where everyone is working in the building. Activities are also tracked for variance, and PPC is monitored to drive continuous improvement in plan reliability.

Value Stream Mapping (VSM) at Rosendin

Rosendin was asked by the owner on the UCSF Cardiovascular project to look for further cost savings. A study team was assembled to explore ideas that might save time and cost.

The team decided to use Value Stream Mapping (VSM) to analyse and document their process improvements. The first study was on the installation of a pendant-hung fluorescent light. First of all, the study team told the installation crews that they were not being assessed. This was a study of the installation of the fitting with a focus on process improvement. They asked the workers not to alter their installation procedure simply because they were now being watched. The first step was to record current practice, and this was followed by an analysis of the potential for improvement. The workers were asked to participate in this. The field crew was excited to be a part of the experiment.

A data capture sheet was devised, allowing for a record of each activity, the time taken and some notes. The group started recording all the steps in the process and identified the workers' activities that might be improved. They also recorded time lost getting materials and tools. The entire process was repeated four or five times to get a good understanding of current practice and the potential for improvement.

The next step was to create a current state process map in which they recorded the work as currently done. This provided a high-level overview of the process showing every step mapped out from beginning to end. The team then analysed each step to identify areas where the process could be improved and waste eliminated. These areas were highlighted on the map, and a technique known as a Kaizen Burst was used to identify potential improvements to these tasks.

The team then created a proposed future state process map which incorporated all the ideas that had been generated. Nine steps in the process were eliminated, reducing the install time from 44 minutes to 22 minutes, a 50 per cent time saving. The refinement of the process required some preparation and prefabrication at the supplier's factory. This required some negotiation with the supplier, who was happy to oblige. This single activity saved the project owner $50,000, while it took less than $2,000 to conduct the study. This result led to further efforts within Rosendin to look at what they could do on their project.

Shortly after this, another group within the company used VSM to improve lighting installation efficiency on a garage project. Once again everyone in the process was encouraged to 'find the waste' within the activity. While the map was not a conventional VSM map, it plotted the process and indicated where there could be savings. With a focus on team collaboration, waste elimination and process improvement, the project attracted a lot of interest across the business unit as this was the first exposure of most of the staff to lean construction.

As with the UCSF team, they began by observing, timing and recording all the steps while two men installed a limited number of fixtures. The simple VSM shown in Figure CS11.1 illustrates these observations.

The following morning, the team met to take the process further. It was made clear that this was an occasion to express positive criticism, discuss initial findings and identify anything where there may be an opportunity for improvement. The following summarizes the suggestions from this team revue.

1   *Open the box*: All fixtures arrived pre-addressed from the factory (using RFID technology) and each fixture was installed at a set location once complete. It was decided to keep each light in its box to ensure the address stickers stay with each fitting and to eliminate unnecessary travel time by using rolling carts on which the fixtures can be stored. The rolling cart was designed to also carry the drawings showing the address for each light, and a helper was to stage the lights ahead of the installer a few fixtures at a time.
2   *Install the back-plate*: This activity was eliminated from the fixture installation and moved to the first pass across the floor for a few reasons:
    a   When the deck is stripped, the tabs on the deck box are bent down. They have to be removed or beaten back flat into the concrete. Also, any nails left by the concrete contractor close to the fixture box have to be removed.

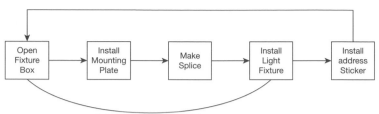

Return to box to retrieve sticker

	Sample 1	Sample 2	Sample 3	Sample 4
Step 1 – Open box	0:35	0:37	see below	see below
Step 2 – Install Mountain plate	2:50	2:40	3:16	3:08
Step 3 – Make Splice	3:01	2:32	2:15	1:48
Step 4 – Install Fixture	1:40	1:12	:40	1:48
Step 5 – Install Sticker	2:00	2:05	2:51	2:32
Total time	10:06	9:06	8:22	9:16

Observations
- Trash removal?
- Can the fixtures be placed near the install area?
- Can battery tools be used to lessen fatigue?
- During the first pass, tap down box support tabs, make splice and add #14 pigtail. The fixture tail is too short to make the splice
- Can push nuts be used instead of wirenuts? Can wirenuts be used for in-slab splice and push nuts for fixture connection?
- Make sure men use clever cutter, no knives
- Only tools needed for this activity, no extra unnecessary tools
- Wait for reshore removal to allow easier path of travel
- For ladder install, can the apprentice hold the fixture while plate is installed, when he hands the fixture to the JW can he then return for the next fixture and place the stickers?

FIGURE CS11.1 Team value stream map of lighting installation

b   During first pass, the MCJ in the deck is spliced, also #14 stranded pigtails were added to make splicing the fixture much easier. The hole in the back-plate was very small making it difficult to push #10 wires up inside the box, while holding the fixture and trying to slide it up tight.

c   It was decided that if the pigtails were to be spliced on, the back-plate should be installed as well.

d   Finally, at locations where there are joints in the plywood framing, the concrete can be slightly out of level, this alignment problem is exaggerated when the back-plate and fixture are installed. It was decided to install fender washers at this time before the back-plate is installed, thereby preventing rework later on.

3   *Make the splice*: It was recognized that the pigtails were too short. It was difficult to splice a #14 to 2 or 3 #10 wires, while holding the fixture up. Long #14 pigtails were added to each box to reduce not only the installation time, but the potential for pinched or smashed conductors. The use of push nuts/Wago's was considered, but it was felt that the #14 stranded might be very difficult to push into the connectors; however, this was tried and ratchet cable strippers were trialled as well.

4   *Install the fixture*: By making the #14 pigtails and back-plate installation prior to installing fixtures, installation should be a SNAP. It was decided to use all 6 foot rolling platform ladders, creating a much smoother and safer way for the men to work at this deck height. Also 12V Milwaukee battery drills were introduced to reduce fatigue by holding the 20V drills all day long. The 12V drills fit right in a tool pouch or hang from a tool belt.

5   *Install the fixture address sticker*: Currently the installer has to return to the table where drawings are kept, place the sticker on the drawing at the fixture location then

return to the fixture to install the sticker on the fixture. However, with a mobile installation trolley, the helper could place the sticker on the fixture before handing the assembly to the installer, and when the installer climbed back down to move to the next location, he installs the sticker on the drawings. This is a critical activity.

Net results: The process was made more streamlined, greatly reducing travel time by making the operation completely mobile. The team was still to examine trash removal once installation is complete, perhaps with a rolling laundry bin and breaking the boxes down. Once again the time saving was significant at 12 minutes saved per fixture.

TECHNICAL INNOVATION

Rosendin staff are constantly encouraged to look for innovative ways to improve their operations; this has included cloud-based computing, prefabrication, improving mobile work stations and adopting advancements in Business Information Modelling (BIM). The company also pushes its technology partners to provide it with innovative solutions. The company has been recognized for its leadership both within the construction community as well as by high-tech organizations. Two recent examples follow.

Rosendin Electric wins a Gold Vision Award in 2014 from Constructech for new mobile material and tool management app

Rosendin Electric was selected for a Gold Vision Award for developing a new mobile app that works with the Autodesk®BIM 360™ Field platform to streamline material and tool management for field personnel.

Rather than manually listing materials or equipment and generating paper or email requisitions, the material and tool management app allows the field team to use mobile devices to order conduit, wire, fixtures and other materials. The requisition is automatically routed for approval and fulfilment by purchasing, the equipment yard, tool manager and other departments. The app is designed to check inventory before generating a purchase order and every order has a QR code that tracks the materials from shipping to the final installation location. The app has even been extended to support Rosendin Electric's prefabrication unit where assemblies are built, coded for tracking and shipped to the jobsite.

With the deployment of this app, the company has been able to dramatically reduce the time required to process materials and equipment orders. It has also reduced shipping and receiving errors and provides excellent materials tracking. Furthermore, the app has been designed to integrate with the existing document repository software and mobile technology already in use so no additional software or training is required.

Rosendin Electric earns spot on the 2013 InformationWeek 500 List of Top Technology Innovators for third year in a row

In 2013, InformationWeek recognized Rosendin Electric for its innovative integration of wireless technology, consumer devices and cloud services to improve productivity on the jobsite. Rather than carrying the volumes of paper previously required, field employees now use iPads to check drawings, 3D models and job specifications. Using mobile devices streamlines on-site processes and allows for constant, real-time communication with the entire project team. This approach increases QA/QC inspection accuracy and substantially reduces the time to identify and resolve any design and installation issues. Rosendin Electric has also harnessed the same technology to streamline procurement. Large shipments from suppliers or the company's prefabrication facility can now be tracked using QR and bar codes, saving time and resources.

Prefabrication

All opportunities for labour savings are explored by Rosendin Electric and improving productivity through prefabrication is used wherever possible. Prefabrication is either conducted at an off-site facility or in a controlled environment on the jobsite. Pre-assembling products enables Rosendin to meet and often exceed customer project schedules, control jobsite waste and improve overall process quality on a project. Prefabrication capabilities include:

1 Palletized Unit Wiring;
2 Duct Banks Fabricated Above-Ground;
3 Pre-Assembled Conduit Stubs;
4 Transformer Lug Kits;
5 Pre-Cut, Pre-Bent Overhead Feeder Conduits;
6 Wireway, Wiremold Pre-Assembled;
7 Fire Alarm Box/Device Assemblies;
8 Pre-Assembled Wall Rough-Ins;
9 Overhead Racks/Supports.

Site workplace optimization

Opportunities for improving productivity in the workplace are achieved through the development of mobile work stations, carts for moving materials and storage of parts on site, and the use of rolling ladders platforms.

FIGURE CS11.2 Examples of carts developed for the site after 5S

BIM

Rosendin Electric has been successfully designing, modelling, coordinating, prefabricating and constructing in the 3-Dimensional world for ten years. Over this time, the company has been an early adopter of new technology and has developed the people and methods needed for successful BIM implementation.

The company is nationally and internationally recognized as an industry leader, driving innovation though people, technology and process. Maintaining strong relationships with NECA, IBEW, AGC and other construction organizations and industry leading software development companies provides the company with the ability to leverage the appropriate technology and processes for its customers' needs, very early in the project.

Rosendin has a team of nearly 60 full-time BIM modellers and coordinators at the corporate and regional level. Field teams are supported by Rosendin Electric Engineering, Information Technology, Operations, Prefab and Project Management. Staff are trained and assessed routinely in the latest software applications needed to meet both company and customer project objectives.

BIM seamlessly integrates the prefabrication department and field operations, allowing model information to move electronically from the model to prefab before being shipped to the field.

These capabilities are creating and driving data from the estimate to the model, providing operations with the most current model information to build from. They are also driving data from the model directly to prefab, reducing costs and improving schedule durations and helping to better manage risks. Field tracking of RFIs, changes in scope, and material procurement, shipping and logistics tracking in real-time are all reducing the possibility of error and improving productivity.

SUSTAINABILITY

Rosendin Electric recognizes that sustainable design and construction practices are important to the future of the environment. A Certified Green Business Provider through the County of Santa Clara, the company offers LEED® Accredited Professionals and Green Associates with the ability to assist with design, estimating and project management. Many Rosendin Electric projects have been recognized by the USGBC for reaching and surpassing LEED® standards in all levels of certification.

In addition to providing design and construction services, which optimize building performance and improve occupant satisfaction, Rosendin Electric also implements sustainable practices throughout its offices, warehouses, prefabrication locations and project sites. Through its technology innovations and mobile device apps, the company has made many 'paper-based practices' almost entirely obsolete.

The company has a nationwide group of employees, the Green Team, who meet regularly and produce a monthly newsletter to facilitate and foster these practices.

Solar power

Rosendin Electric has established itself as a leading EPC builder of mid- to large-scale solar photovoltaic (PV and CPV) systems throughout the United States, Guam, Puerto Rico and Canada. With over 820 MW of project installation experience to date, more than 200 MW in construction and 200+ MW in various stages of development, the company is able to offer turn-key expertise and EPC capabilities to develop the most efficient and cost-effective solar solutions for its customers.

Rosendin's experience and unique capabilities allow it to take on challenging solar PV projects in a variety of different locations and environments, ranging from commercial, hospitals, schools, universities and government facilities to utility grade solar farms. Project sizes range from 100 kW to more than 400 MW systems, utilizing the best applicable CPV, PV or thin film technology in any mix of rooftop, ground mount, single and dual-axis tracker and/or canopy-based installations.

Wind energy

For more than a decade, Rosendin Electric has recognized renewable energy as an integral part of the future of power generation. Since the inception of its Wind Energy Division, Rosendin Electric has installed more than 10,000 megawatts (MW) of clean, renewable wind energy throughout the US. They provide electrical contracting 'turn-key' services, including:

1 Design-Build Services;
2 Substations/Switchyards;
3 Vertical Tower Wiring;
4 Underground Collection Systems;
5 Overhead Collection Systems and Transmission Lines;
6 Fiber Optic SCADA Networks;
7 Substation SCADA Design and Communications Integration;
8 System Power Factor Correction, Upgrades, and Modifications.

SAFETY

Safety philosophy

In the words of Marty Rouse, CSP, CHST, STS-C Vice President of Safety:

> Rosendin Electric's commitment to develop and implement a Safety and Health Programme is unparalleled in our industry. It begins with one simple statement:

Our people are our most valuable asset and should not be exposed to injury or illness as a result of their employment.

Rosendin Electric employees must be treated with respect and we go to great lengths to listen to their suggestions, comments, and concerns. It is our employees who make our Safety and Health Programme a success.

To communicate this commitment to all our employees, Rosendin Electric has adopted and implemented the Injury, Illness and Impact Free (I-3 Free) Programme. The Programme begins with a paradigm shift in the way we view safety. Rather than being a priority, safety is viewed as a value; one we are not willing to compromise.

I-3 Free is a culture where the whole organization continuously improves itself in an effort to completely eliminate at-risk behaviours. At Rosendin Electric, we place this value above all other project management disciplines and strive to provide our people with the tools, equipment and working conditions that will help them do their best safely.

BIBLIOGRAPHY

Akers, P. *2 Second Lean: How to Grow People and Build a Fun Lean Culture*, FastCap Press, USA, 2012.
Heath, C. and Heath, D. *Switch: How to Change Things When Change is Hard*, Random House, New York, 2010.

CASE STUDY 12

The development of the Costain Way

INTRODUCTION

Founded in 1865, Costain is a major British engineering and construction group delivering integrated consulting, project delivery and operations and maintenance services, with a proud history of landmark projects, including the Trans-Iranian Railway in the 1930s, the Channel Tunnel in the 1990s and, more recently, London's Crossrail. Costain meets essential national needs by providing world-class engineering and technology-led solutions to blue-chip customers in the UK's energy, water and transportation markets.

The Costain business employs over 4,000 people, with turnover more than £1bn, and is split into two core operating and reporting divisions, Infrastructure and Natural Resources.

The Infrastructure division delivers engineering solutions in:

- **Rail:** Delivering end-to-end asset lifecycle solutions across the entire railway, from major station projects to multidisciplinary rail projects.
- **Highways:** Providing end-to-end highways services and delivering technology-led solutions for customers.
- **Power:** Playing a significant role in safely delivering the energy infrastructure that will satisfy the UK's future needs.

The Natural Resources division delivers engineering solutions in:

- **Water:** Providing engineering solutions to UK water utility companies across the asset lifecycle.
- **Nuclear:** Creating the facilities to manage nuclear waste safely using the latest technology-led solutions.
- **Oil and Gas:** Providing full lifecycle engineering services in the development, design, delivery and maintenance of Oil and Gas infrastructure.

In addition to the two divisions listed above, Costain has a non-core 50 per cent participation in a land development joint venture in Spain whose assets include land held for development, two golf courses and a 600-berth marina.

Other major projects undertaken by or involving the company have included the Dolphin Square apartments in London completed in 1937, Dubai International Airport completed in 1960, the Deep Water Harbour at Bridgetown, Barbados completed in 1961, the Thames Barrier completed in 1984, the Tsing Ma Bridge in Hong Kong completed in 1997, the Cardiff Bay Barrage completed in 1999 and St Pancras Station in Central London.

In 2010 the Company was named by one of the UK's most influential construction publications, *New Civil Engineer*, as Contractor of the Decade. Since then the Costain name has received further confirmation of its status as a leading national brand – the company has been named as the most admired construction company in Britain.

Costain has also performed well in Britain's 'Most Admired' league table during recent years. The table, organized by *Management Today* magazine and sponsored by BSI, comprises 250 top names, including Rolls-Royce, Royal Dutch Shell, J Sainsbury and Unilever, and is compiled from feedback involving senior directors voting for the companies they most admire. Costain has been number one choice in the Construction – Heavy category and overall ahead of a number of other leading well-known organizations.

COSTAIN VISION, STRATEGIC PRIORITIES AND VALUES – 'ENGINEERING TOMORROW'

Costain's vision is 'To be the UK's top engineering solutions provider' with the overall key objective to meet the needs of stakeholders. Quality and responsibility are at the core of everything they decide and do. Costain cares about their natural and built environment, their relationships with all of their stakeholders, and fundamentally about the future of the UK. Costain's 'Engineering Tomorrow' strategic focus is to enhance growth and market position by providing innovative and sustainable solutions to increasingly complex and large-scale national infrastructure needs.

As Costain aims to be one of the UK's top solutions providers offering a full service from front-end engineering consultancy and design, through construction to back-end care and maintenance, innovation is at the heart of the business and enables Costain to provide the solutions demanded by their customers. Costain cares about all stakeholders and responsibility is at the core of the business. Awards for quality, innovation and safety are regularly won.

An underpin for all of this is the 'Costain Way'.

THE COSTAIN WAY

Original context

- In the mid-1990s the Company had a very challenging time, nearly going under.
- By 2002, 6 out of 7 jobs were running well but the failure rate was still too high. A really big failure could threaten the Company again.
- A review of Customer Satisfaction Surveys showed what customers wanted: on-time delivery, improved design management and better management of the 2nd tier suppliers. Consequently:
    - working groups were established, led by top management, making the case for change;
    - training was deployed to the entire business (CEO attended all sessions);
    - 'Turn Around' was achieved in 18 months;
    - many of the myths around failure were challenged.
- In 2005 a similar journey was achieved with the Oil and Gas business.

Development of the approach

Business process improvement has evolved into the 'Costain Way', which has been a natural progression and evolution over the last twenty years.

The 'Costain Way' is the Costain Group's Business Assurance System, a risk-based, integrated management system that provides instructions and advice on how to promote best practice across the group.

It contains the required standards, guidance, best practices and standard forms for all the activities undertaken by everyone across the Costain Group – the Costain Way of Working. The Costain Way is fast, easy to access, easy to search and is simplified to tell people what they must do and how to do it, with guidance and tools.

The objectives of 'The Costain Way' is to provide assurance:

1 that Costain activities are compliant with appropriate legislation and codes of practice;

2 to customers, employees and stakeholders that Costain systems, procedures and processes are effective at mitigating identified risks;

3   that customer expectations are understood, communicated and effectively delivered;
4   that management controls are consistently applied across the group;
5   that performance is reviewed, validated and continually improved.

Fundamentally the Costain Way has one purpose – to help manage and reduce risk across the business. The Costain Way operates and supports an integrated risk-based management system and they were the first to achieve registration by BSI to the ISO 9001:2015 standard in September 2015.

The 2015 version of ISO 9001 has actually moved closer to the way Costain operates. It is refreshing as it very much mirrors the way we do business and the way we plan on working in the future.

(Tony Blanch, Business Improvement Director, Costain Group)

Over the last twenty years there has been an evolution from a compliance auditing process to a holistic business improvement process, specifying requirements, providing guidance and tools to everyone within Costain regardless of location and role.

Historically Costain used a list-based 'Compliance Audit' to measure process compliance, and over time a clear link between compliance and project margin became established. However, there were a number of drawbacks identified from the use of a compliance audit system:

- compliance lists are not exhaustive;
- having the paperwork was no guarantee of practices being implemented on site.

Average scores had increased to over 95 per cent significant steps in the evolution but the challenge was that, although 95 per cent may seem good, there is still 5 per cent non-compliant:

- It was demonstrated that it was this 5 per cent that resulted in many of the subsequent failures.
- Costain had developed a 'culture of green' – hiding true poor performance.
- Significant cultural shift was needed to overcome some internal resistance – the changes eased by a relatively 'compliant' business and a centrally led structure.

Compliance audit process to 2008

A Compliance Audit process was in place prior to 2008 which provided an assurance function. However, while this did drive compliance to existing agreed processes, it was realized that it did little to drive and further improve processes/performance and responsiveness to customers.

Business Improvement Assessment from 2008

In 2008 Costain's chief executive defined the brief to devise a performance measurement process that equated 'good' projects with a score of approximately 50 per cent and enabled projects to clearly see what they had to do to improve performance.

The Business Improvement Assessment (BIA) is based on the Excellence Model (EFQM) and measures the five enablers, which are broken down into 33 project requirement criteria (see Figure 3.4 from Chapter 3).

Internal assessors visit projects at regular intervals to ask a series of questions which help them measure how well each criteria is being applied.

Assessments take one day and contract leaders take part in assessments of other managers' projects. This spreads best practice still further.

A core team of ten assessors have been trained to award points out of 10 for:

- how well a project plans its approach;
- how well it implements its approach;
- how well it checks (and reacts) to the effectiveness of its approach.

The BIA report is presented graphically showing scores for each of the 'enabler' criteria together with traffic lights showing the 'results' of Safety, Quality, Cost and Time. The assessors report also includes recommendations on what else the project can implement in order to improve further.

What happens after the report is produced?

Regardless of the score, the project is required to produce a 'Business Improvement Plan' to address the key areas for improvement.

Wherever a project scores 7 out of 10 or better on any individual criteria, the evidence is collected by the assessor as examples of best practice or innovation and posted on the Costain intranet, so that all projects can see how improvements can be achieved.

An internal cross-sectional work group used the Excellence Model enablers of Leadership, Strategy, People, Partnerships and Resources and Processes to design a scoring profile which would reward 'successful' projects with approximately 50 per cent and only truly 'excellent' examples with 70 per cent or more.

In addition to the enablers, measurement criteria were developed around four headline results (safety, quality, cost and time). This enabled 'results' to be included as 'traffic lights' showing the current status of the project.

Current model/approach – The Costain Way (2014 onwards)

Further evolution was driven by feedback from the employees for:

- the management system to be faster – so they have created a separate database in a new area of their intranet to speed up search times;
- it to be easier to find information, telling them what they must do and how – so they developed a new interface.

The Costain Way is intended to be intuitive; people do not need lots of training to grasp the basics and it has been made available as an app for use on tablets and other smart devices. There is directorial responsibility for the consistent use and evolution of the Costain Way.

Tony Blanch, Director of Business Improvement, has a team of ca 60 (6 people in central roles) – the rest are 'Quality Managers' assigned to individual projects engaged in both quality and improvement. These report to the project managers but have a separate escalation route if required.

An independent team of 'experienced' quality inspectors provide additional confidence in the projects: nine people doing two inspections per month. The business also has an internal audit function which is far more than a traditional audit group. Accountability for quality remains very clearly with the project managers.

A simplified IMS has been designed for use in the field and use of an interactive process map enables access to be gained to required information. There is a click on a step in the process maps to access information related to the Contract Lifecycle, Group Shared Services or 'Your Personal Conduct'.

- **Contract Lifecycle**: all procedures applicable for contracts, from identifying and securing opportunities, consultancy, design, construction, through to operations and maintenance.

- **Group Shared Services**: all procedures relevant to group functions e.g. Human Resources, Accounts, IT, etc.
- Personal Code of Conduct for all employees; e.g. expenses, absence.

The Costain Way is comprised of the following types of documents: POLICIES, MUST DO, HOW TO, TOOLS:

Policies
The Board's statement of intent in relation to a legal, regulatory or specified requirement which outlines the way Costain intends to conduct its affairs.

Must Do's
Costain have simplified over 200 policies into 23 (one-page statements of board intent) – the rest of the 'rules' have been moved in to the 'must dos'.

How to's
Each process in the Costain Way has an owner, who is responsible for its content. The owner will regularly review the process to ensure it remains effective. Costain have also appointed a custodian of the Costain Way, who is responsible for making sure that it is easy to use, consistent, controlled and regularly updated.

BENEFITS

The results of the hundreds of BIAs undertaken show a clear correlation between the BIA score at six weeks into a project and the successful delivery of the project at completion. If the project team adopts and meets all of the criteria in the BIA, it will almost certainly deliver within the budgeted costs.
Further benefits are:

- It provides a useful framework for the Contract Leadership.
  - Small projects can compare themselves directly against larger ones using the same model.
  - A full range of criteria required for sustainable success is considered in the contract's overall rating.
  - It provides senior management with an early warning indication of potential contract outcome.
  - There is now scope for improvement for those projects were already scoring 100 per cent compliance by traditional systems of measurement/assessment/audit.
  - It provides heads of department with a measure to gauge how well their particular discipline is being implemented (e.g. innovation, training, branding, etc.).
- Improvement initiatives are now more focused.

AREAS OF FUTURE DEVELOPMENT

The Costain Way is being developed so that it can be downloaded as an app and made available on mobile devices. This is particularly important for those working on remote sites where they have no access to the internet. Further developments will follow as Costain simplify and automate key processes, which will ensure compliance and maintenance of comprehensive records.

The next steps include:

- concentration on 'Behavioural Quality';
- renewed focus on the supply chain and partnering (formal / informal BS11000) – better assurance of their delivery;
- development of an integrated improvement strategy.

REVIEW QUESTIONS

1 Evaluate the 'Costain Way' and explain why you would recommend its adaption for application in companies across the construction supply chain.
2 How could the assessment against the EFQM Excellence Model be applied in other parts of the world in the construction sector. Are other 'excellence' frameworks, such as Baldrige, suitable for this type of application?
3 Provide a detailed critique of the strategic priorities set out in Table 12.1.

ACKNOWLEDGEMENT

The authors gratefully acknowledge the involvement and assistance of Tony Blanch, Director of Business Improvement, Costain.

CASE STUDY 13

Re-engineering timber floors in the Australian housing sector: an example of process innovation

PROJECT BACKGROUND

This case study is about process innovation, in contrast to product innovation, in the housing sector in Australia. By this distinction we refer to innovative changes in *who* does certain work and *when* it is done rather than any fundamental change to *what* is done. This type of change is harder to conceive and more difficult to implement than straightforward product innovation, which is generally entirely within the control of a single company. It is also hard for the initiating company to capture the financial benefits of construction process innovation and, hence, the motivation for process-based investment is weak. Yet fundamental changes in the way that we procure buildings require a rethink of both the products and the processes of the sector – simple product innovation simply alters the parts we build with but not the entire sequence or process.

The timber products division of CSR, an international construction materials group headquartered out of Sydney, Australia, decided in the early 1990s to improve the competitiveness of their timber floor system against the main competition, concrete slab on ground.

At that time both the timber and concrete industries in Australia were in an advertising war, each was taking out one-page advertisements in the major newspapers to convince the public that their particular flooring solution was superior.

CSR went to the ACCI, a construction research group at the University of New South Wales under the leadership of Marton Marosszeky, and asked the group to first of all establish the truth regarding the costs of the various flooring systems and to help the company develop a more efficient solution. The company simply wanted to increase the market share of timber floors in the Australian residential market against concrete raft-slabs.

One focus of the case study discusses the factors in construction that are essential for process innovation to be successful. The intellectual and financial capital necessary to re-engineer the supply chain is examined along with the challenge for organizations to capture the benefits of innovation.

Since the project CSR has sold its timber business; however, the development project spurred a number of similar innovations and there are several alternative techniques for building timber floors in the market which use the logic of the innovation described in this case study.

HOUSING CONSTRUCTION IN AUSTRALIA

In Australia, brick veneer houses comprise structural timber frames and trusses that are clad with an external veneer of brickwork. Normally, they have ceramic or concrete tiles on the roof. The ground floor is either of suspended timber construction or a concrete raft slab on ground. The traditional and still most common way of building a suspended timber floor is to construct brick piers, engaged to the external brick veneer and free-standing internally, at approximately 1.8 metre centres in both directions throughout the subfloor area. Hardwood timber 100 x 75mm (4 x 3 inch) bearers are run first, then 100 x 50mm (4 x 2 inch) hardwood joists, and finally the frame is sheeted with 18mm (3/4 inch) particle-board sheeting. The competing concrete slab construction consists of a 100mm (4 inch) slab with thickened ribs around the edge and under internal walls.

On a sloping site, in the case of traditional timber floor construction, the site is left unaltered and the external subfloor walls and internal brick piers create a flat platform for the floor construction. In the case of the concrete slab, a bulldozer creates a flat platform with cut and fill, and the house is partly built on the ground and partly on piers that pass through the fill to the same stratum as is exposed on the high side of the cut.

This case study is an example of process innovation in the construction of timber residential floors. The ideas were developed specifically for brick veneer houses on sloping sites; however, they are applicable to this type of construction on any site. The construction supply chain in the cottage construction sector in Australia is highly fragmented. Before the timber floor of a cottage has been completed, at least six different parties have already been involved in the process.

A surveyor sets out the reference marks and the builder sets profiles for the actual wall dimensions. An excavator digs the trenches and a subcontractor fixes the reinforcing steel and pours the concrete footings. Subcontract bricklayers build the brickwork up to the floor level and subcontract flooring-carpenters build the timber frame and fix the timber floor sheeting. These processes normally take between 1.5 to 3 weeks to complete.

By comparison, the competing concrete raft-slab is a simple arrangement between the builder and subcontractor – a concreter. This subcontractor arranges for the levelling of the site and the drilling of pier holes, erects edge formwork, places a plastic sheet moisture barrier and reinforcement and pours and finishes the concrete slab. This process is faster and cheaper than the fragmented timber alternative – one subcontractor completes the entire work package. The traditional process takes proximately 1 to 1.5 weeks to complete.

Normal product innovation in building construction is ubiquitous and relatively well understood. A manufacturer conducts market research to test the feasibility of a new product idea and invests in an innovation process. The product is then developed and manufactured. If it finds a successful niche in the market, and performs well financially, the product is deemed a success.

Building construction as a whole, however, is highly fragmented and each party is involved in improving their particular segment of the process. The head contractor increasingly acts as a broker, organizing the work through a chain of subcontractors who are expected to take full responsibility for the development, improvement and delivery of their segment of the whole. Under this arrangement, no one looks at the optimization of the entire process.

There is also a major disincentive for contractors to invest in process improvement. Improvements gained by them cannot be quarantined from their competitors because of the fragmented nature of the supply chain. Any competitive benefits gained are almost immediately lost as competitors on competing projects copy any innovation. Hence process innovation has to come from the manufacturing sector.

Overall the research and development created a process for building a timber floor in normal brick veneer construction that was more than 20 per cent faster and cheaper than the traditional alternative. To introduce the new process to the market was as much of a challenge as the process development itself, because it involved forming an organization that could deliver the entire work package. The new entity was required to act as a change agent in the market till the new processes were adopted industry-wide.

INNOVATION PLAN

Step one: Interviews were conducted with 50 tradesmen and builders to identify their perceptions and experiences regarding the differences between the timber and concrete floor construction.

Step two: A detailed field research to identify all the factors that impact on the efficiency of individual tasks and the flow of activities in traditional timber construction as well as of the competing concrete slab-on-grade process was carried out. This was conducted using detailed work-study methods, which involved documenting observations and taking measurements on six timber and six concrete slab sites.

Step three: An international review of timber floor construction techniques was undertaken to discover solutions that had been developed elsewhere for the problems observed.

Step four: Based on findings from the three preceding steps of the research, a novel method for timber floor construction was developed and validated through discussion with builders.

Step five: A full-scale prototype was constructed and every aspect of the process was evaluated.

Step six: As the findings were positive, the company developed an implementation strategy.

EXPECTED OUTCOME

The development process that led to the new timber construction system was informed by a number of key findings.

1 The time difference between the timber and concrete flooring processes was due to periods of tradesmen inactivity in the timber construction process, primarily because of the number of parties involved. This finding led to the idea that a single point of responsibility was needed for the entire timber floor work package.

2   It was identified that waste in reinforced concrete, when footings are cast in stepped trenches on a sloping hillside, is of the order of 43–66 per cent. It was also found that survey set-out tended to be inaccurate on steeper sites, though it was perceived that flattening a sloping site for a concrete slab had some negative consequences (the need for retaining walls and drainage provision resulted in a lower profile for the building). Hence these findings overall led to the idea that sites should be benched, perhaps in about 1.2 metre (4 ft) benches. This had the positive benefits of reducing waste in footing construction and reducing the subfloor brickwork while minimizing the negative consequences of fully benching a site.

3   The fact that the brickwork in a timber floored house is constructed in two stages and the bricks are delivered in two lots. On most sites there is insufficient space to store all the bricks for the building at the outset, especially on sloping sites. This has led to the practice of having the first delivery of bricks simply for the subfloor, and later, after the frame is erected, another for the walls. Problems occur due to colour differences in the batches of bricks delivered at different times, even though they are from the same brickyard. These problems simply can't be avoided in a clay brick manufacturing process. This led to a preference for leaving all the brick construction for a single stage, after the floor, frames and trusses had all been erected.

The solution involved benching the site, cutting accurate footing trenches to avoid concrete waste, then casting concrete strip footings and pads from a stiff concrete mix and standing precast concrete piers into the wet concrete. This process had precedence in Melbourne where experienced workers could stand the piers vertically to accurate levels in the wet, footing concrete. The piers were temporarily braced to the footings at the high side of the site and timber bearers, joists and flooring were then assembled on the piers. All this work was undertaken by a single organization.

BENEFITS/RESULTS

A number of interesting lessons were learned from this case study.

In a single development exercise more than 20 per cent was cut from the time and cost of a multidisciplinary work package. This gives one indication of the scope of benefits that can be captured through a simple process re-engineering exercise.

In the traditional supply chain, the participants are usually only capable of improving their part of the process. Overall process change normally lies beyond the scope and ability of the single, traditional process participants. Such change requires the involvement of organizations, which are first of all able to make the necessary investment and then to capture the benefits. This project was conducted for a major international manufacturer. It had the intellectual and financial capital to conceive of and lead the project. Their benefit is in increased sales of their product.

Increasingly head contactors act as brokers in the construction process, coordinating other parties to plan and execute the work. If they engage in the kind of innovation described in this case study, they cannot capture and hold the benefits for any significant time as their innovation rapidly becomes common knowledge, spread through an extremely fragmented supply chain which enables their competitors to copy their innovations. The industry becomes more efficient, prices fall but the benefit for the individual contractor in terms of increased margin is limited and very short term.However in an international context, actors in any market cannot afford to be complacent. In a global market, head contractors and manufacturers in other, more competitive markets can simply move in to less efficient markets and take significant market share before the home industry can catch up. There are ample precedents for this in other industries.

A major challenge for the construction industry is to develop models of process innovation that enable innovators to capture the benefits on their investment.

REVIEW QUESTIONS

1 Evaluate the 'Innovation Plan' used in this case study and explain why it is particularly process- rather than product-based.
2 Suggest other areas of the construction industry in which this approach could be applied to develop processes.
3 How does the approach described in this case study compare and contrast with 'business process re-engineering' and 'continuous process improvement'?

CASE STUDY 14

Highways England

BACKGROUND

From 1st April 2015 Highways England became a public sector company owned by the UK Government. The primary responsibility of Highways England is for the operation, maintenance and improvement of the 7,000 km of strategic road network which comprises England's motorways and trunk roads.

Provision of new infrastructure is carried out by Highways England major projects directorate (MP). This employs contractors, usually through Early Contractor Involvement. Maintenance of the network is undertaken by the network development and delivery directorate (NDD) which employs 12 managing agent contractors, various design build finance and operate companies, framework contractors and technology managing agent contractors. In 2009/10 Highways England, then Highways Agency spent £1bn on capital projects and more than £1.5bn on maintenance and other activities.

To support improvement of value for money, a new Lean Division was set up in April 2009. This new division was established in response to the successful piloting of Lean and Six Sigma methodologies on several major projects, particularly on the M6 extension from Carlisle to Guards Mill, on which a saving of £4.7m was made. It was decided by the Highways England Board that there should be an active focus on three work stream areas. These were 'Managed Motorways', 'other Major Projects' and 'Maintenance and Renewals'. It was recognised that the implementation of Lean had significant potential benefits in terms of cost, time, and delivery of quality to the road-using public and other stakeholders.

In its 2010/11 business plan Highways England was charged with delivering £114m of efficiencies. Given the current economic position and the government's drive to reduce the financial deficit there has been considerable pressure to further reduce costs and improve value for money, thereby achieving 'more for less'.

The Highways England Lean development strategy had clear objectives, both for the immediate future (Figure CS14–1) and the longer term. They included the

1 delivery of increased Value for Money (VfM) to road users.
2 time compression to enable major schemes and other key Highways England processes to be delivered faster.

3   Realization of tangible and auditable benefits in terms of:
    – target cost and final out–turn cost;
    – provision of sustainability;
    – quality of service and infrastructure;
    – delivery of capability across the Highways community;
    – delivery of measurable efficiency improvements;
    – significant cultural shift towards continuous improvement;
    – development of an industry standard for lean construction;
    – generation of a talent pipeline;
    – evolution of Highways England to become a more agile and responsive organisation.

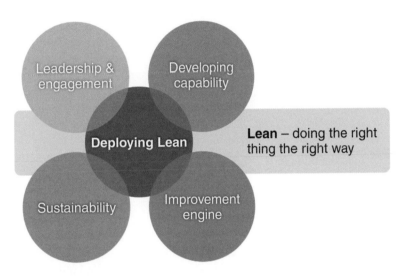

FIGURE CS14.1 Overview of HE's early strategy for deploying Lean improvement across its supply chain

Auditable savings of £100 million were achieved between 2010 and 2015 and a methodology to share good practice via Knowledge Transfer Packs was established. The Return On Investment (ROI) has historically been approximately 25:1 and to underpin these advances a series of guides were published outlining a standardised approach to techniques such as collaborative planning, visual management and benefits realisation. Supplier organisations were encouraged to develop their own Lean programmes and progress was assessed by the then Highways Agency Maturity Assessment Tool (HALMAT) now Highways England Lean Maturity Assessment (HELMA) which was based on industry standard models.

IMPLEMENTATION STRATEGY 2015 TO 2020

The purpose of Lean in Highways England

The Strategic Outcomes sought are:

• Supporting economic growth;
• Safe and serviceable network;

- More free flowing network;
- Improved environment;
- More accessible and integrated network.

The Lean objective

The objective of Lean at HE is 'to get value to flow at the pull of the Customer and then improve each day'.

To achieve this Highways England staff:

- focus on understanding and delivering customer value;
- work with its suppliers and across the HE directorates to improve end to end business processes and get value to flow through value streams;
- work with teams to engage them in continuously improving their own local processes and to do this throughout the entire business.

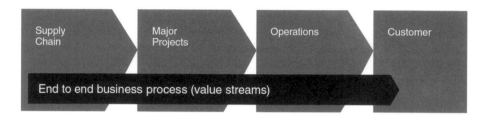

FIGURE CS14.2 End to end improvement

The objective is for Highways England and its supply chain to routinely work together to continuously improve safety, customer experience and efficiency.

There is a structured approach to *increasing Lean maturity* using the Highways England Lean Maturity Assessment (HELMA) .

The five supporting themes are described in the HE house of Lean:

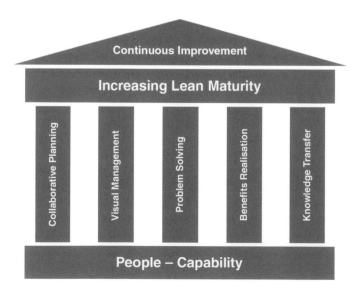

FIGURE CS14.3 Highways England house of Lean

The objective is only realized by developing the Lean capability of HE people and those of its suppliers.

HIGHWAYS ENGLAND 2020 AMBITION

That by 2020 Highways England:

- will realise a £250m contribution toward the Road Period 1 efficiency target of £1.2bn using Lean techniques;
- staff will use their Continuous Improvement skills every day. They will be empowered to see and act on opportunities to improve their own and the organisation's performance;
- staff and delivery partners will be accountable for continuously improving performance in safety, customer satisfaction and efficiency;
- will work with its delivery partners, suppliers, stakeholders, and customers to routinely use collaborative planning as an enabler to genuine collaboration; and
- will see productivity improvement demonstrated by year on year reduction in cost unit rates.

The key enablers

These are designed to enable HE achieve its five strategic Outcomes:

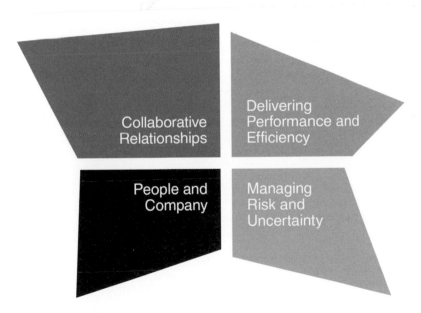

FIGURE CS14.4 Highways England key enablers

COLLABORATIVE RELATIONSHIPS

To strengthen and develop collaborative relationships with suppliers and stakeholders by:

- establishing Collaborative Planning as standard practice across our business;
- using Collaborative Planning to achieve ownership, alignment and commitment to a shared plan;

- helping the supply chain to assess their own Lean maturity and provide guidance to future development;
- sharing Lean efficiencies through knowledge transfer and learning;
- working with our supply chain on process improvement;
- embedding Lean into the Highways England Supply Chain Strategy and Value Chain plans;
- developing supplier capability in the use of Lean principles, tools and techniques.

PEOPLE AND COMPANY

To support Highways England's people strategy by:

- Accountable leadership using Continuous Improvement cells to empower team members to own and improve performance;
- Leaders owning and improving their end to end processes, enhancing value and reducing waste;
- Capable employees providing training and support for people in the use of Lean principles, tools and techniques;
- Customer focused delivery helping our people to identify and understand the requirements of their key customers using Lean 'Voice of the Customer' principles;
- Performance visibility helping leaders develop customer focussed performance indicators.

DELIVERING PERFORMANCE AND EFFICIENCY

To support and facilitate Highways England to improve performance and efficiency by:

- understanding and defining customer value;
- using Lean principles, tools and techniques to deliver performance improvement in:
  - safety;
  - quality;
  - process cycle time;
  - staff time;
  - staff engagement;
  - financial savings; and
  - carbon emission;
- demonstrating that Lean efficiencies have contributed £250m towards the £1.2bn target.

MANAGING RISK AND UNCERTAINTY

To support and facilitate Highways England in its management of risk and uncertainty by:

- establishing a clear definition of customer value to enable focus on 'doing the right thing the right way';
- using Lean tools to minimise subjective decision making;
- establishing standardised work principles;
- reducing process variation to deliver predictable outcomes;
- training all staff in problem solving techniques;
- using Collaborative Planning to improve programme certainty;
- applying 'no errors forward' thinking and methods;
- using performance cells to create a culture of ownership and rapid response to emerging issues.

CUSTOMER OPERATIONS SENIOR LEADERSHIP TEAM

After extensive research it was noted that the major reason for failure in deploying Lean in other businesses was due to insufficient senior level leadership. Consequently, the Lean team gave a high priority to engaging with the leadership both within Highways England and across its supply chain.

To help improve team planning, performance management and problem solving across Customer Operations Senior Leadership Team, HE supported the use of visual management techniques based on Continuous Improvement (CI) cells. This methodology has introduced more evidence based customer focused discussions with a clearer alignment toward the organization's objectives.

Building the Lean capability of leaders, teams and individuals is the key to ensuring that the growth in Lean activity and the realization of benefits is achieved. Over 200 practitioners were trained both across the supply chain and within the agency from mid-2009 to early 2013. These practitioners were expected to carry out a Lean project following the Define, Measure, Analyze, Improve, Control, Measure and Transfer (DMAICT) methodology, in parallel with their training. Waves of one-day champion training were also conducted so that projects received support from informed senior managers.

Each project went through the DMAICT stages to ensure that they were all clearly scoped, based on real and accurate evidence, and that the improvements which were implemented did make a difference. Controls were put in place, ensuring that the improvements would be sustained. In addition, each practitioner had to produce a Knowledge Transfer Pack (KTP) summary of the project to ensure that benefits are captured and lessons learned are shared among the Lean Practitioner community.

Although a key priority was to improve problem-solving capability using the DMAICT Lean sigma methodology, Lean was seen as a philosophy rather than a set of tools and techniques. However, as the tools and techniques enable the embedment of the Lean philosophy at a practical level, training was developed to teach their use.

The Lean division also trialled Lean visual management, using an information center as a focus for the team to communicate on a daily basis. Collaborative planning, based on the Last Planner® System—a technique which allows all parties on site to plan together and improve the stability of the programme—was implemented, with the possibility of finding ways to deliver early. Guides to visual management and collaborative planning were created by Highways England for use by suppliers. In addition, the use of 6S (sort, set in place, sweep, standardize, sustain and safety) was deployed to improve the efficiency and improve safety of the workplace either for the major project sites or maintenance depots.

Each implementation project was required to undergo a benefits realization analysis through which benefits were calculated and signed off by the Lean practitioner who led the project, the Highways England technical manager for the relevant work stream, the supplier contract manager/director and the Highways England project sponsor/manager. The benefits were categorized as improvements in: cost, time, safety, sustainability, culture and other benefits. On the basis of a guide provided Highways England, a standard form to calculate benefits was developed and used by all practitioners.

To share lessons learned, a Knowledge Transfer Pack (KTP) was developed for each completed Lean improvement project. These packs are stored in the Supply Chain area of Highways England's website to enable future Lean practitioners or other supply chain staff to learn from previous projects; and, if possible, to better them. This has become a hub of Lean project learning, feeding into the HE's Managing Down Cost toolkit and sharing information with suppliers about new ideas and processes.

Quarterly seminars, attended by the Lean team and affiliated practitioners, plus others who wish to share and gain knowledge, are now held for the wider 'Lean community'.

Presentations by members and guests of the community form much of the day, and table discussions are used to exchange ideas. Visits to best practice offices and factories in other sectors have also been arranged for the Lean community, to encourage learning from organisations which are further ahead on the Lean journey.

DEVELOPING LEAN WITHIN THE HE SUPPLY CHAIN

HE was one of several highways clients who were keen to see the Civil Construction industry embrace Lean to help foster a culture of continuous improvement for mutual advantage and in doing so improve efficiency and provide 'more for less' for the benefit of the taxpayer.

As a part of its strategy to encourage the take-up of Lean by its supply chain, Highways England developed a self-assessment maturity model (initially HALMAT and recently simplified to HELMA) to support supplier development in Lean and continuous improvement practices. This was designed to enable organisations to assess the maturity of their practices in ten key operational areas.

This framework provided a structured method for Highways England to carry out the moderation of self-assessments. This has proven to be valuable in that:

- scoring is made more consistent across the supply chain;
- areas of best practice can be identified;
- exchange of ideas and moderator feedback can assist with developing action plans.

The HELMA/SLCA (Simplified Lean Culture Assessment) framework was designed for use across the whole of Highways England supply chain, covering the top tier companies and a wide range of different activities from design, consultancy, construction, production, maintenance, etc.

A matrix with five levels of maturity was designed to provide evidence of achievement in ten key areas. The following figure illustrates the stages of maturity, and the ten areas of practice to be assessed are set out in the table below. This tool was incorporated into the 'Motivating Success Toolkit' and used by HE to identify which suppliers to engage with in the future.

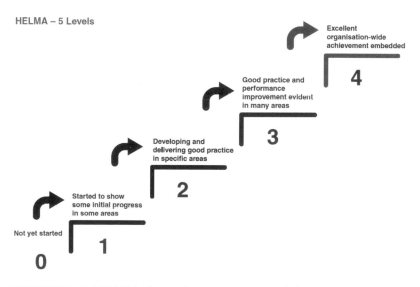

HELMA – 5 Levels

Excellent organisation-wide achievement embedded — 4

Good practice and performance improvement evident in many areas — 3

Developing and delivering good practice in specific areas — 2

Started to show some initial progress in some areas — 1

Not yet started — 0

FIGURE CS14.5 HELMA's 5 step improvement model

This methodology is based on 'HELMA', which is an evolution of 'HALMAT', version 2 (2012) which itself based on previously published material, including *Lean Enterprise Self-Assessment Tool, LESAT*, developed by MIT / University of Warwick and *Lean Aerospace Initiative 1998–2003*, University of Warwick.

TABLE CS14.1 HELMA self-assessment scoresheet *(note all areas carry equal weighting)*

HELMA area	Key questions
1.0 Integration of Lean in business strategy	How explicitly is Lean integrated within the overall business strategy?
2.0 Lean leadership and engagement	How engaged are senior management with the Lean journey? How are the leaders demonstrating commitment and leadership?
3.0 Deployment management/ Lean infrastructure	How are you driving Lean within the business from strategy to implementation? How do you undertake Lean deployment governance, planning and review? How do you measure progress?
4.0 Understanding customer value	How do you understand internal and external customer value? What measurements do you have in place? How do you link customer value back into your delivery process?
5.0 Understanding of process and value streams	How widely are processes and value streams understood? How do you undertake in process measurements? How do you establish pull and flow? How do you identify and eliminate waste?
6.0 Use of methodologies and tools	What is the range of Lean tools and methodologies that you use? How widely are they understood and practised?
7.0 Organizational coverage, activity and capability	What is the depth and breadth of Lean within your organization? How many of your people are involved in Lean activity? Which organizational areas are delivering improvements? How are you developing Lean capability within the business? What training has been delivered, to what level and what coverage?
8.0 Performance improvement, benefits realization & delivery	How do you measure, capture and report Lean benefits? What improvements have you delivered in the last 12 months? How have you recorded and shared best practice and benefits with the sector supply chain and HE for example, do you report benefits on the HE Lean Tracker and complete a knowledge transfer pack?
9.0 Lean collaboration, climate and culture	How would you describe your existing culture? What is your understanding of your desired Lean culture and climate? How do your people work together, with clients, suppliers and partners? How do you plan and manage cultural change?
10.0 Supplier maturity	How do you engage with your suppliers and partners on Lean? What is their level of maturity? What benefits can you achieve through greater Lean collaboration?

In order to find improvement-oriented suppliers, HE pre-qualifies its suppliers for high-value projects using its Strategic Alignment Review Tool (StART). While Lean terminology is deliberately left out of StART, its approach and scoring was designed to identify companies that are already practicing elements of continuous improvement. The StART process begins with an introductory briefing, after which the companies complete an extensive self-assessment process. Once this is done a team from HE conduct a site visit and audit, and provide a feedback report including the company's score.

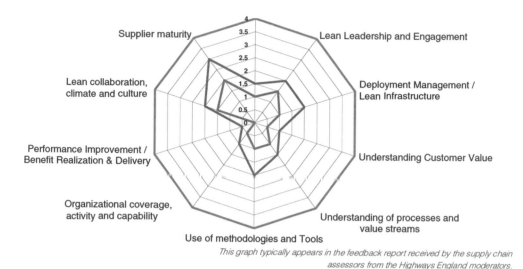

FIGURE CS14.6 Plot of typical supplier assessment results

From there, Highway England provides its suppliers with extensive training opportunities and learning resources – such as workshops, seminars, webinars, and online materials on innovative new construction practices – all at Highway England's expense.

The underlying philosophy of Highway England's overall strategy is to create a community of companies practicing Lean at a high level. Having its suppliers use a common vocabulary and set of management tools makes it easier to share improvements and innovations, and to work together on those areas that cut across the work of many suppliers. HE has also built an extensive online library of articles, helpful templates for the various Lean tools, write-ups of successful Lean projects and innovative construction techniques (many with instructive YouTube videos), and other resources.

The 'Major Projects' side of Highway England's work, which accounts for the almost a 50 per cent share of its budget illustrates how the organisation has rolled out its program. Highway England selected five 'tier 1' suppliers through a bidding process weighted 70 per cent on quality and 30 per cent on cost. Rather than bidding on specific projects, the cost side of the bids were for the rates (including 'normal' profits) they would charge for specific classifications of labour, equipment, materials and overhead – the components that make up every project. Large multi-year contracts were then assigned to each Tier 1 supplier under contractual conditions which encouraged Lean practices. Historical performance data from similar projects were used as a starting point to negotiate each contract price based on existing standards. A scheme which shared savings rewarded suppliers for improving on those standard rates and for sharing their learnings with each other. All savings from the negotiated base price are split 40/60 between HE and the contractors, the 60 per cent going into a pool to be shared by all 'tier 1' suppliers. The improved performance standard becomes the new base standard for the next round of contracts.

An important aspect of the division of savings is that the contractor creating the innovation and making the saving keeps one-third of the saving, and the remaining two-thirds are shared between the other four contractors in the pool. This incentivises both innovation and collaboration between the contractors in the pool. It means that every contractor has a stake in the improvements the other contractors make, and they deploy these improvements on their own jobs as quickly as possible.

The drive for efficiency is ongoing and Highway England are now moving rapidly towards a philosophy which embraces Building Information Modelling (BIM) and 'Factory thinking'. This allows all the aspect of the Lean thinking to be deployed. Also since 2014 the Highway England Executive decided, based on the huge success of the supply chain deployment, to embrace Lean thinking across the entire business of 3,600 employees.

EXAMPLES OF LEAN PROJECTS

Example 1: Lean sigma 'DMAIC' blacktop from Dishforth to Leeming

Encouraged by the success of Lean at the M6 Guards Mill project, which saved £4.7m, both the HE and Carillion/Morgan Sindall Joint Venture were very keen to explore process improvement at A1 Dishforth to Leeming. One of the first steps was to establish capability on site. A Lean project team and a steering group were formed, made up of Lean practitioners with an overseeing master 'black belt'.

In October 2009, the scheme hosted a 'Recognise' workshop. This gave staff the opportunity to identify opportunities that lay ahead in the project. One concern was the logistical challenge; it was anticipated that delivering the scheme would require some 250,000 wagon loads of deliveries, averaging 60 vehicles per hour over the construction period. Following considerable discussion, the Lean project team set about challenging the blacktop laying rates, with the intention of improving upon the 1,000 tonnes per day allowed for in the construction programme.

In line with DMAIC methodology, the first steps were to Define the problem. This involved looking at the many factors behind the variation in asphalt laying rates, as well as identifying the common causes of delay. Daily outputs were then Measured over a two-month period, and further Analysed with the aid of the statistical software Minitab.

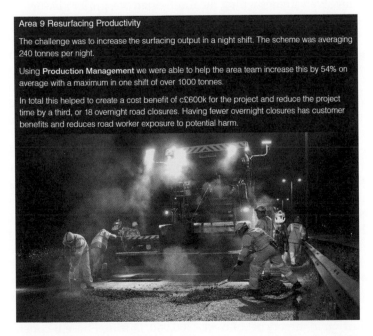

FIGURE CS14.7 Lean driven productivity improvement in paving

This assisted in highlighting inefficiencies, as well as demonstrating bottlenecks within the process. A time and motion study further showed that on a 'perfect day', faced with continuous supply of asphalt, the blacktop gangs' capability could be doubled to 2,000 tonnes per day. It was clear the project's success hinged on establishing this continuous supply.

Improvements were then implemented. These included:

- the use of multiple batching plants within close proximity of the site, each producing the same type of asphalt rather than regularly changing;
- contingency plans to prevent plant breakdown, including night time maintenance and text alerts should an incident occur;
- changes in working patterns, i.e. changing sequencing so that all pavement joints were cut the previous evening in order to prevent morning delays;
- first delivery on site at 7am, with continuous supply up to 5pm; and
- by using the old tank road through MOD Catterick instead of the A684, and therefore it was possible to extend daytime delivery times so that laying could be continuous from 7am to midnight. The most achieved in any 24-hour period was 3,500 tonnes.

With the above improvements in place, daily outputs could exceed the 2,000 tonne output, with the end result that the programme's critical path was reduced by 17 days. Through the accompanying reduction in preliminaries and labour and plant costs requirement a total saving of £602,000 was made.

Example 2: collaborative planning

On the M53 Bidston Moss Viaduct Strengthening project, over £1m of savings were realised because of Lean deployment on the scheme. Through collaborative planning in the design and construction phases, a culture of continuous improvement was developed across the project.

The £89.9m scheme involved strengthening the multi-span box girder viaduct together with refurbishment works in order to restore the network to full capacity. Costain was awarded the contract. Detailed design and investigation was undertaken during Phase A of the Early Contractor Involvement Scheme (ECI) prior to main construction starting.

Collaborative planning was established early in the design phase and deployed on all further activities. Integrated project meetings, weekly in design and daily for construction were used to set programme targets, establish which tasks in the plan were achieved and to identify key barriers to progress. This process provided the PPC (commitment reliability).

In design, collaborative planning drove better communication and transparency helping to improve commitment reliability by 34 per cent. This helped the project meet an accelerated design programme and complete Phase A one month ahead of programme.

Commitment reliability in construction improved from 78 per cent to 90 per cent. In addition, trends affecting production on site were identified and dealt with in a timely manner. This approach ensured that the root causes of any problems were treated rather than a simple reaction to the symptoms.

Design efficiency (commitment reliability) chart.

Collaborative planning during Phase A and advanced construction works provided increased confidence of production capability enabling rates for certain activities to be reduced realising a reduction in target cost of over £300,000. During Phase B construction improved commitment reliability led to further labour efficiencies exceeding £380,000.

In construction, weekly collaborative planning was also used to improve short term look-ahead planning. All key supply chain partners were engaged in examining the whole programme and challenging themselves to exceed targets by either reducing the time needed to complete individual tasks, or by running tasks in parallel. This resulted in commitment reliability for look-ahead planning improving from 59 per cent to more than 85 per cent. As a result of collaborative planning, significant programme benefits were realised, including programme savings in excess of £400,000.

In parallel with these collaborative planning activities, standard work processes were also mapped and reviewed to identify and implement improvements.

Example 3: people and process

Departures from Standard

Departures from "Standard" may be granted by the Highways England if a proposed design is submitted which does not comply with current standards, but can be demonstrated to yield significant benefits and the adverse impacts, if any, remain tolerable. However, with multiple stakeholders involved along the process, there is opportunity for delays to occur. These delays can pose a risk to scheme delivery if they occur on the critical path. The objective of this project was to reduce delay and waste in the process. Both traditional Lean methods and data analysis were used to find common causes of delay and waste.

A workshop was held at which participants represented all the stakeholders along the process. All participants were invited to use a common 'fishbone' diagram and the practice of 5 Whys. Doing so helped to expose reasons why delay might be occurring and did so in a way that represented the perspectives of all the participants.

To investigate the causes of delay in greater depth, analysis was also undertaken of a year's worth of departures data. From looking at a year's worth of records (financial year 2008/2009), concentrating on road geometry departures, a rejection rate of 26 per cent was identified as contributing to delay within the process. Rejection, or not getting the application 'right first time' represented waste within the system, especially since departures had to be resubmitted and delays in receiving approvals could impact the timing of the scheme associated with the proposed design.

Reasons for rejection were investigated, and common causes identified. It was discovered that common causes of rejection tended to be basic errors such as not attaching a risk assessment (30 per cent of all rejected departures), not including an accident summary and commentary (22 per cent of all rejected departures) or not discussing cost and other benefits (21.4 per cent of all rejected departures). More complex errors also occurred, but the likelihood of these occurring was lower.

In response to these findings, the HE team that managed departures created a filter by which those departures that contain the most basic errors were immediately returned to the submitting designer. This gave designers a chance to correct these errors and resubmit without the departure having to undergo further processing before review and certain rejection by the technical specialist. This saved the time of a valuable staff resource, avoided delay in addressing a simple omission and the consequential delay to other departures that the technical specialist had to process. Whereas, previously, internal reporting of the departures processes concentrated on time taken to deal with departures, a move towards also reporting the percentage of departures that were 'right first time' according to subject area helped to monitor the quality of departure submissions and identify waste which could be reduced. This made better use of the capacity within the process and freed time within HE and its supply chain partners.

Example 4: Lean visual management

The first trial of Lean visual management took place in early 2009 on a bridge repair project (Lodge Lane) on the M6 Junction 23. It was undertaken by the Area 10 Managing Agent Contractor (MAC 10). Boards were put up in the site office, showing daily regularly updated information for the following categories: health and safety, minutes of the last meeting, site attendance per contractor, planned inductions, general comments, weekly measures, project plan and delay, project cost information, continuous improvement activity (including problem solving sheets), and traffic management for the week ahead. Meetings of ten minutes were held every day, dealing with each of the categories one by one.

Because the boards acted as a focal point for team communication, all important issues could be captured there and discussed as a team during the daily meetings, saving time that otherwise might have been spent having to transfer knowledge via several informal discussions. Rather than solving problems informally, they could instead be addressed using a formal problem-solving technique that used the knowledge and abilities of all attendees and sought to put in place countermeasures to the root causes of the problems. As an overall result, teamwork improved and all staff had a clearer idea of what needed doing, who was doing what, and how well the work was progressing.

Since the trial took place, all sites on the MAC 10 construction management framework have employed Lean visual management. The practice has proven very popular on site, to the extent that subcontractors have introduced it elsewhere, not just on highways projects. Furthermore, Lean visual management boards and daily meetings are the first experience that many people have of Lean, so this is also a good way of demonstrating the benefits of Lean, and promoting a continuous improvement culture. Visual management has also been trialled on routine maintenance works and in design offices.

CONCLUSION

The above four examples were taken from over from the many hundreds held within the HE's project tracking system to illustrate its strategy for Lean deployment. Over the last seven years, Highways England has saved hundreds of millions of pounds by using a different strategy for dealing with its suppliers. The traditional approach used to ensure that fair value is received in purchasing, is based on rules around the procurement process. These are designed to create arms-length market-type transactions. Requests for proposals are issued, and typically the provider submitting the lowest qualified bid is selected. When there are many qualified bidders and the product or service is relatively standard and easily measured, this approach makes sense.

Experience has shown that many of the products and services purchased by Highways England do not fit this mould, and in this case the traditional purchasing approach does not guarantee good value. As the Highways England experience has demonstrated, public sector managers have to change how they work with the suppliers of these types of products and services. In order to get the best value for money, they must be willing to invest in increasing their suppliers' ability to improve, both in the short term and the long term.

Highways England estimates that for every £1 it has invested in its Lean supply-chain initiative (e.g. training, resources, personnel, etc.) it has received more than £25 back in savings so far, and the improved methods promise to generate additional savings well into the future. And these are all savings that would not have occurred under the old purchasing paradigm.

The final Figure gives an overview of Highways England's Lean implementation stra-
-tegy, goals and outcomes.

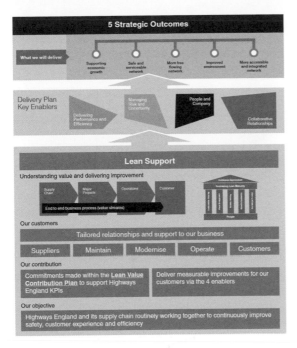

FIGURE CS14.8 Overview of Highways England Lean strategy

ACKNOWLEDGEMENT

The authors acknowledge the contribution of Derek Drysdale, the former head of the Lean construction initiative within Highways England, and the support of Paul Doney, the current head of the initiative and Lucia Fullalove and Andrew Wingrove (HE Lean practitioners).

REFERENCES

Highways England documents:
Lean Support to Highways England 2015 to 2020
Lean Maturity Assessment (HELMA)
Knowledge Transfer Packs for visual management, precast RCB installation
Drysdale D, (2013) *Introducing Lean improvement into the highways agency supply chain*, Proc, 21st IGLC Conference Fortaleza, Brazil.

INDEX

Numbers in italics indicate tables or figures